USING AUTOCAD®

• • • • • • • • • • • • • • • • • • • •

Release 13 for Windows

Also available for AutoCAD Release 13 from Autodesk Press and Delmar Publishers

Basic AutoCAD Books

AutoCAD: A Problem-Solving Approach r13 DOS, by Sham Tickoo
  ISBN 0-8273-6015-0

AutoCAD: A Problem-Solving Approach r13 Windows, by Sham Tickoo
  ISBN 0-8273-7432-1

AutoCAD: A Visual Approach r13 DOS/Windows—Series, by Steven Foster and others
  (Call for individual module ISBNs)

AutoCAD r13 Update Guide for DOS and Windows, by Sham Tickoo
  ISBN 0-8273-7433-X

The AutoCAD Tutor for Engineering Graphics, by Alan Kalameja
  ISBN 0-8273-5914-4

Harnessing AutoCAD r13 for DOS, by Tom Stellman, GV Krishnan, Robert Rhea
  ISBN 0-8273-6822-4

Harnessing AutoCAD r13 for Windows, by Tom Stellman, GV Krishnan, Robert Rhea
  ISBN 0-8273-7199-3
  3-hole punched binding ISBN 0-8273-7224-8

The Illustrated AutoCAD Quick Reference for DOS, by Ralph Grabowski
  ISBN 0-8273-6645-0

The Illustrated AutoCAD Quick Reference for Windows, by Ralph Grabowski
  ISBN 0-8273-7149-7

Using AutoCAD r13 for DOS, by James Edward Fuller, edited by Ralph Grabowski
  ISBN 0-8273-6824-0
  3-hole punched binding ISBN 0-8273-6972-7

Customizing Books

Customizing AutoCAD r13, by Sham Tickoo
  ISBN 0-8273-7501-8

Maximizing AutoCAD r13, by Rusty Gesher, Mark Middlebrook, Tony Tanzillo
  ISBN 0-8273-7993-5

Call your local representative for more information and a complete listing of all that Delmar Publishers offers the AutoCAD user!

USING AUTOCAD®

Release 13 for Windows

James Edward Fuller

Edited by
Ralph Grabowski

Press

I(T)P® An International Thomson Publishing Company

Albany • Bonn • Boston • Cincinnati • Detroit • London • Madrid • Melbourne
Mexico City • New York • Pacific Grove • Paris • San Francisco • Singapore • Tokyo
Toronto • Washington

NOTICE TO THE READER

Publisher does not warrant or guarantee any of the products described herein or perform any independent analysis in connection with any of the product information contained herein. Publisher does not assume, and expressly disclaims, any obligation to obtain and include information other than that provided to it by the manufacturer.

The reader is expressly warned to consider and adopt all safety precautions that might be indicated by the activities herein and to avoid all potential hazards. By following the instructions contained herein, the reader willingly assumes all risks in connection with such instructions.

The publisher makes no representation or warranties of any kind, including but not limited to, the warranties of fitness for particular purpose or merchantability, nor are any such representations implied with respect to the material set forth herein, and the publisher takes no responsibility with respect to such material. The publisher shall not be liable for any special, consequential, or exemplary damages resulting, in whole or part, from the readers' use of, or reliance upon, this material.

Trademarks
AutoCAD® and the AutoCAD® logo are registered trademarks of Autodesk, Inc.
Windows is a trademark of the Microsoft Corporation.
All other product names are acknowledged as trademarks of their respective owners.

Cover Design: Michael Speke
COPYRIGHT © 1996
By Delmar Publishers Inc.
Autodesk Press imprint
an International Thomson Publishing Company
The ITP logo is a trademark under license.

Printed in the United States of America

For more information, contact:

Delmar Publishers
3 Columbia Circle, Box 15015
Albany, New York 12212-5015

International Thomson Publishing Europe
Berkshire House 168-173
High Holborn
London, WC1V 7AA
England

Thomas Nelson Australia
102 Dodds Street
South Melbourne, 3205
Victoria, Australia

Nelson Canada
1120 Birchmont Road
Scarborough, Ontario
Canada, M1K 5G4

International Thomson Editores
Campos Eliseos 385, Piso 7
Col Polanco
11560 Mexico D F Mexico

International Thomson Publishing GmbH
Konigswinterer Strasse 418
53227 Bonn
Germany

International Thomson Publishing Asia
221 Henderson Road
#05-10 Henderson Building
Singapore 0315

International Thomson Publishing—Japan
Hirakawacho Kyowa Building, 3F
2-2-1 Hirakawacho
Chiyoda-ku, Tokyo 102
Japan

All rights reserved. Certain portions of this work copyright 1993, 1992, 1991, 1989, 1988, 1986. No part of this work covered by the copyright hereon may be reproduced or used in any form or by any means—graphic, electronic, or mechanical, including photocopying, recording, taping, or information storage and retrieval systems—without the written permission of the publisher.

1 2 3 4 5 6 7 8 9 10 XXX 02 01 00 99 98 97 96

Library of Congress Cataloging-in-Publication Data
Fuller, James Edward.
    Using AutoCAD release 13 for Windows/James Edward Fuller: edited by Ralph Grabowski.
      p.   cm
    Includes index.
    ISBN 0-8273-7499-2
    1. Computer graphics.   2. AutoCAD for Windows.   I. Title.
T385.F844   1996
620'.0042'02855369—dc20

96-20291
CIP

BRIEF CONTENTS

Introduction	xxvii
Chapter 1—AutoCAD Quick Start	1-1
Chapter 2—Why CAD?	2-1
Chapter 3—Components of a CAD System	3-1
Chapter 4—Your Computer	4-1
Chapter 5—The Disk Operating System	5-1
Chapter 6—General AutoCAD Principles	6-1
Chapter 7—Getting In and Around AutoCAD	7-1
Chapter 8—Drawing Setting Up	8-1
Chapter 9—Drawing	9-1
Chapter 10—Editing	10-1
Chapter 11—Constructing Multiview Drawings	11-1
Chapter 12—Constructing Sectional and Patterned Drawings	12-1
Chapter 13—Text, Fonts, and Styles	13-1
Chapter 14—Layers	14-1
Chapter 15—Introduction to Dimensioning	15-1
Chapter 16—Dimension Styles and Variables	16-1
Chapter 17—Dimensioning Practices	17-1
Chapter 18—Plotting Your Work	18-1
Chapter 19—Inquiry and Utility Commands	19-1
Chapter 20—Intermediate Draw Commands	20-1
Chapter 21—Isometric Drawings	21-1
Chapter 22—Intermediate Edit Commands	22-1
Chapter 23—Intermediate Operations	23-1
Chapter 24—Advanced Operations	24-1

Chapter 25—Viewports and Working Space 25-1
Chapter 26—Attributes 26-1
Chapter 27—External Reference Drawings and OLE Objects ... 27-1
Chapter 28—Customizing AutoCAD 28-1
Chapter 29—Customizing Menus and Icons 29-1
Chapter 30—Introduction to AutoCAD 3D 30-1
Chapter 31—Viewing 3D Drawings 31-1
Chapter 32—The User Coordinate System 32-1
Chapter 33—Drawing in 3D 33-1
Chapter 34—Introduction to Solid Modeling 34-1
Chapter 35—Constructing Solid Primitives 35-1
Chapter 36—Creating Custom Solids 36-1
Chapter 37—Modifying Solid Objects 37-1
Chapter 38—Creating Composite Solid Models 38-1
Chapter 39—Realistic Rendering 39-1
Chapter 40—Introduction to AutoLISP 40-1
Chapter 41—Programming in AutoLISP 41-1
Chapter 42—Programming Toolbar Macros 42-1

Appendices

A—Professional CAD Techniques A-1
B—Command Summary B-1
C—System Variables C-1
D—ACAD Prototype Drawing Settings D-1
E—Hatch Patterns E-1
F—AutoCAD Linetypes F-1
G—Tables G-1
H—Glossary H-1

Index X-1

CONTENTS

INTRODUCTION	**xxvii**
CHAPTER 1—AUTOCAD QUICK START	**1-1**
INTRODUCTION TO USING AUTOCAD	1-1
Starting AutoCAD	1-1
Drawing in AutoCAD	1-2
Starting a New Drawing	1-3
Changing the Pull-Down Menu	1-3
Using Pull-Down Menus	1-5
Blip Marks	1-9
Using Icons	1-9
Summary	1-11
Exiting a Drawing	1-12
Tutorial	**1-14**
USING AUTOCAD	1-24
CHAPTER 2—WHY CAD?	**2-1**
TRADITIONAL DRAFTING TECHNIQUES	2-1
BENEFITS OF CAD	2-2
Accuracy	2-2
Speed	2-2
Neatness and Legibility	2-2
Consistency	2-2
Efficiency	2-2
APPLICATIONS OF CAD	2-3
Architectural	2-3
Engineering	2-4
Interior Design	2-5
Manufacturing	2-5
Yacht Design	2-6
Business	2-6
Entertainment	2-6
Other Benefits	2-7
Chapter Review	**2-7**

CHAPTER 3—COMPONENTS OF A CAD SYSTEM 3-1
COMPONENTS OF A CAD SYSTEM 3-1
THE COMPUTER 3-1
    Categories of Computers 3-2
    Components of a Computer 3-2
PERIPHERAL HARDWARE 3-3
    Plotters 3-3
    Printers 3-4
    Display Systems 3-5
    Input Devices 3-5
SUMMARY 3-6
Chapter Review **3-6**

CHAPTER 4—YOUR COMPUTER 4-1
INTRODUCTION 4-1
DISK DRIVES 4-1
    Floppy Disks 4-2
    Hard Drives 4-2
DISK CARE 4-3
    Write-Protecting Data 4-4
FILE DIRECTORIES AND PATHS 4-5
Chapter Review **4-6**

CHAPTER 5—THE DISK OPERATING SYSTEM 5-1
THE DISK OPERATING SYSTEM 5-1
CHANGING ACTIVE DRIVES—DOS 5-2
CHANGING ACTIVE DRIVES
AND DISPLAYING FILES—WINDOWS 5-3
DISPLAYING FILES—DOS 5-4
FILE NAMES AND DIRECTORY LISTINGS—DOS 5-5
WILD-CARD CHARACTERS 5-6
    Displaying Files in Directories 5-7
    Making a Directory Current—DOS 5-7
CREATING AND DELETING DIRECTORIES—DOS 5-8
DELETING FILES 5-8
COPYING FILES 5-9
    Copy Disk to Disk 5-9
    Copying from Directory to Directory 5-9
    Copy Files and Renaming 5-9

COPYING DISKS	5-10
RENAMING FILES	5-11
FORMATTING DISKS	5-11
DISPLAYING FILE CONTENTS—DOS	5-12
Chapter Review	**5-12**

CHAPTER 6—GENERAL AUTOCAD PRINCIPLES ... 6-1

OVERVIEW	6-1
TERMINOLOGY	6-2
Coordinates	6-2
Display	6-3
Drawing Files	6-3
Limits	6-3
Units	6-3
Zooming/Panning	6-3
USE OF THE MANUAL	6-4
Keys	6-4
Command Nomenclature	6-4
Using AutoCAD Note Boxes	6-5
Chapter Review	**6-5**

CHAPTER 7—GETTING IN AND AROUND AUTOCAD ... 7-1

BEGINNING A DRAWING SESSION	7-1
THE DRAWING EDITOR	7-2
Screen Coordinates	7-2
Mode Indicators	7-2
Toolbar Indicators	7-3
The Command Prompt Area	7-3
ENTERING COMMANDS	7-3
Keyboard	7-4
Screen Menus	7-4
Pull-Down Menus	7-4
Tablet Menus	7-4
SCREEN MENUS	7-5
Root Menu and Submenus	7-6
Pull-Down Menus	7-6
KEYBOARD KEYS USED IN AUTOCAD	7-8
Command Line Entry Keys	7-8
Toggle Keys	7-8
Other Keys	7-9
REPEATING COMMANDS	7-9

SCREEN POINTING	7-9
Showing Points by Window Corners	7-9
DIALOG BOXES	7-10
Displaying Dialog Boxes by Command	7-11
Using a Dialog Box	7-12
ICON MENUS	7-15
Selecting from an Icon Menu	7-15
STARTING A NEW OR EXISTING DRAWING	7-15
Starting a New Drawing	7-16
Editing an Existing Drawing	7-19
SAVING AND DISCARDING DRAWINGS	7-20
Saving Drawings	7-20
Discarding Your Work	7-22
Exercises	**7-23**
Chapter Review	**7-24**
CHAPTER 8—DRAWING SETUP	**8-1**
SETTING UP A DRAWING	8-1
Setting the Drawing Units	8-2
Setting the Angle Measurement	8-4
SETTING DRAWING LIMITS	8-7
SCALING YOUR DRAWINGS	8-8
CHECKING THE DRAWING STATUS	8-10
Exercises	**8-11**
Chapter Review	**8-11**
CHAPTER 9—DRAWING	**9-1**
GETTING STARTED	9-1
DRAWING LINES	9-2
Drawing a Line	9-2
Line Options	9-3
Canceling the Line Command	9-4
DRAWING LINES BY USING COORDINATES	9-4
Absolute Coordinates	9-5
Relative Coordinates	9-7
Polar Coordinates	9-8
DRAWING POINTS	9-9
Point and Sizes	9-9
DRAWING CIRCLES	9-11
Drawing a Circle with Center and Radius	9-12
Drawing a Circle with Center and Diameter	9-12

Drawing a Circle by Designating Two Points	9-13
Drawing a Circle by Designating Three Points	9-13
Drawing Circles Tangent to Objects	9-14
Drawing Tangent Three-Point Circles	9-15
DRAWING ARCS	**9-16**
Drawing Three-Point Arcs (3-Point)	9-17
Start, Center, End (S,C,E)	9-17
Start, Center, Included Angle (S,C,A)	9-18
Start, Center, Length of Chord (S,C,L)	9-19
Start, End, Included Angle (S,E,A)	9-20
Start, End, Radius (S,E,R)	9-20
Start, End, Starting Direction (S,E,D)	9-21
Center, Start, End Point (C,S,E)	9-21
Center, Start, Included Angle (C,S,A)	9-21
Center, Start, Length of Chord (C,S,L)	9-21
Line/Arc Continuation (Contin)	9-22
Manual Entry Method of Constructing an Arc	9-22
USING OBJECT SNAP	**9-22**
Object Snap Modes	9-24
Methods of Using Object Snap	9-29
CLEARING THE SCREEN WITH THE REDRAW COMMAND	**9-31**
Redrawing "Transparently"	9-31
REGENERATING A DRAWING	**9-32**
ZOOMING YOUR DRAWINGS	**9-32**
Using the Zoom Command	9-33
Transparent Zooms	9-40
Panning Around Your Drawing	9-41
Scroll Bars	9-42
Aerial View	9-43
DRAWING AIDS AND MODES	**9-45**
Placing a Grid on the Drawing Screen	9-46
Snapping to a Grid	9-47
Drawing with Orthogonal Control	9-49
Setting the Dragmode	9-49
XLINES AND RAYS	**9-50**
The Ray Command	9-50
The Xline Command	9-51
UNDOING AND REDOING OPERATIONS	**9-53**
Undoing Drawing Operations	9-53
Redoing a Drawing Operation	9-53
Exercises	**9-54**
Chapter Review	**9-66**

CHAPTER 10—EDITING ... 10-1
INTRODUCTION ... 10-1
USING OBJECT SELECTION ... 10-2
   Methods of Selecting Objects for Editing ... 10-2
   Changing the Items Selected ... 10-9
DESIGNATING ENTITY SELECTION SETTINGS ... 10-9
   Using Ddselect to Set Selection Mode ... 10-9
   Setting the Pick Box Size ... 10-12
PRESELECTING OBJECTS FOR EDITING ... 10-13
   Creating Groups of Objects ... 10-13
ERASING OBJECTS FROM YOUR DRAWING ... 10-17
RESTORING ERASED OBJECTS ... 10-17
MOVING OBJECTS IN THE DRAWING ... 10-18
Tutorial ... **10-19**
MAKING COPIES OF DRAWING ENTITIES ... 10-23
   Making Multiple Copies ... 10-24
PARTIALLY ERASING WITH THE BREAK COMMAND ... 10-24
CONNECTING OBJECTS WITH A FILLET ... 10-26
   Filleting Two Lines ... 10-26
   Filleting Polylines ... 10-26
   Filleting Arcs, Circles, and Lines Together ... 10-27
   Filleting Circles ... 10-28
Tutorial ... **10-29**
Exercises ... **10-33**
Chapter Review ... **10-39**

CHAPTER 11—CONSTRUCTING MULTIVIEW DRAWINGS ... 11-1
MULTIVIEW DRAWINGS ... 11-1
   Orthographic Projection ... 11-1
Tutorial ... **11-6**
   Drawing Multiviews with AutoCAD ... 11-9
   Auxiliary Views ... 11-9
CONSTRUCTING AUXILIARY VIEWS WITH AUTOCAD ... 11-11
   Showing Hidden Lines in Multiview Drawings ... 11-11
Tutorial ... **11-12**
   Loading Linetypes ... 11-15
   Writing Linetypes ... 11-17
Exercises ... **11-18**
Chapter Review ... **11-22**

CHAPTER 12—CONSTRUCTING SECTIONAL AND PATTERNED DRAWINGS ... 12-1

SECTIONAL VIEWS ... 12-1
- Types of Sections ... 12-2
- Crosshatching Sectional Views ... 12-4

CREATING SECTIONAL VIEWS IN AUTOCAD ... 12-5
- Using the Bhatch Command ... 12-6
- Selecting the Hatch Boundary ... 12-8
- Advanced Hatch Boundary Definitions ... 12-10
- Using the Hatch Command ... 12-14
- Hatch Command Boundary Definition ... 12-14

Tutorial ... 12-15
- Defining Your Own ... 12-17
- Pattern Alignment ... 12-17

Tutorial ... 12-18

CREATING A POLYLINE BOUNDARY ... 12-20

CREATING SOLID AREAS ... 12-21

Exercises ... 12-22

Chapter Review ... 12-25

CHAPTER 13—TEXT, FONTS, AND STYLES ... 13-1

THE USE OF TEXT IN GRAPHIC DRAWINGS ... 13-1
- Text Standards ... 13-1
- AutoCAD Text Components ... 13-2

THE TEXT COMMAND AND ITS OPTIONS ... 13-3
- Placing Justified Text ... 13-4

Tutorial ... 13-5
- Additional Alignment Options ... 13-6
- Selecting Different Text Styles ... 13-8
- Rotating Text ... 13-8

PLACING MULTIPLE TEXT LINES ... 13-9

DRAWING TEXT DYNAMICALLY ... 13-10
- Placing Dynamic Text ... 13-10

CREATING PARAGRAPH TEXT—THE MTEXT COMMAND ... 13-11
- Using the MText Dialog Box ... 13-13

CHANGING MTEXT ... 13-16

TEXT STYLES ... 13-19
- Creating a Text Style ... 13-20
- Text Styles by Dialog Box ... 13-24
- Using Text Styles ... 13-26

SPECIAL TEXT CONSIDERATIONS 13-26
    Text and Dtext Codes 13-27
MTEXT CODE CHARACTERS 13-27
REDRAWING AND REGENERATING TEXT FASTER 13-30
    Effects of Qtext on Plotting 13-31
    Setting Qtext with a Dialog Box 13-31
Exercises **13-33**
Chapter Review **13-35**

CHAPTER 14—LAYERS 14-1
LAYERS 14-1
    Using the Layer Command 14-3
    Controlling Layers with a Dialog Box 14-7
    Controlling Layers with the Toolbar 14-12
Exercises **14-13**
Chapter Review **14-14**

CHAPTER 15—INTRODUCTION TO DIMENSIONING 15-1
DIMENSIONING IN AUTOCAD 15-1
DIMENSIONING COMPONENTS 15-1
    The Dimension Line 15-1
    Extension Lines 15-3
    Dimension Text 15-3
    Dimension Tolerances 15-4
    Dimension Limits 15-4
    Alternate Dimension Units 15-5
    Leader Lines 15-5
    Center Marks and Center Lines 15-6
    Changing the Look with Dimension Variables 15-6
ENTERING DIMENSIONING MODE 15-6
DIMENSIONING COMMANDS 15-7
DIMENSION DRAWING COMMANDS 15-7
    Placing a Linear Dimension Line 15-8
    Continuing the Dimension String 15-10
    Placing a Vertical Dimension Line 15-12
    Dimensioning Angled Surfaces 15-12
    Creating Baseline Dimensions 15-14
    Dimensioning Angles 15-16
    Dimensioning Circles and Arcs 15-17
    Dimensioning the Radius of a Circle or Arc 15-19
    Placing Center Marks 15-19
    Ordinate Dimensioning 15-21

DIMENSION EDITING COMMANDS	15-22
Restoring Dimension Text to Its Default Position	15-22
Changing Dimension Text	15-23
Obliquing Dimension Extension Lines	15-23
Relocating the Dimension Text	15-25
Rotating Dimension Text	15-26
DIMENSION UTILITY COMMANDS	15-27
Displaying the Dimension Status	15-27
Changing the Dimension Text Style	15-27
DEFINITION POINTS	15-27
ARROW BLOCKS	15-27
Separate Arrow Blocks	15-28
Exercises	**15-29**
Tutorial	**15-37**
Chapter Review	**15-43**

CHAPTER 16—DIMENSION STYLES AND VARIABLES ... 16-1

DIMENSION STYLES AND VARIABLES	16-1
SETTING DIMENSION VARIABLES	16-2
Dimension Line Geometry	16-3
Dimension Format	16-10
Dimension Annotation (Text)	16-14

CHAPTER 17—DIMENSIONING PRACTICES ... 17-1

DIMENSIONING PRACTICES	17-1
Placing Dimensional Information in a Drawing	17-2
Constructing Dimension Components	17-2
Dimensioning 3D Objects	17-5
Dimensioning Mechanical Components	17-9
Exercises	**17-12**

CHAPTER 18—PLOTTING YOUR WORK ... 18-1

PLOTTING OVERVIEW	18-1
Printer Plots	18-1
Plotter Plots	18-2
PLOTTING THE WORK	18-2
PLOTTING FROM THE COMMAND LINE	18-17
Exercises	**18-21**
Chapter Review	**18-21**

CHAPTER 19—INQUIRY AND UTILITY COMMANDS ... 19-1
INQUIRY AND UTILITY COMMANDS ... 19-1
ID SCREEN COORDINATES ... 19-2
LISTING DRAWING INFORMATION ... 19-2
LISTING DRAWING DATABASE INFORMATION ... 19-3
COMPUTING DISTANCES ... 19-4
CALCULATING AREAS IN YOUR DRAWING ... 19-6
   Methods of Calculating Areas ... 19-6
FILE UTILITIES ... 19-8
AUTOCAD OPERATING SYSTEM COMMANDS ... 19-10
TIME COMMAND ... 19-10
Exercises ... 19-11
Chapter Review ... 19-12

CHAPTER 20—INTERMEDIATE DRAW COMMANDS ... 20-1
DRAWING ELLIPSES ... 20-1
   Specifying an Ellipse by Axis and Eccentricity ... 20-2
   Specifying an Ellipse by Axis and Rotation ... 20-3
   Specifying an Ellipse by Center and Two Axes ... 20-3
CONSTRUCTING ISOMETRIC CIRCLES AND ELLIPSES ... 20-5
CONSTRUCTING AN ELLIPTICAL ARC ... 20-6
TWO TYPES OF ELLIPSES ... 20-6
   Properties of Ellipses ... 20-6
DRAWING SOLID-FILLED CIRCLES AND DOUGHNUTS ... 20-7
OFFSETTING ENTITIES ... 20-8
   Constructing Parallel Offsets ... 20-9
   Constructing "Through" Offsets ... 20-10
CHAMFERING LINES AND POLYLINES ... 20-11
DRAWING POLYGONS ... 20-12
   Inscribed Polygons ... 20-13
   Circumscribed Polygons ... 20-14
   Edge Method of Constructing Polygons ... 20-14
   Constructing Polygons ... 20-15
BLOCKS AND INSERTS ... 20-15
   Combining Entities into a Block ... 20-16
   Creating a Drawing File from a Block ... 20-17
   Inserting Blocks into Your Drawing ... 20-18
   Inserting Blocks with a Dialog Box ... 20-23
   Multiple Insertions ... 20-26

| Exercises | 20-29 |
| Chapter Review | 20-35 |

CHAPTER 21—ISOMETRIC DRAWINGS ... 21-1
ISOMETRIC DRAWINGS	21-1
PRINCIPLES OF ISOMETRICS	21-3
ENTERING ISOMETRIC MODE	21-4
SWITCHING THE ISOPLANE	21-5
DRAWING IN ISOMETRIC	21-6
ISOMETRIC CIRCLES	21-7
ISOMETRIC TEXT	21-7
ISOMETRIC DIMENSIONING	21-8
Exercises	21-9
Chapter Review	21-14

CHAPTER 22—INTERMEDIATE EDIT COMMANDS ... 22-1
CHANGING ENTITY PROPERTIES	22-1
Changing Properties	22-2
Changing Properties with a Dialog Box	22-2
Changing Entity Points	22-5
CHANGING PROPERTIES WITH THE CHPROP COMMAND	22-7
ARRAYING OBJECTS IN THE DRAWING	22-7
Constructing Rectangular Arrays	22-7
Rotated Rectangular Arrays	22-9
Constructing Polar Arrays	22-9
MIRRORING OBJECTS	22-9
Mirrored Text	22-10
DIVIDING AN ENTITY	22-11
Using Blocks to Divide	22-11
USING THE MEASURE COMMAND	22-12
EXPLODING BLOCKS	22-12
TRIMMING ENTITIES	22-13
Trimming Polylines	22-14
Trimming Circles	22-15
EXTENDING OBJECTS	22-15
Using Extend with Polylines	22-16
LENGTHENING LINES	22-17
ROTATING OBJECTS	22-18
Rotating from a Reference Angle	22-19
Rotating an Object by Dragging	22-20

SCALING OBJECTS	22-20
Changing Scale by Numerical Factor	22-20
Changing Scale by Reference	22-21
STRETCHING OBJECTS	22-21
Stretch Rules	22-22
UNDOING DRAWING STEPS	22-23
General Notes	22-25
EDITING WITH GRIPS	22-26
Enabling Grip Editing	22-26
Using Grips for Editing	22-29
Grip Editing Commands	22-30
Exercises	**22-36**
Chapter Review	**22-49**
CHAPTER 23—INTERMEDIATE OPERATIONS	**23-1**
SETTING THE CURRENT COLOR	23-1
STORING AND DISPLAYING DRAWING VIEWS	23-3
STORING AND DISPLAYING DRAWING SLIDES	23-4
Making a Slide	23-4
Viewing a Slide	23-5
Slide Libraries	23-6
Slide Shows	23-6
Purging Objects from a Drawing	23-7
RENAMING PARTS OF YOUR DRAWING	23-8
Using a Dialog Box to Rename	23-9
PRODUCING AND USING POSTSCRIPT IMAGES	23-9
Exporting a PostScript Image	23-10
Importing a PostScript Image	23-12
Displaying a PostScript Fill	23-14
Exercises	**23-16**
Chapter Review	**23-17**
CHAPTER 24—ADVANCED OPERATIONS	**24-1**
DRAWING POLYLINES	24-1
USING THE PLINE COMMAND	24-2
DRAWING ARCS WITH POLYLINES	24-3
POLYLINE EDITING	24-5
CONSTRUCTING MULTIPLE PARALLEL LINES	24-8
Using the MLine Command	24-8

Editing Multilines with MlEdit	24-10
Defining Multilines with MlStyle	24-11
CONSTRUCTING SPLINE CURVES	24-14
Spline Anatomy	24-14
Spline Segments	24-15
Effects of Edit Commands on Splines	24-15
THE SPLINE COMMAND	24-16
EDITING SPLINES WITH SPLINEDIT	24-17
POLYLINE VERTEX EDITING	24-18
EXCHANGE FILE FORMATS	24-22
Drawing Interchange File Format	24-22
Drawing Interchange Binary Files	24-23
DRAWING FILE DIAGNOSTICS	24-24
DRAWING FILE RECOVERY	24-24
Exercises	**24-25**
Chapter Review	**24-30**
CHAPTER 25—VIEWPORTS AND WORKING SPACE	**25-1**
USING VIEWPORTS IN AUTOCAD	25-1
The Current Viewport	25-2
Drawing Between Viewports	25-2
Setting Viewport Windows	25-2
Redraws and Regenerations in Viewports	25-5
WORKING SPACES	25-5
MODEL SPACE AND PAPER SPACE	25-6
Switching to Paper Space	25-6
Paper Space Icon	25-6
Switching to Model Space	25-7
MVIEW COMMAND	25-7
Mview Options	25-7
CREATING RELATIVE SCALES IN PAPER SPACE	25-9
Tutorial	**25-10**
Exercise	**25-12**
Chapter Review	**25-14**
CHAPTER 26—ATTRIBUTES	**26-1**
ATTRIBUTES	26-1
Suppression of Attribute Prompts	26-1
Tutorial	**26-2**

CONTROLLING THE DISPLAY OF ATTRIBUTES	26-6
EDITING ATTRIBUTES	26-7
Individual Editing	26-8
Editing Attributes with a Dialog Box	26-9
Global Editing	26-10
Visible Attributes	26-10
ATTRIBUTE EXTRACTIONS	26-10
Using a Dialog Box to Extract Attributes	26-11
Creating Template Files	26-12
Exercises	**26-13**
Chapter Review	**26-16**

CHAPTER 27—EXTERNAL REFERENCE DRAWINGS AND OLE OBJECTS 27-1

OVERVIEW	27-1
XREF COMMAND	27-3
Attach (Adding an External Reference)	27-3
? (List External Reference Information)	27-4
Bind (Bind an Xref to the Drawing)	27-5
Detach (Remove an Xref from the Drawing)	27-5
Path (Change Path to an Xref)	27-6
Reload (Update External References)	27-6
Overlay (Unrepeated Xrefs)	27-7
XREF LOG	27-7
XBIND COMMAND	27-7
Exercise	**27-8**
OBJECT LINKING AND EMBEDDING	27-9
Tutorials	
Placing an OLE Object in AutoCAD	27-12
Placing an AutoCAD Drawing as an OLE Object	27-16
CAUTIONS WITH OLE	27-18
Chapter Review	**27-19**

CHAPTER 28—CUSTOMIZING AUTOCAD 28-1

SETTING SYSTEM VARIABLES	28-1
Using Setvar While in a Command	28-2
DISPLAYING BLIP MARKS	28-3
SETTING APERTURE SIZE	28-3
CONTROLLING DRAWING REGENERATIONS	28-4

SETTING THE VIEW RESOLUTION	28-4
REDEFINING COMMANDS	28-5
Exercise	**28-5**
Chapter Review	**28-6**

CHAPTER 29—CUSTOMIZING MENUS AND ICONS ... 29-1

CUSTOM MENUS	29-1
SIMPLE MENUS	29-2
SCREEN DISPLAY	29-2
MULTIPLE MENUS	29-3
SUBMENUS	29-4
LINKING MENUS	29-4
MULTIPLE COMMANDS IN MENUS	29-5
LOADING MENUS	29-7
TABLET MENUS	29-7
PULL-DOWN MENUS	29-8
ICON MENUS	29-10
Selecting Slides from Libraries	29-12
Designing Icon Boxes	29-12
CUSTOMIZING THE TOOLBAR	29-13
THE TBCONFIG COMMAND	29-14
Chapter Review	**29-18**

CHAPTER 30—INTRODUCTION TO AUTOCAD 3D ... 30-1

INTRODUCTION TO AUTOCAD 3D	30-1
HOW TO APPROACH 3D	30-2
3D THEORY	30-2
Coordinate System	30-2
3D Versus Perspective	30-4
Clipping Planes	30-4
Exercise	**30-6**
Chapter Review	**30-7**

CHAPTER 31—VIEWING 3D DRAWINGS ... 31-1

VIEWING 3D DRAWINGS	31-1
METHODS OF VIEWING 3D DRAWINGS	31-1
SETTING THE 3D VIEWPOINT	31-2
Creating a View by Coordinates	31-2
Creating a View by Axes	31-2

SETTING VIEW BY DIALOG BOX	31-3
DYNAMIC VIEWING	31-4
Dview Options	31-4
CAMERA AND TARGET POSITIONING	31-7
Positioning the Camera and Target	31-8
Setting the Distance to View From	31-11
PANNING THE VIEW	31-12
ZOOMING THE VIEW	31-13
Twisting the View	31-14
Setting Clipping Planes	31-14
PRODUCING HIDDEN LINE VIEWS	31-16
PRODUCING SHADED IMAGES	31-17
Shading Types	31-17
Chapter Review	**31-20**
CHAPTER 32—THE USER COORDINATE SYSTEM	**32-1**
THE USER COORDINATE SYSTEM	32-1
THE UCS ICON	32-3
Head-On Indicator	32-3
CHANGING THE UCS	32-4
PRESET UCS ORIENTATIONS	32-10
UCSFOLLOW SYSTEM VARIABLE	32-10
UCSICON COMMAND	32-11
DDUCS COMMAND (UCS DIALOG BOX)	32-12
Changing the Current UCS	32-13
Listing UCS Information	32-13
Deleting a UCS	32-13
Chapter Review	**32-13**
CHAPTER 33—DRAWING IN 3D	**33-1**
DRAWING IN 3D	33-1
EXTRUDED ENTITIES	33-2
Elevation	33-2
Thickness	33-3
Setting Elevation and Thickness	33-3
Tutorial	**33-4**
Changing Existing Entities	33-6
DRAWING IN COORDINATE SYSTEMS	33-6
Tutorial	**33-7**

CREATING SOLID 3D FACES 33-16
   3DFACE Command 33-17
   Placing 3D Faces 33-17
Tutorial **33-18**
3D POLYGON MESHES 33-19
   Mesh Density 33-19
3DMESH COMMAND 33-20
   Constructing a 3D Mesh 33-20
PFACE COMMAND (POLYFACE MESH) 33-21
RULESURF COMMAND (RULED SURFACE) 33-22
   Constructing Ruled Surfaces 33-23
TABSURF COMMAND (TABULATED SURFACE) 33-23
   Constructing Tabulated Mesh Surfaces 33-24
REVSURF COMMAND (REVOLVED SURFACE) 33-25
   Creating Revolved Surfaces 33-26
EDGESURF COMMAND (EDGE-DEFINED SURFACE) 33-28
   Constructing Edge-Defined Surfaces 33-29
3D OBJECTS 33-29
   Box 33-29
   Cone 33-30
   Dome 33-31
   Dish 33-32
   Sphere 33-33
   Torus 33-34
   Wedge 33-35
Summary **33-36**
Chapter Review **33-36**

CHAPTER 34—INTRODUCTION TO SOLID MODELING 34-1
OVERVIEW 34-1
   Who Uses Solid Modeling 34-2
   Drawing with Solids 34-2
Chapter Review **34-4**

CHAPTER 35—CONSTRUCTING SOLID PRIMITIVES 35-1
SOLID PRIMITIVES 35-1
   Drawing Solid Primitives 35-2
   Drawing a Solid Box 35-3
   Creating a Solid Cone 35-6

Creating a Solid Cylinder	35-10
Creating a Solid Sphere	35-12
Constructing a Solid Torus	35-13
Constructing a Solid Wedge	35-17
Chapter Review	**35-20**

CHAPTER 36—CREATING CUSTOM SOLIDS ... 36-1

CREATING SOLIDS	36-1
Creating Solid Extrusions	36-2
Creating a Solid Revolution	36-5
Tutorial	**36-11**
Chapter Review	**36-15**

CHAPTER 37—MODIFYING SOLID OBJECTS ... 37-1

MODIFYING SOLID SHAPES	37-1
Creating Solid Intersections	37-2
Subtracting Solids	37-3
Joining Solid Objects	37-4
Chamfering a Solid	37-5
Filleting a Solid	37-7
Chapter Review	**37-8**

CHAPTER 38—CREATING COMPOSITE SOLID MODELS ... 38-1

DRAWING A COMPOSITE SOLID MODEL	38-1
Beginning the Model	38-2
Displaying Your Model	38-15
Summary	**38-16**

CHAPTER 39—REALISTIC RENDERING ... 39-1

THE RENDER COMMAND	39-1
Your First Rendering	39-2
Advanced Renderings	39-3
CREATING LIGHTS—LIGHT	39-6
COLLECTING LIGHTS INTO SCENES—SCENE	39-10
APPLYING MATERIALS AND BACKGROUNDS—RMAT AND REPLAY	39-12
SAVING RENDERINGS—SAVEIMG	39-16
Improving the Speed of Rendering	39-18
THE RENDERUNLOAD COMMAND	39-19
Chapter Review	**39-19**

CHAPTER 40—INTRODUCTION TO AUTOLISP ... **40-1**
USING AUTOLISP ... 40-1
Why Use AutoLISP? ... 40-2
Using an AutoLISP Program ... 40-2
Chapter Review ... **40-6**

CHAPTER 41—PROGRAMMING IN AUTOLISP ... **41-1**
AUTOLISP BASICS ... 41-1
Arithmetic Functions ... 41-1
AutoLISP and AutoCAD Commands ... 41-2
SETQ Function ... 41-3
Getting and Storing Points in AutoLISP ... 41-3
Placing Command Prompts within AutoLISP ... 41-3
Using Notes within a Routine ... 41-3
WRITING AND USING LISP ... 41-4
Using the LISP Routine ... 41-5
Summary ... **41-5**
Chapter Review ... **41-6**

CHAPTER 42—PROGRAMMING TOOLBAR MACROS ... **42-1**
USING MACROS ... 42-1
Why Use Macros? ... 42-1
Toolbar Macro Basics ... 42-2
Summary ... **42-9**
Chapter Review ... **42-10**

APPENDICES
A—PROFESSIONAL CAD TECHNIQUES ... A-1
B—COMMAND SUMMARY ... B-1
C—SYSTEM VARIABLES ... C-1
D—ACAD PROTOTYPE DRAWING SETTINGS ... D-1
E—HATCH PATTERNS ... E-1
F—AUTOCAD LINETYPES ... F-1
G—TABLES ... G-1
H—GLOSSARY ... H-1

INDEX ... **X-1**

INTRODUCTION

With more than 1,000,000 users around the world, AutoCAD offers engineers, architects, drafters, interior designers, and many others a fast, accurate, extremely versatile drawing tool. Welcome to the first book about how to *use* AutoCAD and make it the productivity tool for you!

Now in its 8th edition, *Using AutoCAD for Release 13 Windows* makes using AutoCAD a snap by presenting the user with easy-to-master, step-by-step tutorials through all the commands of AutoCAD. Designed to lead the novice through the basics of AutoCAD to more advanced features like customizing, 3D, and AutoLISP, *Using AutoCAD* introduced the method of using menus and submenus in the side margins of the text to illustrate the steps the user must take to execute a command, and allow space for additional notetaking.

WHAT'S NEW IN THIS EDITION:

- New text design for easy access to key topics and exercises.
- Competency-based objectives start off each chapter to keep you on track with the new CADD Skill Standards.
- New illustrations! Clearly display what you will see on the screen as you perform each command.
- Fully updated to Release 13 for Windows 95. New sections added include:

Rays	Xlines	Groups	Mtext	OLE v2
MtProp	Lengthen	MLine	MlEdit	Toolbox
MlStyle	Spline	SplinEdit	elliptical arc	DdStyle

- An option for SIMPLE FLEXIBILITY: Each chapter of *Using AutoCAD* is written to take you step-by-step through the many features of AutoCAD. However, not everyone follows the same path when learning the software; that's why we offer a separate, non-bound, 3-hole punched, alternative of this edition. Chapter page numbering is self-contained, so you may "customize" the direction this text takes you by arranging the chapters in the order that suits you best.

FEATURES:

Pull-Down Menus

Pull-down menus throughout the text offer an easy way for users to follow their progress from the menu bar.

Notes

> ***NOTES:*** highlight programming and user hints on working effectively with the AutoCAD commands.

Tutorials

Tutorials are found throughout each chapter outlined by a box. They serve to reinforce groups of topics learned and range in complexity from brief to chapter-long.

A disk icon points out if you are to use a file found in the accompanying work disk.

Exercises and Chapter Review

All exercises are pulled to the back of each chapter for easy access after reading the concepts of the chapter. Review questions test key chapter concepts.

CONVENTIONS:

Command lines are indented: **User response is boldfaced** *(Author instructions are italicized within parentheses.)*

Key icons are displayed when instructed to use:

Enter	[ENTER]	Backspace	[BACKSPACE]	F1	[F1]
Control	[CTRL]	Escape	[ESC]		

WE WANT TO HEAR FROM YOU!

Many of the changes to the look and feel of this new edition came by way of requests from and reviews done by users of our previous editions. We'd like to hear from you as well! If you have any questions or comments, please contact:

The CADD Team
c/o Delmar Publishers
3 Columbia Circle, PO Box 15015, Albany, NY 12212

DRAW FROM EXPERIENCE:

After learning to use AutoCAD with our text, you *must* have some drawings of which you are especially proud. Would you like to share your experience with the next generation of *Using AutoCAD* users? Send us your drawing files with your name, affiliation, address, phone number, and a brief description of your "experience" using AutoCAD to create this drawing. With your permission, we'll incorporate as many as we can into the next edition of *Using AutoCAD*.

ABOUT THE AUTHORS:

Ralph Grabowski, Delmar Publisher's consulting editor, is a frequent contributor to *CADENCE* magazine, a member of the Review Board of *InfoWorld* magazine and former Senior Editor at *CADalyst* magazine. Mr. Grabowski has written about AutoCAD since 1985, and is the author of over two dozen books on computer-aided design. He now publishes the *CAD++ VRML* newsletter, the international reference for CAD and virtual reality modeling developers.

James E. Fuller earned his degree in architecture from the University of Tennessee. He now holds an NCARB national certificate and is owner of the successful Fuller & Associates Architecture firm. Mr. Fuller is author of numerous articles and several best-selling books.

CHAPTER 1

AUTOCAD QUICK START

The new user needs an opportunity to experience the "feel" of the AutoCAD program. The concept and operation of graphic design software can be unique to a first-time user. This "quick start" chapter gets your feet wet and introduces you to some of the features to be discussed in detail in other chapters. After completing this chapter, you will be able to:

- Start and exit AutoCAD.
- Create a new drawing.
- Practice placing entities into a drawing.
- Examine the basic 3D capabilities of AutoCAD.

INTRODUCTION TO USING AUTOCAD

Welcome to *USING AUTOCAD*! This chapter is specifically designed to acquaint you with the AutoCAD drawing program from the start. Subsequent chapters will cover the subject more thoroughly. Let's get started!

Starting AutoCAD

To start AutoCAD Release 13 for Windows, you must have Microsoft Windows loaded and running. If it is not open, double-click with the left button of your mouse on the program group AutoCAD R13. This window opens and you will see several icons inside it. Double-click on the icon labeled AutoCAD R13.

Your computer will now load the AutoCAD program. After an opening screen is displayed, you will see a drawing screen. This screen is referred to as the drawing editor and should be similar to Figure 1-1.

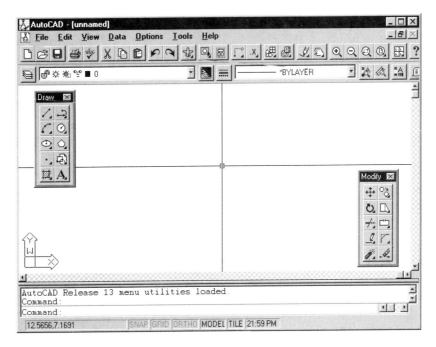

FIGURE 1-1 AutoCAD Drawing Screen

Drawing in AutoCAD

AutoCAD is used with a pointing device. This device is usually a digitizing tablet or mouse. Move the pointing device and notice how a *crosshair* moves around the screen. The intersection of the crosshair is used to specify points on your drawing. From this point forward, we will refer to the pointing device as the *mouse*.

AutoCAD draws or edits objects by using *commands*. Commands are words such as *line*, *circle*, *arc*, and *erase* that describe the object to be drawn or the operation that you wish to perform.

At the bottom of the drawing editor is an area known as the "command prompt area" (refer to Figure 1-1). This area lists the commands you have entered. You should see the word Command: on that line now. You can specify commands from menus or by typing them from the keyboard.

Starting a New Drawing

Let's continue by typing (or "keying-in") a command to start a new drawing. Type **NEW** from the keyboard. Notice how the command is listed on the command line. It doesn't matter whether the commands are typed in uppercase or lowercase letters. If you make a mistake, just use the backspace key to back up and retype. After you have typed **NEW**, press [ENTER]. Pressing the Enter key "sends" the command to AutoCAD. You should now see a *dialog box* similar to the one in Figure 1-2.

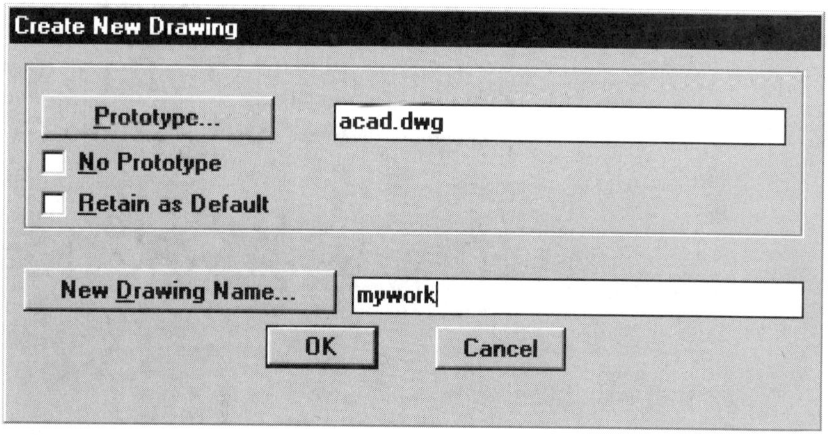

FIGURE 1-2 Dialog Box Used to Create a New Drawing

Type **MYWORK** from the keyboard. The text will appear at the blinking bar next to the "New Drawing Name..." box. Now move the mouse and notice how the crosshairs have turned into an arrow pointer. Move the pointer to the OK box and press the mouse button. AutoCAD will initialize a new drawing screen.

If you wish to type any command from the keyboard, you must first have a "clear" command line. The line must have the Command: prompt, without anything after it. If there is other text, just clear it by entering [ESC] from the keyboard before typing in the new command.

Changing the Pull-Down Menu

Every version of AutoCAD has had a set of pull-down menus that give you access to most of the commands. However, AutoCAD Release 13 for Windows is different. When you use it for the first time after installing it, Release 13 for Windows has an abbreviated set of pull-down menus.

For the purpose of following along in this book, you need to switch to the "full" menu. If this has not already been done with the copy of AutoCAD you are using, then follow these instructions:

1. Type the word "menu".
2. AutoCAD displays the Select Menu File dialog box (Figure 1-3).

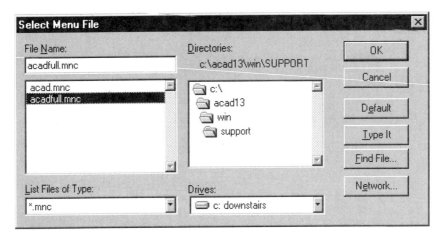

FIGURE 1-3 Selecting the Full AutoCAD Menu File

3. With your mouse, click on the downward-pointing arrow in the bottom left-hand corner box labeled "list files of type." A list of options appears.
4. Click on the *.mnu option. The names of different **mnu** files appear in the file selection box.
5. With your mouse, click on Acadfull.mnu. You have selected the full menu for AutoCAD.
6. Click on the OK button. A warning box appears. This is harmless. Click on the OK button in this box. AutoCAD loads the full menu into itself.

You are now ready to explore AutoCAD's user interface.

Using Pull-Down Menus

Let's enter a command from the *pull-down menu*. Move the crosshairs above the top of the screen on the menu bar that extends the width of the drawing area (Figure 1-4).

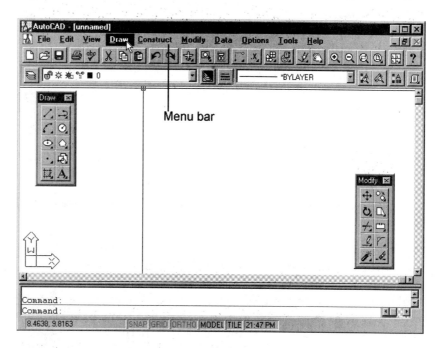

FIGURE 1-4 Accessing Pull-Down Menus

As you move the arrow across each word in the menu bar, it is *highlighted*. Highlight the word Draw and press the left-hand mouse button. A *pull-down menu* extends downward into the drawing area (Figure 1–5).

1-6 USING AUTOCAD

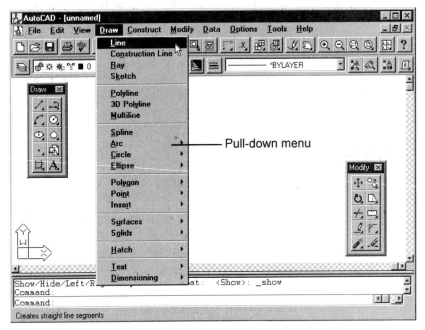

FIGURE 1-5 The Draw Pull-Down Menu

Pull-down menus contain commands that can be selected with the pointing device. Let's pick the Line command from the Draw menu. Move the pointer down and highlight the word Line, then press the mouse button. Notice that the word "Line" appears on the command line at the bottom of the screen. After the Line command is displayed, the words "line From point:" appear. This is called a *prompt*. AutoCAD always tells you what it expects on the command line. AutoCAD is now asking for the point the line starts *from* (the From point). Move the crosshairs into the screen area and enter a point at approximately the location shown in Figure 1-6.

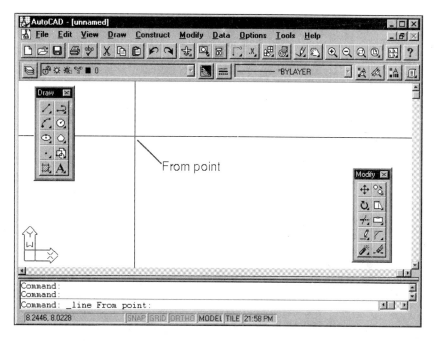

FIGURE 1-6 Drawing a Line

You are now prompted for a "To point". Notice how the line "sticks" to the intersection of the crosshairs. As you move the crosshairs around the screen, the line will stretch and follow. This is called *rubber banding*. Move the crosshairs and enter a point at approximately the point shown in Figure 1-7.

1-8 USING AUTOCAD

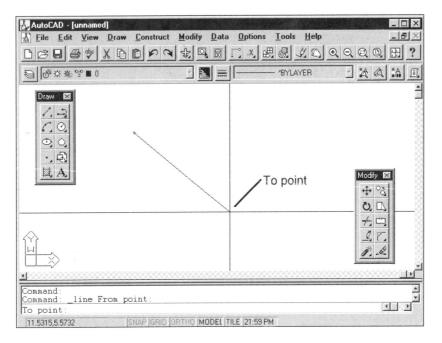

FIGURE 1-7 Designating the Endpoint of the Line Segment

Continue to enter lines as shown in Figure 1-8.

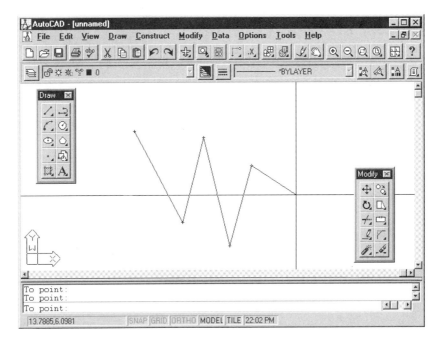

FIGURE 1-8 Drawing Multiple Line Segments

After you have entered the last point, click the right mouse button to end the Line command. Alternately you may enter [ESC] from the keyboard to terminate any command. Some commands terminate automatically, while others must be terminated when you are finished. The Line command remains active so you can draw as many line segments as you desire without having to reselect the command repeatedly.

Blip Marks

If you look closely at the endpoints of the lines, you will see some small crosses. These are called *blip marks*. These temporary blips are *not* part of your drawing. They are displayed for reference purposes at points you have entered. Let's remove the blips from our drawing. Highlight the menu bar and pull down the View menu. Select the Redraw All command. AutoCAD will "redraw" the screen and remove any blip marks.

Using Icons

AutoCAD Release 13 for Windows places a strong emphasis on toolbars and icon buttons. Earlier versions of AutoCAD for Windows had a single floating toolbox and a separate fixed toolbar.

Release 13 is much more flexible. It has dozens and dozens of toolbars; each can be floating or docked. If you were to open them all, AutoCAD's 50 toolbars would practically obscure the entire AutoCAD window!

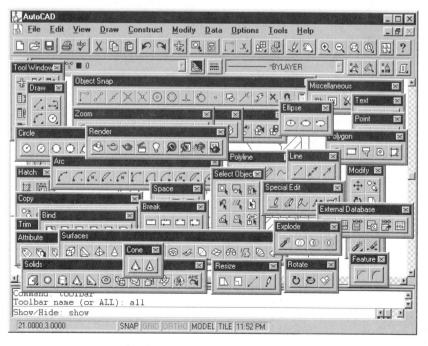

FIGURE 1-9 The Result of Opening Every Toolbar

Instead, AutoCAD normally displays just four toolbars. These are (1) the Standard toolbar (top of the window); (2) the Object Properties toolbar (just below); (3) the Draw toolbar (floating); and (4) the Modify toolbar. These four hold most of the everyday commands you use with AutoCAD.

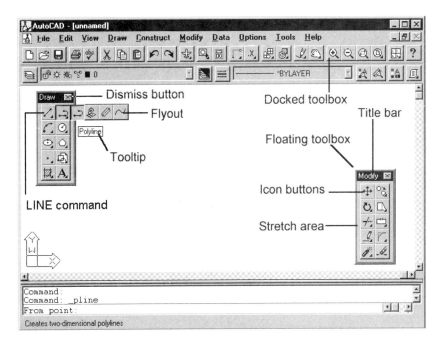

FIGURE 1-10 The Parts of a Toolbar

Using a Toolbar Button. Every button on a toolbar holds a command. For example, the first button of the Draw toolbar is the Line command for drawing line segments. The symbol of the line with two dots at either end is called an "icon" for the Line command. When you cannot remember the purpose of a button, move the cursor over the button and wait for a second or two. AutoCAD displays a "tooltip," a one-word description of the button's purpose.

To draw a line with the icon button:

1. Move the cursor over the Line button.
2. Click the mouse's left button.
3. At the command line, you should see AutoCAD prompting you, as follows:

   Command: _LINE From point:

4. You can now draw some lines or press [ESC] to exit the Line command.

Accessing Flyouts. Some buttons have additional buttons hidden in them. You recognize these by the small black triangle in the corner of the button. The triangle indicates the presence of a "flyout." The flyout is a group of buttons that "flies out" from a single button.

To access a flyout:

1. Move the cursor over any button with the tiny triangle.
2. Press the left mouse button and wait a second or two.
3. AutoCAD displays the flyout buttons.
4. Without letting go of the mouse button, move the cursor over the buttons on the flyout.
5. As the cursor passes over a button, it changes slightly to give the illusion of being depressed.
6. When you reach the button you want to use, release the mouse button.
7. AutoCAD starts the command associated with the button. In addition, the button you accessed moves to the "front" of the line and appears on the toolbar.

Manipulating the Toolbar. The toolbar has many controls hidden in it. You can move, resize, dock, and dismiss the toolbar. Here's how:

1. To move a toolbar, move the cursor over the title bar. Click the mouse button and drag the toolbar to another location.
2. To dock the toolbar means to move it to the side of the drawing area, such as the two docked toolbars at the top of the window in Figure 1-10. To dock the toolbar, move it all the way against one of the four sides of the drawing area. To make the toolbar float, grab it by the edge and drag it away from the edge of the drawing area.
3. To resize (or stretch) the toolbar, move the cursor over one of the four edges of the toolbar. Press the left mouse button, then drag the toolbox into a new shape.
4. To dismiss (get rid of) the toolbar, click on the tiny button with the X. The toolbar disappears.

Summary

- AutoCAD drawings are constructed by using commands.
- Commands can be entered from the keyboard, pull-down menus, screen menus, toolbars, and from a digitizer template (you will learn about this later in another chapter).
- The command line shows your command activities and displays prompts that tell you what input AutoCAD expects from you.
- The command line must be "clear" before typing a new command from the keyboard. You can use either uppercase or lowercase when entering a command from the keyboard.
- The Line command is used to draw lines. The Circle command is used to draw circles. Erase objects with the Erase command and restore them with the Oops command. Use the Redraw command to clear the temporary blip marks from your drawing.
- You cancel a command by pressing the [ESC] key.

Exiting a Drawing

How you exit your drawing will depend on the command you use. The following four sections outline the possibilities. Choose the one you want and follow the instructions. If you intend to continue with the tutorial, select one of the last two choices.

Discard the Drawing and Quit AutoCAD. If you don't want to keep your drawing and you wish to stop work now, enter the Quit command (you can do this from the keyboard). AutoCAD will display a dialog box (see Figure 1-11) that asks you to confirm your choice. Move the pointer to and select the "No" box. AutoCAD will not record your work to disk and will return you to the Windows desktop.

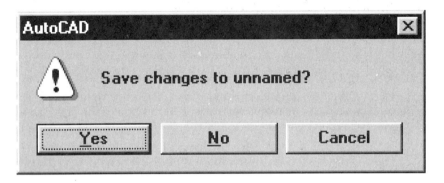

FIGURE 1-11 Drawing Modification Dialog Box

Save the Drawing and Quit AutoCAD. If you want to save your work to disk and quit AutoCAD, enter the End command. AutoCAD will save your work under the name "MYWORK" and return you to the Windows desktop.

Discard the Drawing and Remain in AutoCAD. If you do not want your drawing saved, but would like a new AutoCAD screen, use the New command.

We used the New command to start this drawing. You will need to specify a name for the new drawing. If you will be constructing the following tutorial drawing, enter **TUTOR** as the drawing name.

Save Your Work and Remain in AutoCAD. If you want to keep your work and remain in AutoCAD, enter the Saveas command. A dialog box will be displayed on the screen. Verify that the drawing name "MYWORK" is listed in the File box, then select OK. Next, use the New command to start the new drawing name. If you will be constructing the following tutorial, enter **TUTOR** as the drawing name.

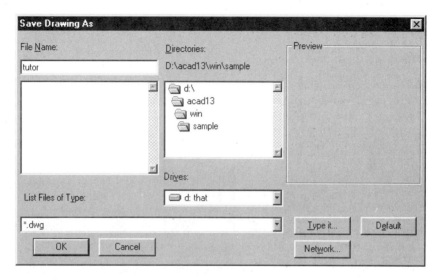

FIGURE 1-12 Saving Your Drawing

Now let's practice a little bit using AutoCAD to get a "feel" for the program.

TUTORIAL

Type New from the keyboard. AutoCAD will display the *Create New Drawing* dialog box. Respond to the "New Drawing Name..." box with the drawing name "TUTOR". Figure 1-13 shows the drawing you will construct.

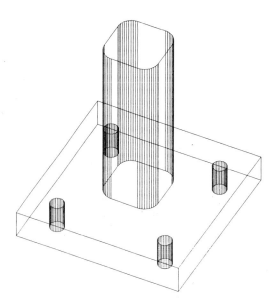

FIGURE 1-13 Finished Drawing

GETTING STARTED

The following is a listing of items to be typed and "entered" to complete the drawing. The items you type will appear at the bottom of the screen on the "command line". The word Command: appears at this location now. Your response is shown boldfaced in this tutorial. You may type the items in either uppercase or lowercase. Be sure to press the Enter key after each response. Items enclosed in parentheses are to be done and not typed. And, if ENTER is shown, press the Enter key on the keyboard.

If you mess up, just press **U** and then press the [ENTER] key. This will "undo" the previous step. You may use it several times to undo each step in reverse order. In order to enter a new "command", the prompt on the command line must say Command:. If it does not, press the [ESC] key. This cancels the current command, and places AutoCAD ready for the next command.

Let's first set up the size of the drawing area.

> Command: **Limits**
> Reset Model space limits:
> ON/OFF <Lower left corner> <0.0000,0.0000>: [ENTER]
> Upper right corner <12.0000,9.0000>: **12,10**

Let's continue by zooming out to display the entire screen. "Zoom" is the way your drawing is enlarged or reduced. Since we changed the size of the work area with the Limits command, we need to zoom out to show the entire work space.

> Command: **Zoom**
> All/Center/Dynamic/Extents/Left/Previous/Vmax/Window/ <Scale (X/XP)>: **A**
> Regenerating drawing.

This will not have a visible effect on the drawing screen.

We will now display a grid with a spacing of 1.

> Command: **Grid**
> Grid spacing (X) or ON/OFF/Snap/Aspect <0.0000>: **1**

Let's set the snap so that our crosshair cursor moves at increments of 1. The crosshair cursor is moved by the input device (mouse, digitizer pad, or cursor keys) and is the way you show AutoCAD where you want to place points when drawing.

> Command: **Snap**
> Snap spacing or ON/OFF/Aspect/Rotate/Style/<1.0000>: **1**

Drawing the Baseplate

Let's draw the baseplate. We are looking down in plan and in 2D view. We will change into a 3D view later.

>Command: **Box**
>Center/<corner of box><0,0,0>: **3,2**
>Cube/Length/<other corner>: **9,8**
>Height: **1**

Your drawing should look similar to Figure 1-14.

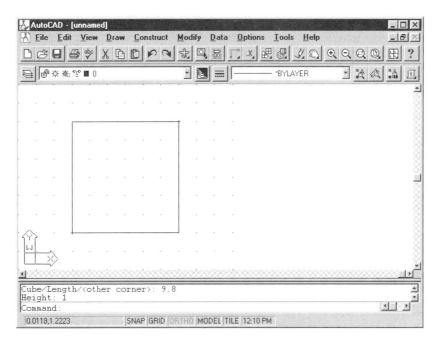

FIGURE 1-14 Baseplate

Drawing the Shaft

Let's now draw the shaft.

    Command: **Box**
    Center/<corner of box><0,0,0>: **5,4,1**
    Cube/Length/<other corner>: **7,6,1**
    Height: **6**

Your drawing should look similar to Figure 1-15.

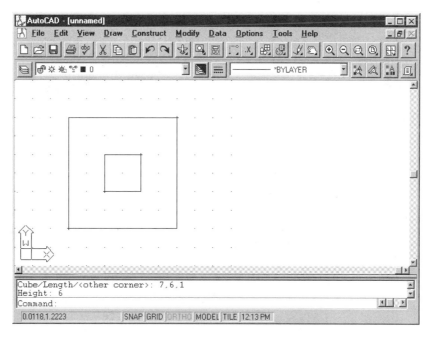

FIGURE 1-15 Baseplate and Shaft

Drawing the Holes

Let's now draw a hole.

> Command: **Cylinder**
> Elliptical/<center point><0,0,0>: **4,3**
> Diameter/ <Radius>: **.25**
> Center of other end/<Height>: **1**

We will now copy the hole to the other parts of the baseplate. Let's use a copy option called *Multiple*. This will allow us to make multiple copies easier.

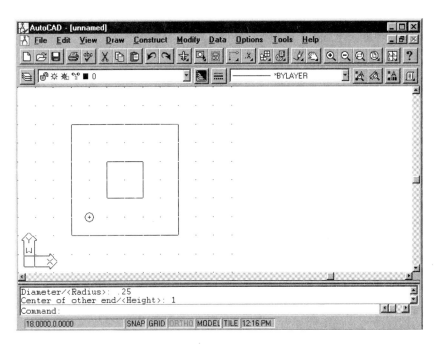

FIGURE 1-16 Drawing the First Circle

Command: **Copy**
Select objects: **L** [ENTER]
1 found
Select objects: [ENTER]
<Base point or displacement>/Multiple: **M**
Base point: *(Place cursor and click at center of circle.)*
Second point of displacement: **8,3**
Second point of displacement: **8,7**
Second point of displacement: **4,7**
Second point of displacement: [ENTER]

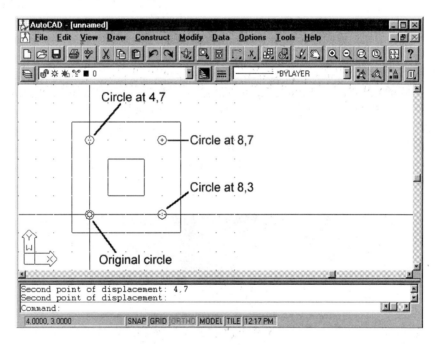

FIGURE 1-17 Copying the Circles

Creating the Holes

You have drawn four cylinders to represent the holes but they are solid cylinders. You need to remove the cylinders from the baseplate so that they become holes. We will now use the Subtract command to remove the cylinders from the baseplate.

Command: **Subtract**
Select solids and regions to subtract from...
Select objects: *(pick the baseplate)*
1 found Select objects: *(press Enter)*
Select solids and regions to subtract...
Select objects: **fence**
First fence point: *(place cursor and click at center of circle)*
Undo/<Endpoint of line>: *(click at center of second circle)*
Undo/<Endpoint of line>: *(click at center of next circle)*
Undo/<Endpoint of line>: *(click at center of last circle)*
Undo/<Endpoint of line>: *(press Enter)*
4 found
Select objects: *(press Enter)*

The baseplate won't look any different after the cylinders are subtracted from the baseplate until we remove the hidden lines at the end of this tutorial.

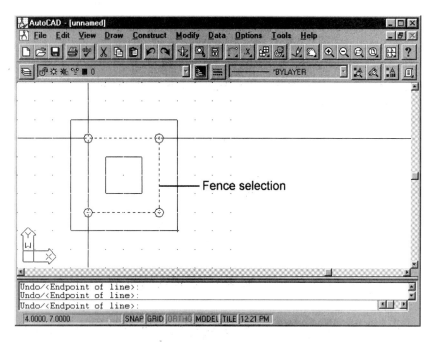

FIGURE 1-18 Subtracting Cylinders from Baseplate

Viewing the Drawing in 3D

Let's have some fun and view the drawing in 3D. (You learn more about these capabilities in Chapter 30.)

Command: **Vpoint**
Rotate/<View point> <0.0000,0.0000,1.0000>: **.5,-1,1**
Regenerating drawing.

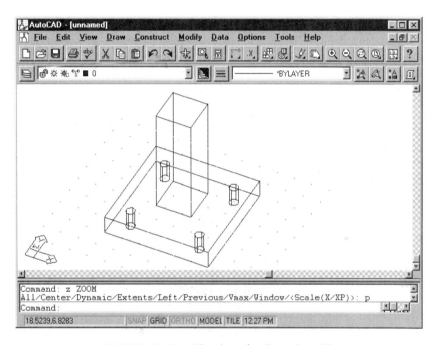

FIGURE 1-19 Viewing the Drawing 3D

Filleting the Corners

Let's use the Fillet command to round the corners of the shaft.

> Command: **Fillet**
> (TRIM mode) Current fillet radius = 0.0000
> Polyline/Radius/Trim/<Select first object>: *(pick one vertical edge of the shaft)*
> Enter radius <0.0000>: **.5**
> Chain/Radius/<Select edge>: *(pick another vertical edge of the shaft)*
> Chain/Radius/<Select edge>: *(pick another vertical edge)*
> Chain/Radius/<Select cdge>: *(pick last vertical edge)*
> Chain/Radius/<Select edge>: *(press Enter)*
> 4 edges selected for fillet.

After a few seconds, AutoCAD fillets the four corners of the shaft.

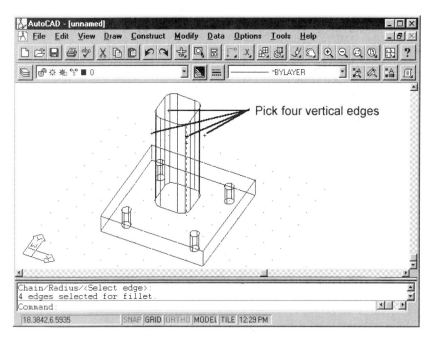

FIGURE 1-20 Filleting the Shaft

Removing Hidden Lines

You are seeing the 3D object in wireframe view. Let's remove the hidden lines. Removing the hidden lines takes a small amount of time.

> Command: **Hide**
> Regenerating drawing.
> Hiding lines 100% done.

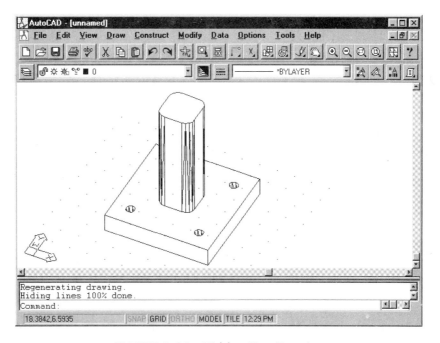

FIGURE 1-21 Hidden Line Drawing

Save or discard your drawing in the manner you learned earlier in this chapter.

USING AUTOCAD

So, this is a quick start for using AutoCAD. Welcome to the rest of *Using AutoCAD*. To start you in the right direction while working through this text, its best to know what lies ahead. Chapters 2 through 4 give you an overview of CAD and your CAD system; Chapter 5 introduces you to DOS (the disk operating system) and Windows; Chapter 6 begins to explain in more detail some of the concepts you learned in this chapter. From there the rest of the chapters take you step-by-step through the program. Before long you'll be using AutoCAD like a pro—enjoy!

> **NOTE:** You can move chapter by chapter or use the table of contents or index to find a specific topic. Each chapter is page-numbered separately.

CHAPTER 2

WHY CAD?

Computer-aided design can be used to do many jobs more effectively. This chapter reviews the benefits and applications of CAD. After completing this chapter, you will be able to:

- Explain the benefits of using computer-aided design.
- Identify some standard applications of computer-aided design and the benefits of using CAD for these disciplines.
- Recognize the availability of information services.

TRADITIONAL DRAFTING TECHNIQUES

The necessity to perform drawings as a means of communication has been present throughout history. While other types of work have benefited from technology, traditional drafting techniques have remained relatively unchanged.

The introduction of Computer-Aided Drafting (CAD) has had an impact on the industry that is greater than all the previous changes combined. The acceptance of the graphics world of CAD and its capabilities has been phenomenal. CAD presents advantages that are undeniably superior to traditional techniques.

BENEFITS OF CAD

Computer-Aided Drafting is a more efficient and versatile drafting method than traditional techniques. Some of the advantages are:

Accuracy

Computer-generated drawings can be drawn and plotted to an accuracy of up to fourteen decimal places of the units used. The numerical entry of critical dimensions and tolerances is more reliable than the traditional methods of manual scaling.

Speed

The ability of the CAD operator to copy, array items, and to edit his work on the screen speeds up the drawing process. If the operator customizes his system to a specific task, the work speed can be greatly increased.

Neatness and Legibility

The ability of the plotter to produce exact and legible drawings is an obvious advantage over the traditional methods of "hand-drawn" work. The uniformity of CAD drawings, which produce lines of constant thickness, print quality lettering, and no smudges or other editing marks, is preferred.

Consistency

Since the system is constant in its methodology, the problem of individual style is eliminated. A company can have a number of draftspersons working on the same project and produce a consistent set of graphics.

Efficiency

The CAD operator must approach a drafting task in a different manner than he would when using traditional techniques. Since the CAD program is capable of performing much of the work for the operator, the job should be preplanned to utilize all the benefits of the system.

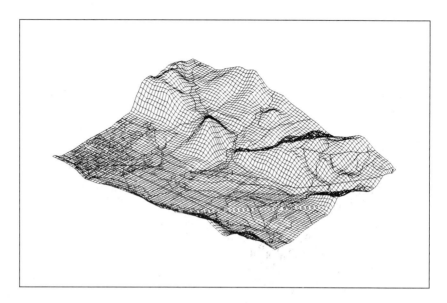

FIGURE 2-1 AutoCAD Drawing (Mapping) *(Courtesy Autodesk Inc.)*

APPLICATIONS OF CAD

Applications of CAD are now present in many industries. The flexibility of the many programs now available has had a major impact in the manner in which various tasks are performed. Some applications of CAD which are now being used are:

Architectural

Architects have found CAD to be one of the most useful tools that has ever been available to them. Designs can be formulated for presentation to a client in a shorter period of time than is possible by traditional techniques. The work is neater and more uniform. The designer can use 3D modeling capabilities to assist him and his client to better visualize the finished design. Changes can be performed and resubmitted in a very short time.

The architect can assemble construction drawings using stored details. Data base capabilities can be used to extract information from the drawings and perform cost estimates and bills of materials.

FIGURE 2-2 AutoCAD Drawing (Architectural) *(Courtesy Autodesk Inc.)*

Engineering

Engineers use CAD in many ways. They may also use programs which interact with CAD and perform calculations that would take more time using traditional techniques.

Among the many engineering uses are:

Electronics engineering	Mechanical engineering
Chemical engineering	Automotive engineering
Civil engineering	Aerospace engineering

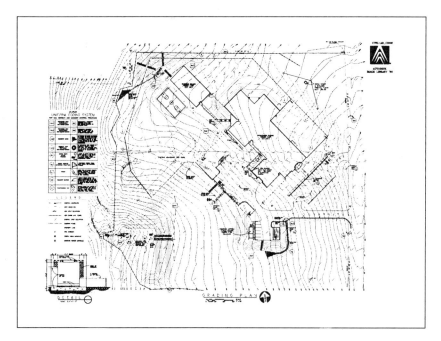

FIGURE 2-3 AutoCAD Drawing (Site Planning) *(Courtesy Autodesk Inc.)*

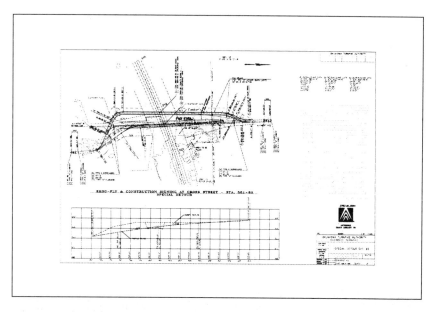

FIGURE 2-4 AutoCAD Drawing (Civil Engineering) *(Courtesy Autodesk Inc.)*

Interior Design

AutoCAD is a valuable tool for interior designers. The 3D capabilities can be used to model interiors for their clients. Floor plan layouts can be drawn and modified very quickly. Many third-party programs are available that make 3D layouts of areas such as kitchens and baths relatively simple.

Manufacturing

Manufacturing uses for CAD are many. One of the main advantages is integrating the program with a database for record keeping and tracking purposes. The ability to maintain information in a central database simplifies much of the work required in the manufacturing process.

Technical drawings used in manufacturing can be constructed quickly and legibly.

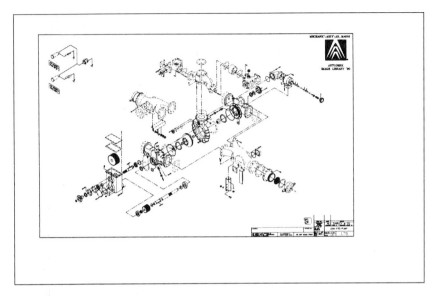

FIGURE 2-5 AutoCAD Drawing (Parts Assembly) *(Courtesy Autodesk Inc.)*

Yacht Design

Yacht and ship designers use AutoCAD to create the complex drawings for their unique designs.

Business

The business use of CAD has been increasing. Visual aids of all types are being used on an increasing basis. Advertising agencies as well as other businesses are finding CAD to be an invaluable aid in many types of graphic work otherwise known as *presentation graphics*.

Other business uses of CAD are for workflow charts, organizational charts, and all types of graphs.

Entertainment

Since the entertainment field is largely based in the electronic media, the use of CAD graphics is a perfect match. Television stations and networks use electronic graphics in place of the traditional artwork that must be photographed and converted. The weather graphics you view daily are a form of CAD graphics. Movie makers are turning to forms of electronic graphics for manipulations and additions to their work.

The introduction of AutoCAD related programs such as Animator Pro and 3D Studio has brought life-like realism and animation to many entertainment applications.

FIGURE 2-6 AutoCAD Drawing (3D Drawing and Rendering) *(Courtesy Autodesk Inc.)*

OTHER BENEFITS

Using a computer to draft and design also makes available a whole new world of information technology. Many systems now incorporate electronic mail software networks (e-mail) and access to electronic bulletin board systems through the use of a modem. Commercial online services such as CompuServe, as well as the worldwide network of the Internet, can be accessed through a computer as well, to gather information on numerous topics, putting you in touch with others using AutoCAD. There are many books published to teach you the ins and outs of getting "online".

CHAPTER REVIEW

1. If drawings are a means of communication, how has CAD helped in this process?

2. Name some applications of CAD.

3. Why is CAD a more flexible method of drafting than traditional drafting techniques?

4. Why would an office using several CAD stations produce more consistent work than a traditional office?

5. Why is CAD more accurate than traditional methods of scaling?

6. How has the entertainment industry been helped by CAD?

7. How can programs that interact with CAD be useful for engineering purposes?

8. What benefits does CAD provide for the user over traditional drafting techniques?

9. Does the construction or editing capabilities of CAD provide the greater increase in production speed?

10. Has CAD altered the traditional drafting industry significantly?

CHAPTER 3

COMPONENTS OF A CAD SYSTEM

The many parts of the CAD computer system can be confusing to the uninitiated user. Chapter 3 reviews the hardware components used in CAD drawing. After completing this chapter, you will be able to:

- Identify the categories of computers used in CAD.
- Identify the advantages and disadvantages of the various types of peripheral equipment unique to CAD systems.
- Demonstrate the proper care and use of CAD equipment.

COMPONENTS OF A CAD SYSTEM

CAD systems are comprised of several pieces of equipment that perform various functions. Equipment that is added to the basic computer is called peripheral equipment. The following is a description of each of the items that could be used with a CAD system.

THE COMPUTER

The computer is the central part of the CAD system. The peripheral equipment is connected to the computer. There are several types of computers that can be used with a CAD system. Let's look at the categories of computers.

Categories of Computers

Computer systems are divided into three main categories. These are:

>Personal Computers
>Workstations
>Mainframe Computers

Personal Computers. The personal computer is the type you often see on a desktop. These small, versatile machines are referred to as personal computers because they are mostly designed for use by one person at a time (although today's personal computers can be "networked", allowing use by more than one operator at a time). Personal computers usually consist of a case which contains the central processing unit and one or more disk drives, a display device, and a keyboard.

Workstations. Workstations are larger, faster and more expensive than personal computers. They allow many users to work on a single computer at one time. This class of computer can run larger and more sophisticated programs than personal computers.

Mainframe Computers. This is the largest type of computer. Mainframes are capable of processing a large amount of data. Mainframes are used by government and companies which handle large amounts of data.

Components of a Computer

A computer is made up of several parts that are essential to its operation. In addition, components can be added that speed up and/or enhance the operation of the computer. Let's look at some of the components that make up a computer.

System Board. The system board (sometimes referred to as the "motherboard") is an electronics board that holds many of the computer's chips and boards. The central processing unit, memory chips, ROM chips, and others are mounted on this board. In addition, the board contains what is referred to as "expansion slots". These are slots in which you can mount add-on boards, such as display and disk drive adapters.

Central Processing Unit. The central processing unit (or CPU) is the center of activity of the computer. It is here that the software program you are using is processed. After processing, instructions are sent out to the display, printer, plotter, or other peripheral. With few exceptions, all information passes through the central processing unit.

Physically, the central processor is a computer chip mounted on the system board. There are several types of processing chips. In a microcomputer, the type of CPU used determines what is known as the "class" of computer.

The CPU is manufactured by Intel, AMD, Cyrix, IBM, and others. Each CPU is designed with more capacity than the previous. The 80386 (referred to as a '386) and the 80486 (referred to as a '486) are 32-bit processors, which process software much faster than 8- and 16-bit processors. Today's Pentium ("586") is a part 32-bit and part 64-bit processor.

Memory. Computer memory can be divided into two categories; ROM (read-only memory) and RAM (random access memory). ROM memory is contained on pre-programmed chips on the system and is used to store basic command sets for the computer.

Random access memory (RAM) is the memory that is mostly referred to when computer memory is being discussed. Random access memory is used to temporarily store information in your computer. It is temporary storage because all data in RAM is lost when the computer is turned off.

Software programs (such as AutoCAD and others) have a minimum requirement for the amount of RAM necessary to run the program. This amount is given in kilobytes. A common reference might be "640K", meaning 640 kilobytes. You may see a reference to "megabytes". A megabyte is 1,000 kilobytes. In reality, these numbers are rounded, since a megabyte actually contains 1024 kilobytes.

Math Coprocessor. A math coprocessor works in tandem with the CPU. The CPU is very good at processing text-based instructions sets that are the heart of many programs, but it is not as efficient when asked to calculate numeric information. When a math coprocessor is installed, it allows the CPU to "hand off" math calculations to the coprocessor, resulting in faster operation of programs that are essentially math based. CAD programs, and other graphics oriented programs are heavily math based and benefit greatly from math coprocessors. Note that in order to benefit from the coprocessor, the software program must be written to use it.

Coprocessors are named according to the class of computer they are designed for. They are named the 80387 and 80487. The 80486 DX and Pentium CPU contains an integral coprocessor. The 486 SX requires a math chip. In addition to Intel, other companies such as Weitek manufacture coprocessors.

PERIPHERAL HARDWARE

Peripheral hardware consists of add-on devices which perform specific functions. A properly equipped CAD station consists of several peripheral devices. Among these are:

Plotters	Displays
Printers	Input devices

Plotters

Plotters are used to produce a "hard copy" of your work. The three main types of plotters used most often are pen plotters, printer plotters, and electrostatic plotters.

Pen Plotters. Pen plotters can produce the most pleasing type of plot. The pen plotter uses technical ink pens to draw on vellum, mylar, or other suitable surfaces.

Some pen plotters can use other types of pens such as ballpoint, pencil, and marker-type pens.

The plotter is controlled by signals sent by the computer to the plotter. The pen movement results from the response to the signals by servo or stepper motors. These motors control both the pen movement and the up and down motion of the pen. The accuracy (resolution) of the plotter is determined by the interval which the servo motors are capable of moving.

Pen plotters are produced in two major types: the flatbed and the rollerbed plotter.

Flatbed Plotters. A flatbed plotter moves the pen over a stationary sheet of paper. The pen carriage moves in both the X and Y direction. Since a flatbed plotter must have a surface large enough to contain the paper sheet, large sheet capacity flatbed plotters can take up a large amount of space.

Rollerbed Plotters. A rollerbed plotter moves both the paper and the pen. The paper is placed over a drum, or roller. The paper is then "rolled" back and forth over the roller while the pen is moved along the other direction. The combination of the two motions, along with the up and down pen movements, create the drawing. While most rollerbed plotters use standard paper sizes, some are capable of using roll paper.

Printer Plotters. Many CAD programs allow the use of a dot-matrix printer to produce a plot. Although a dot-matrix print doesn't contain the same fine resolution and line quality as a pen plot, it provides an easy, inexpensive manner to produce check plots, or plots in which a high degree of quality is not necessary.

Electrostatic Plotters. Electrostatic plotters produce plots very quickly. The advantages of electrostatic plotters are a short plotting time and a relatively high resolution plot. The disadvantage is the high initial equipment cost. Electrostatic plotters are very useful for companies that plot large volumes of drawings.

Printers

Printers provide a means of producing a hardcopy of text information. There are four main types of printers: dot-matrix, letter quality, laser, and ink jet.

Dot-Matrix Printers. Dot-matrix printers produce letters and graphics by impacting a ribbon with tiny striker pins. These pins are contained in a head which thrusts the pins outward in the designated pattern. Dot-matrix printers are fast and relatively inexpensive to operate.

Dot-matrix printers can be used as printer plotters if they are equipped with graphics capabilities. The advantage is the relatively low cost of equipment and operation. The disadvantage is the low quality of plot for most dot-matrix printers. Some specialized dot matrix printers, however, can plot relatively high resolution drawings up to C-size (18" × 24").

Laser Printers. Laser printers create prints similar to a pen plot. They use a laser-copy process that produces sharp, clear prints. Although some laser printers have excellent graphics capabilities, they are relatively expensive to purchase when compared with other types of printers. Another disadvantage is the restriction to B paper size.

Ink Jet Printers. Ink jet printers create letters and graphics by spraying a fine jet of ink onto the media. Ink jet printers can produce graphic plots of good quality. Advantages are the ability to create good quality plots in a reasonable time and the capability of some ink jet printers to use color. One disadvantage is the less than optimum quality.

Display Systems

Displays are used to view the work while in progress. Displays are often referred to as monitors or CRTs (cathode ray tubes). The quality of the image on the display is determined by its resolution. The resolution is controlled by the number of dots (pixels) contained on the screen. It is these pixels that make up the image. A 640×480 resolution display contains 640 pixels horizontal and 480 pixels vertical on the screen. Many professional CAD users prefer higher resolution displays such as 1024×768 or 1280×1024.

In order to display higher resolution, both the software program and the graphics board contained in the computer must be capable of displaying the higher resolution. The monitor must also be matched to the graphics board.

CAD drafters use both monochrome and color display systems. The advantages of monochrome displays are the lower cost, and in many cases, a clearer display. The advantages of color systems are the ability to color code drawing layers and entities and the ability to produce more realistic rendered models.

Input Devices

Just as a word processing program requires a keyboard to input the individual letters, numbers, and symbols, a CAD program requires an input device to create and manipulate drawing elements. Although many programs allow input from the keyboard arrow keys, an input device speeds up the drawing process. The most common input devices used for CAD drawing are:

    Digitizers

    Mice

    Track balls

Digitizers. A digitizer is an electronic input device that transmits the X and Y location of a cursor which is resting on a sensitized pad. Digitizers may be used as a pointing device to move a point around the screen, or as a tracing device for copying drawings into the computer in scale and proper proportion.

The points are located on the pad by means of a *stylus* (similar in appearance to a pencil), or a *puck* (alternately referred to as a cursor). Digitizers use a fine grid of wires sandwiched between glass layers. The cursor is then moved across the pad and the relative location is read and transmitted to the computer.

In "tablet" mode, the digitizing pad is calibrated to the actual absolute coordinates of the drawing. When used as a pointing device, the tablet is not calibrated.

Digitizers are available in several sizes. Small pads may be used to digitize drawings larger than the pad surface. This is accomplished by moving the drawing on the pad and recalibrating. However, this can be very annoying if you frequently work with large scale drawings to be digitized.

Mouse. A mouse is an input device that is used for pointing only. The name comes from its appearance. There are primarily two types: a mechanical mouse and an optical mouse.

A mechanical mouse has a ball under the housing. The rolling ball transmits its relative movement to the computer through a wire connecting the mouse to the computer.

An optical mouse uses optical technology to read lines from a special pad to sense relative movement. Although an optical mouse uses a pad, it is not capable of digitizing.

Track Balls. A track ball can be thought of as an inverted roller ball mouse. A ball is moved by rolling it with the palm of the hand. In the same way as a roller ball mouse, relative movement of the ball is translated to the screen.

SUMMARY

A large number of peripheral devices are available for use by the CAD operator. The reason for choosing each peripheral is as diverse as the number of devices and users. In choosing the proper peripherals for you, the following criteria should be considered:

1. What will the primary use for the device be?
2. Does the software you plan to use support the device?
3. How much money are you willing to spend? (Large digitizing pads can cost several thousand dollars).

CHAPTER REVIEW

1. What are the three main categories of computers?

2. What is peripheral hardware?

3. What are the advantages of pen plotters over dot-matrix plots? Of dot-matrix plots over pen plotter plots?

4. What does the term *resolution* mean?

5. How does a mouse differ from a digitizing pad as an input device?

6. Can the performance and/or speed of a computer be altered or enhanced?

7. What is the center of activity of the computer?

8. Explain the difference in RAM and ROM memory.

9. How are coprocessors categorized?

10. Discuss the advantages and disadvantages of electrostatic plotters.

11. Compare laser and ink jet printers.

12. How do an optical mouse and a mechanical mouse differ?

13. What is a track ball?

14. How is a stylus used with a digitizing pad?

15. Would a complete CAD system with necessary peripherals be appropriate for architectural as well as manufacturing applications? Why?

CHAPTER 4

YOUR COMPUTER

In order to use a CAD program effectively, knowledge of basic computer principles is essential. Chapter 4 introduces the new computer user to the various principles of computer use. After completing this chapter, you will be able to:

- Identify the basic parts of the personal computer.
- Demonstrate the proper care and operation of storage media.
- Identify, create, and use directory structure and batch files.

INTRODUCTION

In order to operate your AutoCAD program best, it is helpful to understand your computer. There are several aspects of computer use to become familiar with. This chapter is an overview of computer operation. The next chapter discusses the disk operating system, covering the basic operations you need to compute effectively. Let's start by looking at an important part of your computer: disks and disk drives.

DISK DRIVES

Disk drives are identified either as hard drives or floppy drives. Floppy drives use disks that are referred to as floppy disks. They are called floppy disks because the original disks (5¼") are bendable (although they might damage if bent). The more popular "floppy" disks (3½") have hard shells instead of the vinyl shell. There are also removable drives available, such as SyQuest and Bernouli.

CD-ROMs have become a popular means of software distribution for programs that require many disks. Release 13 comes on 27 disks — or one CD-ROM. Using a CD-ROM requires a CD-ROM drive, which is usually a separate peripheral. Some newer computers include a CD-ROM drive in the computer case.

Floppy Disks

Floppy disks used with personal computers are found in two sizes: 5¼" and 3½". The 5¼" disks have the vinyl "floppy" shell. They typically hold either 360 kilobytes or 1.2 megabytes of data files. The 1.2 megabyte disk is referred to as a high-density disk.

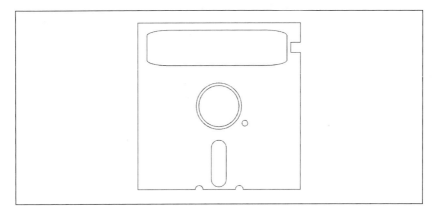

FIGURE 4-1   5¼" Diskette

The second, more predominate size is the 3½" disk. These disks use the plastic shells and are made in two capacities. They hold 720 kilobytes in the standard format, or 1.44 megabytes in the high-density format.

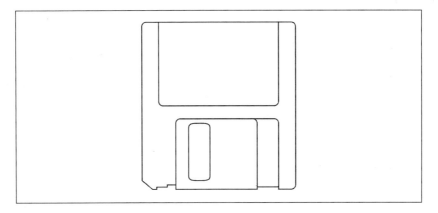

FIGURE 4-2   3½" Diskette

Hard Drives

Hard disks are usually non-removable drives. These drives are typically installed inside the computer case. Hard disks (alternately referred to as hard drives) are manufactured in different capacities. The lower-end capacity is usually 120 megabytes, although these have mostly been replaced by 500 megabyte capacity disks at the higher end. The size can range up to a thousand megabytes (a gigabyte).

DISK CARE

Since all your work is recorded to disk, care of the computer disk is very important. If a disk is damaged, you may lose your files! Frequent back-ups (copying files to a second diskette or computer tape) and proper handling of your disks can minimize the possibility of file loss.

Hard disks are installed inside the computer and are not handled. This may eliminate the danger of improper handling, but does not prevent damage. A hard drive can be damaged by shock. If you move the computer, be sure the power is off and move it gently. If the hard drive does not have self parking heads, use the included software to park them before moving the machine. This moves the heads to a sector that does not have data stored on it. If the hard drive is on, do not move or tilt the computer.

Floppy disks are especially subject to damage from handling. A 5¼" disk has an open area in its jacket where the writing head is positioned when the disk is in the drive. Touching the disk surface through this opening leaves oil from the skin on the disk, possibly leaving it unreadable.

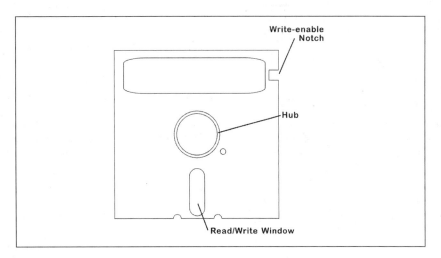

FIGURE 4-3 5¼" Diskette

Dust and smoke can also leave particles on the disk surface that can prevent the drive from reading the disk properly. Heat and cold can cause the disk material to expand or contract, causing problems. Magnets scramble the data on the disk's tracks. Spilling a liquid onto a disk leaves a residue. If the liquid is hot or cold, it can cause temperature damage.

The smaller hard-shelled 3½" disks are less prone to handling damage. The read/write surface is protected by a sliding door. The purpose of this door, of course, is defeated if you open it to look or touch. Otherwise, all the perils of a 5¼" disk apply to the 3½" disk.

CD-ROMs are relatively sturdy forms of carrying data in a "read only" capacity. However, they too must be handled with care not to touch or scratch the surface from which the laser reads the data.

Write-Protecting Data

Normally, a disk can be read from or written to. If you do not want to be able to write to a disk, you can "write-protect" the disk. This allows the disk to be read, but not written to.

The method of doing this depends on the type of disk you are write-protecting. The 5¼" disk usually comes with a set of write-protect tabs. These are small stickers that are placed over the write-enable notch on the disk. This notch can be found on the upper right side of the disk. Place the sticker over the notch, wrapping it around to the back of the diskette. To allow disk writing, simply remove the sticker.

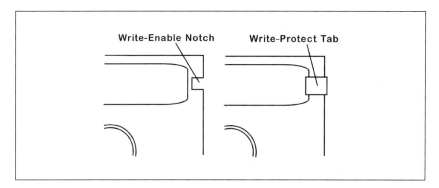

FIGURE 4-4 Write-Protecting a 5¼" Diskette

The 3½" hard shell disk has a small slide switch in one corner of the disk. Sliding the switch alternates between write-protecting and write-enabling.

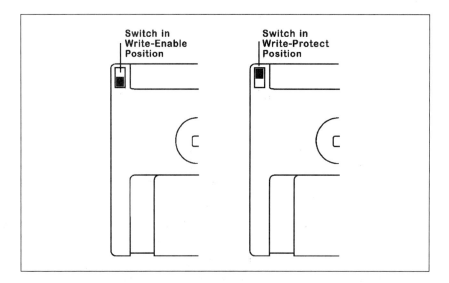

FIGURE 4-5 Write-Protecting a 3½" Diskette

FILE DIRECTORIES AND PATHS

As you add files to your disk, file management becomes an increasing problem. Imagine trying to find a single file out of hundreds on your disk! Computer files are usually able to be categorized, either by files generated by a single program (such as AutoCAD), or by jobs (such as drawing files for the Smith Widget Company). This is similar to standard office files. If you file work, you place it in a file that is designated for that type of information. You wouldn't throw it into a pile. Placing files indiscriminately on a disk is like throwing it into a pile.

You can make "file drawers" for your work. They are called *directories*. You can also make files for the drawers. They are called "subdirectories". This electronic equivalent of a filing system is used to organize files on disks.

To properly plan your filing system, you should outline an overview of your programs and data files. Let's look at an example. We have several software programs we wish to use. They are AutoCAD, Lotus, WordPerfect, and dBase. We also would like to keep work files in a separate place.

The following outline shows the software programs in separate directories (file drawers) and work files in separate subdirectories (file folders).

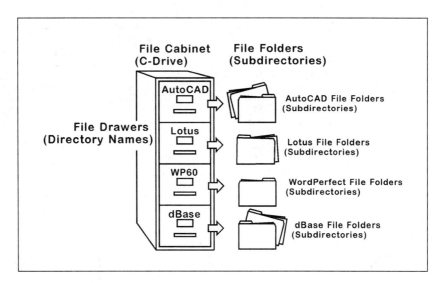

FIGURE 4-6 File Drawers and Folders

Note also that subdirectories can be created several "layers" deep. This is convenient when your outline breakdown of files requires subcategories.

This type of file management is standard for hard drives that can record large amounts of data. The methodology for doing this is found within the disk operating system. We will study this in the next chapter.

CHAPTER REVIEW

1. List two ways of creating batch files.

2. What sizes floppy disks are available? What space capacities are they usually available in?

3. How might you avoid complete loss of a file, other than proper disk maintenance?

4. What does the term *write-protect* mean? How is this accomplished?

5. Why is file management important?

6. What makes a hard shell floppy disk less prone to damage?

7. What may happen to the information on a disk if it passes through a magnetic field?

8. Is write protection of a disk permanent?

9. What would be a good initial approach to file management?

10. What is a directory within a directory called?

11. List 5 improper care procedures when handling floppy disks.

CHAPTER 5

THE DISK OPERATING SYSTEM

The disk operating system (DOS) is the most basic level of computing. The new user must achieve a basic level of understanding of DOS and Windows in order to perform fundamental computing operations. After completing this chapter, you will be able to:

- Understand DOS and Windows.
- Demonstrate proper file management techniques using DOS and Windows commands.
- Format a floppy disk.
- Change directory paths.

THE DISK OPERATING SYSTEM

To use AutoCAD and other Windows software, your computer must have a program which translates between it and the computer. This program is called the Disk Operating System, or DOS for short. Windows is a graphical user interface for DOS.

DOS and Windows can be thought of as an umbrella program under which all other programs can be run. You must first start DOS before beginning other programs. If DOS (or Windows 95) is installed on your hard drive, it will be loaded automatically when you start your computer. If you are running Windows v3.1 or Windows for Workgroups, then you have to start it by typing "Win" at the DOS prompt:

C>win [ENTER]

When you start your computer, it looks for certain files. If they are present, they are loaded. You may notice a disk operating system message and/or version number when you start your computer.

5-2 USING AUTOCAD

The disk operating system is originally supplied on floppy disk or CD-ROM and is often included in the purchase price of your computer.

DOS designates which disk drive is currently being worked on (or "active") by showing a "drive specifier". Disk drives are identified by a letter, such as "A", "B", or "C". The active, or current drive is shown by a letter, followed by a "greater than" symbol. For example, if the A-drive is active, it is designated on the computer screen by:

    A>

The letters designating the drives have a significance. Your computer first looks in the A-drive for the DOS files, then in the first hard drive.

In all computers, drives "A" and "B" are floppy drives, while drives "C" and above are hard drives.

The position of the drives in the computer can differ, according to the computer cabinet design. The following illustrations show the typical setup for different cabinet designs.

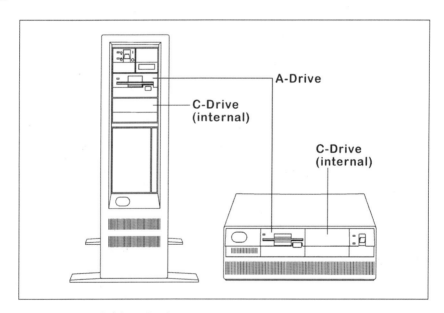

FIGURE 5-1 Cabinet Designs

CHANGING ACTIVE DRIVES — DOS

The active drive can be easily changed. If you are in DOS (that is, not currently in a program), you will notice that the active drive is shown on the screen. For example, if the active drive is now the C-drive, the bottom line on the screen shows:

    C>

To change to another drive, simply enter the drive letter, followed by a colon (:). For example, if the C-drive is current, enter the "A:", followed by pressing [ENTER].

   C>A:

The last line of the screen will now display:

   A>

The A-drive is now the active drive!

CHANGING ACTIVE DRIVES AND DISPLAYING FILES — WINDOWS

There is less worry about the active drive under Windows, since Windows tends to take care of drives and subdirectories. Still, there may be times when you need to change the active drive, such as when your computer is hooked up to a network.

Changing the active drive is most commonly done with the File Manager (see Figure 5-2). To change from the C: drive to the E: drive, simply click on the icon of the E-drive (a hard drive connected via network, in this case). The File Manager displays the subdirectories and files on drive E:.

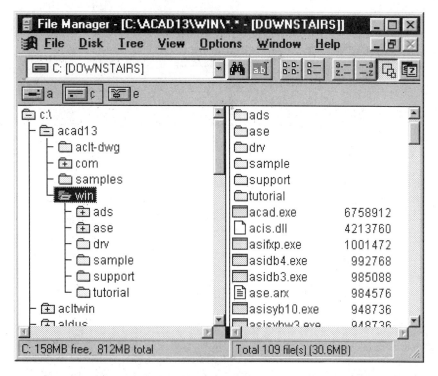

FIGURE 5-2 The Windows File Manager

To change to a different subdirectory, simply click on a subdirectory name in the left window (double-click on the subdirectory name when it appears in the right window). If the subdirectory folder has a + sign, that means there are more subdirectories below it.

To display files in different ways, select the View option of the File Manager's menu bar. This menu lets you sort and view files in a number of different ways. You can sort the directory listing by file name, extension, size, and date. I find it helpful to customize the File Manager's toolbar to give me instant access to the four file sorting options.

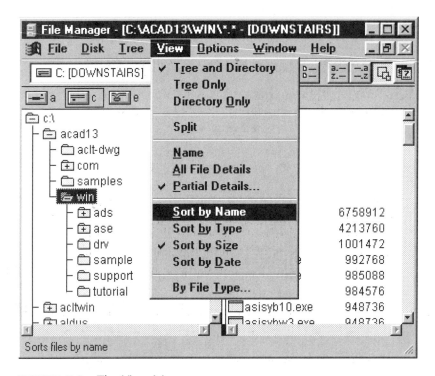

FIGURE 5-3 The View Menu

The View menu also lets you change the display of files. The choices are (1) just the file name and extension; (2) some details (size, date, time, attributes); (3) all these details; or (4) files selected by type. The figure shows the partial details (file name and size) sorted by size.

When a change happens to a subdirectory, Windows does not always automatically update the display. This happens when files are created or deleted, or when you insert a different diskette in drive A:. To refresh the display of files, press function [F5].

DISPLAYING FILES — DOS

The Directory (Dir) command is used to display a listing of files on the disk. If you wish to display a list of files on the active drive, simply enter:

DIR

You can show the files on any drive by adding the drive specifier after the Dir command. For example, if you want to display the files on the disk in the B-drive, enter:

DIR B:

You can do this from any active drive.

If you have a lot of files on the disk, they will *scroll* by before you have a chance to read them. You can place a "pause" in the directory by entering:

DIR /P

This will cause the directory of files to scroll up on the screen until the screen is filled, then pause. Press any key once to continue the scrolling until the screen is filled again. You can remember this by thinking of the "P" option as the "Pause" option.

You can pause the scrolling of the listing at any time by pressing [CTRL], then striking [S]. You can remember this by thinking of the "S" key as the "Stop" key.

If you noticed the screen when you displayed the files with the Dir command, you observed that the listing is one file per line. You can display a "wide" directory with the "W" option. It is used in the same manner as the pause option. To display a file listing in wide format, enter:

DIR /W

You can remember this by thinking of the "W" option as the "Wide" option.

You may mix the options and obtain a wide listing with a pause by entering:

DIR /W/P

FILE NAMES AND DIRECTORY LISTINGS — DOS

The following is an example of a directory listing. Notice the difference between files and directory listings.

```
C:\> DIR
    Volume in drive C has no label
    Directory of C:\
    COMMAND      COM             23456      1-23-95
    DOS                          <DIR>      1-23-95
    AUTOEXEC     BAT               128      2-13-95
    ACAD         BAT                28      3-21-95
    ACAD                         <DIR>      3-21-95
    PIPESTAR                     <DIR>      4-12-95
    SIDEKICK                     <DIR>      4-30-95
    MENU                         <DIR>      (the current date)
```

FIGURE 5-4 File Listing

Also notice the file names. The names are listed, then a three letter code is listed after them. These are called *file extensions*. File names can have up to eight letters and characters (some characters are exempted). The file extension can be up to three letters or characters long. The file extensions usually denote the type of file it is. For example, an AutoCAD drawing file has a "DWG" extension. If you refer to the file name in DOS, the correct notation is the file name, followed by a period (.), then the file extension. For example, an AutoCAD drawing named "WIDGET1" would be written "WIDGET1.DWG". Some programs create their own file extensions, while you must create file names in other programs. Note that a file extension is not required if you have the option of creating the extension.

WILD-CARD CHARACTERS

When using the directory and other DOS and Windows commands, you can use "wild-card" characters to specify files. The two wild-card characters are the question mark (?) and the asterisk (*).

The question mark can fill in for any single character. For example, you may want to display a directory of all the files that are listed as "CAR_.DWG", where the underlined space can be any letter. If you enter:

DIR CAR?.DWG

you could display a listing of files such as:

CAR1.DWG
CAR2.DWG
CART.DWG
CARD.DWG

The question mark wild card can be used in either the file name or extension and can be used to represent as many letters as desired.

The asterisk is used to represent all the characters on either side of the period in a file name. For example, if you enter:

DIR *.DWG

you will display all the files with the ".dwg" file extension. Alternately, if you enter:

DIR FLPLAN.*

you will display all the drawings named "FLPLAN", regardless of their file extension (or even if no file extension exists).

Displaying Files in Directories

If you use directories, you must specify the directory under which you wish to list the files. For example, if you wish to see the files under the "ACAD12" directory, enter:

DIR \ACAD12

Notice the use of the backslash (\). The backslash used before the name indicates that the file name is a directory. If you had a subdirectory named "DWGS" under the ACAD12 directory and wished to see the contents of that subdirectory, you would enter:

DIR \ACAD12\DWGS

If you are in a directory (we will learn how to make a directory active later), and wish to see the contents of any other directory, the process is the same. If, however, you are in a directory and wish to see the contents of the root directory, you simply place a backslash after the DIR command. For example, the \ACAD11\DWGS directory is current and you wish to see the contents of the root directory. Enter the following:

**DIR **

Under Windows, the File Manager displays files in directories.

Making a Directory Current — DOS

There may be times that you wish to make the current directory active. This might be necessary to run a program that is copied in that directory. For example, you may want to change to a directory named "WRITER" to start a word processing program. To change to the directory, type:

CD \WRITER

To return to the root directory, enter the change directory (CD) command, followed by a space and backslash.

**CD **

Since directories and subdirectories can be created several levels deep, there may be times that you want to "back up" one level. To do this, enter:

CD..

CREATING AND DELETING DIRECTORIES — DOS

Now that you know how to navigate through directories and subdirectories, it is time to learn how to create and delete them.

To create a directory, use the Make Directory (MD) command, followed by a space and the directory name you wish to create. For example, to create a directory named "DWGS", enter:

MD DWGS

To create a subdirectory, first enter the directory you want the subdirectory to be created under (using the CD command discussed earlier) and use the MD command to create the subdirectory while in the directory.

To remove a directory, use the Remove Directory (RD) command. For example, to remove a directory named "OLDFILES", enter:

RD OLDFILES

Note that you can not remove a directory if it contains files or a subdirectory. You must first delete (we will learn how to do this later) files and/or remove subdirectories if you wish to remove the directory. If you attempt to remove a directory that contains files or subdirectories, DOS will display an error message.

If you wish to remove a subdirectory, change (CD) to the level "above" the subdirectory and use the RD command. Like directories, if files or sub-sub directories exist, you must first delete them before removing the subdirectory.

DELETING FILES

As you work, you will create files that you no longer wish to have on your disks. You can use the Delete (Del) command to remove files. The Delete command is the same as the Erase command. Delete is typically used since the command can be accessed by typing the shorter **Del** entry.

If you wish to delete files that are in a directory or subdirectory, it is recommended that you first change to that directory or subdirectory before deleting the file. This eliminates the possibility of a fatal error.

> **NOTE:** Unless you are familiar with utility programs that can "unerase" files, your deleted files will not be recoverable. Be sure before you proceed! "Undelete" is now a standard utility with DOS 5 and 6.

Let's look at how we would erase a file. The file we wish to erase is named "PLAN.DWG". To delete the file in DOS, enter:

DEL PLAN.DWG

Note that you could delete all the drawing files in a directory or subdirectory (if, of course, you were really sure you wanted to) by using the asterisk wild-card character and entering:

DEL *.DWG

Under Windows, click on the file name in the File Manager, then press the Del key.

COPYING FILES

The Copy command is used to copy files from disk to disk or to different locations on the same disk.

There are some rules to be aware of. You cannot have two files of identical names in the same directory. You must include the file extension when referring to files to be copied. If you are copying a file from a directory or subdirectory, you must include the path to the file.

Let's look at some examples of file copying.

Copy Disk to Disk

To copy a file from one disk to another, use the following DOS format:

COPY A:FILENAME.EXT B:

This would copy a file from drive A to drive B.

To copy a file under Windows, click on the file name in the File Manager. Then drag the file name to the drive icon on the toolbar.

Copying from Directory to Directory

To copy from or to directories under DOS, you must specify the directory paths.

COPY C:\ACAD\PLAN.DWG D:\DWGS

This would copy a file named "PLAN.DWG" in a directory named "ACAD" on the C-drive to a directory named "DWGS" on the D-drive.

To copy a file to another directory under Windows, click on the file name in the File Manager. Then drag the file name to the subdirectory name while holding down [CTRL].

Copy Files and Renaming

You can copy a file and rename the file in a single step in DOS but not in Windows.

COPY A:PLAN.DWG B:SCHEME1.DWG

This would copy the drawing named PLAN.DWG from the A-drive to the B-drive and rename the file SCHEME1.DWG. The original file name from the A-drive would remain unchanged.

This is a good method for copying a file to the same directory. As mentioned earlier, you can not have two files by the same name. You can, however, have two identical files, each with a different name. The following is an example of copying a file to the same directory with a new name.

COPY PLAN.DWG PLAN2.DWG

Note that we did not have to list a drive specifier, since we were both copying from and to the default drive.

> ***NOTE:*** If you copy a file to a disk that already has a file by the same name, that file will be replaced by the new file of the same name!

COPYING DISKS

It is often desirable to copy (or back up) an entire disk. The Diskcopy command is used to do this in DOS. Diskcopy is somewhat different than the Copy command. You could use the following format to copy all the files from the A-drive to the B-drive:

COPY A:*.* B:

This would copy all the files from the A-drive to the B-drive one by one. If the disk in the B-drive had existing files, the new files would simply be added in addition to the existing files.

Diskcopy, however, makes an exact copy of the original disk. If there are files on the target disk, the files will be removed before the new ones are copied.

Diskcopy is a DOS program. Because of this, you must be in the directory where the DOS files reside before you can use Diskcopy. The following format is used for Diskcopy.

DISKCOPY A: B:

This would duplicate the disk in the A-drive with the disk in the B-drive.

Not everyone, however, has two floppy drives. If you only have one floppy (it will be an A-drive), the following format can be used:

DISKCOPY A: A:

DOS will prompt you to place either the Target disk (the disk to copy to) or the Source disk (the disk to copy from) in the disk drive. You will have to swap disks as DOS prompts you to do on the screen.

Note that you can not use Diskcopy between two drives and/or disks that are not the same capacity.

Under Windows, use the File Manager: select Copy Disk from the Disk item on the menu bar. Then follow the instructions in the dialog boxes.

RENAMING FILES

You can rename a file by using the DOS Rename (Ren) command. Use the following format:

REN FILE1.EXT FILE2.EXT

For example, to rename the file SPROCKET.DWG to COG.DWG, enter:

REN SPROCKET.DWG COG.DWG

In Windows, use the File Manager's Rename command (found in the File item of the menu bar).

FORMATTING DISKS

Before you can use a disk, it must be formatted. Formatting installs the tracks, sectors, and other items needed to make your disk usable. This is achieved by using the Format command. Format, like Diskcopy, is a DOS program. You must be in the directory where the DOS files are located before you can use it.

A simple format can be started by using the following format:

FORMAT A:

This will format the disk in the A-drive.

> **NOTE:** The Format command is destructive to any existing files.

You can create a "bootable" disk (one that has the necessary DOS files to start your computer) by using the "/s" (system) option. Note the following format:

FORMAT A: /S

This option copies files (both visible and hidden) to the formatted disk, making it a "bootable" disk. You do not need to use this option for a disk that is not used to start the computer.

Hard drives are also formatted. The procedure for doing this, however, is more complicated and potentially damaging unless performed by a knowledgeable operator.

Some types of drives use special options for formatting. The product information with the drive gives any special instructions you may need.

In Windows, use Format Disk command, found in the Disk menu of the File Manager.

DISPLAYING FILE CONTENTS — DOS

If a file is written with ASCII text, its contents can be displayed in DOS by using the Type command. To display a file, use the following format:

TYPE FILENAME.EXT

If the file is not ASCII, you will get some very unusual looking characters. If the file you display is more than one screen in length, you can use [CTRL] [S] to stop the screen. Use [CTRL] [S] again to continue the scrolling.

CHAPTER REVIEW

1. What function do DOS and Windows perform?

2. What are wild-card characters? List them.

3. When can a subdirectory not be removed?

4. What would you type at the DOS prompt to rename a drawing file named FLPLAN as FLOORPL?

5. What do the following DOS commands stand for: MD, CD, DEL, REN, RD?

6. How would you return to the root directory from a subdirectory?

7. What would you type in DOS to make the current directory active?

8. How does a /P affect the DIR command?

9. How can you stop the scrolling of file listings?

10. How does DOS indicate the active drive?

11. How can you display a listing of files?

12. Why is it necessary to format a disk before using it?

13. What is a *bootable* disk?

14. What would the effect be of formatting a disk containing files?

15. Is the procedure for formatting a hard disk different from that of a floppy disk?

CHAPTER 6

GENERAL AUTOCAD PRINCIPLES

The operation of a CAD program is unique among other types of software programs. Chapter 6 introduces the user to the principles of the AutoCAD program. After completing this chapter, you will be able to:

- Understand the concepts of a computer-aided design program.
- Understand the AutoCAD screen layouts.
- Understand the terms used in this text and in AutoCAD.
- Understand the functions of special keyboard keys.

OVERVIEW

AutoCAD is a powerful computer-aided drafting package. A computer-aided drafting system is to drafting what a word processor is to writing.

Your drawings are displayed on a graphics monitor screen. This monitor takes the place of the paper in traditional drafting techniques. All the additions and changes to your work are shown on the monitor screen as you perform them.

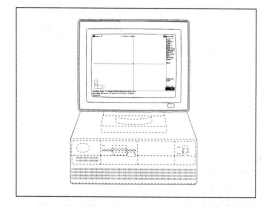

FIGURE 6-1 Graphics Monitor

A drawing is made up of separate elements consisting of lines, arcs, circles, strings of text, and other elements supplied for your use. These elements are called *entities* or *objects*.

The commands are selected from a screen menu or from the keyboard. A *menu* is a list of items from which you may choose what you want. AutoCAD provides for menu selections to be placed on the screen and on a digitizing pad. The next chapter illustrates the parts of the AutoCAD drawing screen.

Entities are placed in the drawing by means of commands. Each command is performed by choosing an entity or function from the menu. You are then asked to identify the parameters of the command. After identifying all the information which AutoCAD requests, the entities or changes are shown on the screen.

TERMINOLOGY

This manual contains terms and concepts that you need to understand in order to use AutoCAD properly. Some of the terms you need to know now are briefly listed in this chapter. If you need other help, refer to the Glossary and special Command Summary, in the Appendix. In addition, the index contains all the terms used and the pages on which they are explained.

Coordinates

The *Cartesian coordinate system* is used in AutoCAD. The diagram shown below illustrates the system. The *X-axis* is represented by the horizontal line. The *Y-axis* is represented by the vertical line. Any point on the graph can be represented by an X and Y value shown in the form of (X,Y). The normal position of the AutoCAD screen is overlaid on the axis.

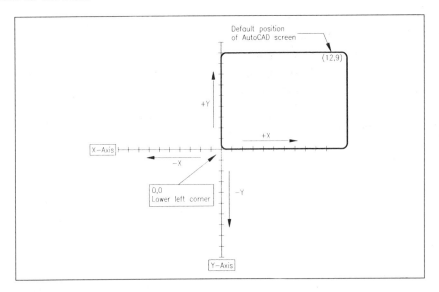

FIGURE 6-2 Coordinate System

The intersection of the X and Y axis is the (0,0) point. This point is normally the lower left corner of the drawing editor. (You may, however, specify a different lower left corner).

The AutoCAD 3D system uses a third axis called the *Z axis*.

Display

In this manual, the term display refers to the part of the drawing that is currently visible on the monitor screen.

Drawing Files

The *drawing file* is the file which contains the information used to describe the drawn graphics image. A drawing file automatically has a file extension of .DWG added to it. A file extension is a three character suffix that is sometimes placed after the period which follows the file name.

Limits

When using AutoCAD, you draw in a rectangular area. The borders of this area are called the *Limits*. You may draw anywhere in these limits, but not outside them. You may set the limits to whatever size you wish. You may also change them at any time. The limits are described by (X,Y) coordinates for the lower left corner and the upper right corner. Your drawing limits may be thought of as the "sheet size" that you are drawing on.

Units

The distance between two points is described in *units*. The units for AutoCAD may be set to any of the following:

1. Scientific
2. Decimal
3. Engineering
4. Architectural
5. Fractional

With the exception of Architectural units, one unit may be equal to whatever form of measurement (for example, feet or meters) you wish.

Zooming/Panning

You will often want to enlarge a portion of the drawing to see the work in greater detail, or to reduce it to see the entire drawing. The *Zoom* command facilitates this. AutoCAD's zoom ratio is about one trillion to one!

The *Pan* command allows you to move around the drawing while at the same zoom level.

USE OF THE MANUAL

Each chapter contained in this manual is designed to build on the previous chapters. Many sections contain tutorials which you should follow. The problems found at the end of a chapter are specifically designed to use the commands covered in the manual to that point.

This method allows your learning to be self-paced. Remember, not everyone will grasp each command in the same amount of time.

Keys

Several references are made to different keyboard keys in this manual. The following keys will be referred to in this manual. Note that the keys may be located in a different location on some keyboards.

Control key: Some commands are executed by a multiple keystroke. The Control key is labeled [CTRL] and is used in conjunction with another key.

Flip Screen: On single screen configurations, the text and graphics screens may be alternately displayed by using the Flip Screen key. On most keyboards, the [F1] key is used for this.

Fast Cursor: When using the arrow keys to move the screen cursor, the movement interval may be adjusted using the fast and slow cursor keys. On IBM-compatible models, the [PAGE UP] and [PAGE DOWN] keys perform this function.

Command Nomenclature

When a command sequence is shown, the following notations are used:

UNDERLINED: An underlined response designates user input. This is what you enter. Entry may be made from either the screen menu or the keyboard.

<DEFAULT>: Entries enclosed in brackets are the default values for the current command. The default value will be executed if you press [ENTER].

ENTER POINT: When prompted to "enter point", you should enter a point on the screen at the designated place. You will usually be shown a point on a drawing. The points will be designated, such as "point A".

RETURN or ENTER: Means to press the [ENTER] key after the entry. (You will need to do this after each input).

CHOOSE or SELECT: Make the desired choice.

You may use either uppercase or lowercase in response to any command inquiry.

When a response option has one letter capitalized, the capital letter for that response is all that is required to be entered. For example, if an option is "eXit", simply entering **X** is sufficient for choosing that option.

The menus shown with tutorials serve to guide you through the complex menu structure. The boxed part of the menu represents the correct choice to make from the screen menu.

Using AutoCAD Note Boxes

Note boxes are provided throughout the text. These note sheets contain helpful hints in using AutoCAD.

> **NOTE:** Be sure to read the note boxes in each chapter for helpful hints for using AutoCAD.

FIGURE 6-3 Note Box

CHAPTER REVIEW

1. The area on the screen in which you draw is surrounded by borders called _____.

2. Some commands prompt for additional information in order to continue execution of the command, and in some instances a default setting can be chosen. If a default setting is present, how can it be identified and what must be done to select it?

3. What are the drawing elements used in computer-aided drafting called? Give three examples of drawing elements.

4. For each of the following options, list the required response you would enter at the prompt:

   Close: _____

   eXit: _____

   circle: _____

   LAyer: _____

   Edit vertex: _____

5. Where can the commands used to draw and edit be selected from?

   _____

   _____

6. When using the [F1] key to facilitate the flip screen function, in what way will the display be altered and what information is presented by using this option?

   _____

   _____

   _____

7. Using the figure below, label each portion of the axis indicated. Next, draw a rectangle on the figure to represent the drawing screen orientation relative to the coordinate axis.

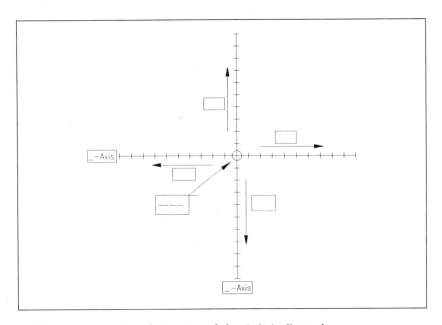

FIGURE 6-4 Label Each Portion of the Axis Indicated

8. To enlarge or reduce all or a portion of your drawing on the screen, what command would you use?

9. What term is used to refer to the portion of the drawing that is currently visible on the screen?

10. What can be done to increase or decrease the increment of movement in the screen cursor when the arrow keys are used?

11. What is the increment of length in which a drawing is constructed?

12. When using this manual, where would you look for extra help or information about commands listed?

13. What happens when you press the [CTRL] key?

14. If you have zoomed in to a portion of the drawing for close examination, how can you move to another part of the drawing while staying in the same zoom?

15. What is used to place entities into a drawing?

16. What is the only option that designates the units a drawing is constructed in?

CHAPTER 7

GETTING IN AND AROUND AUTOCAD

In order to use the AutoCAD program, you must learn how to start the program and manage the files. After completing this chapter, you will be able to:

- Create a new drawing.
- Save or discard your work.
- Demonstrate file management commands.

BEGINNING A DRAWING SESSION

Your drawing sessions with AutoCAD will begin by starting the AutoCAD program. The program is started by double-clicking on the AutoCAD icon. If there is no icon in Windows 95, click on the start button, then select Programs. Click on AutoCAD R13.

After you have started AutoCAD, you will see the drawing editor. The drawing editor is the screen where your drawing activities will take place.

THE DRAWING EDITOR

Once in the drawing editor, you are presented with a drawing area in which to perform your work. You will notice that the top, bottom, and sides of the screen contain information. Let's look at these areas.

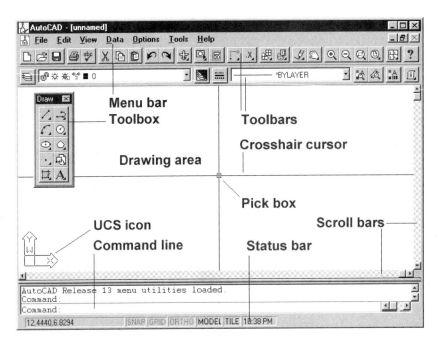

FIGURE 7-1 AutoCAD Drawing Screen

Screen Coordinates

The bottom left of the screen contains two sets of numbers which represent the X and Y coordinates of the crosshair. The type of numerical read-out is controlled by the Units command. The constantly updating coordinates may be switched between absolute coordinate display, polar coordinate display, and off by using [CTRL] [D] or [F6]. If the coordinates do not update as you move the cursor, they are toggled off.

Mode Indicators

The bottom of the screen displays the mode indicators. This includes the following:

Snap: Indicates whether the snap mode is on or off.
Grid: Indicates whether the grid is on or off.
Ortho: Indicates whether the orthogonal mode is on or off.
Model: Indicates whether AutoCAD is in model or paper space.
Tile: Indicates whether Tilemode is on or off.
Time: The time, as provided by the computer's clock.

GETTING IN AND AROUND AUTOCAD **7-3**

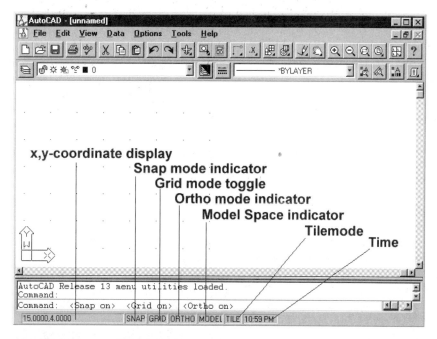

FIGURE 7-2 Mode Indicators

If the mode is listed as bold black letters, it is currently active or "on". You change the status by double-clicking on the word.

Toolbar Indicators

The top of the screen displays indicators on the toolbar. This includes:

Layer status, color, and name: Lists the status, color, and name of the current layer. The four little icons prefixing the layer name indicate whether the layer is unlocked/locked, thawed/frozen, viewport thawed/frozen, and on/off. The small square prefixing the layer name indicates the current color.

Linetype and name: Lists the look and name of the current linetype.

The Command Prompt Area

The bottom of the screen contains the prompt area. This area is where your commands are listed and the resulting prompts are displayed. If you become confused as to the present status of AutoCAD, the prompt line is the place to look for the answer.

ENTERING COMMANDS

Commands may be entered in AutoCAD in several different ways. Let's look at the methods you can use to enter commands in AutoCAD.

Keyboard

The most basic method of input is the keyboard. You may type the command directly from the keyboard. Your input will be displayed on the prompt line. If you make a mistake, simply backspace to correct. After typing, you must press [ENTER] or the spacebar to activate the command. AutoCAD doesn't care if you use uppercase, lowercase, or a combination of each.

In order to type a command from the keyboard, the command line must be "clear". That is, another command must not be in progress. If the command line is not clear, you can clear it by using [ESC].

Screen Menus

The most basic method of command entry is from the screen menu. Selecting items from the menu will have the same effect as typing them from the keyboard, except you don't have to press Enter.

Pull-Down Menus

Selecting items from a pull-down has the same effect as selecting from the screen menu.

Tablet Menus

AutoCAD allows the use of tablet menus. These are printed templates that are placed on a digitizing tablet. The user may customize the tablet menu for a particular application. Items chosen from a tablet menu respond in the same manner as those chosen from the screen menu.

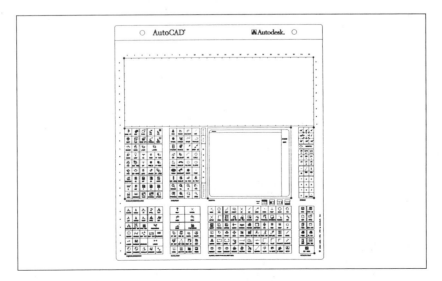

FIGURE 7-3 Tablet Menu Supplied with Release 13

SCREEN MENUS

On the right side of the screen, you will see the optional on-screen menu. This menu and its associated submenus contain all the AutoCAD commands. In the Windows version, this menu is turned off.

Items are "entered" from the menu in different ways, depending on the input device which you are using.

Mouse or digitizer: Move the cursor to the far right of the screen to "light up" the menu items. Move up or down to select the desired command and press the pick button.

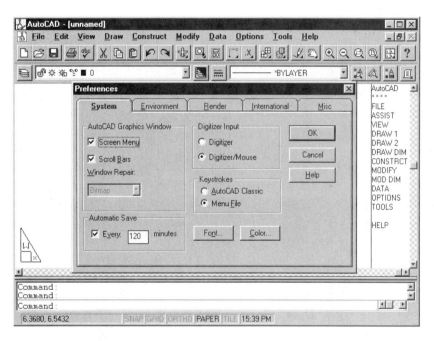

FIGURE 7-4 Turning on the Screen Menu with the Preferences Dialog Box

Keyboard: Press the Menu Cursor key (some computers such as IBM and most IBM-compatibles use other keys such as [INSERT]) to light up the menu items and move up and down the menu using the arrow keys on the keypad. Select the desired item by pressing [ENTER].

To turn on the on-screen menu, type in the PREFERENCES command and press [ENTER]. In the resulting dialog box, select the "System" tab near the top. In the upper left you will see a boxed-off section titled "AutoCAD Graphics Window". With your mouse, click with the left button with the pointer over the words "Screen Menu". An "X" appears in the box to the left. Click "OK" to accept the PREFERENCES dialog box. The screen menu will appear in the right side of the drawing editor.

Root Menu and Submenus

The menu which is initially displayed on the screen is the ROOT MENU. This menu is made up of several items which represent Submenus. Each submenu, in turn, contains several commands that are associated with it. When you enter each submenu, a new menu will appear which lists the options associated with the previous menu. The commands listed in the menus have colons (:) after them. Commands that are followed by three periods (such as PLOT...) prompt a dialog box. This means that selecting this command will cause a dialog box to be displayed on the screen. You will use this dialog box to enter information. Use of dialog boxes is covered later in this chapter.

Each submenu contains options which allow you to "jump" between other submenus. In most cases, you will have the choice to move either to the Last (previous) menu or the Root menu.

The illustration below shows the menu path taken from the root menu to the Line command (contained in the Draw submenu) and back to the root menu. Notice that the commands are followed by a colon and command "modifiers" are not.

The commands are arranged in a logical manner under headings in the root menu. For example, the Line, Circle, and Arc commands are located under the DRAW heading. The Erase command is located under the MODIFY menu. After some practice, the location of each command within the screen menu system will become familiar. It takes a bit of practice to become proficient in the use of nested menus, but once mastered, it speeds up the drawing process.

Pull-Down Menus

The pull-down menu is displayed under the title bar. Clicking on one of the listings will cause a pull-down menu to appear on the screen. You may select a command or function from the pull-down menu in the same manner as a screen menu.

GETTING IN AND AROUND AUTOCAD **7-7**

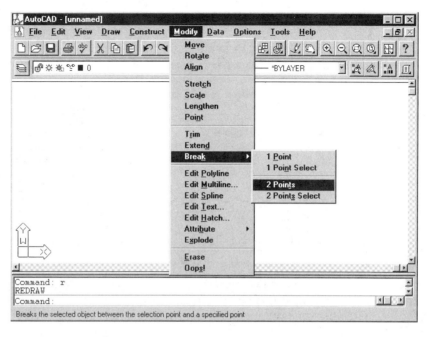

FIGURE 7-5 Pull-Down Menu

If you pull down a menu and do not want to select an item, you can exit the menu by clicking on the pull-down menu.

Pull-down menus are also arranged in a logical manner. In many cases, however, they do not parallel the screen menus. Figure 7-6 shows the sequence used to pull down and select an item from a pull-down menu.

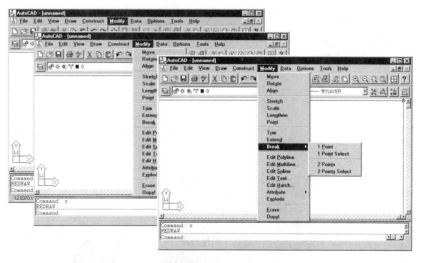

FIGURE 7-6 1. Select Command Category
2. Click to Pull-Down Menu
3. Select Command from Menu and Click

KEYBOARD KEYS USED IN AUTOCAD

AutoCAD provides special keys to aid in toggling modes on and off and for correcting errors in command entry. Modes may be toggled on and off even if you are currently in a command.

Command Line Entry Keys

The following keys can be used to edit your keyboard input at the command line.

Backspace: The backspace key removes one character at a time from the command line. You may use the backspace at any time before you Enter the contents of the command line.

CTRL-H: Used the same as the backspace key.

CTRL-X: Cancels all the characters on the command line.

ESC: Cancels the current command and returns the command line. You may enter [ESC] at any time. If entered while an operation is in progress, it will terminate the command at its present point.

Toggle Keys

Computer keyboards use function keys that are interchangeable with the normal toggle keys. The function keys are designated as "F" keys. For example, "F9" refers to the function nine key.

F1: Calls up the help window.

F2: Toggles between the graphics and text windows.

CTRL-T or F4: Toggles tablet mode on and off. You must first calibrate the pad before you can toggle the Tablet mode on.

Function Key

CTRL-E or F5: Toggles to the next isometric plane when in ISO mode. The planes are activated in a rotating fashion (left, top, right, and then repeated).

CTRL-D or F6: Toggles the screen coordinate display on and off.

CTRL-G or F7: Toggles Grid on and off.

CTRL-O or F8: Toggles Ortho mode on and off.

CTRL-B or F9: Toggles Snap mode on and off.

CTRL-V: Switches to the next viewport.

CTRL-A: Toggles group mode on and off.

Using a control key combination

Other Keys

INS: Turns on the menu cursor. You only need to use this key if you are using the keyboard arrow keys to move the cursor. This is the light bar that is used to highlight screen menu items.

HOME: Turns the crosshairs on. After some types of command entries, the crosshairs are not visible on the screen. The crosshairs will also be redisplayed if you use an arrow key from the key pad to move the crosshair.

UP ARROW: Moves the crosshairs up.

DOWN ARROW: Moves the crosshairs down.

LEFT ARROW: Moves the crosshairs left.

RIGHT ARROW: Moves the crosshairs right.

PgUp: "Speeds up" crosshair movement by increasing the interval which it moves with each press of an arrow key.

PgDn: "Slows down" crosshair movement by decreasing the interval which it moves with each press of an arrow key.

REPEATING COMMANDS

If you would like to repeat the previous command given to AutoCAD, press [ENTER] (or Return) when the "Command" prompt appears. The last command entered will repeat. The Spacebar can also be used for the same purpose.

Some commands require that you enter additional information. The Zoom command, for example, prompts you for the type of zoom you desire. In most cases, the first letter of your choice will be adequate (such as responding with a **W** for "window" when prompted by the Zoom command).

SCREEN POINTING

Points, distances and angles may also be entered simply by "showing" AutoCAD the information on the screen. Entering two points could indicate the distance and angle requested.

On some commands (such as Line) the absolute coordinate display converts to a relative distance and angle display. Using this as a method of measurement is quite suitable in most circumstances, although actual numerical entry using coordinates is more accurate.

Showing Points by Window Corners

Some commands require input which specifies both a horizontal and a vertical displacement. Both points may be shown at one time by requesting a "window" and using the *X* and *Y* distances from the lower left corner and the upper right corner as the displacement.

For example, the window in Figure 7-7 displaces a value of (5,3) from the lower left corner of the window.

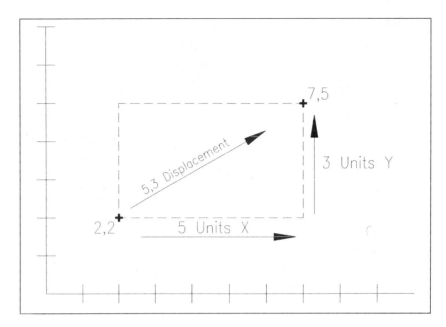

FIGURE 7-7 Using a Window to Show Displacement

DIALOG BOXES

Certain commands allow you to enter or select information from a dialog box. Note that not all display systems are capable of showing dialog boxes.

Figure 7-8 shows the dialog for controlling layer settings.

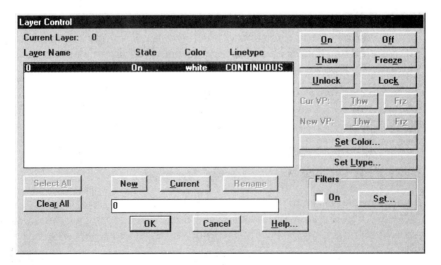

FIGURE 7-8 Layer Control Dialog Box

Displaying Dialog Boxes by Command

Many commands automatically display dialog boxes. Others are invoked by command. The following is a list of commands used to display a dialog box. Each command starts with DD (Dynamic Dialogue).

DDEMODES: Used for setting entity properties of linetypes, colors, etc.

DDLMODES: Controls layer settings.

DDRMODES: Drawing modes and aids such as snap, grid, etc.

DDATTE: Attribute editing.

DDUCS: Dialogue control for the user coordinate system.

DDEDIT: Dialogue control for attribute definitions.

DDIM: Controls dimension settings.

DDGRIPS: Controls Autoedit grips and settings.

DDSELECT: Determines entity selection modes (pick box size, etc.).

DDOSNAP: Sets object snap modes and aperture size.

DDINSERT: Used for block insertion.

DDRENAME: Renames various named items.

DDATTEXT: Used to extract information from a drawing file for use with another program.

DDATTDEF: Used to create attribute definitions.

DDUNITS: Sets drawing units, coordinates, and angles.

DDCHPROP: Controls the layer, color, linetype, and thickness of selected objects.

DDCOLOR: Select current color.

DDLTYPE: Loads and selects linetypes.

DDMODIFY: Modifies object properties.

DDPTYPE: Select point style.

DDUCSP: Select a preset UCS.

DDVIEW: Creates and restores views.

DDVPOINT: Select viewpoint.

DDSTYLE: Select 2 text style from a font file (Release 13c4 only).

The display of dialog boxes can either be on or off. The Filedia system variable is used to turn the display of dialog boxes on or off. If the Filedia system variable is set to 0 (off), you will be prompted for input on the command line.

If the Filedia is off, you can override by placing a tilde (~) in response to a command request if a dialog would normally be available.

Using a Dialog Box

When a dialog box is displayed, the crosshairs are replaced by an arrow pointer. The arrow pointer is used to select items from the dialog box.

Some dialog boxes contain sub-dialog boxes. If a sub box is displayed, you must reply to the prompt or select Cancel to continue.

Many dialog boxes contain a Help button. Selecting the help button will display a help box that explains the purpose and use of the dialog box.

You can move a dialog box by moving the pointer over its title bar and holding down the button while moving the box. Moving dialog boxes is useful if you need to see a screen object that is under the box.

Figure 7-9 shows a typical dialog box, with the parts labeled.

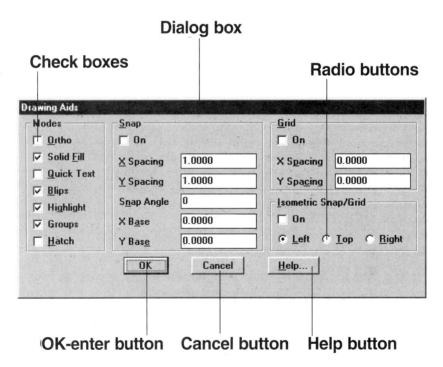

FIGURE 7-9 Typical Dialog Box with Check Box and Radio Button

Let's look at the parts of a dialog box.

Scroll Bars. A list box can contain more entries than can be displayed at one time. Scroll bars are present on some list boxes that are used to move (scroll) the items up or down.

To move the displayed entries up or down one item, move the arrow to either the up or down arrow and click.

GETTING IN AND AROUND AUTOCAD **7-13**

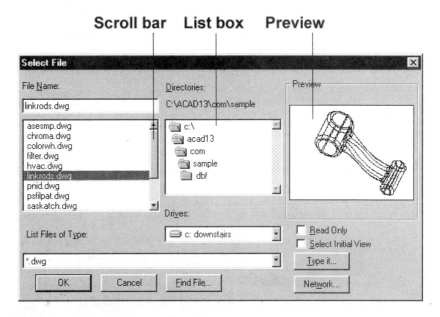

FIGURE 7-10 List Box and Scroll Bars

If you move the arrow on the slider box and click, you can move the box up or down with the arrow. When you click again, the entries will be redisplayed at the new location. Note that the position of the slider box is relative to the position of the displayed items. Thus, if there are a lot of items in the list, a relatively small movement of the slider box will scroll several items. If you click the slider bar and then wish to cancel before clicking the second time, entering [ESC] will cancel, returning the list to the original location.

Buttons in Dialog Boxes. Buttons are used to select items. The buttons can be selected with the pointer.

Buttons that have a heavy border are the default. Selecting OK from the dialog box will automatically select the default buttons.

Three periods (...) after a button prompts a sub-dialog box.

Buttons with two arrow pointers (<<) refer to an action that is required in the graphics screen, such as selecting an object.

If a button is "grayed out" it is not available at that time.

The following is a summary of the buttons used in dialog boxes.

Check button: Boxes following items that show a check mark if selected. These are mostly used to "turn on" a function. If an X mark is present, the function is "on".

Radio button: A radio button selects between a series of items, only one of which can be "active" at a time.

List box: A list box contains a list of items that you can choose. List boxes contain scroll bars that allow you to scroll among the available choices.

Drop-down List Box: A drop-down list box is a type of list box, except it "drops down" when an item with an arrow follows it. An example of a drop-down list box is shown in Figure 7-11.

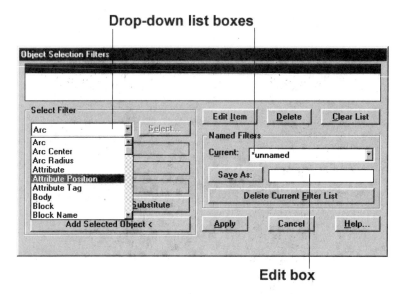

FIGURE 7-11 Dialog Box List Box and Edit Box

Edit box: Edit boxes contain a single line of text that can be edited. To edit the text in the box, click in the box. A cursor bar will appear at the text. You can move the cursor bar with the arrow keys on the keyboard.

Image tile: Some dialog boxes contain an image tile. This is a small window that displays a selected item such as a drawing file or line type.

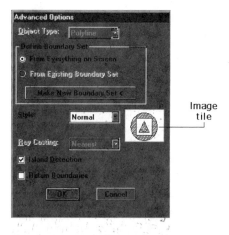

FIGURE 7-12 Dialog Box Image Tile

ICON MENUS

Icon menus are displayed on the page as graphic images instead of words. You can select the image you wish with the screen pointer. Figure 7-13 shows a sample icon menu.

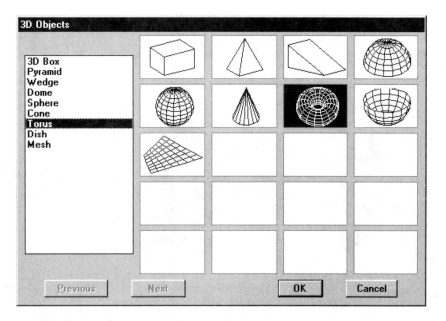

FIGURE 7-13 Icon Menu

Icon menus are prompted by certain command selections. For example, if you select Surfaces and 3D objects from the Draw pull-down menu, an icon box displaying 3D objects will be displayed.

Selecting from an Icon Menu

When an icon box appears, the crosshairs change to an arrow pointer. Each selection has a box next to it. If you move the pointer to the box, an outline appears around the selection. Clicking on the selection box will select the item.

An icon box is cleared from the screen by selecting an item, or pressing [ESC]. Any other keyboard activity is suppressed.

STARTING A NEW OR EXISTING DRAWING

When you start AutoCAD, you will either begin a new drawing, or call up an existing drawing you have previously saved. Let's look at how to perform these functions.

Starting a New Drawing

The New command is used to start a new drawing. Type New at the keyboard and press [ENTER]. AutoCAD will display a *dialog box*. Enter the new drawing name from the keyboard. The drawing name will be displayed in the dialog box (see Figure 7-14).

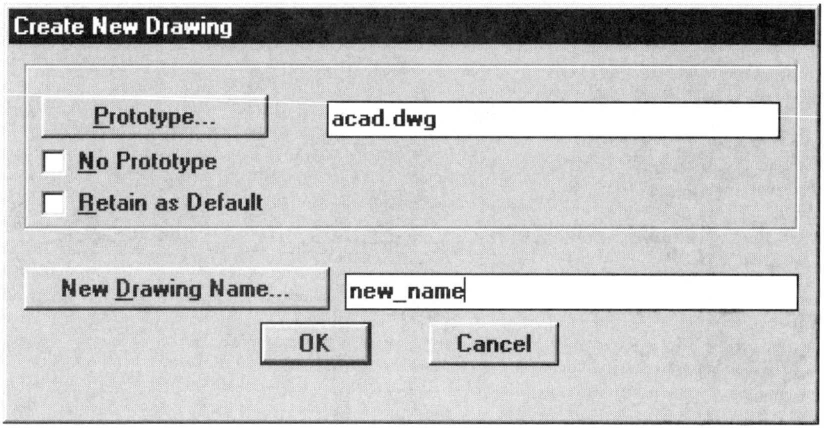

FIGURE 7-14 Starting a New Drawing

Place the pointer over the OK button and click. The drawing editor will return and you will be ready to start your drawing session!

Prototype Drawings. When you begin a new drawing, AutoCAD uses a standard list of settings for the drawing. These settings are taken from a *prototype drawing* named "ACAD". A prototype drawing is a separate drawing containing settings and, optionally, drawing elements, that your new drawing can use as a template. Your new drawing will be exactly equal to the prototype drawing.

Although the default prototype drawing is ACAD, you can use any drawing as a prototype drawing. For example, you may want to create a drawing that contains all the settings you normally use and has a title block already drawn. You could then use this prototype drawing to create a starting point for all your new drawings. Each new drawing that used this prototype would have the same settings and the title block at the moment it was created. It is not necessary, however, to use a named prototype drawing. If you do not select a prototype drawing, AutoCAD will assume the standard settings. Note that you can change any of the settings, regardless of the prototype drawing. The following sections explain how to use prototype drawings with AutoCAD.

Selecting a Prototype Drawing. You select a prototype drawing when a new drawing is started. When you enter New, the *Create New Drawing* dialog box is displayed. Enter the new drawing name, then click on the "Prototype..." box. A new dialog box will paste over the top of the existing box.

FIGURE 7-15 Selecting the Prototype Drawing

This is the *Prototype Drawing File* dialog box. Select a prototype drawing from the files list and click on the OK button. You will now be returned to the *Create New Drawing* dialog box. Click on OK to start the new drawing.

An alternate method of creating a new drawing with a prototype is to specify the new drawing to be "equal" to another drawing. To do this, simply enter the new drawing name in the *Create New Drawing* dialog box, then place an equals sign (=) followed by the prototype drawing name.

For example, if you wish to start a new drawing named "WIDGET" and use a drawing named "PROTO1" as the prototype drawing, enter;

WIDGET=PROTO1

as the drawing name in the dialog box as shown in Figure 7-16.

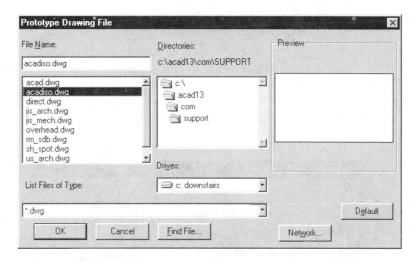

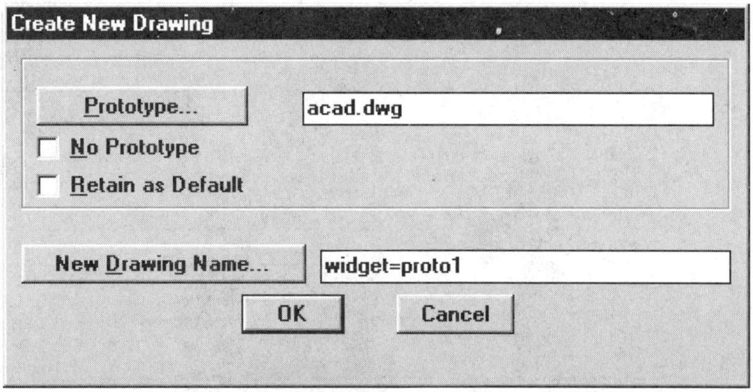

FIGURE 7-16 Creating a Prototype Drawing

Configuring a Default Prototype Drawing. You may wish to retain a drawing as the default prototype. This would allow you to use the default prototype drawing as the template for all new drawings without having to select it each time you start a new drawing. To retain a drawing as the default prototype, click on the Prototype... box in the *Create New Drawing* dialog box and select the drawing you want to use as the prototype from the *Prototype Drawing File* dialog box that is displayed. Next select OK, then click on the Retain As Default box. The following figure shows the sequence.

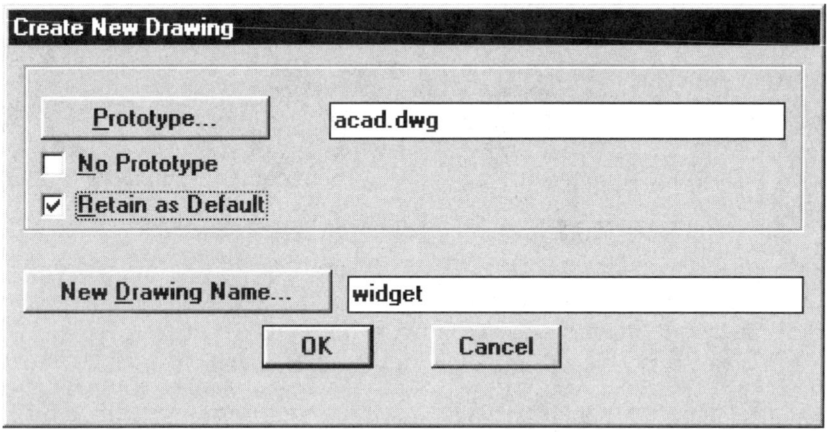

FIGURE 7-17 Selecting a Default Prototype Drawing

Starting a New Drawing With No Prototype. If you do not want to use a prototype drawing, click on the No Prototype box in the *Create New Drawing* dialog box. This sets the drawing values to the default settings.

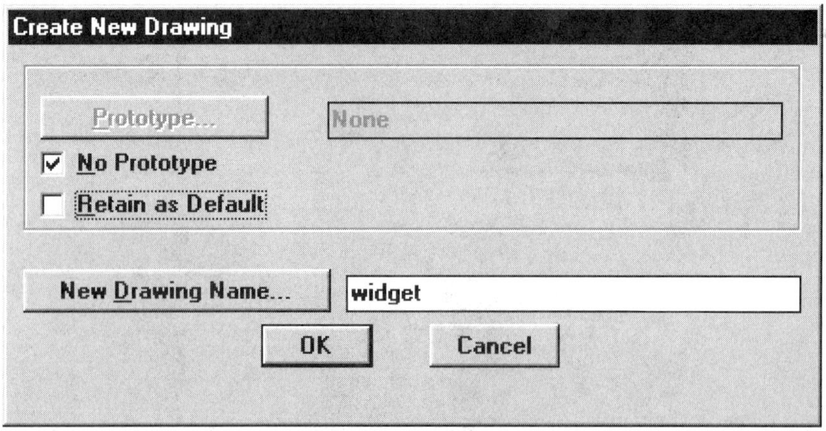

FIGURE 7-18 Starting a New Drawing Without a Prototype

Editing an Existing Drawing

Many times a drawing that was previously saved must be edited. In order to edit an existing drawing, you must first "open" it. To open an existing drawing, use the Open command.

When you enter Open from the keyboard, an *Open Drawing* dialog box is displayed.

Figure 7-19 shows the *Open Drawing* dialog box.

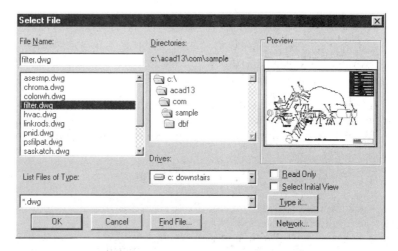

FIGURE 7-19 Open Drawing Dialog Box

To open a drawing, select the drawing from the files listing. When you select the drawing, it will be entered for you on the file line. Click on the OK button to open the drawing. The following figure illustrates the sequence to open an existing drawing file.

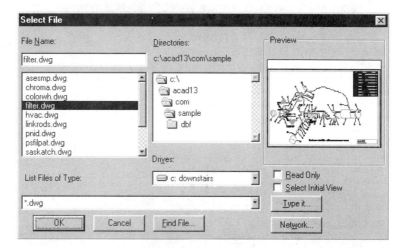

FIGURE 7-20 Opening an Existing Drawing

If you wish to choose a different directory, *double click* on the desired directory in the directories listing. Double clicking means to rapidly click the mouse button twice on the entry you desire.

SAVING AND DISCARDING DRAWINGS

Either during a drawing session or when you have completed a session, you will want to either save or discard the work you have just completed. If you save your work, it will be recorded to the hard drive on your computer. A saved file is recorded with the name you specified when the drawing was originated as new.

You may also save your work and remain in the drawing file. It is good practice to save your work to disk periodically. If you experience a power outage or your computer hangs up, you will lose the work completed since you last saved to disk.

If you choose to discard your work, only the part of the drawing performed since the last save will be discarded. If you have not saved any of the work since the drawing was created, discarding the work will delete all instances of the drawing.

Saving Drawings

There are four commands you can use to save drawings. The following listing explains how each method works.

SAVE: Use the Save command if you wish to save your work and remain in AutoCAD. This is useful if you want to save your work at periodic intervals and continue to work. When you use Save, a dialog box is displayed. You may select the current file name, or a new file name to save the drawing under. Selecting a new file name functions the same as the Saveas commands that follows.

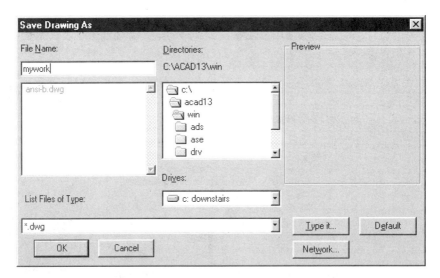

FIGURE 7-21 Dialog Box for Saving a Drawing with a New Name

SAVEAS: The Saveas command is used in the same manner as the Save command, except the drawing is saved in its current state under a new drawing name. When you use Saveas, the *Save Drawing As* dialog box is displayed.

If AutoCAD is set to suppress dialog boxes (using the Filedia system variable), a prompt will appear on the command line:

Save current changes as <default>:

Using the Saveas command creates a new file that contains the drawing in its current state. Note, however, that the original drawing has not been saved. If you wish to save the original drawing you are working in, use the Save command.

If you enter the name of a drawing that already exists, AutoCAD displays the following message box:

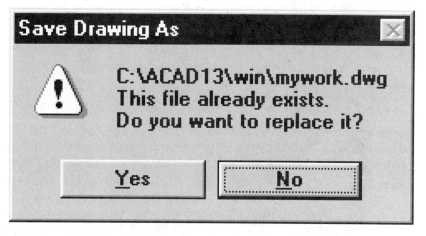

FIGURE 7-22 Message Box

If you want to replace the existing drawing with the new one, click on YES. If you do not, click on the NO button.

> **NOTE:** Selecting YES will write over and delete the existing file.

QSAVE: The Qsave command is functionally the same as the Save command, except the dialog box is not displayed. The drawing is automatically saved to disk under the name specified when the drawing was created.

END: The End command is used to record the work to disk and exit the AutoCAD program. This is one way to exit the AutoCAD program (see Quit for the other way).

NOTE: Although there are four methods of recording your work to disk, they do not function in the same manner. The Save, Saveas, and Qsave commands "clean up" the drawing file by omitting items marked for deletion (previously erased, etc.). This results in a file size that is smaller. The End command does not perform a cleanup. If you wish to compact your drawing file size and exit AutoCAD, use the Save or Qsave command, then use End, Exit, or Quit (the Quit command is covered later in this chapter) to terminate your drawing session.

Saving Your Drawing Automatically. If you wish, you can set a time interval for AutoCAD to save your work automatically. The Savetime system variable is used to set the time interval.

Discarding Your Work

Occasionally you will create a drawing that you do not wish to keep. If you want to exit the current drawing, discard the changes, and start a new drawing, use the New command. AutoCAD will display a message box that the drawing has been changed and ask you what you want to do.

Click on the box you want. The *Create New Drawing* dialog box will be displayed next.

If you want to exit the drawing, discard the changes, and exit the AutoCAD program, use the Quit or Exit command. A message box is displayed asking you whether you wish to save the changes, discard the changes, or cancel the Quit command and reenter the command. The message box is shown in Figure 7-23.

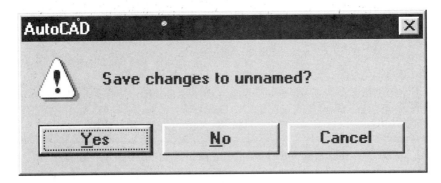

FIGURE 7-23 Drawing Modification Dialogue Box

EXERCISES

1. Move the cursor around with your input device (mouse, digitizer, etc.). Do the coordinates at the lower left corner of the screen move? Press [F6]. Move the cursor again and watch the coordinates.

2. Press the [F8]. Watch the status line to see the display show the ORTHO mode turn on and off. Press [F9] for SNAP mode.

3. Move the crosshairs to the top of the drawing area and into the menu bar.

    Move the arrow from left to right over the entries. Position the arrow over the name "DRAW" and click. A pull-down menu will extend down into the drawing area. You select items the same way as you do from the side screen menu. Move the cursor back to the name "DRAW" and click. The pull-down menu will disappear.

    Select the "DRAW" pull-down menu again and select "LINE". Draw some more lines.

4. Clear the command line with [ESC]. Type "FILEDIA" and [ENTER]. Enter a "1" in response to the prompt. This turns on the dialog boxes.

    From the menu bar, select "DATA", then "LAYERS...". Do you see the *Layer Control* dialog box?

    Let's enter a new layer. Move the cursor. Notice the arrow pointer. Click in the text entry box. Type **MYLAYER**. Move the arrow to the New box and click. Notice how the new layer name is now listed in the layer names listing. Make your layer current by clicking on the layer name, then in the Current box. Now select the OK box at the bottom of the dialog box. Look at the layer name in the Layer toolbar near the top of the screen. Is the layer "MYLAYER" current?

5. Press [F2]. Do you see the text screen? Press [F2] again to return to the drawing screen.

6. Select LINE again. Draw some lines and use [ESC] to cancel. Press the enter key. Did the Line command repeat? Clear the command line with [ESC] again. Press the spacebar. The Line command will repeat again. Either [ENTER] or the spacebar can be used to repeat a command.

7. Clear the command line again. Enter a **?** from the keyboard and press [ENTER]. You should see the *Help* dialog box. To get more detailed information on a topic or to access a different section of the Help menu, position the cursor over any text that is green and click with the left mouse button. Familiarize yourself with using the Help dialog box.

CHAPTER REVIEW

1. Is it necessary to add a .DWG extension to the drawing name when you create a drawing file from the main menu?

2. When you are ready to plot a drawing, how is the distinction made to let AutoCAD know that you desire a printer plot instead of a pen plot, or vice versa?

3. What would be some of the benefits of using prototype drawings?

4. What are mode indicators? Where are they located when turned on?

5. A linkage of menus and submenus progressing from one to another is referred to as what?

6. Must you exit AutoCAD to copy or delete a file?

7. List the control and/or function key associated with the following definitions:

   turns Ortho mode on and off _____

   used to turn on or redisplay the crosshairs _____

   removes characters from the prompt line one at a time _____

   allows you to toggle between graphic and text screens _____

   toggles Tablet mode on and off _____

   turns on the menu cursor _____

   cancels all the characters on the command line _____

   toggles to the next isometric plane in iso mode _____

   used the same as the backspace key _____

   cancels the current command and returns to the command line _____

   toggles the screen coordinate display on and off _____

   toggles the grid on and off _____

   toggles snap on and off _____

8. After executing a command, and you have returned to the command prompt, what happens when you press [ENTER]?

9. How is the Help facility invoked?

10. What is the purpose of the prompt area?

CHAPTER 8

DRAWING SETUP

Before starting any drawing, you should set the type of measurement units to be used, and the size of the workspace. After completing this chapter, you will be able to:

- Perform drawing set up.
- Set the type of measurement used in your drawing.
- Set the size of the workspace in the current units.
- Set units and limits that will match the size of plot media and scale.
- Display the status of the current drawing.

SETTING UP A DRAWING

If you begin a drawing on a traditional drawing board, you start by determining the scale and size of the drawing. You would not draw the first line before choosing the paper size, or before selecting the scale at which you would construct the drawing.

AutoCAD uses settings that somewhat parallel these decisions. You could, as you have seen, just start drawing with AutoCAD's default settings. The problem is that these settings are not suitable for every type of drawing. Consider the difference between architectural and engineering (decimal) scales, or the differences between drawing on 8½" × 11" and 36" × 48" media.

To properly use AutoCAD, you must learn to perform some basic settings. In this chapter, we will learn the following concepts and commands:

Units: Setting AutoCAD to draw in specified units such as architectural, engineering, decimal, or scientific.

Limits: Setting the actual "real world" drawing area.

Scaling: Setting units and limits to match a desired plot scale.

Status: Displaying the settings in the drawing.

Setting the Drawing Units

Since CAD programs are utilized in many different types of work, the units in which distances are measured can be many. An engineer, scientist, or architect may require different notations for coordinates, distances, and angles. AutoCAD provides the capabilities for each through the Units command. Setting the units is the first step in setting up a new drawing.

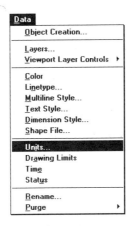

To set the units format for your drawing, select "Units..." from the Data pull-down menu. AutoCAD will display a dialog box that you will use to set the type of units. Figure 8-1 shows the dialog box.

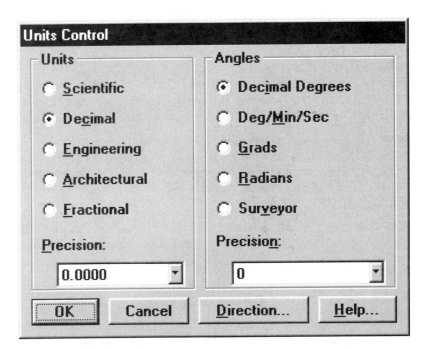

FIGURE 8-1 Units Control Dialog Box

The first "section" of the dialog box is titled "Units". Let's look at the available choices.

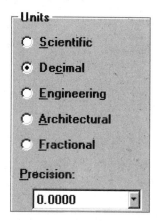

FIGURE 8-2 Units Settings

There are five selections:

Scientific: AutoCAD provides scientific units for use in projects that deal with very large dimensions. An example of scientific units represents 21.75 units as 2.175E+01.

Decimal: Decimal units are common in engineering drawings. An example of decimal units would be 21.75.

Engineering: Engineering units are used for applications such as civil engineering and highway construction. In engineering units, 1' 9¾" would be shown as 1'-9.75".

Architectural: Architectural drawings use the feet, inches, and fractions style of showing measurements. For example, 1' 9¾" is an example of architectural units.

> **NOTE:** The Architectural format designates that one unit equals one inch.

Fractional: AutoCAD can also be set to display fractional units. An example of fractional units would be 21¾.

To set the desired units, simply click on the corresponding radio button next to listing.

Setting the Precision for the Units. After you have selected the format, you should set the precision for the coordinates and distances. For example, if you choose Architectural format, you will need to determine the smallest fractional denominator. To do this, use the popup list box to change the number in the precision box.

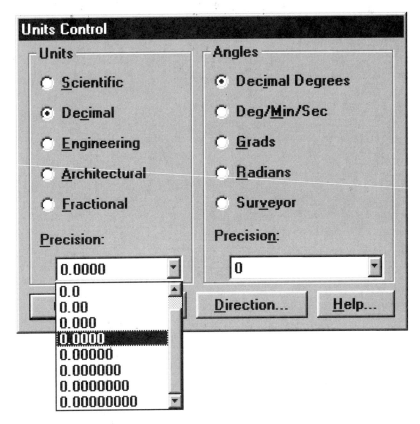

FIGURE 8-3 Setting Units Precision

Setting the Angle Measurement

After selecting the format and the precision for the units, set the method of angle measurement. The angle measurement setting determines the type of coordinate listing (at the top of the drawing editor screen), and the type of input you enter from the keyboard.

The type of angle measurement is set in the right box of the *Units Control* dialog box.

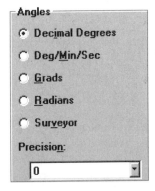

FIGURE 8-4 Angle Settings

AutoCAD can display five types of angle measurement. The following is a listing of the different types of measurement and an example of each.

Decimal degrees: Using decimal degrees will instruct AutoCAD to display whole degrees as whole numbers and partial degrees as decimals (45.0000).

Degrees/minutes/seconds: Degrees can be represented as whole degrees, with partial degrees represented as "minutes", and partial minutes as "seconds". One degree contains 60 minutes and one minute contains 60 seconds. AutoCAD uses a "d" to represent degrees, a (') to designate minutes and a (") to represent seconds. For example, 22 degrees, 14 minutes, 45 seconds is listed as 22d14'45".

Grads: AutoCAD can also specify angles in grads. An example is 35.0000g.

Radians: For mathematical work, you can use radians to display angle. An example is 0.678r.

Surveyor's units: Surveyor's units are used to designate property lines. They are a form of the degrees/minutes/seconds type of format, except that they are relative to compass points. An example is N22d15'32"W, with "N" and "W" representing North and West, respectively.

To set the angle measurement for your drawing, select the radio button next to the type of angle measurement you wish to use.

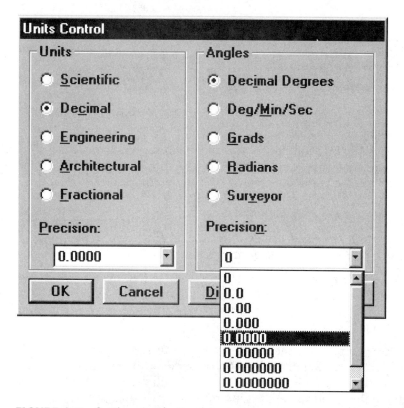

FIGURE 8-5 Setting Angle Precision

Setting the Precision for the Angle Measurement. After you have specified the angle format, set the precision of the angle format. You may select up to eight decimal places. The angle precision is set at the angle precision box and in the same manner as explained before.

Setting the Zero-Angle Direction. The zero-angle direction is set by first selecting the "Direction..." button. The resulting dialog box appears as follows.

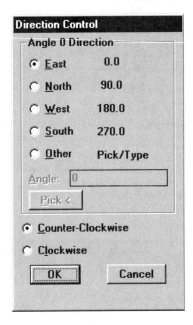

FIGURE 8-6 Setting Zero-Angle Direction

AutoCAD, by default, sets the angle zero to the right (east). This setting makes all angles that have a direction of zero read to the right. You may set the zero angle to display in any direction you wish (this will not affect text in the drawing, since you set the text angle relative to this setting).

For example, you may want the zero angle to be straight up on the screen. Selecting "North 90.0" would achieve this. To set the desired zero angle direction, select the radio button next to the desired listing.

After you have selected the zero angle direction, you can designate whether AutoCAD measures angles in a clockwise or counterclockwise direction. As you become familiar with the operation of some commands, you will better understand the effect of this setting. The default setting is counterclockwise.

Once you have set the type of unit measurement and the method of angle definition, the settings will remain in effect throughout the remainder of the drawing. If you save your drawing and return, the settings will remain intact. It is only necessary to set the units when you are beginning a new drawing.

DRAWING SETUP **8-7**

You can alternately use a prototype drawing to preset the drawing units.

> ***NOTE:*** You may, if you wish, "show" AutoCAD the new angle for "0".
>
> To show AutoCAD the zero direction, click the Other radio button, then click the Pick button, then select any point on the screen. Next, move the cursor in the direction of the desired angle and enter a second point.
>
> This can be especially useful if you must work in reference to a certain angle.

SETTING DRAWING LIMITS

Setting the *Limits* of your drawing allows you to determine the size of working area which you need to draw in. This can be important, since the area needed to construct different types of drawings can be dramatically different. For example, a grid map showing an area of a city or a detailed section of a watch would obviously require different limits.

In order to better understand how drawing limits work, you should also read the following section on "scaling your drawings".

Limits are set in the current drawing units. For this reason, you should set the units *before* specifying the limits. To set the limits, select the "Drawing Limits" from the Data pull-down menu. The following sequence is shown at the command line.

Command: **Limits**
Reset Model space limits:
ON/OFF/<Lower left corner> <*default*>:

Let's look at each of the command line options.

ON: Turns the limits "on". When limits are on, you may only draw within the area of the limits.

OFF: Turns the limits "off". When limits are off, you may draw either inside or outside the area defined by the limits.

LOWER LEFT CORNER: The default, allows setting of the lower left corner (by coordinates) for the drawing area. If you enter a coordinate for the lower left corner, you will be prompted for the coordinate of the upper right corner.

Command: **Limits**
Reset Model space Limits:
On/Off/Lower left corner <*current*>:
Upper right corner <*current*>:

ENTER: Retains the current (default) value for the lower left corner and prompts you for the coordinates for the upper right corner.

The initial setting of the drawing limits is determined by the prototype drawing. You may, however, change the drawing limits at any time with the Limits command. The following exercise will help you better understand the use of the limits command.

> **NOTES ABOUT LIMITS:**
>
> 1. The screen does not display the new limits after setting. Make a habit of executing a "zoom all" after setting new limits. This displays the actual "drawing page size" on the screen.
> 2. To see the size of your "drawing page", turn on the grid pattern. This will allow you to visualize the "edges" of your paper.

SCALING YOUR DRAWINGS

One of the hardest concepts for a new CAD user to grasp is that of scale. Some of the most-asked questions are:

> How do the drawing limits and the plot size relate?
>
> Does the plot scale affect the size of my limits?
>
> How do I know where the edges of the paper are? I need to see the paper size on the drawing to compose the work.

All you need to answer these questions is just some simple arithmetic.

First of all, with the exception of architectural units, you draw in generic units, not in feet, inches, miles, or anything else. A unit may equal one centimeter or it may equal one mile. AutoCAD really doesn't care what your units equal until you plot.

At the time of the plot, you must specify what one unit is equal to. You do this by telling AutoCAD that one inch (or one centimeter if you are working in centimeters) is equal to x number of units on the paper. For example, if 1" = 10 units and you have determined that one unit would equal a foot when you began drawing, then the scale of the plot would be 1" = 10' (remember, you determined that one unit equals one foot).

Let's suppose that you were planning to plot on paper that measured 36" × 24". How would you set limits that were in true relation to the page size?

First, since you intend to plot at 1" = 10', and one inch on the paper will contain ten units, we can multiply the dimensions of the paper by the units per inch and determine the number of units contained in both dimensions of the paper.

   10 units/inch × 36" = 360 (X limit)
   10 units/inch × 24" = 240 (Y limit)

DRAWING SETUP **8-9**

If you now set limits of 0,0 (lower left) and 360,240 (upper right), the limits will match the page size (assuming again that you intend to plot at one inch equals ten units and you have assigned one unit to equal one foot).

You may, of course, change plotting scale at a later time. If you do, you will have to change the limits if you desire to "match" the limits to the paper size.

Knowing the methodology of calculating the sheet size to the intended plotting scale allows you to set up your drawing to see the available "drawing page" size.

Table 8-1 shows the relationship between drawing scale and plot sheet size. You can use this table to set up the limits for your drawings.

FINAL PLOT SCALE		SHEET SIZE				
		A 11 x 8½	B 17 x 11	C 24 x 18	D 36 x 24	E 48 x 36
	1/16	176', 136'	272', 176'	384', 288'	576', 384'	768', 576'
	3/32	132', 102'	204', 132'	288', 216'	432', 288'	576', 432'
	1/8	88', 68'	136', 88'	192', 144'	288', 192'	384', 288'
	3/16	66', 51'	102', 66'	144', 108'	216', 144'	288', 216'
	1/4	44', 34'	68', 44'	96', 72'	144', 96'	192', 144'
	3/8	29'-4", 22'-8"	45'-4", 29'-4"	64', 48'	96', 64'	128', 96'
	1/2	22', 17'	34', 22'	48', 36'	72', 48'	96', 72'
	3/4	14'-8", 11'-4"	22'-8", 14'-8"	32', 24'	48', 32'	64', 48'
	1	11', 8'-6"	17', 11'	24', 18'	36', 24'	48', 36'
	1½	7'-4", 5'-8"	11'-4", 7'-4"	16', 12'	24', 16'	32', 24'
	3	3'-8", 2'-10"	5'-8", 3'-8"	8', 6"	12', 8'	16', 12'

FINAL PLOT SCALE		SHEET SIZE				
		A 11 x 8½	B 17 x 11	C 24 x 18	D 36 x 24	E 48 x 36
	10	110, 85	170, 110	240, 180	360, 240	480, 360
	20	220, 170	340, 220	480, 360	720, 480	960, 720
	30	330, 255	510, 330	720, 540	1080, 720	1440, 1080
	40	440, 340	680, 440	960, 720	1440, 960	1920, 1440
	50	550, 425	850, 550	1200, 900	1800, 1200	2400, 1800
	60	660, 510	1020, 660	1440, 1080	2160, 1440	2880, 2160
	100	1100, 850	1700, 1100	2400, 1800	3600, 2400	4800, 3600
	Full Size	11, 8.5	17, 11	24, 18	36, 24	48, 36

TABLE 8-1

CHECKING THE DRAWING STATUS

AutoCAD has a multitude of modes, defaults, limits, and other parameters that you will occasionally need to know. The Status command displays the current state or value of each of these. The status command may be used any time you are in the drawing editor. To enter it, choose Data from the menu bar, and then Status.

Command: **Status**

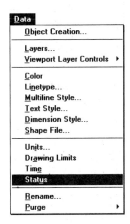

You will be presented with the text window as shown in Figure 8-7.

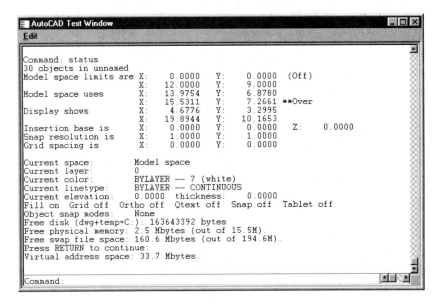

FIGURE 8-7 Drawing Status Screen

EXERCISES

1. Let's suppose that you would like to set limits that have 180.00 units horizontal (X value) and 120.00 units vertical (Y value). First, verify that the units are set to decimal units. Next, from the menu bar, select the Data menu and choose Drawing Limits.

   Command: **Limits**
   Reset Model space limits:
   On/Off/Lower left corner <current value>: **0,0**
   Upper right corner <current value>: **180,120**

   This will set limits that have a lower left corner origin at 0,0 and an upper right corner at the coordinates at 180,120.

   You do not have to retain 0,0 for the lower left corner; however this should be maintained unless you have special circumstances which require it.

2. Calculate drawing limits for a drawing that will be plotted on a 24" × 18" sheet of paper at ¼"=1'-0" scale.
3. Calculate drawing limits for a drawing that will be plotted on a 36" × 24" sheet of paper at 1"=50.00'.
4. Use the Status command to check the limits of your current drawing.

CHAPTER REVIEW

1. How might you obtain a listing of your drawing parameters?

2. How are angles measured in surveyor's units?

3. What choices of unit systems are available under the units option?

4. What might you do to visualize the drawing limits you have set?

5. Can the zero angle position be altered? How?

6. Why should the units of a drawing be set before the limits?

7. What restrictions apply to your drawing when the Limits are turned on?

8. Why would you want to change the zero angle setting from the default position?

9. What is the only exception to drawing in generic units?

10. At what point does AutoCAD need to know the units you have used to construct your drawing?

CHAPTER 9

DRAWING

Most CAD drawing activities involve the most basic drawing commands. After completing this chapter, you will be able to:

- Demonstrate the most basic drawing commands.
- Use display commands to magnify and maneuver around the drawing.
- Operate the commands used to save and record work to disks.

GETTING STARTED

Now that we know how to set up a drawing, let's jump in and start drawing right now. We will be using some basic drawing commands, modes, and assistance. They are:

- **LINE:** Drawing basic line entities.
- **POINT:** Constructing point entities.
- **CIRCLE:** Drawing circle by different methods.
- **ARC:** Constructing multiple types of arcs.
- **OBJECT SNAP:** Connecting precisely to points on an entity.
- **REDRAW:** Clearing the drawing screen of clutter.
- **REGEN:** Regenerating the drawing display from the database.
- **ZOOM:** Enlarging and reducing the view of the drawing.
- **PAN:** Moving the screen around the drawing.
- **GRID:** Placing "grid paper" dots on the screen.

SNAP: Setting a drawing increment.

ORTHO: Forcing cursor movements to be either perfectly horizontal or vertical.

DRAGMODE: Determining whether items are moved "real time" on the screen.

U (single undo): Undoing a single drawing operation.

REDO: Negating a single undo operation.

DRAWING LINES

Lines are drawn with the Line command. The line command is the most basic part of a CAD program. To draw a line, you must first select the Line command.

Drawing a Line

In the menu bar select Draw. You will then be presented with the draw menu. From the menu, select LINE.

You will notice that two things happen. First, the command line prompts you for a "From point". Second, you are presented with a special "Line menu." Before we proceed to draw, let's first talk about "from points" and "to points."

You must first show AutoCAD the point from which you wish to start drawing. Think of this as the place where you put your pencil to the paper. This is the "from point." Now you must determine the point at which the line will end. This is the "to point." Since many drawings contain lines that connect (such as a box consisting of four consecutive and connecting lines), the "to point" prompt is repeated until you terminate the command by choosing CANCEL from the menu, entering [ESC], or pressing [ENTER] a second time. This allows you to enter a series of lines without the inconvenience of invoking the Line command each time.

AutoCAD can draw more than one type of line.

Line: Lines are drawn in the same manner you use when selecting the Line command from the screen menu. The command will repeat until canceled, allowing you to draw several line "segments" without the necessity of reentering the Line command.

Multiline: Allows you to draw up to 16 parallel lines at one time.

Sketch: Sketching is "free hand" drawing.

Construction Line and Ray: Draws infinite construction lines (xlines) and semi-infinite construction lines (rays) that display but are not plotted.

Line Options

There are some drawing aids which are associated with the Line command. These aids provide assistance that is unique to a CAD drawing system.

They are:

1. Continue
2. Close
3. Undo
4. @

Let's continue and look at each option.

Continue. If you terminate the line command, you are free to begin a new line elsewhere. However, if you wish to go back and begin at the last endpoint, you may "reconnect" by using the Continue option. This option is found in the Line submenu.

Close. When you constructed the box, the last line you drew was connected to the beginning point. Aligning the end of the line with the beginning of the first line was a tedious process, wasn't it? The Close option will automatically perform this for you. Let's construct the box again. This time we will connect the lines at the final intersection with the Close option.

Command: **Line**
From point: (*Enter Point 1.*)
To point: (*Enter Point 2.*)
To point: (*Enter Point 3.*)
To point: (*Enter Point 4.*)
To point: **CLOSE**

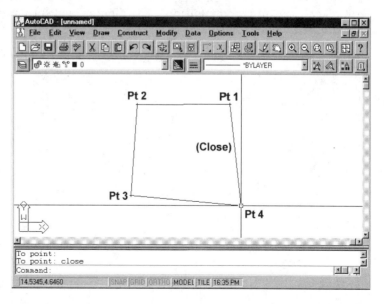

FIGURE 9-1 Close Option

The Close option may be used to return to the start (from point) of any consecutive string of lines. Using Close will terminate the string.

Undo. Sometimes you may draw a line in the series that you wish to erase. Instead of the time-consuming task of erasing and then reconnecting with the continue option, you can "step back" through the series. This is accomplished with the Undo option. Undo takes you back, one line segment at a time. Undo is used while you are still active in the Line command. Since the Undo option does not repeat, you do not have to terminate the command. When you have backed up to the desired point, simply continue the line entries. Notice that the Undo option in the Line command is similar to the U command, except that it can be used on a single line segment while the Line command is active. Using the U command after the Line command undoes all lines drawn by that Line command.

@. The at symbol is the prefix for entering relative coordinates, which you will learn about later this chapter.

Canceling the Line Command

Unless you are using the Segment option in the pull-down menu, you can cancel the command by pressing [ENTER], using [ESC], or by selecting another command from the menu.

DRAWING LINES BY USING COORDINATES

Accurate drawing construction requires that lines and other entities in a drawing be placed in a precise manner.

One method of placing lines and other entities precisely is the use of coordinates. There are three types of two dimensional coordinates: absolute, relative, and polar. Let's look at each type and learn how to use it.

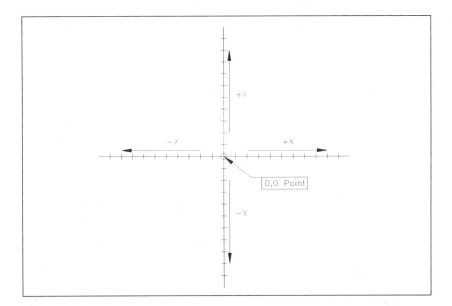

FIGURE 9-2 The *X,Y* Coordinate Grid

Absolute Coordinates

Absolute coordinates specify a point on an X,Y grid, with the 0,0 point as a reference. Let's look at how the X,Y coordinate grid works. Figure 9-3 shows a coordinate grid. The 0,0 point is at the center of the grid.

Notice that the grid has two lines, known as axes. There is an X axis that runs horizontally and a Y axis that runs vertically. The intersection of the X and Y axis is located at the 0,0 point. The increments increase in positive numbers as you move to the right on the X axis and up on the Y axis. Likewise, they decrease to the left in the X axis and down along the Y axis.

The AutoCAD drawing screen can be overlaid on the grid as shown in Figure 9-3. Notice that the 0,0 point is in the lower left corner of the screen. This is the "normal" work screen in AutoCAD. If you remember, when we learned how to set the drawing limits in Chapter 8, we set the lower left drawing corner to 0,0. This places the AutoCAD screen in the upper right quadrant of the grid where all the X and Y values are positive.

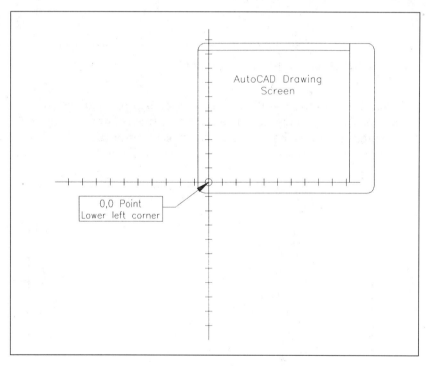

FIGURE 9-3 AutoCAD Screen Superimposed on Coordinate Grid

A single point on the grid can be identified by a coordinate. This coordinate is called an absolute coordinate and is specified by the numeric value of the location where it lines up with the X and Y axes, in respect to the 0,0 point. That designation is listed in an "X,Y" format. Let's look at an example.

Figure 9-4 shows a coordinate grid with the AutoCAD screen placed in the normal location. The point 6,4 is shown on the grid. Notice that it is six units to the right along the *X* axis and four units up on the *Y* axis.

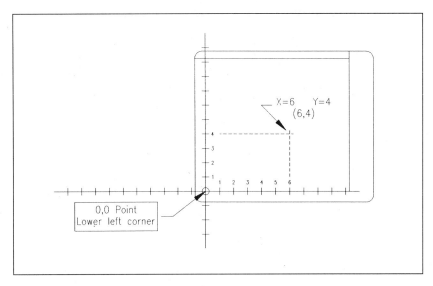

FIGURE 9-4 *X,Y* Coordinate of (6,4)

We can use absolute coordinates to specify points for our commands. Consider the example of a line command. If we select the line command, we are first prompted for the beginning point, then the end point of the line segment. If we enter **2,2** for the beginning point and **5,4** for the end point, we will draw a line as shown in Figure 9-5.

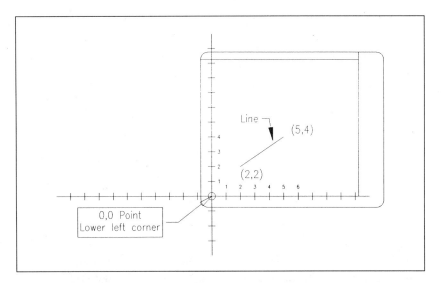

FIGURE 9-5 Drawing a Line with Absolute Coordinates

Relative Coordinates

Relative coordinates are not specified from the 0,0 point, but from any *given point*. Relative coordinates are typically specified from the last point entered. By their nature, relative coordinates can not be the first point entered.

Relative coordinates are specified in the same *X,Y* format as absolute coordinates. The difference is that absolute coordinates describe the *X,Y* distance from the *0,0 point* and relative coordinates describe the distance from the *last point entered*. Let's look at an example.

We can draw the same line we constructed in Figure 9-5 with relative coordinates. Refer to Figure 9-6.

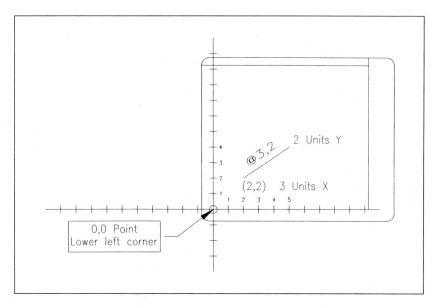

FIGURE 9-6 Drawing a Line with Relative Coordinates

If we choose the Line command, we can specify the first point with the absolute coordinate of 2,2. When prompted for the endpoint of the line segment, we can enter:

@3,2

Notice the format used. The "@" symbol is used to specify a relative coordinate. You can think of this as "at the last point, go 3 units in the *X* direction and two units in the *Y* direction." The "@*X,Y*" format is always used in this manner.

Our line will be drawn three units *X* and two units *Y* from the first point (2,2) as shown in Figure 9-6. Note also that you could have entered the first end point of the line by simply moving the cursor on the screen and entering the beginning point wherever you wished.

Polar Coordinates

Polar coordinates are a type of relative coordinates. Polar coordinates, however, specify a point by defining a distance and angle from the last point. The format is:

@distance<angle

Before we can properly specify an angle, we need to look at the default AutoCAD angle specification. Figure 9-7 shows the angles used by AutoCAD. The default direction for angle zero is to the right (east). The default angle rotation is counterclockwise. The direction for angle zero and the angle rotation direction can be changed with the Units command.

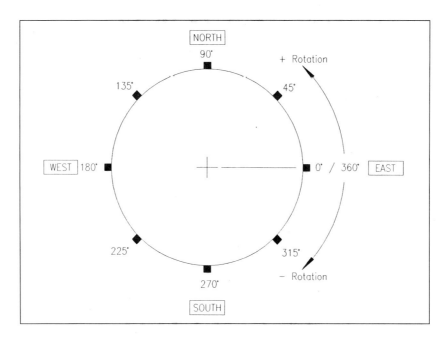

FIGURE 9-7 AutoCAD Default Angle Specification

Now that we know AutoCAD's angle specifications, let's look at an example using polar coordinates. We could specify the endpoints of a line by entering the first endpoint, then entering

@10<30

for the endpoint of the segment. This would draw a line segment that is 10 units and 30 degrees from the first point as shown on the graph in Figure 9-8.

As you proceed through each of AutoCAD's commands you will acquire a perception of the best method of point entry to be used for the task.

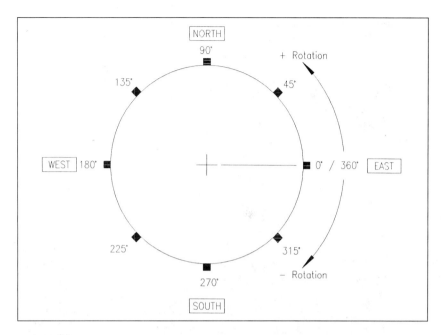

FIGURE 9-8 Using Polar Coordinates

DRAWING POINTS

A point, by default, is a dot (like a single touch of a pen point) that is placed on the drawing. To place the point, enter the Draw menu and select POINT.

Command: **Point**
Point: [ENTER]

Point and Sizes

Points can also be different designs. You can specify the type and size of point you want by using the Pdmode and Pdsize system variables. These variables are accessed through the Setvar command or by selecting the Options menu, followed by Display and Point Style. You may also use Pdmode and Pdsize as commands, typing them directly on the command line. Pdmode designates the type of point drawn and Pdsize controls the size of the point entry.

Let's look at how to specify a particular point type using Pdmode. You may set a Pdmode value of 1 to 4 to select a figure to be drawn at the point. Figure 9-9 shows each setting and its corresponding figure:

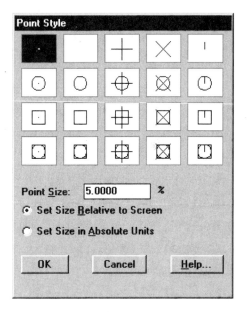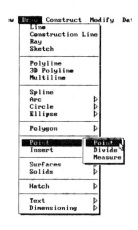

FIGURE 9-9 Point Types

You may also add a circle, square, or both to the point. Each of these figures also has a numerical value. To add one of these to the point entity, simply add the numerical value to the point figure in Figure 9-9. Figure 9-10 shows the values of the circle, box, and both:

Value	Type of Point Drawn
32	○ Circle around the point
64	□ Square around the point
96	◻ Both circle and square

FIGURE 9-10 Adding Point Figures

Adding the values allows several point types. Figure 9-11 shows the possible combinations and their corresponding point entities:

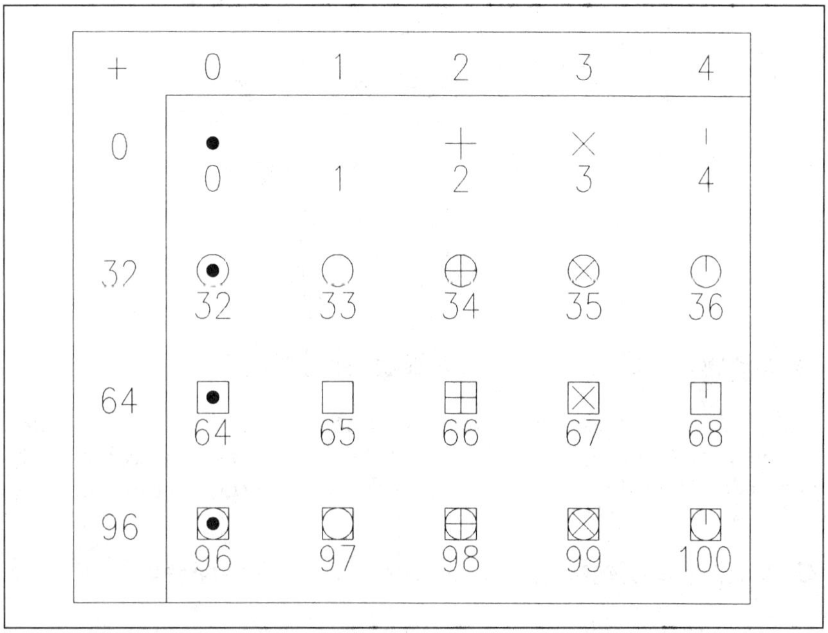

FIGURE 9-11   Point Value Coordinates

Pdsize controls the size of the point entities. Pdmode values of 0 and 1 are not affected by Pdsize. If a positive number is entered, an absolute size for the point entity is specified. If the number is negative, AutoCAD uses the number as a percentage of the screen size. Thus, if you zoom in or out, the size is approximately the same in relation to screen size.

> **NOTE:** Pdmode and Pdsize are retroactive. Setting new Pdmode or Pdsize values will change all existing points to reflect the new size and style.

DRAWING CIRCLES

AutoCAD provides six methods of constructing circles. These are:

1. Center and radius
2. Center and diameter
3. Two-point circles
4. Three-point circles
5. Tangent circles
6. Tangent three-point circles

Let's look at each type of circle construction. It is helpful to read each brief section, then follow the short exercise that follows.

9-12 USING AUTOCAD

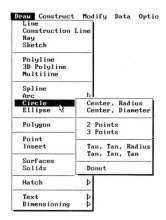

Drawing a Circle with Center and Radius

AutoCAD allows you to enter the center point of the circle and stipulate the radius. The radius may be designated by either entering a numerical value or showing AutoCAD the distance on the screen by entering a point on the circumference of the circle.

Drawing a Circle with Center and Diameter

A circle may be constructed by stipulating the center point and a diameter. Either pick Center,Diameter from the circle menu or enter **D** when prompted for the diameter or radius. You will then be prompted for the diameter.

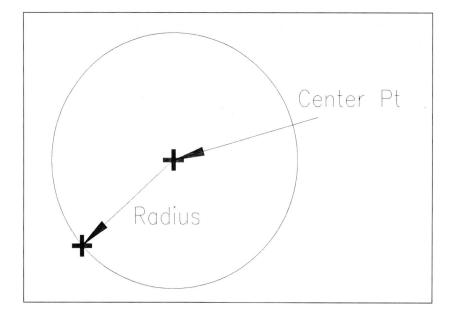

FIGURE 9-12 Circle with Cen,Rad

Drawing a Circle by Designating Two Points

Responding to the "3P/2P/<Center point>" prompt with **2P** allows you to construct the circle by showing AutoCAD two points on the circumference.

Command: **Circle**
3P/2P/TTR/<Center point>: **2P**
First point on diameter: (*Select first point.*)
Second point on diameter: (*Select second point.*)

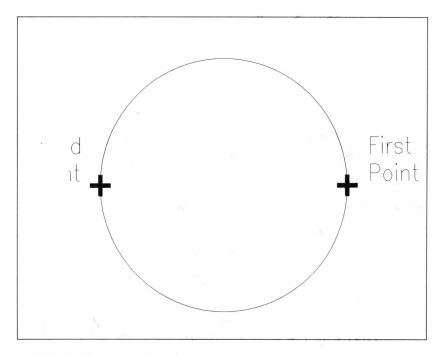

FIGURE 9-13 Two-Point Circle

Drawing a Circle by Designating Three Points

A circle may also be drawn by simply entering three points on the circumference.

Command: **Circle**
3P/2P/TTR/<Center point>: **3P**
First point: (*Select first point.*)
Second point: (*Select second point.*)
Third point: (*Select third point.*)

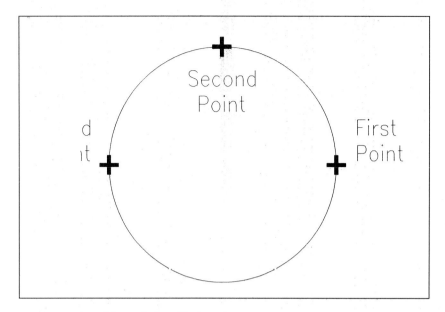

FIGURE 9-14 Three-Point Circle

Drawing Circles Tangent to Objects

AutoCAD allows you to construct circles tangent to lines or circles. You can do this by identifying two lines, two circles, or one line and one circle to construct a tangent circle to, and the tangent circle's radius. To construct tangents, choose the Tan,Tan,Radius option from the Circle submenu.

Command: **Circle**
3P/2P/TTR/<Center point>: **TTR**
Enter Tangent spec: (*Select first circle/line-point "A".*)
Enter second Tangent spec: (*Select second line/circle-point "B".*)
Radius: (*Enter value.*)

The following illustrations show the different effects of the TTR command in constructing circles tangent to lines and/or circles.

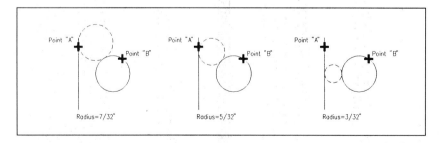

FIGURE 9-15 Using TTR to Construct Tangents

DRAWING **9-15**

Drawing Tangent Three-Point Circles

Three-point circles may be constructed to any combination of lines and circles by specifying each point using the TANgent snap mode option. Let's look at an example of this. Construct a square box of two units on each side as shown in the following illustration. Enter the Circle submenu and choose "3 Points".

Command: **Circle**
3P/2P/TTR/<Center point>: **3P**
First point: **TANGENT** to (*Choose point "A".*)
Second point: **TANGENT** to (*Choose point "B".*)
Third point: **TANGENT** to (*Choose point "C".*)

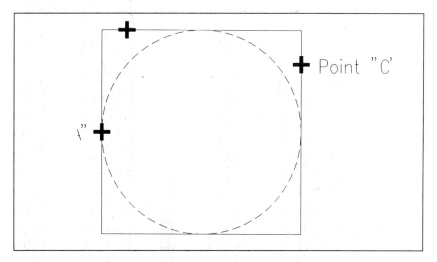

FIGURE 9-16 Constructing Three-Point Tangent Circle

Set up the TANgent option in continuous mode from Running Object Snap from the Options menu.

The following illustrations show 3-point circles constructed using the TANgent object snap method.

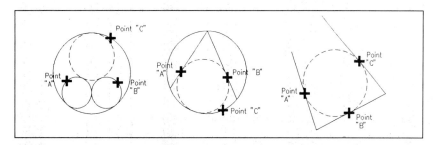

FIGURE 9-17 Using Tangent Object Snap

DRAWING ARCS

Arcs are segments of a circle. There are 11 methods of constructing an arc in AutoCAD. To properly utilize the Arc command, you must have a thorough understanding of the different ways in which it can be used. You may specify an arc in the following ways:

1. Three points on an arc
2. Start, center and end point
3. Start, center and included angle
4. Start, center and length of chord
5. Start, end and included angle
6. Start, end and radius
7. Start, end and starting direction
8. Center, start and end point
9. Center, start and included angle
10. Center, start and length of chord
11. Continuation of a previous line or arc

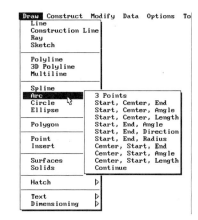

References to "center" mean the radius point of the arc.

If you do not specifically choose any of the above combinations of arc construction, method one (three points on an arc) is used as the default.

Although 11 methods might seem like a lot to learn, you really only have to understand a few principles to be able to use them all.

The arc menu uses letters which designate particular methods of arc construction. The arc command options are designated by the following:

      A—included Angle

      C—Center

      D—starting Direction

      E—End point

      L—Length of chord

      R—Radius

      S—Start point

By using combinations of the abbreviations, you can specify any type of arc construction you wish. For example, to construct an arc using the start, center and end points, the correct choice from the arc menu would be S,C,E.

Let's review each of the methods of arc construction and look at some examples.

Drawing Three-Point Arcs (3-Point)

The "three points on an arc" is the default method of constructing an arc. The first and third points are the endpoints of the arc, while the second point is any point on the arc which occurs between the beginning and end points.

    Command: **Arc**
    Center/<Start point>: *(Enter point "1".)*
    Center/End/<Second point>: *(Enter point "2".)*
    End point: *(Enter point "3".)*

Three-point arcs may be constructed from either direction (clockwise or counterclockwise). The arc will run from the first point toward the second and third points.

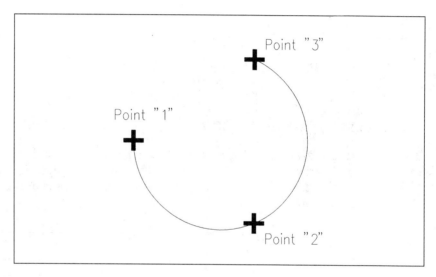

FIGURE 9-18   Three-Point Arc

Start, Center, End (S,C,E)

This method constructs an arc counterclockwise from the start to the specified end point. The arc will be constructed from a radius using the specified center point. The radius will be equal to the actual distance from the center point to the start point. For this reason, the arc will not pass through the specified end point if it is not the same distance from the center.

Relative coordinates and specified angles from the center point may be used if desired.

Some examples of angles constructed with this method are shown in Figure 9-19.

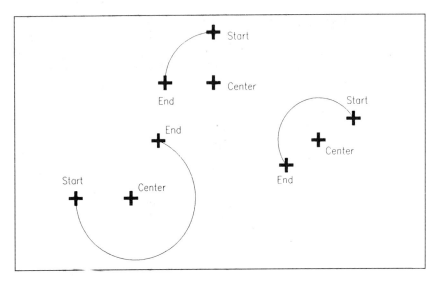

FIGURE 9-19 Arcs with S,C,E

Start, Center, Included Angle (S,C,A)

This method draws an arc with a specified start and center point of an indicated angle. The arc is drawn in a counterclockwise direction if the indicated angle is positive and clockwise if the indicated angle is negative. Examples of arcs constructed by this method are shown in Figure 9-20.

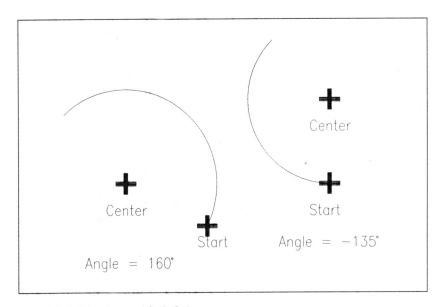

FIGURE 9-20 Arcs with S,C,A

Start, Center, Length of Chord (S,C,L)

A chord is a straight line connecting an arc's start and endpoint (see Figure 9-21).

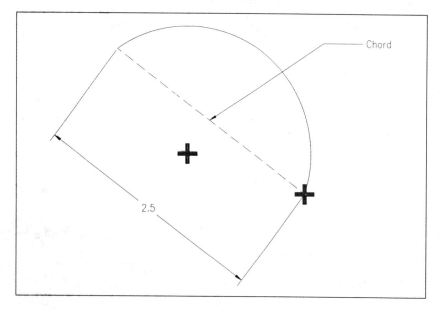

FIGURE 9-21 Arc with a Chord

In some applications, an arc of a specified chord length is required. AutoCAD allows construction of such an arc and allows the user to specify the chord length.

For construction of this type of arc, the chord length is used to determine the ending angle. The arc is drawn in a counterclockwise direction. An example of an arc with a specified chord length is shown in Figure 9-22.

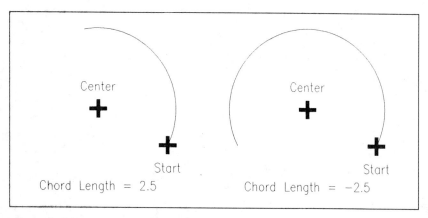

FIGURE 9-22 Arcs with S,C,L

Start, End, Included Angle (S,E,A)

This type of arc is drawn counterclockwise if the specified angle is positive, and clockwise if the specified angle is negative.

Two examples, one using a positive angle, and one using a negative angle are shown in Figure 9-23.

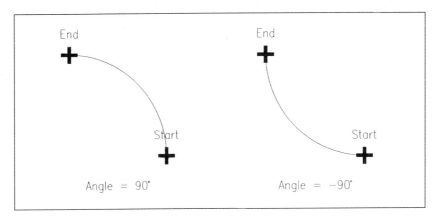

FIGURE 9-23 Arcs with S,E,A

Start, End, Radius (S,E,R)

This type of arc is always drawn counterclockwise from the start point and normally will draw the minor arc. However a negative value for the radius will cause the major arc to be drawn. Two examples of this type of arc are shown in the following illustration.

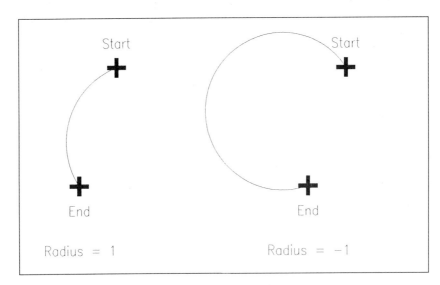

FIGURE 9-24 Arcs with S,E,R

Start, End, Starting Direction (S,E,D)

This method allows you to draw an arc in a specified direction. It will create an arc in any direction, clockwise or counterclockwise, major or minor. The type of arc depends strictly on the relation of direction specified from the starting point. You may specify the direction by using a single point.

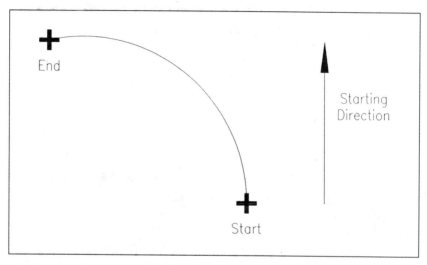

FIGURE 9-25 Arc with S,E,D

Center, Start, End Point (C,S,E)

This construction method is similar to the SCE option, except the order of entry is different.

Center, Start, Included Angle (C,S,A)

This method constructs an arc from a specified center point, using a start point and an included angle that describes the length of the arc. The construction method is similar to the SCA method, except for the order of entry.

Center, Start, Length of Chord (C,S,L)

This method of construction draws an arc that is described from a center point, then a point that designates the start of one end of the arc. The other end is specified by the length of a chord from the first endpoint to a second endpoint. This is similar to the SCL method, except for the order of entry.

Line/Arc Continuation (Contin)

This method allows you to attach an arc to a line or arc previously drawn. To invoke this method, simply respond with a space or Return when prompted for the first point. The start point and direction are taken from the endpoint and starting direction of the previous line or arc drawn.

This method is very useful for smooth connection of arcs to lines and other arcs.

An example of this method is shown in Figure 9-26.

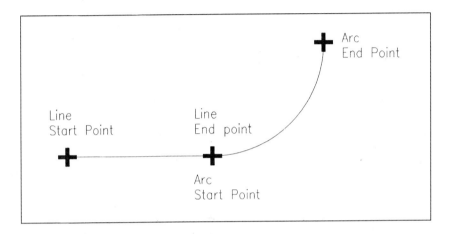

FIGURE 9-26 Line-Arc Combination

Manual Entry Method of Constructing an Arc

In addition to the arc menu's choices for arcs, you may choose your own combination and order of input to construct an arc.

After entering the arc command, simply type one of the letters which designates a particular option (for example, **C** for Center) and press [ENTER]. AutoCAD will then prompt you for the designated information.

You may also use this method even if you have already entered one of the sequences of the arc menu. Doing so will allow you to change the order or even the method of entry.

USING OBJECT SNAP

You have probably noticed in your drawings that you spend a great deal of time lining up points on the screen. Connecting to desired points on lines, circles, arcs, and other entities can be a very tedious procedure.

AutoCAD has provided a drawing aid that makes this process much easier. It is called "object snap".

Object snap provides a "window" that is attached to the intersection of the crosshairs. This window is called an "Aperture." You may stipulate parts of an entity, such as an endpoint or intersection, that can be "captured" if they are within the aperture (sort of like target practice). These points are then treated as though you entered them precisely from the screen.

Object snap can be used "in the middle" of commands that request the entry of a point. We can refer to this use of object snap as "temporary mode" because the object snap will only be effective for one operation. For example, if you are using the Line command and wish to place the endpoint of the currently drawn line segment in the middle of a circle, you would choose object snap mode CENter before placing the point.

Let's look at an example. The following command sequence illustrates how you would connect a line to the center point of a circle as shown in Figure 9-27.

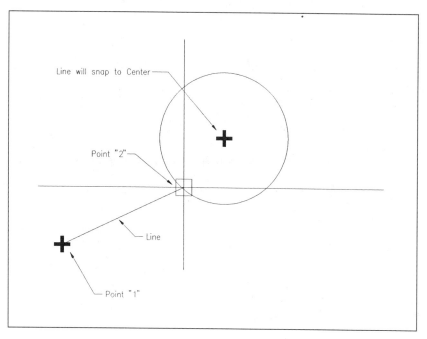

FIGURE 9-27 Object Snap to Center of a Circle

Command: **Line**
From point: (*Point "1".*)
To point: **CEN**
of (*Point "2".*)
To point:

Object Snap Modes

The following modes describe the different object snap capture methods, with an illustration of each application.

CENter: Captures the center point (radius point) of a circle or arc. The aperture must contain a part of the circle or arc in order to identify the entity.

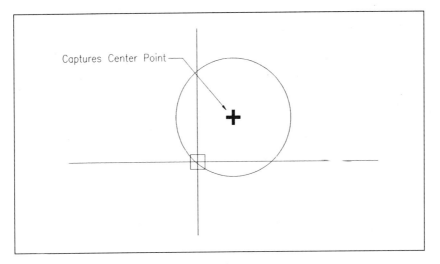

FIGURE 9-28 Object Snap Center

ENDpoint: Causes the nearest endpoint of a line or arc to be captured.

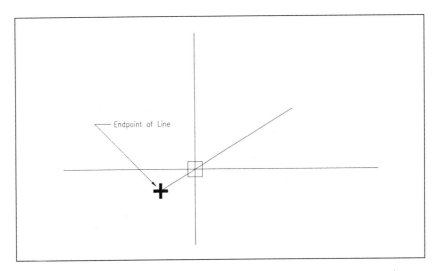

FIGURE 9-29 Object Snap Intersection

INSert: Captures the insert point of a block.

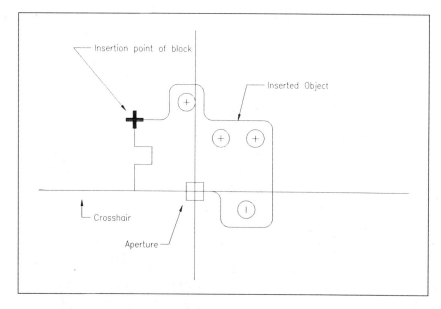

FIGURE 9-30 Object Snap Insert

INTersection: Captures the intersection of two lines, of a line and either a circle or an arc, or the intersection of two circles and/or arcs.

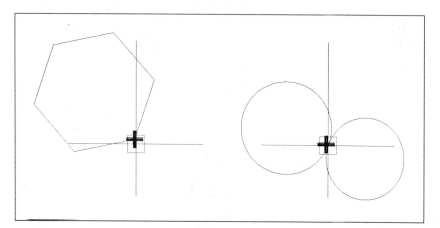

FIGURE 9-31 Object Snap Intersection

APParent Intersection: Capture the "apparent" or projection intersection of two lines that do not physically cross.

MIDpoint: Captures the midpoint of a line or arc.

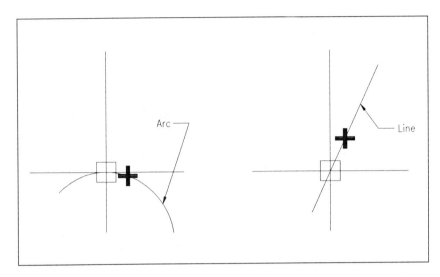

FIGURE 9-32 Object Snap Midpoint

NEArest: This causes the nearest point on a line, circle or arc to be captured. Circles and arcs that are part of a block are not captured.

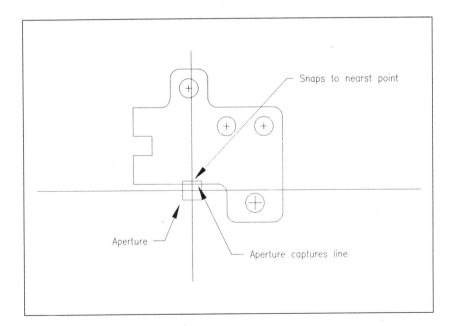

FIGURE 9-33 Object Snap Nearest

NODe: Snaps to a point.

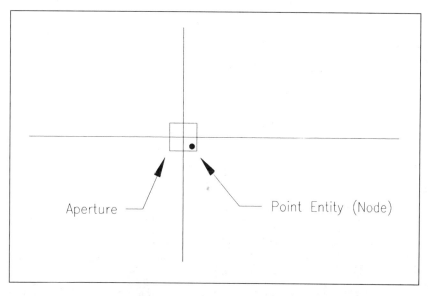

FIGURE 9-34 Object Snap Node

QUAdrant: Snaps to the nearest quadrant point of a circle or arc. The quadrants are the parts of a circle or arc that occur at 0, 90, 180, and 270 degrees. Only the parts of an arc that are visible will be captured. If the circle or arc is part of a rotated block, the quadrant points are rotated with it.

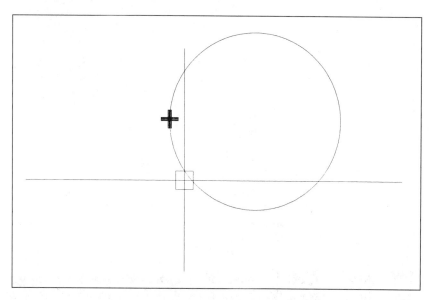

FIGURE 9-35 Object Snap Quadrant

PERpendicular: Snaps to a point on the entity that forms a perpendicular from the last point.

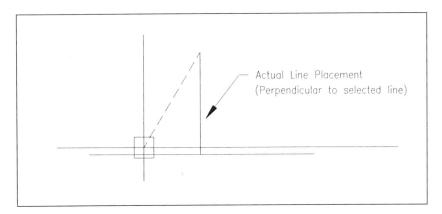

FIGURE 9-36 Object Snap Perpendicular

TANgent: Snaps to a point on a circle or arc that will construct a tangent to the last point entered.

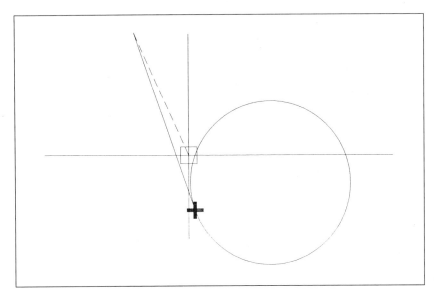

FIGURE 9-37 Object Snap Tangent

QUIck: Quick is a modifier that is entered before choosing each snap function. Choosing Quick will cause the object snap routine to choose the first point that it finds instead of choosing the one that is closest to the intersection of the crosshairs. This may cause problems if there are several entities in the aperture. Quick is used to save "search time" when there are several entities to choose from. If you encounter problems using Quick, cancel the point captured and try again without using Quick. Quick is a modifier that you should use only after you are proficient with the use of the object snap feature.

A menu of object snap modes can be found by pressing the [SHIFT] along with the right button on a two-button mouse.

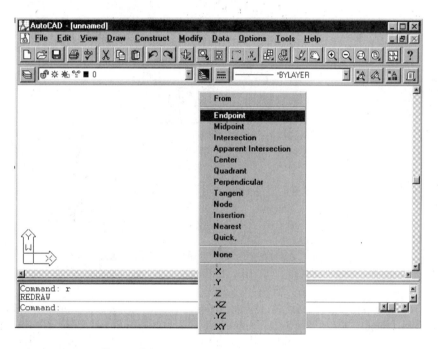

FIGURE 9-38 Cursor Menu

Methods of Using Object Snap

Object snap may be used in two different ways. The first is what we will call *running mode*. When used in this way, you may set the points which will be captured in the aperture whenever AutoCAD prompts for a situation that may require a captured point. At these times, the aperture will appear on the screen, then disappear when it is not required. (The aperture does not affect normal drawing activities.)

The second way to use object snap is what we will refer to as *temporary mode*. In temporary mode, object snap is only used after selecting the mode for a single operation.

Let's look at how to use each mode.

Running Mode Object Snap. To set the object snap modes, select "Running Object Snap..." from the Options pull-down menu. AutoCAD will display a special object snap dialog for setting running mode object snap.

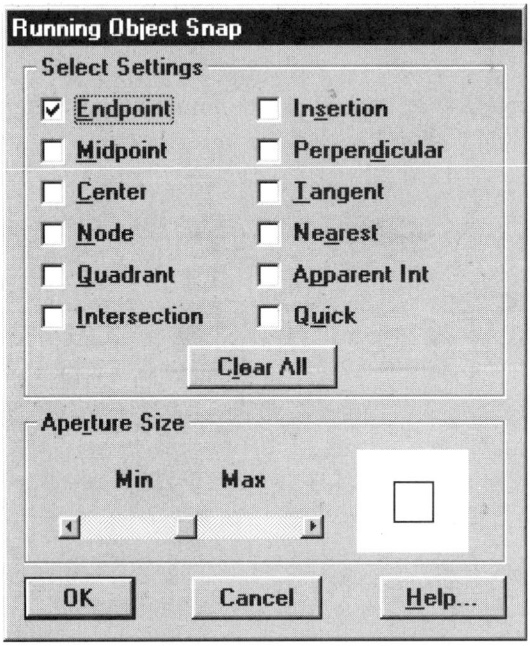

FIGURE 9-39 Object Snap Dialog Box

Select each object snap mode that you want to be in effect in continuous mode. You may choose more than one. Selecting the same mode a second time will "deselect" it. The selected modes will be highlighted and contain a check in the check box next to them.

You may also set the size of the aperture. The slide bar at the bottom of the dialog box is used to set the size. The window to the right of the slide bar shows the actual size of the aperture box.

To clear object snap running mode, deselect the modes from the dialog box.

Temporary Mode Object Snap. Object snap may also be invoked in what we previously referred to as "temporary" mode. This is for those special times when you need an object snap for just one capture.

> **NOTE:** The temporary mode may be used in the middle of other commands without disturbing the current status.

DRAWING 9-31

The object snap (temporary mode) submenu is accessible by pressing the [SHIFT] and the right mouse button.

You may also override the running object snap functions by invoking the temporary method. For example, if you are using running mode object snap and want to enter a point without using object snap, access cursor menu and select NONE. After you enter the point, the running mode will return.

The Status command displays the current object snap features selected.

CLEARING THE SCREEN WITH THE REDRAW COMMAND

After you have been drawing for a while, you begin to build up a number of "markers". While these markers are often helpful in locating points on a drawing, too many of them can be distracting.

AutoCAD provides a method of "cleaning up" a drawing called Redraw. Redraw clears the screen and redraws the entities.

Redraw is also useful when an erase command causes a partial loss of some entity which is to remain. This is common when an erased entity overlaid an entity that remained.

Command: **Redraw**

Some commands, such as zooms, automatically execute a redraw, so it is not necessary to perform the redraw before executing these commands.

Redrawing "Transparently"

A redraw may be executed while another command is active. This is called a "transparent" redraw. To perform a transparent redraw, enter an apostrophe (') before the command. For example, to perform a transparent redraw, enter 'Redraw at a non-text prompt. When the transparent redraw is completed, the previously active command will be resumed. The Redraw command contained in the screen menu will always redraw transparently. This means you can select Redraw from the menu while you are currently in a command without exiting that command.

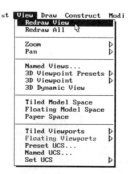

REGENERATING A DRAWING

The Regen command causes AutoCAD to regenerate the drawing display from the database. This differs from the Redraw command that just redraws the screen, in that AutoCAD actually recalculates line endpoints, hatched areas, etc., as well as "cleaning up" erased entities by removing them from the database. To invoke the Regen command, enter:

Command: **Regen**

Several commands may cause an automatic regeneration. Among these are the zoom, pan, and view restore. The regeneration will occur if the new display contains areas that are not within the currently generated area. The Regenauto command controls the automatic regeneration in AutoCAD. See Chapter 28 for an explanation of Regenauto.

You may cancel the regeneration by pressing [ESC]. If the regen is terminated, however, some of the drawing may not be redisplayed. To redisplay the drawing, you must reissue the Regen command. Normally, a redraw is used to "clean up" the screen since a regen takes longer to perform.

> **NOTE:** Regen a drawing after extensive editing to see the "actual" state of the modified drawing.

ZOOMING YOUR DRAWINGS

The Zoom command allows you to enlarge or reduce the view of your drawing. You can think of zoom as a magnifier.

Most drawings would be too small and detailed to work with on a small drawing screen. CAD operators zoom into small areas to show greater detail. Let's consider an analogy to zooming.

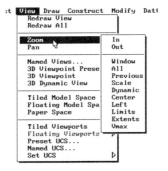

Imagine that your drawing is the size of a wall in a room. The closer you walk to the wall, the more detail you can see. But you can only see a portion of the wall. If you move a great distance from the wall, you may see the entire drawing, but not much detail. Note that the wall does not change size; you just change your viewing distance.

The Zoom command works much in the same way. You may enlarge or reduce the drawing size on the screen, but the drawing does not actually change size. If you zoom closer, you can see more detail, but not all of the drawing. If you zoom out (further away), you can see more of the drawing, but less detail.

Using the Zoom Command

A zoom is performed by the Zoom command:

Command: **Zoom**
All/Center/Dynamic/Extents/Left/Previous/Vmax/Window/<Scale(X/XP)>:

Let's look at the options for the Zoom command.

Zoom Scale. A zoom scale allows you to enlarge or reduce the entire drawing (original size) by a numerical factor. This is the default option. For example, entering a **5** will result in a zoom that shows the drawing five times its normal size. The entire drawing, of course, can not be displayed on the screen in this instance. The zoom will be centered on the previous screen center point.

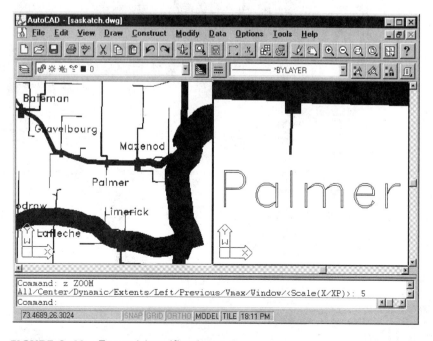

FIGURE 9-40 Zoom Magnification x 5

If the zoom factor is followed by an "X", the zoom is computed relative to the *current* display.

Only positive values can be used in zooms. If you desire a zoom that is smaller than current, use a decimal value. For example, .5 results in a zoom that is one-half normal size.

Zoom All. A zoom all causes the entire drawing to be displayed on the screen. This typically results in the entire area of the limits to be shown. If, however, the drawing extends outside of the current limits, the zoom all will show all the drawing, including the area of the drawing that is outside of the limits.

Occasionally, the zoom all has to generate the drawing twice. If this is necessary, the following will display on the prompt line:

* * Second regeneration caused by change in drawing extents.

If the limits are changed, the entire drawing area will not be shown until a zoom all is performed.

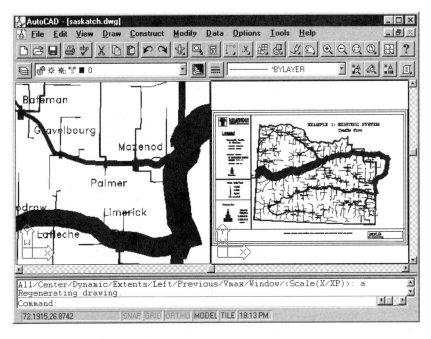

FIGURE 9-41 Zoom All

Zoom Center. The zoom center option allows you to determine the center for the zoom. You are then given the choice for a magnification or a height for the display screen in units.

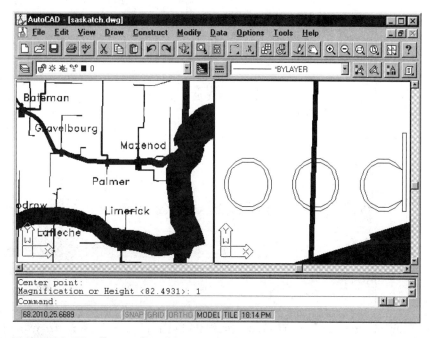

FIGURE 9-42 Zoom Center

Command: **Zoom**
All/Center/Dynamic/Extents/Left/Previous/Vmax/Window/ <Scale(X/XP)>: **C**
Center point: (*Select.*)
Magnification or Height <default>: (*Select.*)

If the magnification value is followed by an "X", the zoom factor will be relative to the current display.

Zoom Dynamic. This option allows dynamic zoom placement. To use this option, enter the ZOOM command and choose "Dynamic" from the screen menu, or enter **D** from the keyboard.

Command: **Zoom**
All/ Center/ Dynamic/ Extents/ Left/ Previous/Vmax/ Window/ <Scale(X/XP)>: **D**

When you choose dynamic zoom, you are presented with a special view selection screen containing information about current and possible view screen selections. The screen shown in Figure 9-43 is representative of a typical display for dynamic zoom.

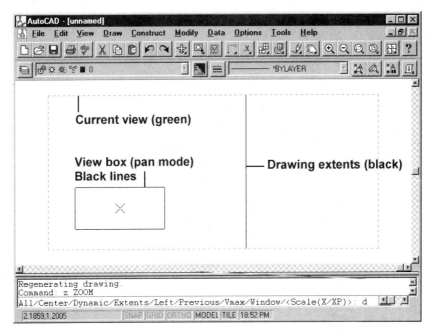

FIGURE 9-43 Dynamic Zoom Display

Each of the viewing windows are labeled and noted as to their respective colors. (Monochrome displays will not, of course, show these colors.)

Let's look at the meaning of each of these "windows".

DRAWING EXTENTS The drawing extents window is the black, solid-line box. The drawing extents can be thought of as the actual "sheet of paper" on which the drawing resides.

CURRENT VIEW WINDOW The current view window is the highlighted (usually dotted and green in color) box. This box defines the screen when you invoked the dynamic zoom command. This box will contain the elements of the drawing that were shown on the screen at that time.

GENERATED AREA The four corner brackets in red define the area of the drawing that AutoCAD has currently generated. Zooms that fall into this area will be zoomed at about half-redraw speed (which is faster than Regen speed). Drawings outside of this area will be regenerated, and thus require more time to zoom. If the generated area is the same size as the drawing extents, the red brackets will be located at the corners of the monochrome display, and this may make the brackets invisible; if it is the same size as the current view window, the brackets will overlay its dotted lines.

VIEW BOX The black, solid-line view box defines the size and location of the desired view. You may manipulate this box to achieve the view you want. The view box is initially the same size as the current view window.

The view box can be enlarged or reduced, and moved to the desired location. A large X is initially placed in the center of the box. This denotes panning mode. When the X is present, moving the cursor will cause the box to move around the screen.

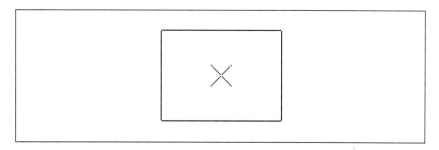

FIGURE 9-44 View Box

Pressing the pick button will cause the X to change to an arrow at the right side of the box. The arrow denotes the zoom mode. Moving the cursor right or left will increase or decrease the size of the view box. The view box will increase and decrease in proportion to your screen dimensions, resulting in a "what you see is what you get" definition of the zoomed area. This differs from the standard "Window" zoom, which works from a stationary window corner and may show more of the screen depending on the proportions of the defined zoom window.

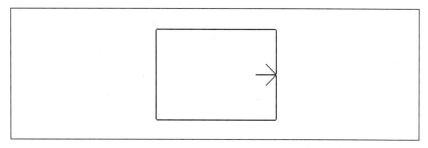

FIGURE 9-45 Changing Size of a View Box

You may toggle between zoom and pan modes as many times as you wish to set the size and location of the view box. When you have "windowed" the desired area, press [ENTER] and zoom will be performed. The area defined by the zoom box is now the current screen view.

Redrawing and Regenerating Dynamic Zooms. Choosing a view box within the current generated area results in the zoom being performed at redraw speed. If you choose a view box outside of this area, the zoom will be performed at the slower regen speed. AutoCAD displays an hourglass at the lower left of the dynamic zoom screen to alert you of the necessity to perform the zoom at regen speed if you move the view box outside of the current generated area.

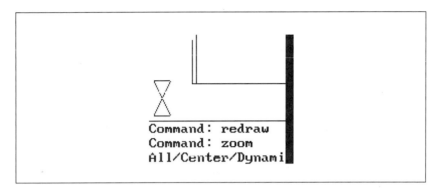

FIGURE 9-46 Hourglass Indicator

In 3D mode, AutoCAD does not generate any area outside the present view. Because of this, if you use the dynamic zoom while in 3D mode, you only choose view windows within the current screen view area. In addition, all zooms in 3D are performed at regen speed, regardless of the location of the view box. When you execute the dynamic zoom command while in 3D, AutoCAD displays a three-dimensional cube at the lower left corner of the screen. This cube has no other purpose than to remind you that you are in 3D.

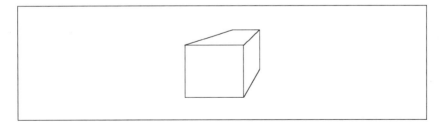

FIGURE 9-47 3D Indicator

Using Dynamic Zoom Without a Pointing Device. Although recommended, it is not necessary to have a pointing device installed to use the dynamic zoom. Simply use the arrow keys to move the view box. If you have used the arrow keys, the Enter key will toggle between pan mode and zoom mode. If you have not used the arrow keys since last toggling with the Enter key, pressing it will perform the zoom at the current location of the view box.

The easy way to perform this is to manipulate the view box to the desired location and size, and press [ENTER] twice.

Zoom Extents. Zoom extents will display the drawing at its maximum size on the display screen. This results in the largest possible display, while showing the entire drawing. Note that areas within the limits that do not contain drawing entities will not be displayed.

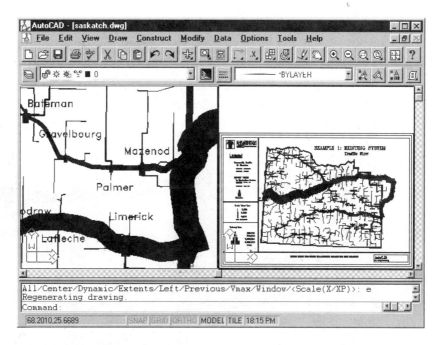

FIGURE 9-48 Zoom Extents

Zoom Lower Left Corner. The zoom left corner is the same as a center zoom, except that you may choose the lower left corner instead of the center. You may use the "X" option in the same manner.

Zoom Previous. The zoom previous option allows you to return to the last zoom you used. This option is often useful if you need to interact between two areas frequently. AutoCAD remembers the last zoom for you. Up to 10 previous zooms are stored. Since the zooms are stored automatically, you are not required to use any special procedure to use them.

Zoom Vmax (Maximum virtual screen). Zoom Vmax zooms to the maximum virtual screen size. The virtual screen size is the pre-generated area of the display. Using the "V" option will allow you to zoom out to the maximum zoom that does not force a regeneration.

Zoom Window. The zoom window command will allow you to determine the area you wish to see in the zoom. The zoom window uses a "window" to specify this area. To use a zoom window, enter:

Command: **Zoom**
All/Center/Dynamic/Extents/Left/Previous/Vmax/Window/<Scale(X/XP)>: **W**
First corner: (*Select.*)
Other corner: (*Select.*)

A box will be displayed around the area to be zoomed.

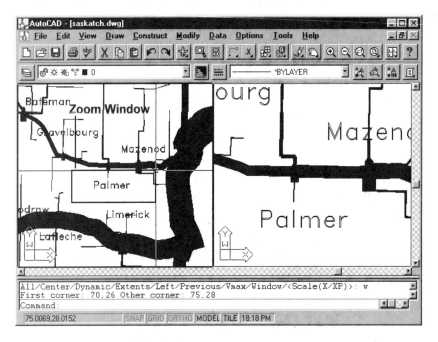

FIGURE 9-49 Zoom Window

Transparent Zooms

A transparent zoom may be executed while another command is active. Enter **'Zoom** at any non-text prompt. The Zoom command found in the screen menu performs a transparent zoom. The following notes apply to using transparent zooms.

The fast zoom mode must be on. This is set with the Viewres command.

You may not perform a transparent zoom if a regeneration is required (zoom outside of the generated area—see "Zoom Dynamic").

DRAWING **9-41**

You may not perform transparent zooms with the zoom all or zoom extents options. These options always force a regeneration.

Transparent zooms cannot be performed when certain commands are in progress. These include the Vpoint command, Pan, View, or another Zoom.

Panning Around Your Drawing

The Pan command is used in conjunction with the Zoom commands.

Many times a CAD operator will zoom into an area of the drawing in order to see more detail. He may want to "move" the screen a short distance to continue work while still in the same zoom magnification. You can think of a pan as similar to placing your eyes at a certain distance from a paper drawing, then moving your head about the drawing. This would allow you to see all parts of the drawing at the same distance from your eyes.

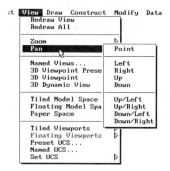

Panning is performed with the Pan command. To use the Pan command, enter:

Command: **Pan**
Displacement: (*Select.*)
Second point: (*Select.*)

You may show a point on the screen for the "displacement" prompt and another point for the "second point" prompt. This has the effect of "dragging" the drawing the specified distance and angle.

This distance and angle is called the "displacement". AutoCAD computes the distance and angle automatically and performs the pan.

The following illustration shows a pan using this method:

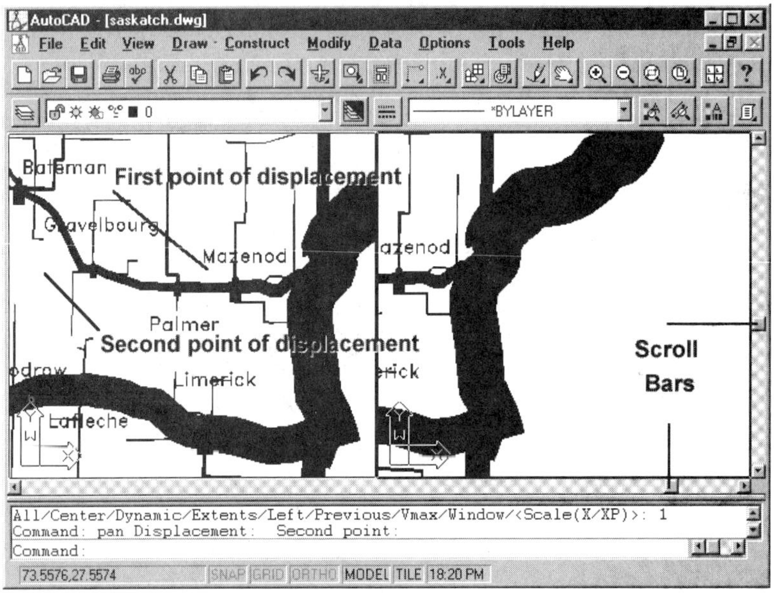

FIGURE 9-50 Panning the Drawing

You may also use relative coordinates to tell AutoCAD how far to move the drawing relative to the screen. For example, using the following coordinates would move the drawing 5 units X and 3 units Y:

Command: **Pan**
Displacement: (*Select a point on the screen.*)
Second point: **@5,3**

Transparent Panning. A pan may be performed transparently while another command is in progress. To do this, enter **'Pan** at any non-text prompt. The same restrictions for transparent pans apply as for transparent zooms. Note that the Pan command in the screen menu performs transparent pans.

Scroll Bars

All Windows versions of Release 13 have a pair of scroll bars, a feature not available in the DOS version. The scroll bars allow you to pan the drawing horizontally and vertically (see Figure 9-51) but not diagonally.

To pan with a scroll bar, move the cursor over a scroll bar and click. AutoCAD transparently pans the drawing. There are three ways to use the scroll bars:

1. Click on the arrow at either end of the scroll bar. This pans the drawing in one-tenth increments.
2. Click and drag on the scroll bar button. This pans the drawing interactively as you move the button. This is also the way to pan by a very small amount.
3. Click anywhere on the scroll bar, except on the arrows and button. AutoCAD pans the drawing by 80 percent of the distance.

Aerial View

AutoCAD has a third alternative to the Zoom/Pan commands and scroll bars for changing the view. Called Aerial View, this window lets you see the entire drawing at all times in an independent window. This is sometimes called the "bird's-eye view." To display the Aerial View, type AV at the Command prompt.

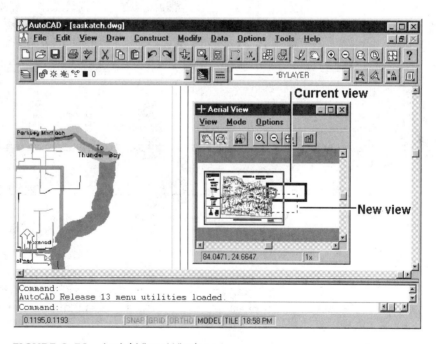

FIGURE 9-51 Aerial View Window

After the drawing appears in the Aerial View window, zooming in is as simple as the Zoom Window command: pick two points. To instantly zoom to another area, pick another two points.

To pan, first click on the hand icon. Then, simply move the rectangle to the new location.

The Aerial View can also work in reverse. In Global mode, it shows a magnified view. This is sometimes known as the "spyglass view." To use Global view, first set the level of magnification. Select Options from the Aerial View menu bar, then Locator Magnification. When the Magnification dialog box appears, change the value from the default of 1 to 2 or 8 or 16. Click the OK button.

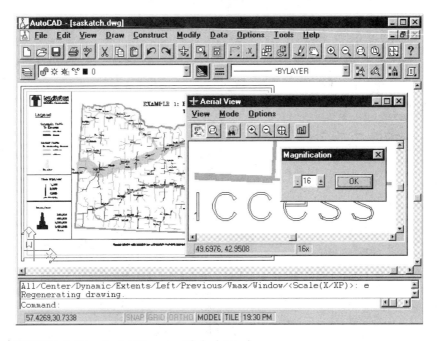

FIGURE 9-52 Aerial View in Global Mode

Move the cursor to the Locate button on the Aerial View tool bar (looks like binoculars). Click and drag the cursor (it now looks like a gunsight) over the drawing in AutoCAD. As you move the locator around, you see a magnified view in the Aerial View window.

Aerial View only works when you configure AutoCAD with a display-list processing device driver. If you try to use the AV command with a regular display driver, AutoCAD responds with an error message.

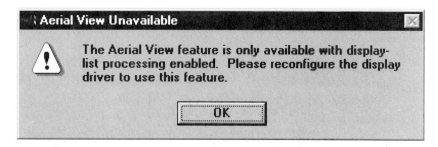

FIGURE 9-53 Error Message

Use the Config command to change display drivers, as follows:

1. Type the Config command.
2. Select option 3, Configure video display.
3. Answer Yes, to select a different video display.
4. Select either WHIP-HEIDI accelerated display driver or Accelerated display driver by Rasterex.
5. Keep pressing Enter until you are back at the drawing editor.

DRAWING AIDS AND MODES

AutoCAD contains many commands that make drawing more accurate, efficient, and easy. In this chapter, we will learn to use the following commands and modes:

    Grid command

    Snap command

    Object snap modes

    Ortho mode

    Dragmode command

    XLine command

    Ray command

Placing a Grid on the Drawing Screen

The Grid command is used to display a grid of dots with a specified spacing. You may determine the spacing of the dots. You may additionally specify the X and Y spacing separately.

The grid dot is for reference purposes and is not part of the drawing. You can not erase the grid markers. The grid will not print or plot.

To display a grid, enter the command:

Command: **Grid**
Grid spacing (X) or ON/OFF/Snap/Aspect <default>:

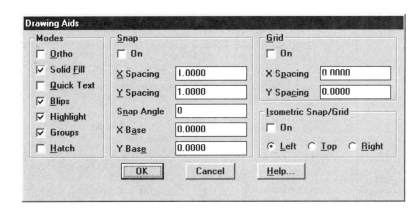

The options are as follows:

Grid Spacing: The grid spacing is the default. If you enter a numerical value, that value becomes the grid spacing. It is not necessary to first enter a letter to designate the grid spacing option.

The grid spacing may be set to a multiple of the snap setting by placing an "X" after the value. For example, if the snap spacing is 1, entering a value of **10X** will display a grid dot at every 10th snap point. A grid spacing of zero will set the grid to exactly match the snap value.

On The On option activates the grid. The previously set grid value is used.

Off The Off option turns off the grid. The current value is stored for later use.

Snap Sets the grid spacing equal to the current snap setting. If the snap resolution is changed, the grid value will be automatically changed to match.

Aspect The aspect option allows you to specify a different X and Y value. For example, you may choose a value of 10 units of spacing for the X value and a value of 5 units of spacing for the Y value.

Command: **Grid**
Grid spacing (X) or ON/OFF/Snap/Aspect: **A**
Horizontal Spacing(X): **10**
Vertical Spacing(X): **5**

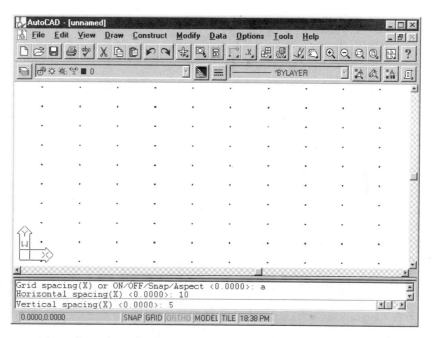

FIGURE 9-54 Grid with Aspect Option

If the grid spacing is too close to display properly, the message:

Grid too dense to display

will appear. This may also happen when you zoom the drawing.

> **NOTE:** Many designers use a grid when laying out or "sketching" design ideas. Setting the grid spacing to a convenient interval allows better visualization of scale.

Snapping to a Grid

Points which are entered on the screen may be aligned to an imaginary grid. This grid is known as the "snap grid". The spacing of this grid is the "snap resolution". If a point is entered that is not exactly aligned with a snap point, the point is forced to the nearest snap point.

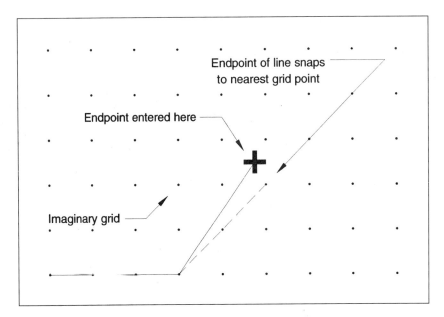

FIGURE 9-55 Using a Snap Grid

You may specify different X and Y spacing for the snap grid. The snap spacing may also be set to coincide with the grid spacing. Unless the snap spacing is set with the grid spacing and the grid is currently displayed, the snap settings are invisible.

To set a snap grid, select SNAP from the Settings/Next menu.

   Command: **Snap**
   Snap spacing or ON/OFF/Aspect/Rotate/Style <default>:

Snap Options.   The options for the snap command are as follows.

   Snap Spacing: The snap spacing is the default. If you enter a numerical value, that value becomes the snap spacing. It is not necessary to first enter a letter to designate the snap spacing option. The value must be non-zero and positive.

   ON: Causes the Snap mode to turn on. The default snap value is determined by the prototype drawing.

   OFF: Turns off the snap. The last value is remembered for use if you turn the snap back on.

   Aspect: Used to set different X and Y snap spacings. If you enter the **A** (Aspect) option, AutoCAD will prompt you for the different X and Y values.

   Rotate: Entering the **R** (Rotate) option will allow you to set a rotation angle for the snap grid and the base point at which the grid will be positioned from. This also has the effect of rotating the crosshairs.

Command: **Snap**
Snap spacing or On/Off/Aspect/Rotate/Style <default>: **R**
Base point <0,0>: (*Select.*)
Rotation angle <default>: (*Select.*)

Style: The Style option is used to choose the format of the snap grid. You may choose either the "standard" or "isometric" format.

Drawing with Orthogonal Control

When you use an input device, such as a mouse or digitizer, drawing lines at true horizontal and vertical angles requires extra effort. The Ortho mode assures that all lines will be orthogonal (either horizontal or vertical).

You can turn on the Ortho mode either by pressing [F8], or by entering the Ortho command:

Command: **Ortho**
ON/OFF: **ON**

Ortho mode is turned off by reentering the command and responding to the prompt with **off**, or by pressing [F8] again. If ortho is on, the status line at the top of the screen will display the word ORTHO.

When the ortho mode is on, all the lines will be forced either horizontal or vertical. If a point is entered that is not true to either, the point will be forced to the nearest true point. That is, if the point is more nearly vertical, the line will be forced vertical.

Setting the Dragmode

AutoCAD allows you to draw many entities by "dragging" them dynamically. For example, a circle radius may be dynamically set by dragging the circle from its center point. The advantage of dragging, for example, is that you can see the circle "grow" as you move the crosshairs away from the center. Many edit and other commands also use dragging in their operations. In addition, "drag" may be selected from some menus to aid in visual placement of points and objects.

The Dragmode command is used to enable or disable dragging. The command sequence is:

Command: **Dragmode**
ON/OFF/Auto <current>:

Setting the mode to OFF disables all drag requests, whether selected from the menu or built into a command macro.

Selecting "Auto" will enable the dragmode for each command that supports it.

Setting Dragmode to ON permits selecting dragmode manually by entering "DRAG" transparently from the command line.

Using Drag on some computers and display systems can be time consuming. Disabling dragmode in these situations can speed up the drawing process.

> **NOTE:** Disabling the Dragmode is useful when complex objects that slow down AutoCAD are being drawn or manipulated.

XLINES AND RAYS

AutoCAD has the ability to create construction lines. You can snap to construction lines to help you create a drawing but construction lines do not plot or print. AutoCAD has two kinds of construction lines created with the Ray and Xline commands. Creating a grid of construction lines is better than using the grid dots, since you can use object snap modes on the construction lines.

The Ray Command

The *ray* is a semi-infinite construction line. The ray starts at a point you specify. The other end of the ray is in "infinity": no matter how far you zoom out, the ray will always appear (whereas other objects in the drawing will grow smaller and eventually disappear). Rays are created with the Ray command, as follows:

Command: **ray**
From point: *(Pick point.)*
Through point: *(Pick point.)*
Through point: *(Press Escape.)*

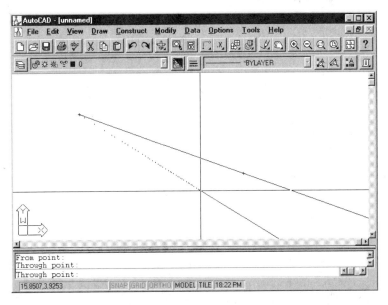

FIGURE 9-56 Drawing Rays (Semi-Infinite Lines)

The Ray command prompts you for the starting point of the ray, and then for a point that the ray passes through. The Ray command automatically repeats itself, using the initial "From point:" for new rays until you press [ESC].

The Xline Command

The Xline is an infinite construction line. The Xline starts in one part of infinity, passes through the drawing area, and continues on into infinity. Like the ray, the Xline will always appear in the drawing, no matter how far out you zoom or whether you pan to AutoCAD's limit.

Xlines are created with the Xline command, which has six options. Like the Ray command, the Xline command keeps repeating its prompts until you press [ESC].

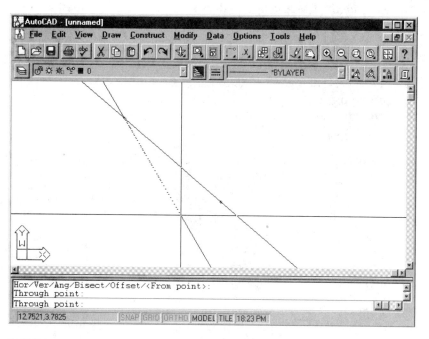

FIGURE 9-57 Drawing Xlines (Infinite Lines)

Default Option. The default option prompts you to pick a pair of through points (an Xline can't actually have a "from point" since it doesn't start anywhere), as follows:

Command: **Xline**
Hor/Ver/Ang/Bisect/Offset/<From point>: *(Pick point.)*
Through point: *(Pick point.)*
Through point: *(Press [ESC].)*

Pick the "from" and "through" points, then press [ESC] to end the Xline command.

Horizontal Option. The Hor option draws a horizontal xline through a single point you pick, as follows:

    Command: **Xline**
    Hor/Ver/Ang/Bisect/Offset/<From point>: **h**
    Through point: *(Pick point.)*
    Through point: *(Press [ESC].)*

Vertical Option. The Ver option draws a vertical xline through a single point, as follows:

    Command: **Xline**
    Hor/Ver/Ang/Bisect/Offset/<From point>: **v**
    Through point: *(Pick point.)*
    Through point: *(Press [ESC].)*

Angled Option. The Ang option draws an xline at a specified angle through a picked point, or by a reference, as follows:

    Command: **Xline**
    Hor/Ver/Ang/Bisect/Offset/<From point>: **a**
    Reference/<Enter angle (0.0000)>: *(Pick point.)*
    Second point: *(Pick point.)*
    Through point: *(Pick point.)*
    Through point: *(Press [ESC].)*

Bisector Option. The Bisect option draws an xline that bisects (half-way between) a pair of intersecting lines, as follows:

    Command: **Xline**
    Hor/Ver/Ang/Bisect/Offset/<From point>: **b**
    Angle vertex point: *(Pick point.)*
    Angle start point: *(Pick point.)*
    Angle end point: *(Pick point.)*
    Angle end point: *(Press [ESC].)*

Offset Option. The Offset option draws an xline parallel to an existing line. You specify either the offset distance or pick a point the xline should go through, as follows:

    Command: **Xline**
    Hor/Ver/Ang/Bisect/Offset/<From point>: **o**
    Offset distance or Through <Through>: *(Enter distance.)*
    Select a line object: *(Pick line.)*
    Side to offset? *(Pick point.)*
    Select a line object: *(Press [ESC].)*

UNDOING AND REDOING OPERATIONS

AutoCAD allows you to undo your work, then to redo it! This is very useful if you have just performed an operation that you wish to "undo". After you undo an operation, you use the Redo command to reverse the undo. Let's take a look at how to do these interesting operations.

Undoing Drawing Operations

The U command will undo the most recent command. You may execute a series of "Undos" to back up through a string of changes. The U command should not be confused with the more complex Undo command, although it functions the same as Undo 1.

Undoing a command will restore the drawing to the state it was in before the command was executed. For example, if you erase an object, then execute the U command, the object will be restored. If you scale an object, then undo it, the object will be rescaled to its original size. The U command will list the command that is undone to alert you of the type of command that was affected. For example, executing the U command after the Scale command would result in the following:

Command: **U**
SCALE

Several commands cannot be undone. Some, Plot, and Wblock, for example are not affected. If you attempt to use the U command after these commands, they will be displayed, but not undone.

Undoing a just completed block command will restore the block and delete the block definition that was created, leaving the drawing exactly as it was before the block was performed.

Redoing a Drawing Operation

The Redo command is the antidote of the U command.

Redo "undoes" the undo. The Redo command must be used immediately after undo commands.

To use Redo, enter:

Command: **Redo**

EXERCISES

1. Let's draw a box using the Line command.

   Command: **Line**
   From point: (*Enter point 1.*)
   To Point: (*Enter point 2.*)
   To Point: (*Enter point 3.*)
   To Point: (*Enter point 4.*)
   To Point: (*Enter point 1.*)

   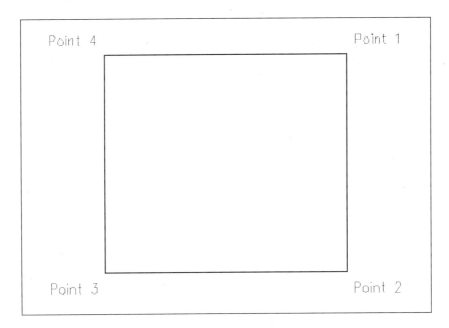

   FIGURE 9-58 Using the Line Command

   You may have noticed that the line "stretched" behind the crosshair. This is called *rubber banding*.

2. Draw lines on grid paper, connecting the points designated by the following absolute coordinates.

   Point 1: 1,1
   Point 2: 5,1
   Point 3: 5,5
   Point 4: 1,5
   Point 5: 1,1

3. Use Figure 9-59 to fill in the missing absolute coordinates. Each side of the shape is dimensioned. Place the answers in the boxes provided.

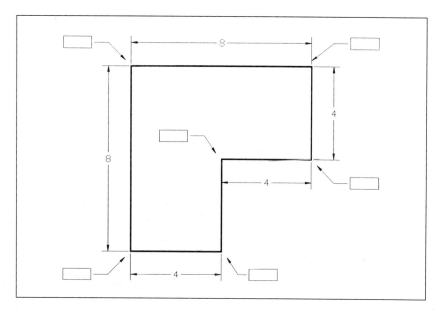

FIGURE 9-59 Fill in the Coordinates

4. List the length of each side of the object in Figure 9-60. (Calculate from the absolute coordinates given.)

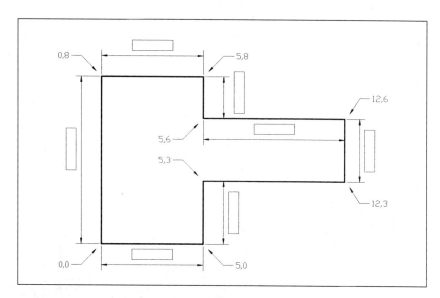

FIGURE 9-60 List the Length of Each Side

5. Use the following coordinates to draw an object.

   Point 1: 0,0
   Point 2: @3,0
   Point 3: @0,1
   Point 4: @-2,0
   Point 5: @0,2
   Point 6: @-1,0
   Point 7: 0,0

6. List the coordinates used to draw the following object. Use relative coordinates.

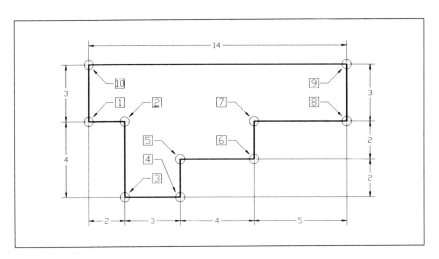

FIGURE 9-61 List the Relative Coordinates

7. Use the following coordinates to draw the object.

   Point 1: 0,0
   Point 2: @4<0
   Point 3: @4<120
   Point 4: @4<240

8. List the coordinates used to construct the following object. Use polar coordinates whenever possible.

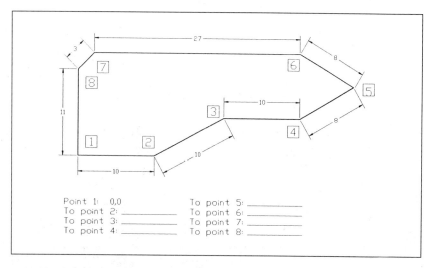

Point 1: 0,0
To point 2: _____
To point 3: _____
To point 4: _____
To point 5: _____
To point 6: _____
To point 7: _____
To point 8: _____

FIGURE 9-62 List the Polar Coordinates

9. Write a list of the coordinates used to construct the following objects. You may use any combination of absolute, relative, or polar coordinates.

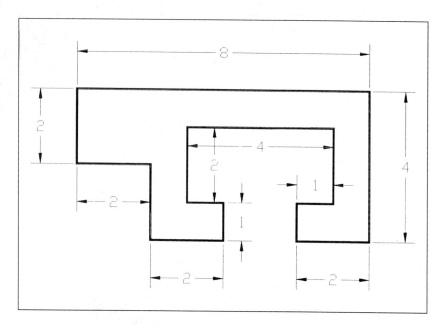

FIGURE 9-63 List the Coordinates

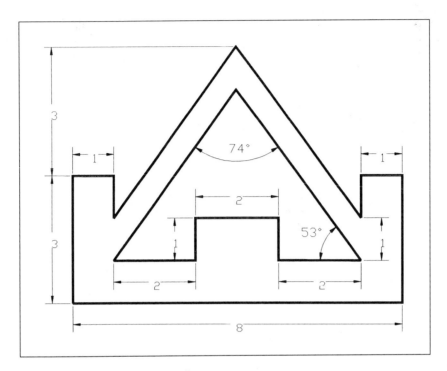

FIGURE 9-64 List the Coordinates

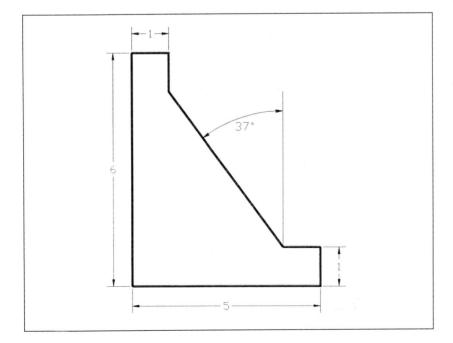

FIGURE 9-65 List the Coordinates

10. Let's set up a point type of 66 and place it in our drawing. First, execute the Setvar command:

    Command: **Setvar**
    Variable name or ?: **Pdmode**
    New value for PDMODE <*default*>: **66**

    Each point entered after setting a new Pdmode will be the type specified. Point entities entered before changing to the new setting will be updated at the next command that causes a regeneration. To place the point we defined in the drawing, enter:

    Command: **Point**
    Point: (*Enter location on screen.*)

11. Let's construct a circle with a radius of 5. From the menu bar, select Draw. From the Draw menu, select Circle. You will be presented a special Circle menu on the screen. You may then select the type of entry desired. Select "Center,Radius".

    Command: **Circle**
    3P/2P/TTR/<Center point>: (*Select a point on the screen.*)
    Diameter/<Radius>: **5**

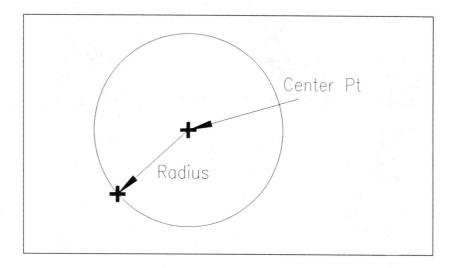

12. Let's suppose you want to construct a circle using a center point and a diameter of 3.

    Command: **Circle**
    3P/2P/TTR/<Center point>: (*Enter point 1.*)
    Diameter/<Radius>: **D**
    Diameter: **3**

    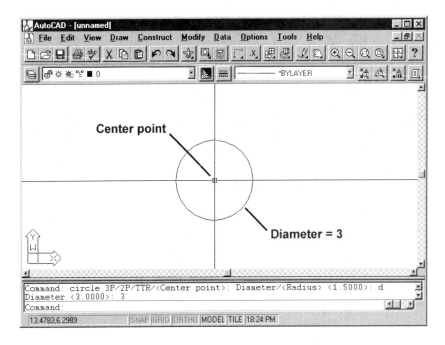

    FIGURE 9-66 Circle with Cen,Dia

13. Start a new drawing named "SNAP". Draw a line, circle, and arc of any size on the screen.

    Select the Line command from the DRAW menu. Hold down the shift key and press the right mouse button to access the Object Snap menu. Select Endpoint from the menu. Notice the aperture box at the intersection of the crosshairs. Place the aperture over one end of the line and click. The new line should "snap" precisely to the endpoint of the existing line.

    Before placing the endpoint of the line, select CENter from the object snap menu (remember to display the object snap menu by pressing [SHIFT] and right mouse button). Place the aperture over any part of the circle circumference and click. Notice how the line snapped to the center of the circle. Repeat the same procedure, except select a part of the arc. The line will snap to the center point of the arc.

14. Draw several more lines, circles, and arcs. Use each of the object snap modes to capture the parts of the entities. Be sure to use TANgent object to construct tangent lines with circles.

15. Type in the Grid command. In response to AutoCAD's prompt, enter a value of **1**. The grid should be visible on the screen.

16. Press [F7]. This key toggles the grid on and off.

17. Enter the Grid command again. In response to the prompt, enter **A** for aspect. Set the horizontal grid to a value of 1, and the vertical grid to a value of .5. The grid should line up with the axis marks you set previously.

18. Let's set up an ordinary snap grid, with a snap resolution of .25. Start a new drawing named "SNAP". Select the Snap command from the Settings/Next menu.

    Command: **Snap**
    Snap spacing or ON/OFF/Aspect/Rotate/Style <default>: **.25**

    Next, select the Line command and enter the line endpoints on the screen. Did you notice the crosshairs "snapping" to a point?

19. Use the Grid command to set a grid with .25 spacing. Use the Line command to set some endpoints. Notice how the crosshairs line up with the grid points.

20. Use [F9] to turn the snap mode on and off. Draw two boxes; one with snap mode on, and one with snap mode off. Notice how the points are easier to line up with snap mode on.

21. Use the Zoom Window command to zoom in on one of the boxes. Move the crosshairs around the area of the box with the snap mode on (use F9 to turn the mode on and off). Notice how the movement of the crosshairs is more exaggerated with the closer zoom. This is because the "screen distance" between the snap points is larger when zoomed in. The actual snap distance (resolution), however, remains unchanged.

22. Select the Snap command again and enter values as shown in the following command sequence.

    Command: **Snap**
    Snap spacing or ON/OFF/Aspect/Rotate/Style <default>: **A**
    Horizontal spacing <default>: **.25**
    Vertical spacing <default>: **.50**

    Next, select the Grid command. Choose the grid option of Aspect and enter the same values. Notice how the spacing for the vertical and horizontal snap resolution is different. It is not necessary for the vertical and horizontal values to be the same. Try drawing another box with the snap on.

9-62 USING AUTOCAD

23. Let's try rotating the snap points and the crosshairs. Select the Snap command again.

    Command: **Snap**
    Snap spacing or ON/OFF/Aspect/Rotate/Style <default>: **R**
    Base point <0.0000,0.0000>: [ENTER]
    Rotation angle <0>: **45**

    Your display should look like the one in Figure 9-67. Notice how the snap resolution and aspect are maintained. The entire snap grid and crosshairs are rotated. Draw a box with the rotated snap grid and crosshairs. This is an excellent method of drawing objects that have many angular lines.

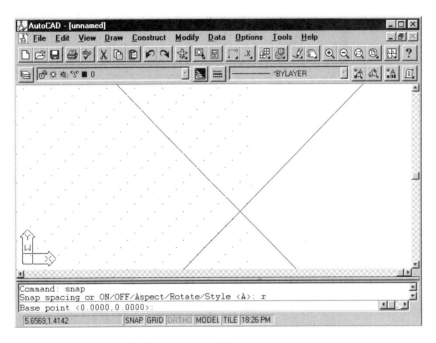

FIGURE 9-67 Snap with Aspect and Rotate

24. The base point can be placed at a particular position, relative to a drawing. Select the Snap command again and choose the Rotate option. When prompted for the base point, select a corner of one of your boxes. Maintain the rotation angle of 45 degrees. Notice how a point falls on the corner of the box. Use the base point position to locate the snap grid where you want it.
25. Start a new drawing named "ORTHO". Select the Line command and draw a four-sided box with horizontal and vertical lines. Do not use the Ortho mode. Now press [F8] to turn on Ortho mode and draw another box. Notice how much easier it is to create straight lines with the Ortho mode.
26. Select line again and enter the first endpoint. Now move the crosshairs around the first endpoint and notice how AutoCAD forces the line to be horizontal or vertical, depending on the location relative to the first endpoint.
27. Use the Snap command to set a snap increment of .25. Be sure Ortho is on and draw another box. Notice how the combination of Snap and Ortho makes drawing the box very easy.

28. Start a new drawing named "DRAG". Draw a circle on the screen. Type "Dragmode" at the Command prompt and turn the mode OFF. Select the Move command and move the circle. Notice how the circle does not move.
29. Enter the Dragmode command again and set it to Auto. Move the circle again. Notice how the circle "drags" with the crosshairs.
30. Select the Dragmode command again and set it to ON. Select the Move command and select the circle with a pickbox and press [ENTER]. Move the circle. The circle will not drag. Repeat the same procedure, except when AutoCAD prompts

    Base point or displacement:

    enter **DRAG** on the command line. Continue the sequence and move the circle. The circle will drag to the new position. Setting the dragmode to ON allows you to select the times you wish to drag an object.
31. Start a new drawing named "TEST". Select the Line command and draw a line segment. Use [ESC] to clear the command line. Enter **U** from the keyboard and press [ENTER]. Did the line segment disappear?
32. Use the Line command to draw several line segments. Select the Line command and draw more line segments. Use the U command to undo the lines drawn with the last line command. Which segments were undone? Press [ENTER] to repeat the U command. What happened?
33. Use the line command to draw two line segments. Before entering the last point, use [F8] to turn on Ortho mode. Notice the ORTHO listing on the status line. Now use the U command to undo the sequence. Is the ortho mode on now?
34. Use [F8] to turn on Ortho. Notice the listing on the status line. Now use the Line command to draw a line segment. Next, use the U command to undo the sequence. Is Ortho mode still turned on? What is the difference between this and the last sequence you performed?
35. Use the Line command to draw several line segments. Use the U command to undo the lines. Now use the Redo command. Did the lines reappear?
36. Enter the Redo command again. What does the prompt line say? Why?
37. Use the commands you have learned to draw the following objects.

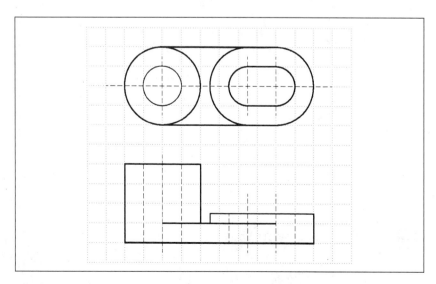

FIGURE 9-68

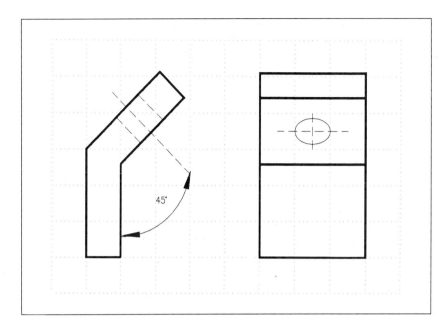

FIGURE 9-69

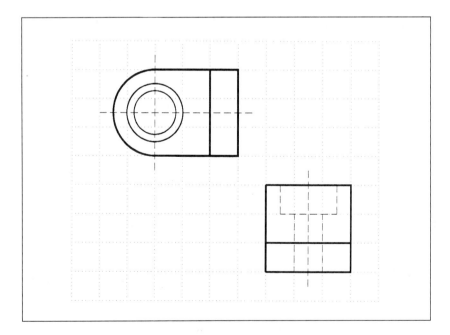

FIGURE 9-70 List Coordinate Entry

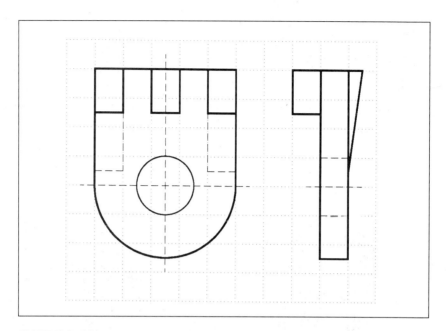

FIGURE 9-71

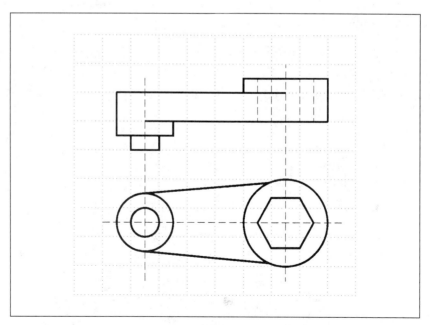

FIGURE 9-72

CHAPTER REVIEW

1. What command allows you to enlarge and reduce the display size of your drawing?

2. When drawing a circle and you are prompted for a diameter or radius, a numerical value can be given, but can the distance be shown by screen pointing as well?

3. Explain the function of the close option under the Line command.

4. Why does AutoCAD offer so many methods of arc construction?

5. What is the purpose of the Pan command?

6. What is the default method of arc construction?

7. What two variables can be altered to change the design and size of points?

8. Why is the close option of the Line command a more accurate method of closing lines than attempting to line up the endpoints manually?

9. Can the U command be used to undo a sequence of commands?

10. Pdmode and Pdsize are retroactive. What does this mean?

11. What is the Vmax option under the Zoom command used for?

12. When an arc sequence is in progress, can you override the preset selections? How?

13. How might you make a backup file of the drawing in progress without ending the drawing?

14. How is the line command terminated?

15. There are six methods of constructing circles. What four circle properties are used in various combinations to comprise these methods?

16. What is the chord of an arc?

17. Can the U command be entered from the screen menus as well as from the keyboard?

18. How do Pan and Zoom commands differ?

19. When a dynamic Zoom is invoked, what information appears on the screen in reference to your drawing?

20. What is a transparent command? How is this option invoked?

21. If the Undo command is used while in a line sequence, does the entire sequence disappear or is each endpoint stepped through backwards?

22. How do redraw and regen differ?

23. What option allows you to enlarge or reduce the original drawing size by a numerical factor for viewing?

24. Relative and polar coordinates use the last point entered as a reference point. What reference point are absolute coordinate entries relative to?

25. In what type of coordinate entry is the @ symbol required?

26. Using Figure 9-73, list the coordinate entry necessary to enter the second point, first using absolute coordinates and then relative coordinates.

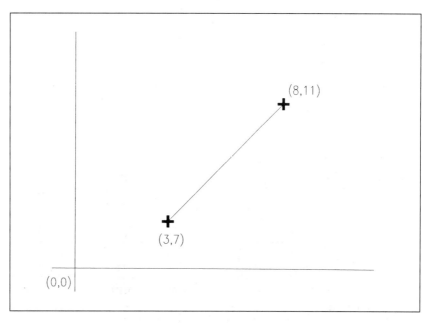

FIGURE 9-73

27. AutoCAD uses a default angle setting that may be altered if the user desires. Where would these setting changes be made?

28. When using polar coordinates, from what point is the distance and angle measured?

29. What method of point entry is entered in the *X,Y* format?

30. What type of coordinate identifies a single point on the coordinate axis?

31. What direction is negative in AutoCAD's default angle specification?

32. List the differences in the three types of coordinate entry, using illustrative examples of each.

33. The coordinate readout at the top of the AutoCAD drawing screen is in absolute coordinates (X,Y), so how could the length of a line resting at an angle be determined if the point entry were from screen pointing rather than through the keyboard?

34. At what point on the coordinate axis is the lower left corner of the AutoCAD drawing screen normally located?

35. Where is the zero angle located in AutoCAD's default angle specification?

36. Given the first point of a line at 12,37, what coordinate entry would you use to locate an endpoint 15 units and 79 degrees from the first point?

37. Using AutoCAD's default angle settings, in which direction does the angle measurement increase?

38. How is the format different for absolute coordinate entry and relative coordinate entry?

39. What is the aperture?

40. What object snap mode would you choose to snap on to a point entity?

41. Can you snap on the midpoint of an arc?

42. Define the Dragmode command.

43. May different X and Y spacing be given to the snap grid? Can it coincide with the grid spacing?

44. What term is given to the dynamic insertion of an object?

45. What mode can provide great accuracy in creating true horizontal and vertical lines?

46. When must the quick modifier be entered? What purpose does it serve?

47. Can you overrun the running object snap mode?

48. When does the Redo command have to be used?

49. How can the results of the last command executed be removed?

50. Although the grid is not a part of the drawing, can it be plotted?

51. Will the U command undo any command just executed?

CHAPTER 10

EDITING

Editing drawings is one of the strengths of CAD. AutoCAD provides an exceptional toolbox of editing functions. After completing this chapter, you will be able to:

- Utilize AutoCAD's basic editing commands.
- Demonstrate the methods of selecting the part(s) of the drawing to be edited with the object selection process.
- Demonstrate the use of AutoCAD's edit commands by using the functions with work problems.

INTRODUCTION

So far you have learned how to use AutoCAD to create a simple drawing with lines, circles, arcs, and points.

From time to time you will have reason to change parts of your work. AutoCAD supplies edit commands to achieve this. In this chapter you will learn about:

OBJECT SELECTION—Selecting the entities to be edited.
SELECT—Preselects objects to be edited.
GROUP—Create a named group of objects.
ERASE—Deleting one or more entities from the drawing.
OOPS—Restoring to the drawing what you just erased.
MOVE—Moving entities around the screen.
COPY—Copying entities already drawn.
BREAK—Removing portions of entities.
FILLET—Making smooth and perfect corners with lines and arcs.

These commands are found in the screen menu under "Modify", and under the "MODIFY" pull-down menu.

The ability to electronically edit your work is a very powerful feature of CAD drafting. Last-minute changes, correction of mistakes or any other reason for change can be accomplished quickly and accurately.

In order to edit a drawing, you must first determine the following:

> Which entity (or entities) would you like to edit?
>
> How would you like to edit them?

After determining which entities you would like to edit, you use the object selection process to isolate them.

USING OBJECT SELECTION

In order to identify which entities are to be edited, you will use the method of identification called "object selection." Object selection is the standard method of entity identification in AutoCAD and is used with most edit commands.

In order to efficiently edit objects, it is essential that you understand the object selection options. As you will see, the edit commands are very easy to use. Efficiency will result from innovative use of the methods used to select objects to be edited. This is especially true of intricate drawings. You may use any combination of selection methods or a single method several times to build the selected set of objects.

Methods of Selecting Objects for Editing

You may select objects for editing by many methods. Let's look at each method.

Selecting by Object Pointing. When you first select an edit command, the cursor is replaced with a small box that we will refer to as a "pick box." If you place the pick box over the object and press [ENTER], the drawing is scanned and the entity that the pick box covers will be selected.

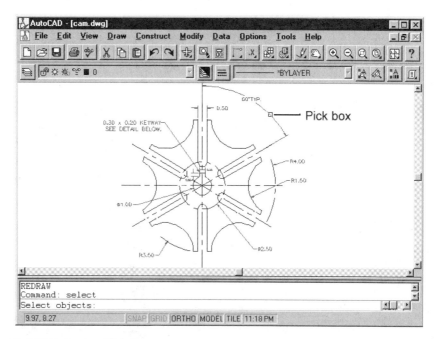

FIGURE 10-1 Pick Box

The pick box can be changed in size to assist in working in drawings of differing complexities. The *Ddselect* dialog box is used to change the pick box size and is covered later in this chapter.

> **NOTE:** Don't place the pick box at an intersection of two objects. This will give unpredictable results, since AutoCAD won't know which entity you desire to choose.

After you select the item(s) to be edited, you will notice that AutoCAD highlights the items which were isolated. Each selected item will become dotted instead of solid. On color monitors they will change color. This helps you to see which items were selected. (The method of highlighting may be different or not present at all on some display systems.)

Using the Multiple Option. Each time you select an entity, AutoCAD will scan the drawing to find that entity. If you are selecting an entity in a drawing that contains a large number of objects, there can be a noticeable delay. The "M" option causes AutoCAD to scan the drawing only once. This can result in a shorter selection time if the drawing is complex. Press ENTER when you finish object selection to begin the scan.

Selecting Objects with a Window. A window may be placed around a group of objects that you wish to select for editing. A *window* is a box that you define by its opposite corners. You may choose the window option by entering a **W** in response to the "Select objects:" prompt.

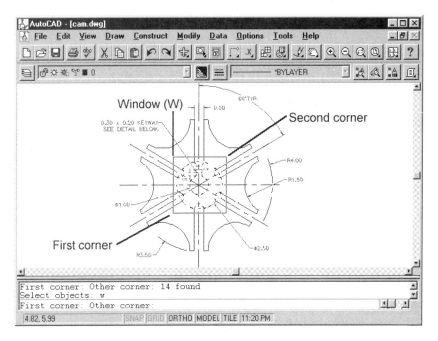

FIGURE 10-2 Selecting with a Window

Only objects that are currently visible on the screen may be selected. Any entity that is entirely in the window box will be selected. If all of an entity is not currently on the screen, it will only be chosen if all parts that are visible are inside the box.

> **NOTE:** Selection of certain objects may be made by placing the box so that all parts of those you want to choose are contained in the box and those which you don't want chosen are not entirely contained. With this method, you may make certain selections in areas of your drawing where objects overlap.

Selecting the Last Object Drawn. The "Last" option selects the last object drawn. If the command is repeated and Last is used again as the object selection technique, the selection will choose the current "last" entity drawn.

Reselecting the Previously Designated Selection Set. The "Previous" option uses the previously selected group of entities for the edit set. This option is useful for performing several edit functions on the same group without the necessity of redefining them.

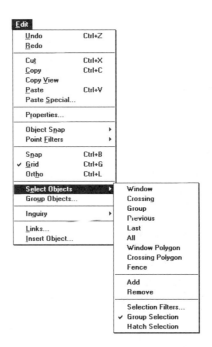

Selecting Objects by Defining a Crossing Window. The "Crossing" option is similar to the Window option. When you use Window, entities which are entirely contained within the window are selected. Using the Crossing option allows you to place a window that will select any entities that are either within or cross through the window. On many display systems, the Crossing window box will be dashed (or highlighted in some other fashion). This distinguishes the crossing box from a standard "window" box.

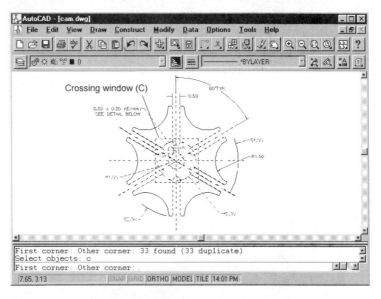

FIGURE 10-3 Selecting with a Crossing Window

Selecting Objects with a Box. Entering **Box** in response to the "select objects" prompt allows you to place either a Crossing or Window box to select objects. If you set the first corner of the box, then move to the right, the resulting box is a Window box and will select objects in accordance with the standard window box method of object selection. Placing the first point of the box, then moving left will designate a crossing box.

Selecting Objects with the Automatic Option. If **AU** is entered in response to the "select objects" a pick box is used for selection. If you select a point with the pick box and an object is found, a selection is made. If an object is not found, the selection point becomes the first corner of a "Box" method of object selection. Move the box to right for Window or the left for Crossing. The automatic method of object selection is excellent for advanced users who wish to reduce the number of modifier selections that must be input.

Selecting Entities by Using a Polygon Window. The Wpolygon (WP) option is used to select entities by placing a polygon window around the desired objects. The polygon window functions the same as the Window option, except that you build a multisided window that surrounds the desired entities. Let's look at a sample command sequence using the Erase edit command.

Command: **Erase**
Select objects: **WP**
First polygon point:
Undo/<Endpoint of line>:

After you enter the first polygon point, you can proceed to build a window by placing endpoints. The window lines rubberband to the cursor intersection, always creating a closed polygon window.

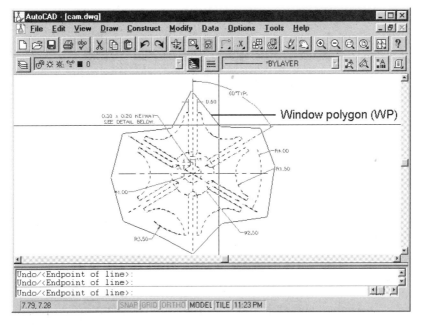

FIGURE 10-4 Selection with a Polygon Window

You can undo the last point entered by entering a **U** in response to the prompt. Pressing [ENTER] will complete the polygon window and complete the process.

The polygon window must not cross itself or be placed directly on a polygon entity.

Selecting Objects by Placing a Crossing Polygon Window. The crossing polygon window (CP) option works in the same manner as the WP option, except the window functions in the same manner as a crossing window.

The crossing polygon is displayed as a dashed line similar to a crossing window. Any entity crossing (touching) the crossing polygon window is selected for editing.

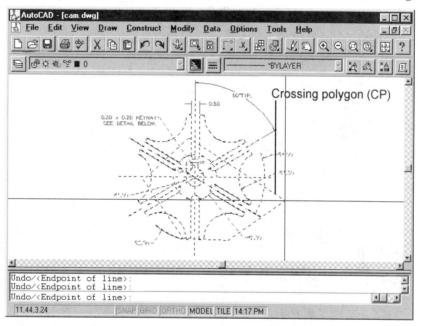

FIGURE 10-5 Selecting with a Crossing Polygon Window

Using a Fence to Select Entities. The Fence (F) option is similar to the crossing and polygon crossing methods, except that AutoCAD uses a crossing "fence line" to select objects.

The fence line is displayed as a dashed line in the same way that a crossing window is displayed. The fence lines are constructed the same as line entities, except that all objects that are crossed or touch the fence line(s) are selected.

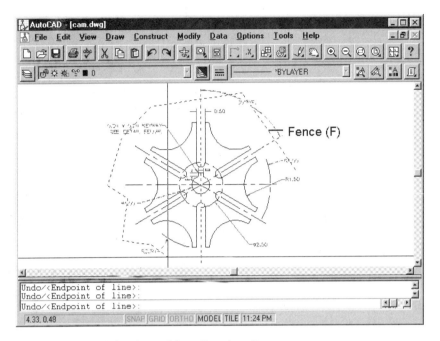

FIGURE 10-6 Selecting with a Crossing Fence

You may consecutively construct as many fence lines as you wish, and may use the "U" option to undo a fence line segment.

Single Entity Object Selection. Selecting "SI" (Single) causes AutoCAD to allow only a single "select objects" prompt, suppressing subsequent "select objects" prompts. The single option can be used in menu strings for an efficient single object selection operation, since it deletes the requirement of a null response (Enter) to end the object selection process.

Selecting all the Objects in the Drawing. Selecting All picks all the entities in the drawing except those contained on frozen or locked layers.

> **NOTE:** If most of the entities in a drawing are to be selected for edit, first designate all the drawing entities with the All option, then use the Remove (explanation follows) option to "deselect" the items to remain.

Canceling the Selection Process. Entering [ESC] at any time during selection will cancel the selection process and restore all previously selected objects to normal. The prompt line will return to "Command:".

Changing the Items Selected

You may also add or remove objects to be edited from the group of selected objects by using "modifier" commands. Modifier commands are entered after you have selected objects with the object selection process, but before you press [ENTER] to accept the designated objects.

Undoing the Last Selected Entity. U (undo) removes the most recent addition to the set of selections. If the undo is repeated, you will step back through the selection set.

Removing Designated Objects from the Selection Set. Entering an **R** will cause the object selection process to begin to remove the next selected objects from the set.

When you start the selection process, you may add objects until you identify every object you wish to edit. "Remove" allows you to begin to remove objects from the selection set that you do not wish to edit. The prompt line will show "Remove objects:" when you are in the Remove mode. You may remove objects from the selection set by any object selection method.

> ***NOTE:*** Many new CAD operators only think of object selection in terms of adding entities to a selection set. It is often the case that you wish to choose a large number of entities, with the exception of one or two entities that are located within the area of the other entities.

Adding Designated Objects to the Selection Set. Entering an **A** will cause the object selection process to add objects to the set. The Add option is usually used to toggle back to the Add mode after the Remove option has set the process to the Remove mode. Add changes the prompt line back to "Select objects:" so you may add objects to be edited to the selection set with the select objects options.

Each of the edit commands requires the object selection process to identify the entities to be edited. Let's examine each command to see how to edit your work.

DESIGNATING ENTITY SELECTION SETTINGS

AutoCAD allows you to set different modes of entity selection. The previous sections covered the basic entity selection methods. As you gain more proficiency, you will want to use other modes of selection.

Using Ddselect to Set Selection Mode

The *Ddselect* dialog box is used to set selection modes. This dialog box is accessed through the Options pull-down menu. From the Options menu, select "Selection". You may alternately type **DDSELECT** at the command line.

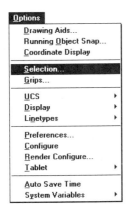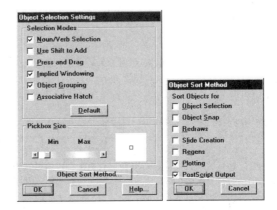

FIGURE 10-7 Ddselect Dialog Box

The dialog contains four selection modes. They are "turned on" by clicking in the check box next to the selection. Let's look at how each mode functions.

Noun/Verb Selection. So far you have learned to edit entities by first selecting the edit command, then the objects to modify with that command. An analogy of this method would be choosing the verb (action represented by the edit command), then the noun (object of the action represented by the selection set). AutoCAD also allows you to reverse this procedure by first picking the object(s) you wish to modify, then the edit command to use. This is called *noun/verb selection*.

The *Ddselect* dialog box is used to set this type of selection. When you set the noun/verb selection to on, AutoCAD places a box at the intersection of the crosshairs. This box is used to select entities in the same way as a pick box.

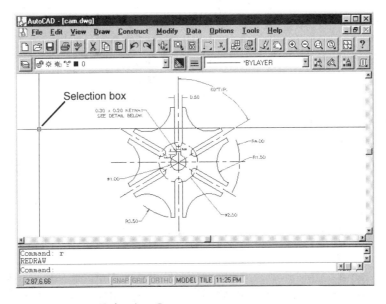

FIGURE 10-8 Selection Box

To edit objects, first select the object(s) to be edited, then choose the edit command. The desired edit command can be either typed at the command line, or selected from the menus. Let's look at how the command sequence would look if we used this method to erase a single entity.

Command: *(Select the Object.)*
Command: **Erase**
1 found

The following edit commands can be used with the noun/verb selection method.

Array	List
Block	Mirror
Change	Move
Chprop	Rotate
Copy	Scale
Ddchprop	Stretch
Dview	Wblock
Erase	Explode
Hatch	

Use Shift to Add. When using the selection process, you have learned to select the objects, then press [ENTER] when you are finished. As you choose each entity, it is automatically added to the selection set.

If you check the "Use Shift to Add" box, each selection will only choose that entity, canceling the choice of the one previously selected. To add entities, you must hold down the shift key on the keyboard. This is similar to the method used by many computer "draw" programs.

Press and Drag. The traditional method of using a window to select entities is to click on one corner, then move the cursor to the other corner and click again.

If you check "Press and Drag", a window is built by clicking on one corner, then moving to the other corner *while holding down the mouse button*. Like the "shift to add" option, this is also similar to the windowing method used by many "draw" programs.

Implied Windowing. You have already learned how to use a selection window or a crossing window to select objects. All you have to do is enter either a **W** or a **C** to invoke the window mode. Checking "Implied Windowing" allows you do this automatically.

A window can be "implied" by clicking on an empty area of the drawing when building a selection set. If AutoCAD does not find an entity within the area of the pick box, then it assumes that you want to use a window. If you move the crosshair away from the area of the first click, a window will be built. If the movement is to the right, a *selection window* will be made. If you move to the left, a *crossing window* will occur.

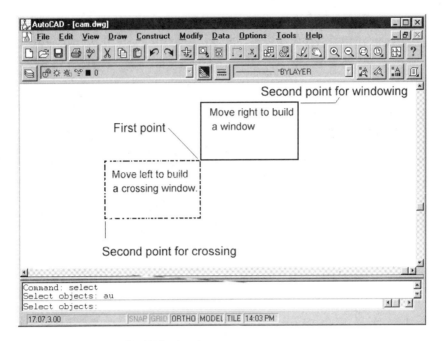

FIGURE 10-9 Implied Windowing

Since the first point entered describes the first corner of the window, be sure to select a position that is desirable.

Resetting to the Default Modes. Selecting the "Default Selection Mode" button resets the selection modes to the original settings. By default, "Use Shift to Add" and "Implied Windowing" are selected.

Setting the Pick Box Size

The pick box size is set from the *Ddselect* dialog box. To change the size, use the slidebar by clicking and holding the button in the slidebar and moving to left to decrease the size and to the right to increase the size. The window to the right of the slidebar shows the actual size of the pick box as you change it.

> **NOTE:** The pick box size can aid in building the selection set. A larger pick box makes selection less tedious, while a smaller pick box allows you to select entities in a crowded area without the necessity of enlarging the drawing area with the zoom command. Change the pick box size as the conditions warrant.

PRESELECTING OBJECTS FOR EDITING

The Select command is used to preselect entities for editing. To execute the Select command, enter:

Command: **Select**
Select objects: *(Select objects to be edited.)*

When next using an edit command, you may simply enter the "previous" option to choose the preselected entities.

> **NOTE:** The select command is especially helpful when edits of several types must be performed on the same group of entities. The target group of entities may be reselected by executing the Previous option when using each successive edit command.

Creating Groups of Objects

A selection set only lasts until a new selection set is created. When you select a group of circles, then later select a group of lines, AutoCAD forgets about the group of circles you selected first.

To overcome this limitation, AutoCAD lets you create any number of "groups." Each group has a name and consists of any selection set of objects. The Group command displays the *Object Grouping* dialog box.

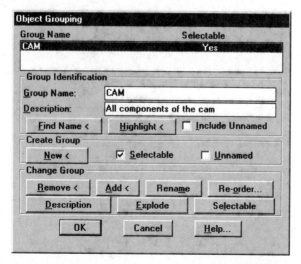 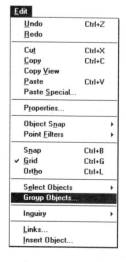

FIGURE 10-10 Object Grouping Dialog Box

Group Identification. The first step to creating a group is to give it a name. You can use any name up to 32 characters long (no spaces). You can also give the group a description up to 64 characters long, such as "The left end of the linkage."

After the groups are created, you can remind yourself of the objects that belong to a particular group by (1) selecting a group name and (2) clicking on the Highlight button. The dialogue box disappears and AutoCAD highlights the objects in the group.

Create Group. The second step to creating a group is to select the objects that will be part of the group. Click on the New button, then use any of the object selection modes to select the objects.

The Selectable check box is very important. When checked (turned on), selecting one object in the group selects the entire group. When unchecked (turned off) selecting an object in the group selects only the object, not the group.

The Unnamed check box determines whether the group is named. When checked (turned on), AutoCAD assigns a name to the group. The name is *A0; the next unnamed group is *A1; and so on. When unchecked (turned off), you give the group its name.

Change Group. After the groups are created, you can change the description and the objects in the group, as follows:

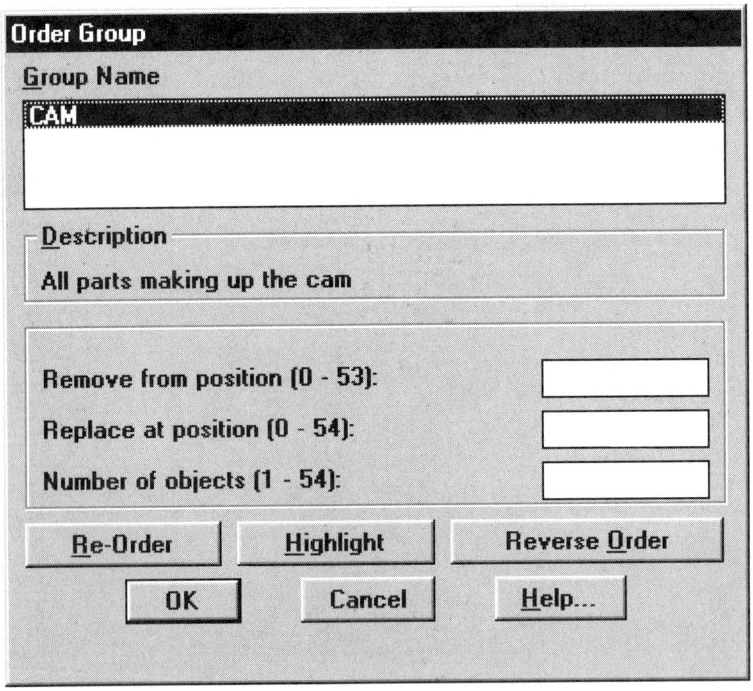

FIGURE 10-11 Order Group Dialog Box

The Remove button lets you remove objects from the group.

The Add button lets you add objects to the group.

The Re-order button lets you change the "order" of the objects in the group. AutoCAD numbers the objects in the group as you add them (the first object in the group has number 0). Changing the order of objects in the group can be important in numerical control and analysis operations.

The Selectable button changes the selectable setting.

The Rename button lets you change the group's name.

The Description button lets you change the group's description.

The Explode button gets rid of the group and name.

If you prefer to work with the Group command, at the command line, precede the command with the negative sign. The command-line version of the Group command has many of the dialog box version's options.

Create Group. The default option of the -Group command creates a new named group, as follows:

Command: **-Group**
?/Order/Add/Remove/Explode/REName/Selectable/<Create>:
Group name (or ?): **lefthand**
Group description: **Left side of the linkage**
Select objects: *(Select objects.)*
8 found
Select objects: *(Press* [ESC]*.)*

You give the group a name (32 characters, no spaces), a description (64 characters, spaces allowed) and then select the objects using any object selection method.

? List Groups. The ? option lists the names and descriptions of the groups, along with their selectability status, as follows:

Command: **-group**
?/Order/Add/Remove/Explode/REName/Selectable/<Create>: **?**
Groups(s) to list <*>:
Defined groups.   Selectable
LEFTHAND   Yes
Description: Left side of the linkage.
RIGHTHAND   Yes
Description: Right side of the linkage

Order Groups. The O option lets you change the order of objects in the group, as follows:

Command: **-Group**
?/Order/Add/Remove/Explode/REName/Selectable/<Create>: **o**
Group name (or ?):
Reverse order/Remove from position <0 - 10>:
Replace at position <0 - 10>:

After specifying the name of the group, the O option reverses the order of objects in the group, or else lets you change the order by (1) first removing an object from its current position and (2) then placing it at a new position.

Add to Group. The A option lets you add objects to the group, as follows:

Command: **-Group**
?/Order/Add/Remove/Explode/REName/Selectable/<Create>: **a**
Group name (or ?):
Select objects to add: *(Pick.)*
Select objects to add: *(Press* [ESC]*.)*

Remove from Group. The R option lets you remove objects from a group, as follows:

Command: **-Group**
?/Order/Add/Remove/Explode/REName/Selectable/<Create>: **r**
Group name (or ?):
Select objects to remove: *(Pick.)*
Select objects to remove: *(Press* [ESC]*.)*

Explode Group. The E option removes a group by exploding it, as follows:

Command: **-Group**
?/Order/Add/Remove/Explode/REName/Selectable/<Create>: **e**
Group name (or ?):

> **NOTE:** The Undo command does not reverse the explode group operation.

Rename Group. The Ren option lets you change the name of a group, as follows:

Command: **-Group**
?/Order/Add/Remove/Explode/REName/Selectable/<Create>: **ren**
Old group name (or ?):
New group name:

Selectability. The S option lets you change the selectability option of a group. When the selectivity option is turned on (YES), selecting a single object in the group selects the entire group, as follows:

Command: **-Group**
?/Order/Add/Remove/Explode/REName/Selectable/<Create>: **s**
Group name (or ?):
This group is selectable. Do you wish to change it? <Y>

Answering **Y** makes the group unselectable (selecting an object in the group only selects the object, not the group).

EDITING 10-17

ERASING OBJECTS FROM YOUR DRAWING

The Erase command is used to remove entities from the drawing. The command sequence for an erase is as follows:

Command: **Erase**
Select objects:

You will notice that the screen crosshair is now a pick box. The pick box can be used to select an object (point method). If you wish to use one of the other object selection options (such as Window), enter the letter for the option after the "Select objects:" prompt.

RESTORING ERASED OBJECTS

The Oops command restores the entities that were last erased from the drawing.

Command: **Oops**

You cannot always Oops backward through a drawing to return objects erased several commands back. Most of the time you will be required to Oops back objects immediately after you erased them. Don't press your luck by executing any commands between the two!

If you find yourself unable to Oops back erased entities, remember that you can undo the erase with the U command.

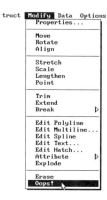

MOVING OBJECTS IN THE DRAWING

The Move command allows you to move one or more objects to another location.

Objects are moved by showing a point to start from and a point to move to. The selected object(s) will then move relative to the specified displacement.

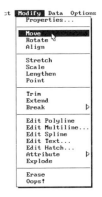

Command: **Move**
Select objects:
Base point or displacement:
Second point of displacement:

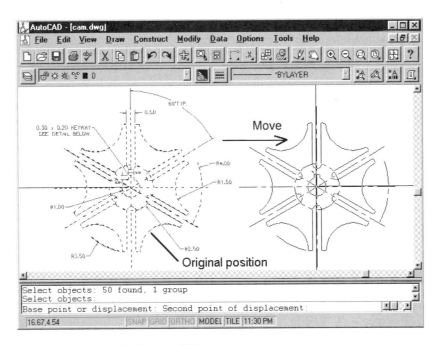

FIGURE 10-12 Moving an Object

> **NOTE:** Your first point need not be on the object to be moved; however, using a corner point or other convenient point of reference on the object makes the displacement easier to visualize.

TUTORIAL

Let's use the erase command to delete some objects from a drawing. We will use a drawing from the work disk named "EDIT1". Start the drawing. You should see the following on your screen.

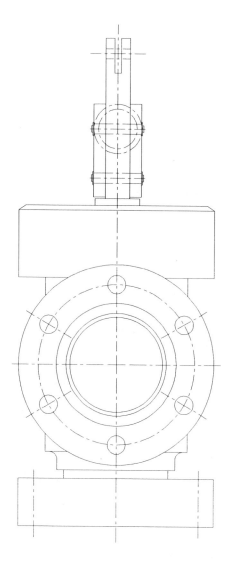

FIGURE 10-13 EDIT1.DWG Exercise

Using a Pick Box. Let's use the Erase command to delete some of the entities. From the Modify menu, select Erase. You should now see a pick box on the screen. Place the pick box over the bottom line of the part as shown in Figure 10-14 and click. The line should now be highlighted. Now select the remainder of the lines as shown in the figure in the same manner. Finally, press ENTER.

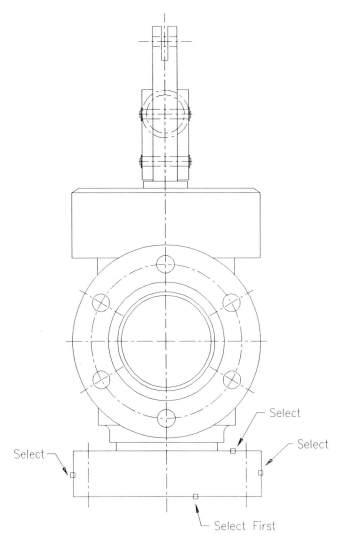

FIGURE 10-14 Selecting Entities to Erase

Using a Window. Let's continue. First, use the U command to undo the erase. Let's now use some of the object selection options.

Select Erase again and enter a **W** in response to the prompt. Refer to Figure 10-15 for the points referenced in the command following sequence.

Command: **Erase**
Select objects: **W**
First corner: *(Select point "1".)*
Other corner: *(Select point "2".)*
58 found
Select objects: [ENTER]

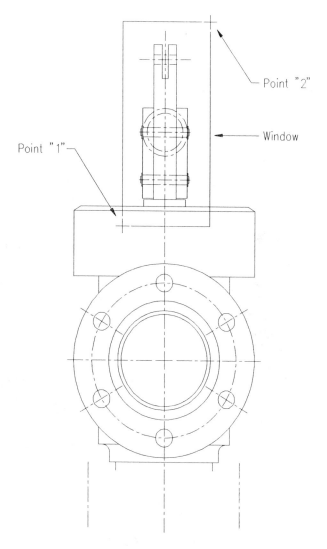

FIGURE 10-15   Erasing with a Window

All the items that were contained in the window were selected. The selected items are highlighted. Notice that entities that extended into the window area, but were not wholly contained within the window, were not selected. Press [ENTER] and the selected objects will be deleted.

Using a Crossing Window. Use the U command to undo the erase. Select Erase again, entering **C** (for Crossing) as the option. Refer to the following command sequence and Figure 10-16.

Command: **Erase**
Select objects: **C**
First corner: *(Select point "1".)*
Other corner: *(Select point "2".)*
61 found
Select objects:

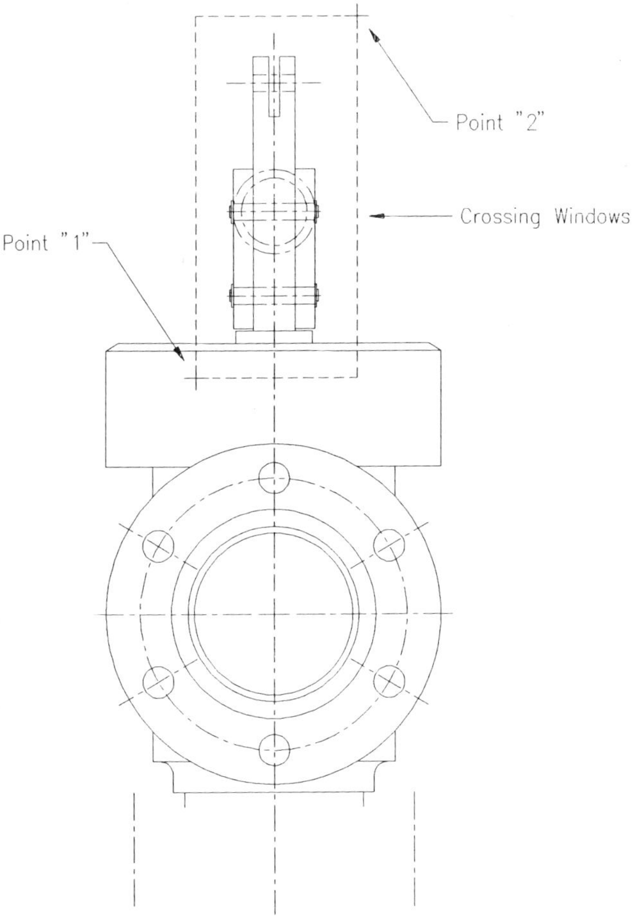

Removing Objects from the Selection Set. Notice that *all* the objects that were touched by the window were selected (as noted by the highlighting). Let's remove some of the objects. After you placed the crossing window, AutoCAD again asked you to "Select objects:" (see the previous command sequence). Enter **R** for Remove. The command sequence will continue:

Remove objects: *(Select one of the horizontal lines.)*
1 selected, 1 found, 1 removed
Remove objects: *(Select the other horizontal line.)*
1 selected, 1 found, 1 removed
Remove objects: [ENTER]

FIGURE 10-16 Using a Crossing Window

The objects you removed from the object selection set were not erased.

We will use this drawing for some additional exercises in this chapter. If you exit the drawing, discard the changes so any edits are not recorded. If you want to practice with some of the edit commands, you can restore the drawing by using the U command to undo the edits.

MAKING COPIES OF DRAWING ENTITIES

The Copy command is used to make copies of existing objects in the drawing.

Use the object selection process to choose the object to be copied. The prompt line will then ask for the displacement from the original object to the location of the new object.

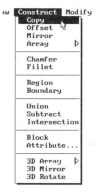

Command: **Copy**
Select objects:
<Base point or displacement>/Multiple:
Second point of displacement:

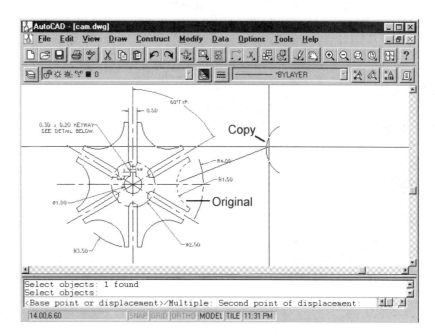

FIGURE 10-17 Copying an Object

> **NOTE:** Use the Copy command freely to repeat items "on the go" as you draw.

Making Multiple Copies

The Multiple option allows placement of multiple copies.

> Command: **Copy**
> Select objects:
> <Base point or displacement>/Multiple: **M**
> Base point:
> Second point of displacement:
> Second point of displacement:

The "Second point of displacement:" prompt will repeat until you cancel the command. The copy will originate from the originally selected object, using the base point you first selected.

PARTIALLY ERASING WITH THE BREAK COMMAND

The Break command is used to erase parts of a line, circle, arc, trace, or 2D polyline.

> Command: **Break**
> Select object:
> Enter first point:
> Enter second point (or F for first point):

To Break an object, select the two points on the object between which the Break is to take place. The object selection process may be used to select the desired object to be broken.

If object pointing is used to select the entity, AutoCAD assumes that point is also the first Break point. If you wish to redefine the first break point, enter **F** in response to the prompt line's request for the second point and you will be prompted again for a first point.

> Command: **Break**
> Select object: *(Select.)*
> Enter second point (or F for first point): **F**
> Enter first point:
> Enter second point:

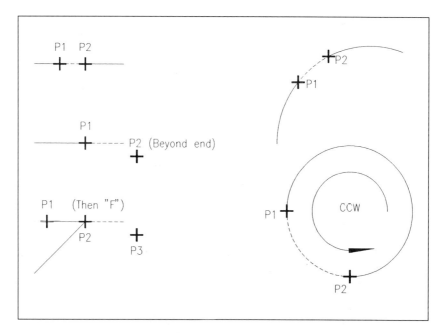

FIGURE 10-18 Using Break

NOTE: The ability to redefine the first point is useful if the drawing is crowded or the break occurs at an intersection where pointing to the object at the first break point might result in the wrong object being selected.

The Break command affects entities in different ways:

Line: The portion of line between the points is removed. If one point is on the line and the other point is off the end of the line, the line is "trimmed off" at the first break point.

Trace: Traces are broken in the exact manner as lines. The new endpoints of the trace are trimmed square.

Circle: A circle is changed into an arc by removing the unwanted piece going *counterclockwise* from the first point to the second point.

Arc: An arc breaks the same as a line.

Polyline: Polylines of non-zero length are cut square (similar to breaks on traces). Breaking a closed polyline creates an open polyline.

Viewport entities: Viewport entity borders can not be broken.

CONNECTING OBJECTS WITH A FILLET

The Fillet command is used to connect two lines or polylines with a perfect intersection or with an arc of a specified radius. Fillet can also be used to connect two circles, two arcs, a line and a circle, a line and an arc, or a circle and an arc.

The two objects do not have to touch in order to perform a fillet, including parallel lines.

Fillet expects you to choose two lines or polylines. After you have chosen the second line, you do not have to press Enter.

Filleting Two Lines

The fillet radius is a default setting which you specify. A zero radius fillet will connect two lines with a perfect intersection.

Filleting Polylines

You may fillet an entire polyline in one operation. If you select the "P" (Polyline) option, the fillet radius is constructed at all intersections of the polyline. If arcs exist at any intersections, they will be changed to the new fillet radius. Note that the fillet is applied to one continuous polyline only.

The following figure illustrates a fillet on a polyline.

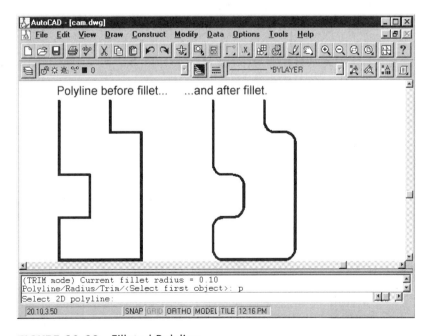

FIGURE 10-19 Filleted Polyline

Filleting Arcs, Circles, and Lines Together

Lines, arcs, and circles can be filleted together. When filleting such entities, there are often several possible fillet combinations. You can specify which type of fillet you desire by the placement of points when you select the objects. AutoCAD will attempt to fillet the endpoint that the selection point is closest to.

Figure 10-20 shows several combinations between a line and arc. Observe the placement of the points used to pick the objects in the middle row, and the resulting fillet shown in the top row.

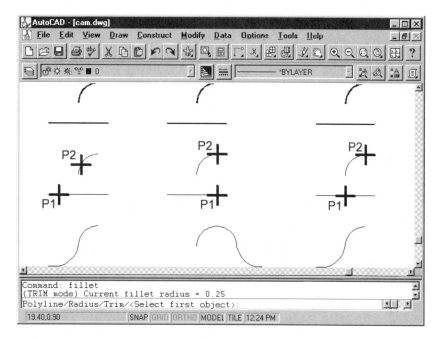

FIGURE 10-20 Filleting Lines and Arcs

If you select two objects for filleting and get undesirable results, use the U command to undo the fillet and try respecifying the points closer to the endpoints you desire to fillet.

Filleting Circles

As with lines and arcs, the result of filleting two circles depends on the location of the two points you use to select the circles. Figure 10-21 shows three possible combinations, each using different selection points.

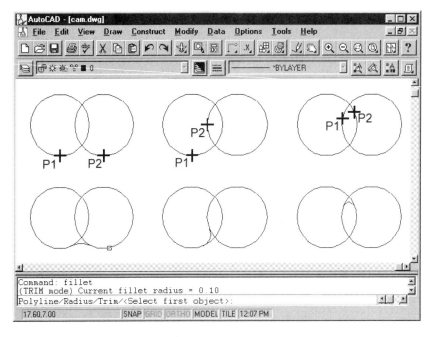

FIGURE 10-21 Filleting Circles

> **NOTE:** If you have several filleted corners to draw, construct your intersections at right angles, then fillet each one later. This allows you to continue your line command, without interruptions, and results in fewer commands to be executed.
>
> Your line intersections can be "cleaned up" by setting your fillet radius to zero and filleting the intersections.
>
> Changing an arc radius by fillet is cleaner and easier than erasing the old arc and cutting in a new one. Let AutoCAD do the work for you!

TUTORIAL

Let's try some edit commands with a drawing on the work disk project named "EDIT7". Figure 10-22 shows the drawing.

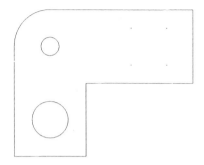

FIGURE 10-22 EDIT7.DWG Work Disk Exercise

Removing the Lower Circles

Suppose that a design change has been initiated and you have been instructed to remove the lower circle from the drawing.

In the Modify toolbar, move

The prompt line should now say "Select objects".

Move the pick box until it intersects on the lower circle. Click on the circle.

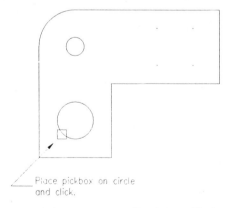

FIGURE 10-23 Erasing a Circle

The circle is now highlighted! The prompt line shows that AutoCAD is ready to add more entities, but let's stop with this one now.

Press ENTER to tell AutoCAD to execute the Erase command on the highlighted object, and watch the object disappear!

Removing the Points

Now, remove the four "Points" on the object by using a window selection:

Command: **Erase**
Select objects: **W**
First corner: *(Select point "1".)*
Other corner: *(Select point "2".)*

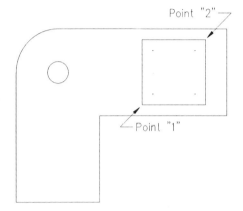

FIGURE 10-24 Erasing the Points

Notice how the window stretched out and followed the crosshair! You may "stretch" the window to any size you wish. It is only "set" when you enter the second corner location.

Restoring the Points

Let's put back the four points you just erased. You could redraw them. Or you could say "Oops! I made a mistake!" Over in the screen menu, you will find the Oops command. Execute it and watch the points return.

Moving a Circle

Now that you have erased the lower circle on your drawing, move the remaining circle to the middle of the object. Refer to Figure 10-25.

Command: **Move**
Select objects: *(Select the circle.)*
Base point or displacement: *(Select point "1".)*
Second point of displacement: *(Select point "2".)*

Copying a Circle

Let's suppose that you have now been directed to add another circle to your drawing which is identical to the remaining circle. (Remember erasing the larger circle earlier?)

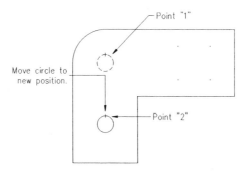

FIGURE 10-25 Moving a Circle

Command: **Copy**
Select objects: *(Select the circle.)*
<Base point or displacement>/Multiple: *(Select point "1".)*
Second point of displacement: *(Select point "2".)*

You now have an exact copy of the first circle.

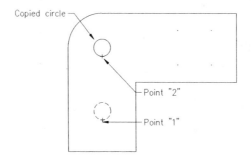

FIGURE 10-26 Copying a Circle

Adding a Notch

Let's add a notch to the object by breaking a line to receive the notched-in area. Refer to Figure 10-27.

Command: **Break**
Select objects: *(Select the bottom line.)*
Enter second point (or F for first point): **F**
First point: *(Select point "1".)*
Second point: *(Select point "2".)*

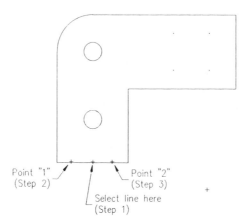

FIGURE 10-27 Breaking a Line

Now, using the Line command, draw in the notch.

Radiusing the Corners

Let's add a radius to each of the two right corners of the object. We will set the radius to .5.

> Command: **Fillet**
> Polyline/Radius/<Select first object>: **R**
> Enter fillet radius <default>: **.5**

Now repeat the Fillet command again.

> Command: **Fillet** (or [ENTER] to repeat the command.)
> Polyline/Radius/<Select first object>: *(Select point "1".)*
> Select second object: *(Select point "2".)*

The corner now has an arc with a radius of .5. Now repeat the fillet on the lower corner.

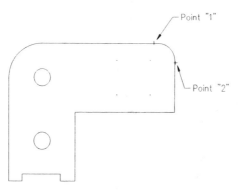

FIGURE 10-28 Radiusing a Corner

EXERCISES

1. Use the work disk drawing named "EDIT1". Use the Erase command to delete some of the entities. Next, issue the Oops command. Did the objects return?
2. Start the drawing named "EDIT2" from the work disk. This exercise is a jigsaw puzzle. The drawing is the same as Figure 10-13. When you start the drawing, it will appear as in Figure 10-29. Use the Move command to move the pieces into position, leaving a small space between them. The most effective method is to move the pieces roughly into position while the drawing is in a zoom-all, then zoom in and fine position the pieces.

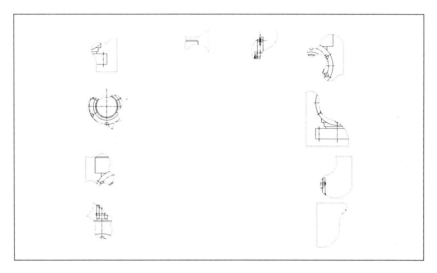

FIGURE 10-29 EDIT2.DWG Work Disk Exercise

3. From the work disk, start the drawing named "EDIT3". Figure 10-30 shows the drawing. Use the copy command to copy the windows from the left side to the right side. Then copy all the windows on the lower level (including those you just copied) to the upper level.

FIGURE 10-30 EDIT3.DWG Work Disk Exercise

When you copy, use the object selection options you think will work best.

4. From the work disk, start the drawing named "EDIT4". The drawing is a site plan as shown in Figure 10-31.

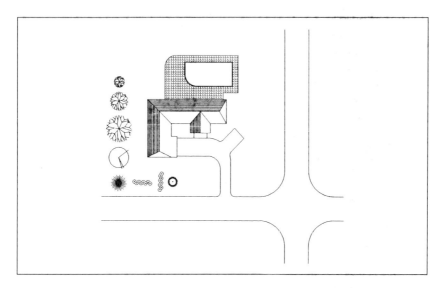

FIGURE 10-31 EDIT4.DWG Work Disk Exercise

Use the Copy command, with the Multiple option, to copy the landscaping items (trees, shrubbery) and create your own landscape scheme.

5. From the work disk, start the drawing named "EDIT5". The following figure shows the drawing.

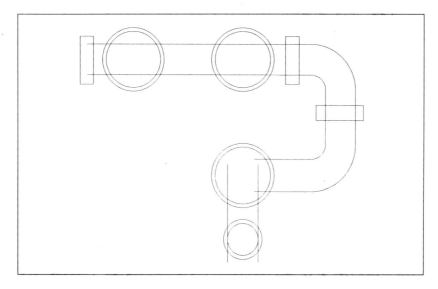

FIGURE 10-32 EDIT5.DWG Work Disk Exercise

Use the Break command as shown in Figure 10-33 to break each of the objects in the drawing, achieving the results shown.

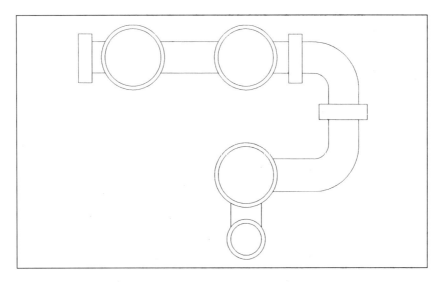

FIGURE 10-33 Completed EDIT5 Exercise

6. Let's look at an example of connecting two lines with a "zero" radius intersection. Refer to the following command sequences and Figure 10-17. First, draw lines similar to those in the illustration. Now let's set the fillet radius to zero.

Command: **Fillet**
Polyline/Radius/<Select two objects>: **R**
Enter fillet radius: **0**

Now, issue the Fillet command again and select the two lines.

Command: **Fillet**
Polyline/Radius/<Select two objects>: *(Select the two lines.)*

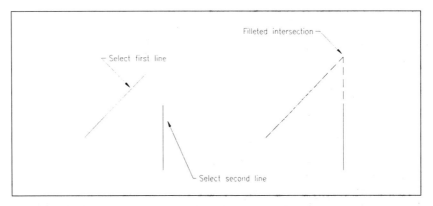

FIGURE 10-34 Fillet 0 Intersection

Notice how the two lines are now connected in a perfect intersection. Let's continue and connect the same two lines with a radiused fillet. We will connect the lines with an arc with a radius of .15. Let's first set the fillet radius to .15.

Command: **Fillet**
(TRIM mode) Current fillet radius: 0.0000
Polyline/Radius/<Select two objects>: **R**
Enter fillet radius <default>: **.15**

The default radius is now set to .15 and will remain until it is changed to another value.

Now, select the Fillet command again. Refer to the following command sequence and Figure 10-35.

Command: **Fillet**
(TRIM mode) Current fillet radius: 0.1500
Polyline/Radius/<select two objects>: *(Select the two lines.)*

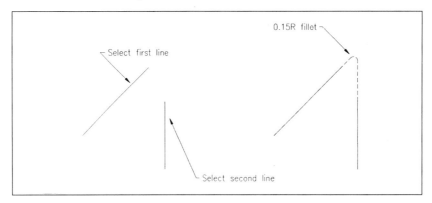

FIGURE 10-35 Fillet Radius 0.15 Intersection

7. Start the drawing named "EDIT6" from the work disk. Perform fillets on the objects to achieve the results shown in Figure 10-36.

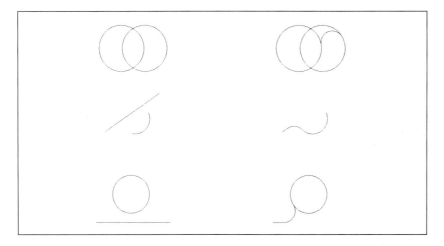

FIGURE 10-36 EDIT6.DWG Work Disk Exercise

8. Additional drawing exercises:

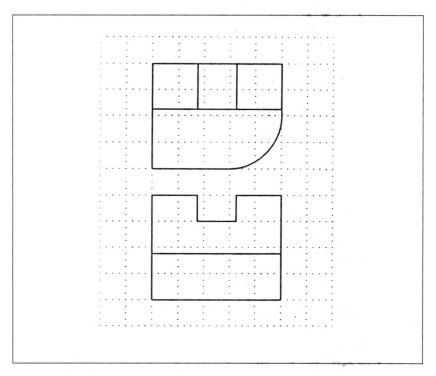

FIGURE 10-37

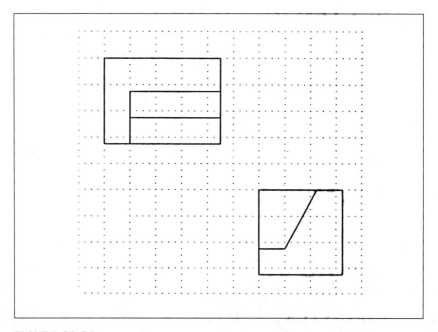

FIGURE 10-38

10-38 USING AUTOCAD

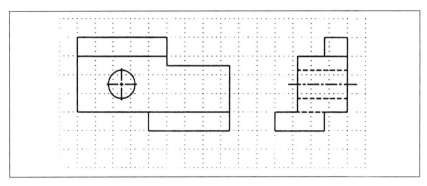

FIGURE 10-39

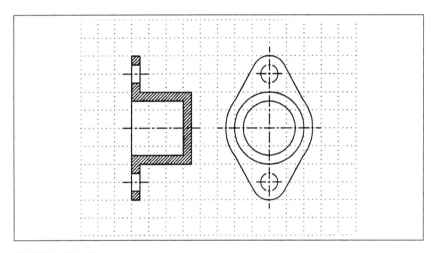

FIGURE 10-40

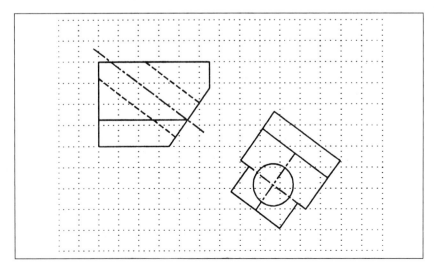

FIGURE 10-41

CHAPTER REVIEW

1. When an edit command is invoked, must a selection option, such as single or window, be entered before an entity is selected?

2. What two changes can a fillet make to an intersection?

3. When choosing a set of objects to edit, other than using the U command, how can you alter an incorrect selection without starting over?

4. If you wish to edit an item that was not drawn last, but was the last item selected, would the "Last" selection option allow you to select the desired item?

5. How do you increase or decrease the size of the pick box?

6. When moving an object, must the base point of displacement be on the selected object?

7. What five entities are affected by the Break command?

8. When using the Multiple option of the Copy command, is each copy relative to the point of the last copy made, or the first base point entered?

9. In breaking an object, what happens if you do not enter an F for selection of the first point?

10. Can elements of different types be filleted (such as a line and an arc), or must they be alike?

11. When entering **box** in response to the select object prompt, how are you then allowed to choose entities?

12. What command will restore entities just erased?

13. What makes it evident that an item has been selected?

14. How can entities be completely removed from the drawing?

15. Once you have begun selecting objects during a command sequence, can you alter your method of selection?

16. When a group of items are selected by using the window option, will an item become a part of the selection set as long as it is partially inside of the window?

17. What happens when you choose the U option during a selection process? Can you use this option more than one time in a row?

___

___

___

18. In the object selection process, how is the box option different from the automatic option?

___

___

___

19. How does a group differ from a selection set?

___

___

___

CHAPTER 11

CONSTRUCTING MULTIVIEW DRAWINGS

Three-dimensional objects are commonly described by multiview drawings. Chapter 11 covers the use of AutoCAD for constructing multiview drawings. After completing this chapter, you will be able to:

- Perform the fundamentals of multiview drawings.
- Manipulate the commands that draw multiview drawings.
- Use the methodologies of using AutoCAD to construct multiview drawings.

MULTIVIEW DRAWINGS

The description of three-dimensional objects by use of flat, two-dimensional drawings is a common drafting practice. An accurate 2D description can be accomplished by drawing the object from several directions, thus *multiple* views.

Orthographic Projection

An *orthographic projection* is a view of an object that is created by projecting a single view onto an imaginary projection plane. Let's look at an example. The face of an object in Figure 11-1 is projected onto the viewing plane.

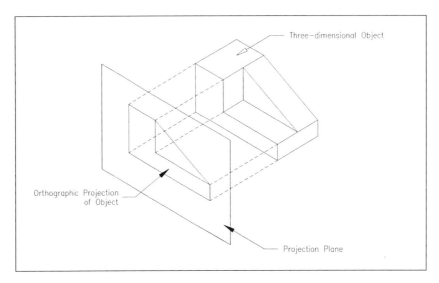

FIGURE 11-1 Orthographic Projection

The image that is projected onto the projection plane represents the true lengths of the edges on the object. The projection plane in this example is parallel to a viewing face on the object. This is referred to as a *normal* view of the object. Normal views are a more accurate method of viewing an object in orthographic projection.

One-View Orthographic Projections. Many thin, simple objects can be described by a single orthographic projection, such as the one in Figure 11-2.

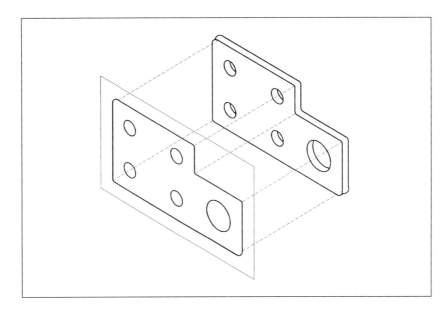

FIGURE 11-2 One-View Orthographic Projection

Two-View Orthographic Projections. More often, a single view cannot adequately describe a three-dimensional object. Faces that do not lie in the same plane can be projected onto the projection plane. The viewer may see the edge lines, but cannot determine the location of the different planes.

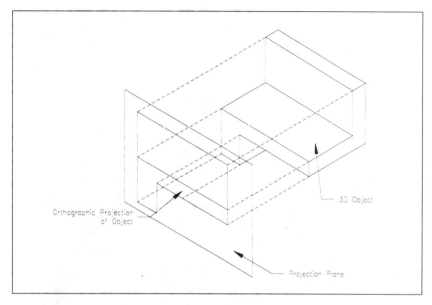

FIGURE 11-3 Orthographic Projection of a Complex Object

Two orthographic views can be used to accurately describe such an object. To do this, two projection planes must be used.

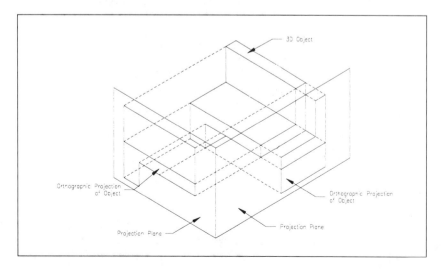

FIGURE 11-4 Two-View Orthographic Projection

The views should show the length, height, and width of the object.

Multiple-View Orthographic Projections. More complex objects may require multiple views to adequately describe them. These views can be projected from several sides of the object. The term *multiview* describes several views of an object.

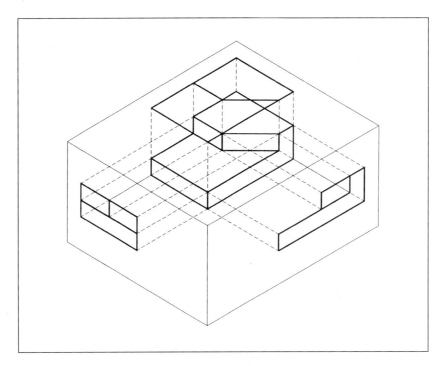

FIGURE 11-5 Multiple-View Orthographic Projection

Positioning Views. In most situations, three views of an object will adequately describe an object. The views are usually labeled as the front, top, and side. The front view is considered the primary view, with the top view positioned above it, and the side view to one side of the front view.

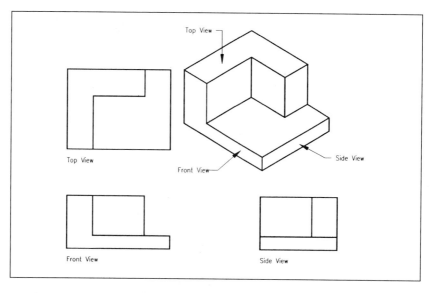

FIGURE 11-6 Three Views of an Object

With this arrangement, the dimensions for the side view can be transferred from the top and front views. Figure 11-7 shows the previous example with imaginary transfer lines shown. Note how the transfer lines from the top view are reflected off a 45 degree miter line.

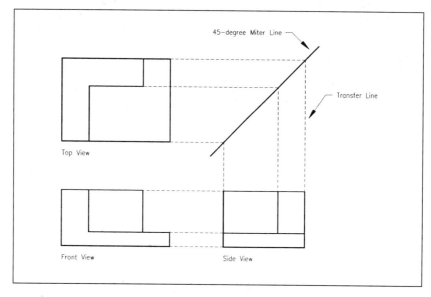

FIGURE 11-7 Transferring Line Lengths

This method can be used as an alternate technique to constructing each line length.

TUTORIAL

Let's construct a simple three-view drawing with AutoCAD. The following figure shows the object we will use.

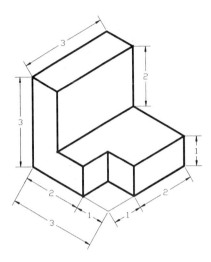

FIGURE 11-8 3D Object (with Dimensions Shown)

Let's begin a new drawing. Enter **New** from the command line. Specify the drawing name "3VIEW". Let's set a snap increment so our crosshairs will only move one unit at a time.

Command: **Snap**
Snap spacing or ON/OFF/Aspect/Rotate/Style <default>: **1**

Now use the Line command to draw the front view of the object. Figure 11-9 (left) shows the dimensions of the front view. Figure 11-9 (right) shows how your drawing should look after you have drawn the front view.

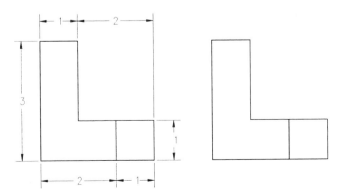

FIGURE 11-9 Dimensioned Front View (left); Completed Front View (right)

Now we will draw the top view. Use the Line command to draw transfer lines to transfer the widths of the object, then draw a line that will serve as the upper edge of the top view as shown in the following illustration.

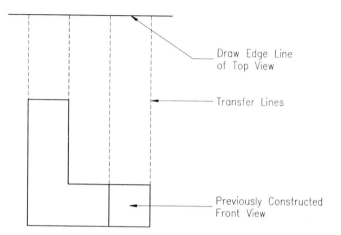

FIGURE 11-10 Transferring Object Widths

Next, we will use a command called Offset to offset the 3-inch width of the object. Following this, we will again use the Offset command to create the "notch". Use the following illustrated sequence.

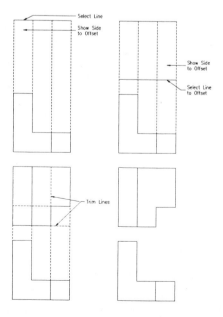

FIGURE 11-11 Illustrated Sequence

Now we will draw the side view. Let's start by placing a miter line so we can transfer the top view dimensions to the side view.

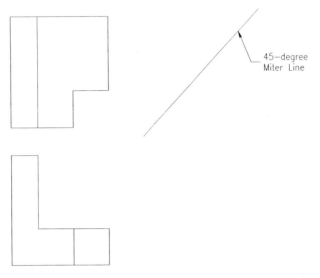

FIGURE 11-12 Placement of Miter Line

Now use the line command to transfer the object's edges from the top and front views as shown in Figure 11-13. Using Ortho mode and object snap will aid you in constructing accurate transfer lines.

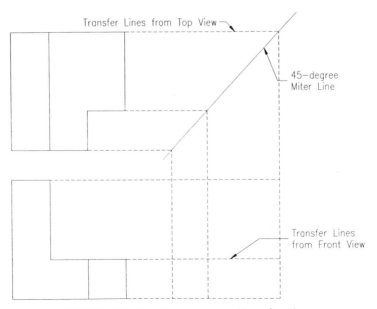

FIGURE 11-13 Constructing Transfer Lines

Now use the Trim command to trim away lines so your drawing looks like the one in Figure 11-14. Be sure to erase any remaining transfer lines.

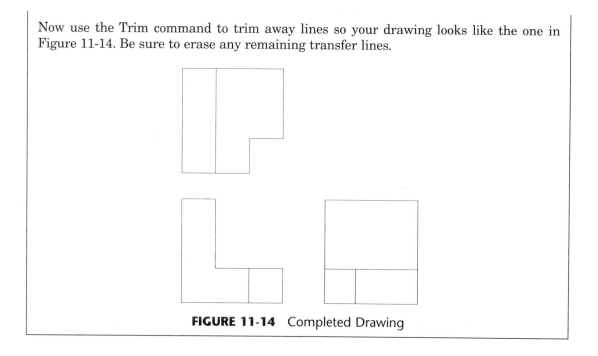

FIGURE 11-14 Completed Drawing

Drawing Multiviews with AutoCAD

AutoCAD is an excellent tool for constructing multiview drawings. The method of construction closely follows that used on the drawing board.

Auxiliary Views

Objects sometimes have angular faces that are not parallel to the projection plane. The object shown in Figure 11-8 contains an angular face that is not truly represented in the top or side view.

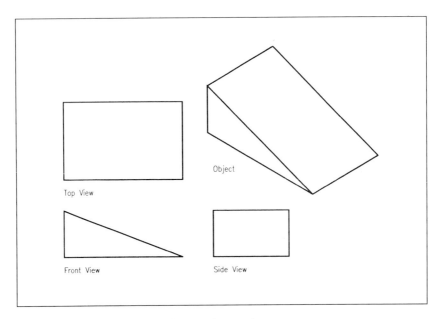

FIGURE 11-15 Object with Angular Surface

We can show the true size of the angular face by adding a projection from the object called an *auxiliary view*. An auxiliary view is one that is projected onto a projection plane that is parallel to the angular surface. You can also think of this as the view you would see if you looked at the object from a point perpendicular to the angular face. Figure 11-16 shows the previous object with an auxiliary view added.

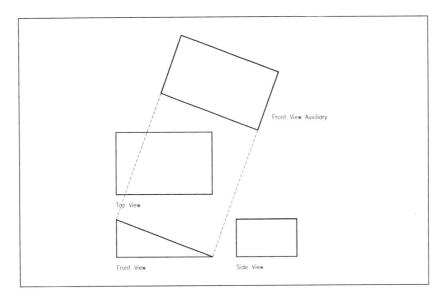

FIGURE 11-16 Auxiliary View

Auxiliary views serve three purposes that cannot be achieved by the normal three views.

- To show the true size of the angular surface.
- To illustrate the true shape of the surface.
- To aid in the projection of other views.

Auxiliary views can be projected from any other view. The name of the view is determined from the view from which it is projected. For example, if you project the auxiliary view from the front view, it is named the *front view auxiliary*.

CONSTRUCTING AUXILIARY VIEWS WITH AUTOCAD

Constructing an auxiliary view with AutoCAD involves drawing the view at an angle parallel to the angular face. There are some tricks that make this process easier. Let's look at how we would draw an auxiliary view.

Showing Hidden Lines in Multiview Drawings

Auxiliary views are line drawings. Solid lines are used to represent the edges of the object. In AutoCAD, solidly drawn lines are referred to as *continuous* lines. Edges that are hidden from view are shown in a linetype referred to as *hidden*.

This is a line that is constructed from a series of short line segments. Let's look at an example. Figure 11-17 shows an object containing edges that are hidden in some views. These edges are defined with the hidden linetype.

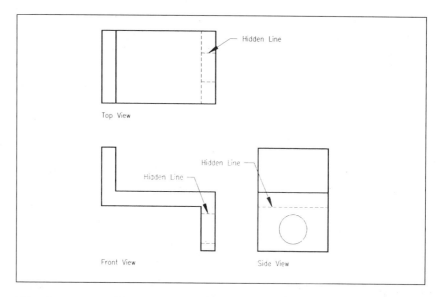

FIGURE 11-17 Object with Hidden Lines Shown

TUTORIAL

We will construct an auxiliary view for the following object. The electronic work disk contains a drawing named "AUX_VIEW" that can be used if you wish to follow along.

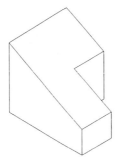

FIGURE 11-18   3D Object with Angular Surface

A simple trick to get started is to copy the angle line from a projected view into the position where the auxiliary view will be positioned. Figure 11-19 illustrates this procedure.

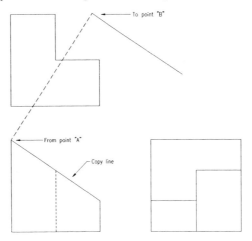

FIGURE 11-19   Copying the Angle Line

Next, use the Snap command to rotate the snap grid. This will not only rotate the snap grid, but will also rotate the crosshairs so we can draw at the proper angle. The following command sequence and Figure 11-20 show how to do this.

Command: **Snap**
Snap spacing or ON/OFF/Aspect/Rotate/Style <default>: **R**
Base point <0.0000,0.0000>: **END**
of *select point "1"*
Rotation angle <0>: **END**
of *select point "2"*

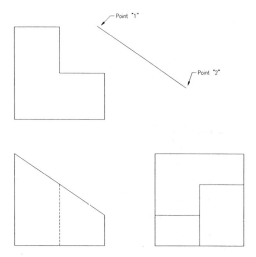

FIGURE 11-20 Rotating the Snap Grid

Notice how using object snap assists in obtaining greater accuracy. If you move the crosshair around the screen, you will notice that it is rotated to the same angle as the line you copied.

Next, use the Line command to complete the drawing. Be sure to use the Ortho mode and the snap increment setting to assist in drawing the object. You may also want to construct some temporary transfer lines to assist in determining the intersections. Figure 11-21 shows the completed view with transfer lines shown dashed.

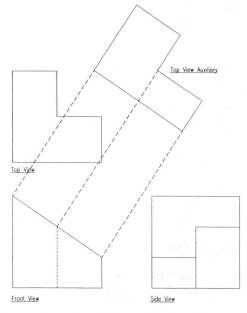

FIGURE 11-21 Completed View with Transfer Lines Indicated

AutoCAD provides several linetypes for your use. Let's look at how to draw different types of lines.

Using the Linetype Command. Different linetypes are used frequently in drawings. Figure 11-22 shows the linetypes provided in AutoCAD.

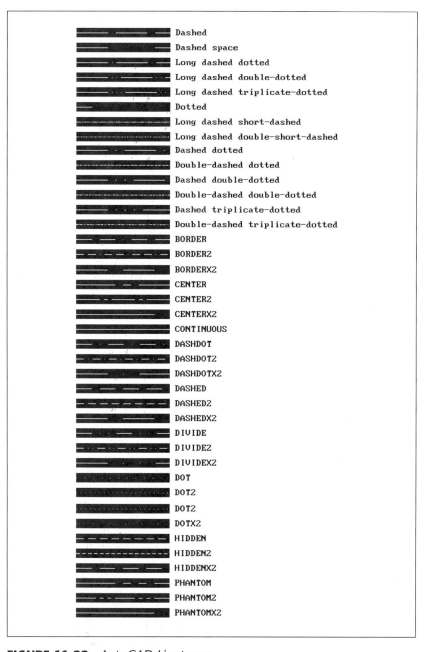

FIGURE 11-22 AutoCAD Linetypes

These linetypes can be added to your drawing by using the Linetype command. Before you can use a linetype, you must first load that linetype. Let's look at how to use the Linetype command to load a linetype.

Loading Linetypes

You must load a linetype before using it. When you use the Layer Ltype command, the linetype is automatically loaded from ACAD.LIN when it is needed.

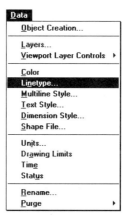

If you explicitly want to load a linetype into your drawing, issue the DDLType command:

Command: **DDLtype**

AutoCAD displays the Select Linetype dialog box. It displays the linetypes currently loaded. To use a linetype, click on it.

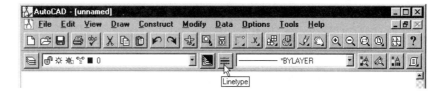

If you need a linetype not loaded, click on the Load button. AutoCAD displays the Load or Reload Linetypes dialog box. The file to search is the file in which the linetype is located. Initially, you will use the ACAD file. It is not necessary to enter the .LIN file extension. Click on the linetype you want to load, then the OK button.

You may use this process to reload current linetypes. If a regeneration is performed, the effect of the new linetype is immediately seen on the screen.

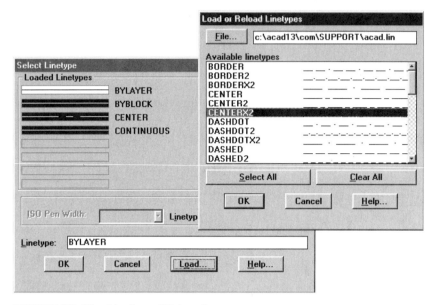

FIGURE 11-23 Linetype Dialog Box

Specifying Linetypes in Other Ways.   Linetypes can be designated in many ways.

1. The Linetype command can be used to designate the current linetype.
2. The linetype can be specified as the default type of line for a layer.
3. The linetype can be assigned during the plotting setup to a specified entity color.

Using the Change Command to Designate Linetypes.   You can use the Change command to designate linetypes for existing entities. Let's look at the command sequence used to change a linetype.

>   Command: **Change**
>   Select objects: *(Select the entities you want to change.)*
>   Properties/<Change point>: **P**
>   Change what property (Color/Elev/LAyer/LType/Thickness) ? **LT**
>   New linetype <*default*>:

Linetype Scales.   Linetypes are constructed of line segments and dots. The line segments are of a specified length of one. The scale of these segments may be adjusted by use of the Ltscale (linetype scale) command:

>   Command: **Ltscale**
>   New scale factor <*default*>:

Entering larger numbers results in longer line segments, while smaller numbers create shorter line segments.

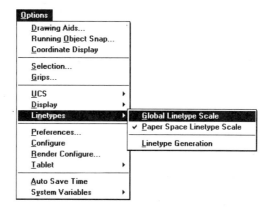

To properly display the new linetype scale, a regeneration must be performed. In certain conditions, AutoCAD will automatically regenerate. If a regeneration does not occur after changing the linetype scale, you can force a regeneration with the Regen command.

If you change linetypes and the line still appears as a continuous line, change the Ltscale to a different setting. It is possible for the linetype scale to be too large or small to display.

Writing Linetypes

The Create option of the Linetype command is used to create a linetype. In order to do this, we must understand how linetype files work.

Linetype descriptions are held in linetype files. The linetype descriptions you received are held in a file named "ACAD.LIN". Each linetype description is described by two lines. The first is a visual description of the line. This description is not actually a part of the description, but is used to display if you enter a question mark (**?**) in response to the Linetype prompt. Do this now and look at the linetype descriptions in the ACAD linetype file.

The second line is the actual description AutoCAD uses to build the line from. This line uses a numerical description to create the line. This line always begins with "A," then continues with the description. The numerical description can be thought of as simple pen-up and pen-down commands, with the numbers as segment lengths. A positive number is a pen-down length, and negative numbers are pen-up lengths.

EXERCISES

1. Let's write a new linetype definition. We will construct a new linetype that could be used for property line definitions. This line will have a long length, followed by two short dashes, then repeat. Let's enter the Linetype command.

   Command: **Linetype**
   ?/Create/Load/Set: **C**
   Name of linetype to create: **PROP**

   A dialog box appears. Enter "ACAD2" in the "File Name" box and click "OK". We have just designated a new linetype file to hold our linetype and named it "ACAD2". We could have just used the file named "ACAD" and appended the linetype description. The command sequence continues with:

   Descriptive text: ———————— — — ————————

   Enter a series of dashes and spaces as shown. It doesn't matter how many. Remember, this is just a visual representation of the line you are about to write a description for. Let's continue with the command sequence.

   Enter pattern (on next line):
   A,1,-.1,.1,-.1,.1-.1

   Press [ENTER] when you are finished. The command sequence will continue:

   New definition written to file.
   ?/Create/Load/Set: **L**
   Linetype to load: **PROP**
   File to search <ACAD@>: **ACAD2**
   Linetype PROP loaded.
   ?/Create/Load/Set: **S**
   New object linetype (or ?) <BYLAYER>: **PROP**
   ?/Create/Load/Set: (*Cancel the command with Esc.*)

   Now use the Line command to draw several lines with your new linetype. If the linetype appears as a solid line, use the Ltscale command to change the linetype scale.

2. Use AutoCAD to draw three views of the following 3D objects.

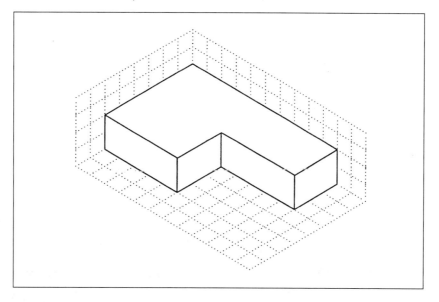

FIGURE 11-24

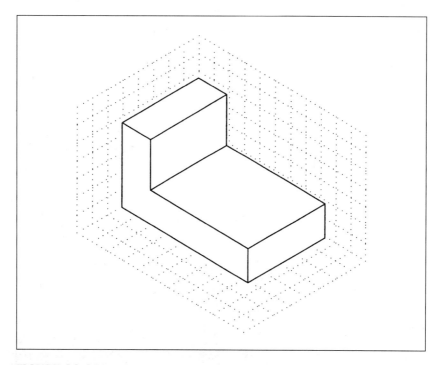

FIGURE 11-25

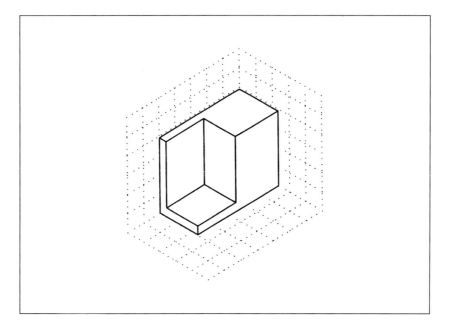

FIGURE 11-26

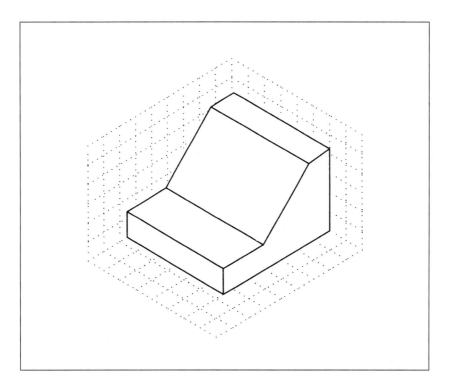

FIGURE 11-27

CONSTRUCTING MULTIVIEW DRAWINGS **11-21**

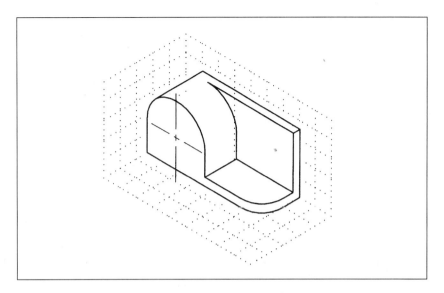

FIGURE 11-28

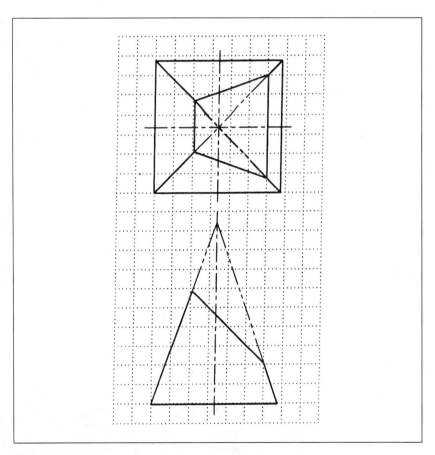

FIGURE 11-29

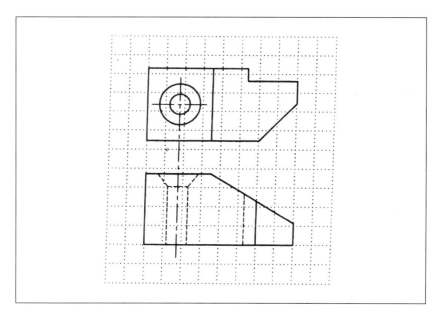

FIGURE 11-30

CHAPTER REVIEW

1. What is an orthographic projection?

2. What type of object can be described by only one projection?

3. What term is used to describe a drawing that contains several views of an object?

4. What are the three typical views used in orthographic projection?

5. What is an auxiliary view?

6. Why would you use an auxiliary view?

7. What do hidden lines show in a drawing?

8. Name two ways a linetype can be designated in AutoCAD.

9. How would you control the length of individual segments in a dashed line?

10. Write a linetype pattern that contains a long line segment, three short segments, then repeats.

CHAPTER 12

CONSTRUCTING SECTIONAL AND PATTERNED DRAWINGS

Sectional views are used to describe objects when other views are not sufficient. Computer aided design is an excellent tool for constructing sectional views. After completing this chapter, you will be able to:

- Use the many types of sectional views and their applications.
- Perform the techniques of constructing the crosshatching that is an integral part of sectional views.
- Create the solid areas that can be used as part of sectional drawings.

SECTIONAL VIEWS

Many parts and assemblies cannot be fully described by use of orthographic projection. It is often helpful to view the object as if it were cut apart. A view of an object or assembly that has been cut apart is called a *section*.

A section is used in many disciplines. The mechanical designer uses sections to show details that cannot be described in other ways.

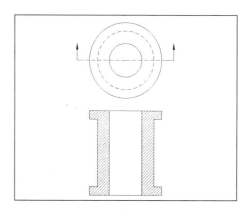

FIGURE 12-1 Section

Civil engineers detail roadway profiles by showing sections through the road.

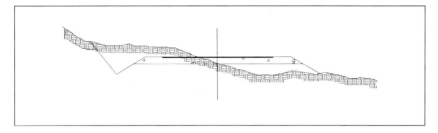

FIGURE 12-2 Roadway Section

Architects use sections frequently. Sections through entire structures show how a building structure is designed. Individual wall sections detail vertical measurements and delineate materials.

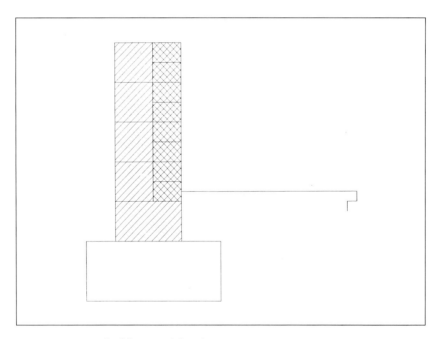

FIGURE 12-3 Architectural Section

Types of Sections

Depending on the desired information to convey, the designer uses different types of sections. Let's study each type of section and look at an example of each.

Full Sections. A *full section* is a section that is cut across the entire object. Full sections are usually cut through the larger axis of the object.

CONSTRUCTING SECTIONAL AND PATTERNED DRAWINGS   12-3

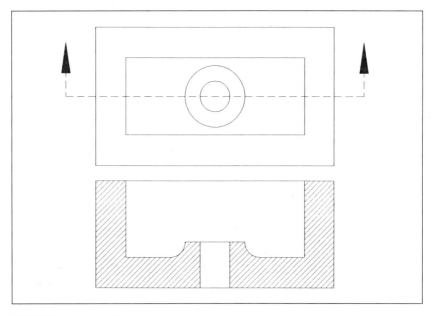

FIGURE 12-4   Full Section

Such a section cut along the longer axis is referred to as the longitudinal axis. If the section is cut along the minor axis, it is referred to as the latitudinal axis.

Parts of the object that are "cut" are shown with crosshatching.

Revolved Sections.   It is often helpful to view a section of a part of an object that is transposed on top of the point where the section was cut. Such a section is referred to as a *revolved* section. Figure 12-5 shows a revolved section.

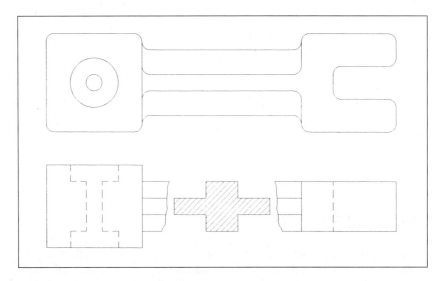

FIGURE 12-5   Revolved Section

Removed Sections. A *removed section* is similar to a revolved section, except the section is not placed at the point that the section was cut.

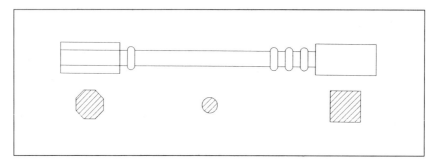

FIGURE 12-6 Removed Section

Offset Sections. *Offset sections* are sections that are cut along an uneven line. Offset sections should be used carefully; changing the cutting plane only to show essential elements.

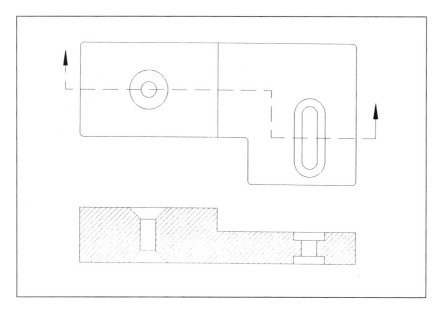

FIGURE 12-7 Offset Section

Crosshatching Sectional Views

Crosshatching refers to placing a pattern within a boundary area. When drawing sections, it is customary to place crosshatching on the faces through the section cuts.

The spacing of the crosshatching should be relative to the scale of the section. The angle of the crosshatching should be oriented 45 degrees from the main lines of the cut area whenever possible.

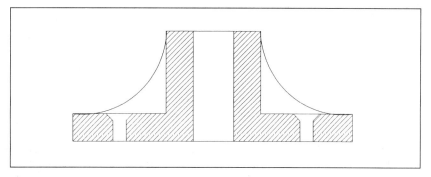

FIGURE 12-8 Section with Crosshatching

In most drafting applications, a simple crosshatch composed of parallel lines is used. In some applications, however, it is acceptable to show different materials with representative crosshatching. Figure 12-9 shows some standard hatch patterns for different materials.

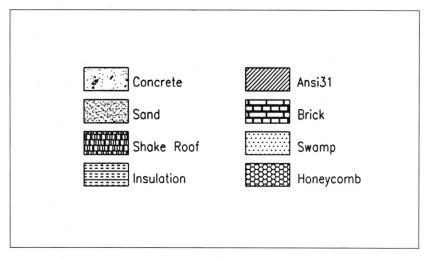

FIGURE 12-9 Typical Material Hatches

CREATING SECTIONAL VIEWS IN AUTOCAD

AutoCAD can be used to create sectional views effectively. You can draw the section, then render the cut areas with hatch patterns that are provided with the AutoCAD program.

12-6 USING AUTOCAD

Hatches differ from other entities in that the entire hatch is treated as one entity. Because of this, you cannot edit a portion of it unless it is exploded. If you identify one line of a hatch to be erased, the entire hatch will be erased. Likewise, an entire hatch may be erased by the Erase-Last command.

Hatch patterns are placed with the Hatch and Bhatch commands. Let's look at how each command is used.

Using the Bhatch Command

The Bhatch (boundary hatch) command creates a boundary around an area, then places an associative hatch pattern within that area. An associative hatch pattern is one that automatically updates when you change the hatch boundary. The Bhatch command is controlled from a dialog box. To use this command, select Constrct/Bhatch from the screen menu, or Draw/Hatch from the pull-down menu.

The Bhatch command displays a "Boundary Hatch" dialog box as shown in Figure 12-10. Let's look at how we can use the dialog box to place a hatch.

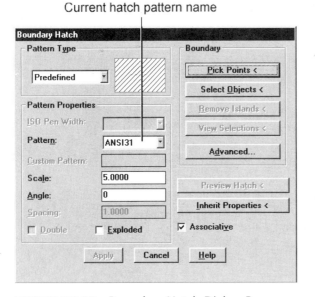
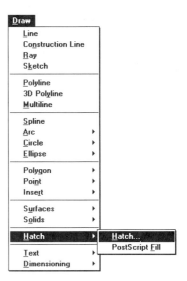

FIGURE 12-10 Boundary Hatch Dialog Box

Selecting a Hatch Pattern and Style. The first step in placing a hatch pattern is to select the pattern you want to use. If a selection has already been made in the current drawing session, its name is displayed to the right of the "Pattern:" drop box.

If you wish to use the named pattern, it is not necessary to select it again. If you want to select a new pattern, select the down arrow. Selecting this button will display the list of available hatch patterns.

If we know the name of the pattern we wish to use, we can click in the text box next to the "Pattern" button and type the name from the keyboard. If a name is already present, we can use the backspace key to erase it and type a new one.

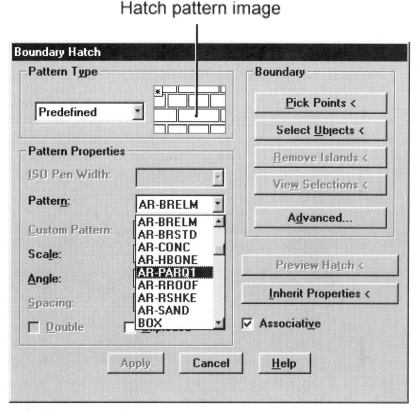

FIGURE 12-11 Hatch Pattern Name

Let's assume that we do not know the name of the pattern we wish to use. Click on the image in the "Pattern Type" area. Each time you click on the image, AutoCAD displays the next hatch pattern.

Selecting the Angle and Scale. Hatch patterns can be drawn at any angle and scale. The angle of the hatch is specified in degrees and is true to AutoCAD's angle specifications (see Chapter 8). The angle of the hatch patterns shown in the dialog box is zero. This is true even if the angle of the lines within the hatch are drawn at another angle.

To set the angle, click in the Angle box and use the keyboard to specify the new angle.

The scale of the hatch pattern is defined as a numerical factor. The default scale is one. Setting the factor as two will create a hatch twice as large, while setting a factor of .5 will create a hatch half the size of the default.

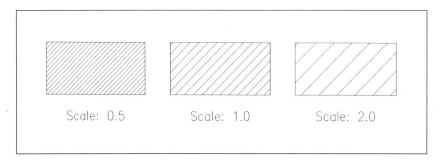

FIGURE 12-12 Hatch Scale Factors

Setting the Hatch Style. Hatch *styles* are used to determine the manner in which AutoCAD will place a hatch within a bounded area. Let's look at an example. Figure 12-13 shows an object that contains hatching in part of an area.

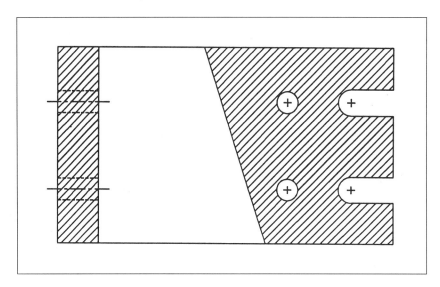

FIGURE 12-13 Hatched Object

Selecting the Hatch Boundary

A hatch must be contained within a boundary that is made up of entities that form a closed polygon. The hatch may be contained by entities such as lines, polylines, circles, and arcs. To place a hatch, the entities that surround the area to be hatched must be identified. There are different ways to identify this boundary area. Let's look at each.

Defining a Boundary by Picking Objects. A hatch boundary can be selected by selecting the objects that surround the area to be hatched. Let's look at a simple example. Figure 12-14 shows a simple rectangle constructed by four line entities.

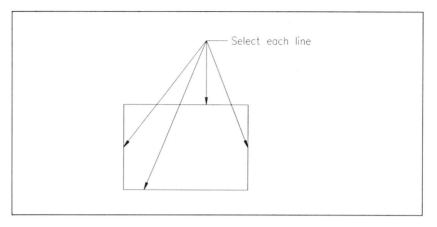

FIGURE 12-14 Selecting the Hatch Boundary

To place a hatch in the rectangle, select the Bhatch command. From the *Boundary Hatch* dialog box, select the "Select Objects" button.

After you pick the Select Objects button, the dialog box leaves the screen. Next, place the pick box over each line and select. Now press [ENTER]. You have just selected the objects that bound the area to be hatched.

> **NOTE:** Use the window or crossing option to select all the sides of the boundary in one operation.

Now you can select the "Apply" button to complete the hatch. If you wish to see how the hatch will look without actually completing the hatch operation, select the "Preview Hatch" button. The proposed hatch will be displayed, with the prompt:

Press RETURN to continue

displayed on the command line.

If you wish to review the objects selected for the hatch boundary, select "View Selections" from the dialog box. The drawing will be displayed, with the boundary objects highlighted.

Defining a Boundary by Selecting the Hatch Area. A simpler way to define a hatched area is to show AutoCAD the area inside a boundary that you wish to be hatched. Let's look at an example. Figure 12-15 shows an object with different areas that could be hatched. First select the Bhatch command. From the dialog box, select the "Pick Points<" button. When the dialog box is removed from the screen, place the crosshair into the area shown in Figure 12-15; click, then press [ENTER]. The dialog box will again be displayed. Select "Apply" and the hatch will be completed.

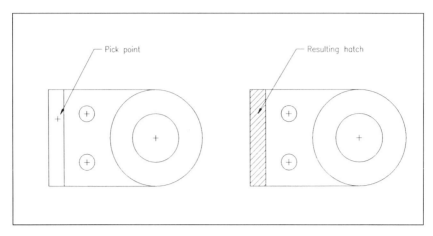

FIGURE 12-15 Picking Inside the Boundary

Advanced Hatch Boundary Definitions

Hatch boundary areas can also be described and stored. Selecting the "Advanced" button will display the *Advanced Options* dialog box.

FIGURE 12-16 Advanced Options Dialog Box

Let's look at each part of the dialog box.

From Everything on Screen. Selecting "From Everything On Screen" will choose all the entities currently visible on the drawing screen. If there is no current boundary stored, this is already checked. If a boundary is already stored, selecting this button will clear the stored set and reselect all the screen entities.

From Existing Boundary Set. When you first define a hatch boundary, that boundary is stored for reuse. If the radio button next to the "From Existing Boundary Set" is checked, you can reuse the stored boundaries.

If you have not previously used a hatch boundary, this button is greyed out.

Make New Boundary Set. Selecting the "Make New Boundary Set" button clears the existing set, then returns you to the drawing screen to select entities that will make up the new set.

To define the area that is to be hatched, a style is used to control the hatch. Let's look at how styles work.

If the area bounded by the identified entities have no objects within them, the hatch is filled with the selected pattern. If, however, the area contains other entities, the hatch works differently. You may choose the manner in which the hatch behaves. The following illustration shows a part called "PART A". Let's see how each hatch style would work.

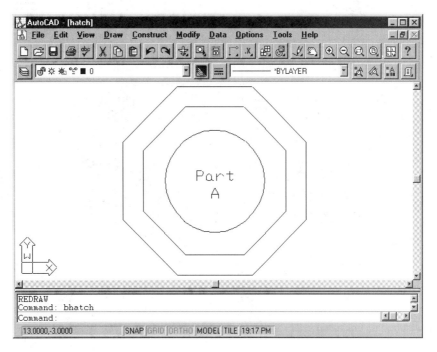

FIGURE 12-17 Part "A" Drawing

Normal Style. The default style of hatching is called Normal. The normal style will hatch inward from the first enclosed boundary, then skip the next enclosed boundary, and hatch the next. Notice how the text is hatched around in Figure 12-18. Text is protected by an invisible window. This ensures that the text is not obscured by the hatch pattern.

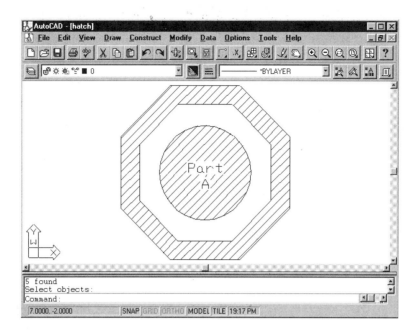

FIGURE 12-18 Normal Style

Outer Style. The Outer style option will cause the hatch pattern to hatch only the outermost enclosed boundary. The hatch continues until it reaches the boundaries and continues no further.

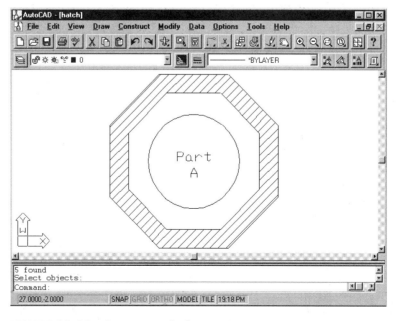

FIGURE 12-19 Outermost Style

Ignore Style. The Ignore style of hatching will hatch all areas that are defined by the identified boundary with no exceptions. This style also hatches through text.

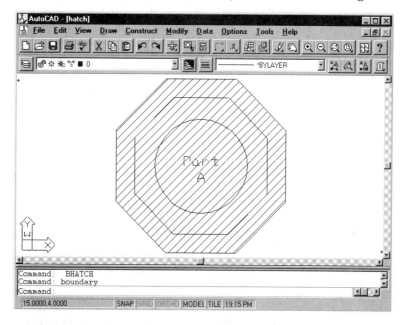

FIGURE 12-20 Ignore Style

Ray Casting. The popup list box next to "Ray Casting:" is used to control the way that AutoCAD finds the boundary objects when you select a boundary with the "Pick Points<" option. When you pick a point inside of a boundary area, AutoCAD, by default, projects a "ray" to the nearest boundary and selects it. The Nearest setting is displayed by default.

You can control the direction that AutoCAD projects this ray by selecting the positive or negative direction in either the X or Y direction. The following illustration shows the areas of valid selection for each setting.

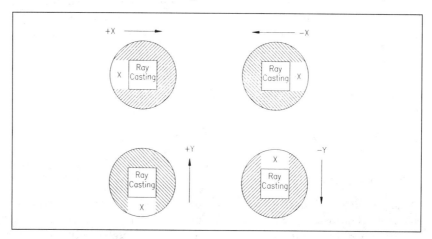

FIGURE 12-21 Valid Pick Areas

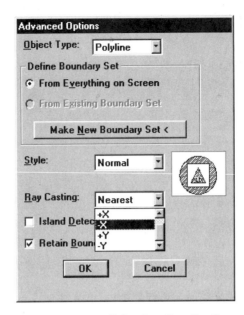

FIGURE 12-22 Selecting Ray Casting

Retain Boundaries. If the hatch produces the desired results and you wish to retain the boundary as a polyline in the drawing, check the "Retain Boundaries" box.

Island Detection. Click the "Island Detection" box if you want BHatch to detect islands (unhatched areas inside a larger area to be hatched).

Using the Hatch Command

The Hatch command is used to place hatches from the command line. The result and principles are the same as a hatch placed with the Bhatch command, but the methodology is somewhat different. The Hatch command places non-associative hatch patterns that do not change when the boundary changes. To use the Hatch command, enter **Hatch** at the command line.

Command: **Hatch**
(? or name/U,style) <default>:

Hatch Command Boundary Definition

Before you can use the Hatch command to place a hatch in your drawing, you must first learn how to identify the hatch boundaries.

A hatch must be bounded by a closed polygon. The entities which create this boundary may be identified by the normal object selection process. There are, however, some tricks which make hatches constructed with the Hatch command work more smoothly.

TUTORIAL

Let's look at a more complex example. Figure 12-23 shows an object that contains three areas that could be hatched. Let's suppose that we wanted to hatch the square and the circle, but not the triangle. Create a quick drawing similar to the one in the illustration.

This is a good example of how hatch styles can be used to control how a hatch pattern behaves. From the "Draw" menu, select "Hatch". In the "Pattern" popup list box, select "ANSI37". Now click on the "Advanced" button. Make sure that "Normal" appears in the "Style" popup list box, and the "Island Detection" box is NOT checked. Click "OK" to return to the previous dialog box.

Next, Select "Pick Points<". Click near the right side inside the box, the triangle, and the circle as shown in Figure 12-23. Click the right mouse button to return to the previous dialog box.

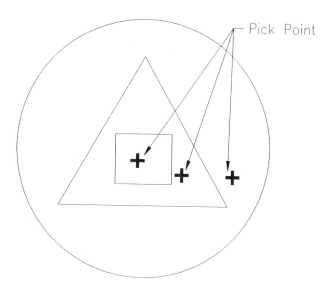

FIGURE 12-23 Selecting Pick Points

Next, select the "Apply" button. Your hatch should look similar to the one in Figure 12-24.

The "Island Detection" option eliminates the need to select multiple points when boundaries are nested inside boundaries. Try turning this option on and only selecting the point within the circle. The result should look like Figure 12-24.

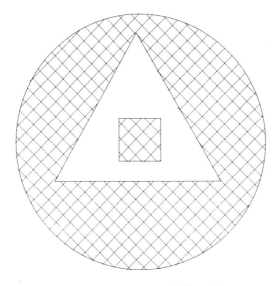

FIGURE 12-24 Hatch Drawing

The entities that bound the area should be perfectly joined at their intersections. If, for example, a line extends beyond the intersection, the hatch may not operate smoothly. In a situation such as that shown in Figure 12-25, the line which runs past the intersection along the full length of the object can cause problems for the hatch.

In order to prevent incorrect hatching, the lines should be broken at the intersection. A good way to do this is to utilize the Break command with the First option to break the entity at the intersection as shown in Figure 12-25.

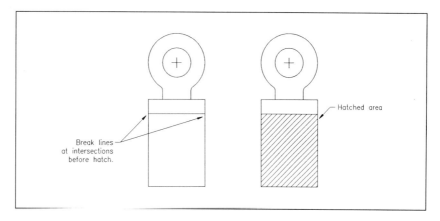

FIGURE 12-25   Intersection Break for Hatch Boundary

Now the entities which define the boundaries properly meet at their endpoints. Before we insert the Hatch, let's review the different styles utilized by the Hatch command. We will do this by creating some hatches with the Hatch command.

Defining Your Own

You may define your own hatches. If you choose the "U" option, AutoCAD will prompt you for the necessary information:

Angle for crosshatch lines <default>:
Spacing between lines <default>:
Double hatch area? <default>:

Simply respond to the prompts with the desired values. You then proceed as normal.

Pattern Alignment

There may be times when you will place hatches adjacent to each other and will want them to line up. AutoCAD compensates for alignment problems by normally using the 0,0 point as the hatch origin for all hatches. This means that the hatches align properly. You can change the origin point by using the variable Snapbase to change the base point.

TUTORIAL

Using the work disk, start the drawing called "HATCH1".

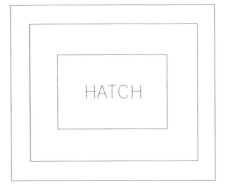

FIGURE 12-26 HATCH1.DWG Work Disk Drawing

We will start by using a Normal hatch style and the hatch pattern called ANSI31.

Command: **Hatch**
Pattern (? or name/U, style) <default>: **ANSI31**
Scale for pattern <default>: **1**
Angle for pattern <default>: **0**
Select objects: **W**

Place the window around the object and press ENTER.

Your drawing should now look similar to the following illustration:

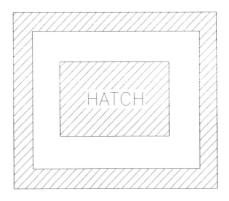

FIGURE 12-27 Normal Hatch Style

CONSTRUCTING SECTIONAL AND PATTERNED DRAWINGS 12-19

Let's now hatch the same drawing using the Outermost style. (Use the Erase-Last command to erase the hatch from your drawing and Redraw, if necessary.)

    Command: **Hatch**
    Pattern (? or name/U, style) <*default*>: **ANSI31,O**
    Scale for pattern:
    Angle for pattern: **0**
    Select objects: **W**

Using the Outermost style, your drawing should look similar to the following illustration:

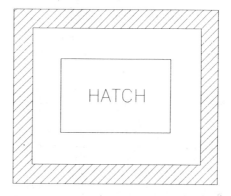

FIGURE 12-28 Outermost Hatch Style

Now, erase the hatch and use the Ignore style.

    Command: **Hatch**
    Pattern (? or name/U, style) <*default*>: **ANSI31,I**
    Scale for pattern:
    Angle for pattern: **0**
    Select objects: **W**

Notice how the hatch has ignored the boundaries and the text.

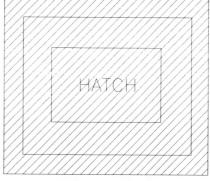

FIGURE 12-29 Ignore Hatch Style

Hatches can be handled more easily if they are put on their own layer. They can also be turned off and frozen to speed redraw time. Be sure that the layer linetype is continuous. Even though the hatch pattern may contain dashed lines and dots, the linetype should be continuous to ensure a proper hatch.

CREATING A POLYLINE BOUNDARY

Since a polyline is a single entity, defining a hatch boundary as a polyline allows you to select a single entity as the boundary. A polyline can be composed of many segments, all functioning as a single entity.

To define an entire boundary as a single polyline, use the Boundary command. When you select Boundary, the *Polyline Creation* dialog box is displayed.

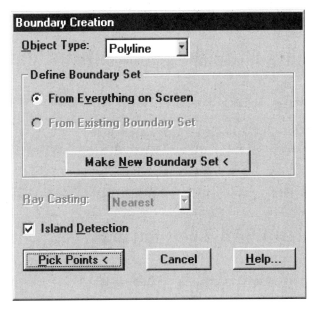 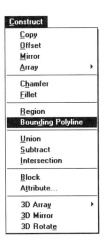

FIGURE 12-30 Boundary Creation

This dialog box is the same as the one displayed under the advanced options selection of the Bhatch command. The boundary is selected in the same manner. The difference is that the hatch is not placed within the boundary; just the boundary is constructed as a polyline.

> **NOTE:** The Boundary command has other options. Converting the outline of an object to a polyline allows you to select all parts of it as a single option when editing.

CREATING SOLID AREAS

The Solid command allows you to fill areas with a solid color. Solids plot as solid ink areas.

> **NOTE:** The solid areas are only displayed if the Fillmode is on.

To use the Solid command, enter:

Command: **Solid**
First point:
Second point:
Third point:
Fourth point:
Third point:
Fourth point:
Third point:
Fourth point:

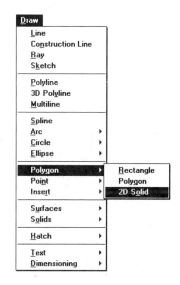

This sequence will continue until you terminate it with [ENTER]. This allows you to create solid polygons with any number of sides.

In some cases, the Solid command won't work properly if the points are not entered in the right sequence. For example, if you enter the points in the fashion shown in the following example, the solid will form a "bow tie".

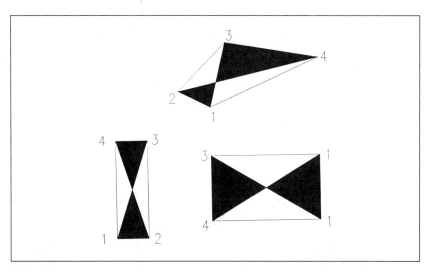

FIGURE 12-31   Bow Tie Sequences for Solid Command

12-22 USING AUTOCAD

The following examples show the sequence to enter points for a correct solid fill. Draw each shape and enter the points as shown.

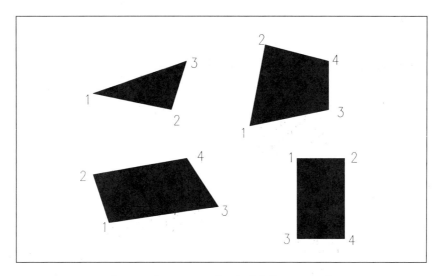

FIGURE 12-32 Proper Sequences for Solid Command

Turn off the fill after using the Solid command to speed up the regeneration process caused by zooms and pans, then turn the fill back on and Regen to redisplay filled solids.

EXERCISES

1. Use the work disk and start the drawing named "SOLIDS". The drawing shows several hot-air balloons. Use the Solid command to place solid areas in some of the balloon areas.

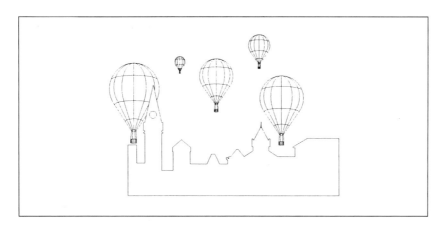

FIGURE 12-33 Solids Exercise

2. Start the drawing on the student work disk named "SOLIDS". This is the same drawing you used in the Solid command exercise. Use the Hatch command to place hatched areas in some of the balloons. Remember to use the Break command to make clean hatch boundaries.

FIGURE 12-34 Hatch Exercise

3. Draw the following items.

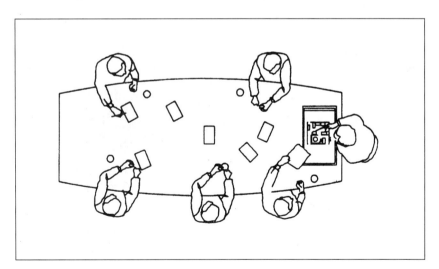

FIGURE 12-35 Use the Sketch Command to Draw the Human Forms and Construct the Drawing

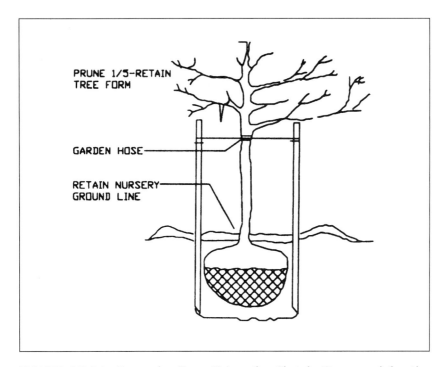

FIGURE 12-36 Draw the Tree, Using the Sketch Command for the Limbs, Then Hatch as Shown

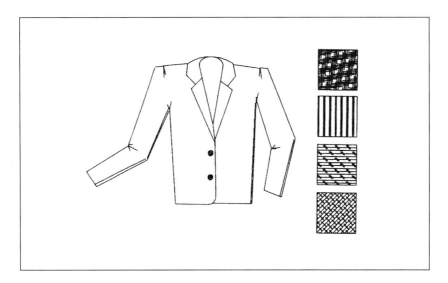

FIGURE 12-37 Draw the Coat and Hatch with a Pattern. Some Suggestions are Shown. Design an Outfit and Hatch to Delineate It

CHAPTER REVIEW

1. Explain what the Fill command does.

   _____
   _____
   _____
   _____

2. If you did not want the solid filled areas in your drawing to plot, what could you do?

   _____
   _____
   _____

3. What is a hatch boundary?

   _____
   _____
   _____
   _____

4. When using the Hatch command, why must a hatch boundary be perfectly constructed?

   _____
   _____
   _____

5. How would you change the origin point of a hatch?

   _____
   _____

CHAPTER 13

TEXT, FONTS, AND STYLES

Drawings are a means of communication that uses both graphics and text. The proper use and placement of text in a drawing is an important aspect of constructing effective drawings. After completing this chapter, you will be able to:

- Manipulate AutoCAD's text capabilities.
- Create text styles that contain fonts and make modifications to those fonts.
- Utilize the methods of placing text in a drawing.

THE USE OF TEXT IN GRAPHIC DRAWINGS

While graphics are used in a drawing to convey information, the total description of an object often requires written words. Written words, or text, are easily placed in a drawing by AutoCAD. The traditional, laborious manner of placing text in a drawing by hand is made much easier with the use of CAD.

Text Standards

The use of text in drawings is usually governed by standards. Many industries use the American National Standards Institute (ANSI) style of letters and numbers. Other companies and offices set standards of their own.

The most important aspect of lettering is that it is clear and concise. The standards should be consistent. In general, note text should be ⅛-inch high. Headings should be ³⁄₁₆-inch high. Note text is typically *left justified*. This means that each line of text is aligned at its left edge.

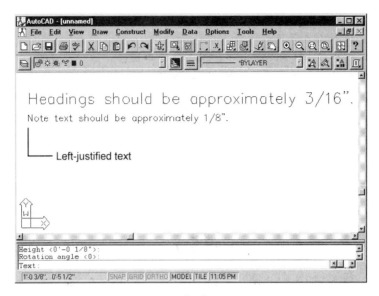

FIGURE 13-1 Text Size Standards

AutoCAD Text Components

AutoCAD provides many text tools for your use. You can use many different styles, create text in any size, slant, compress, rotate, and even tell AutoCAD how long any text line length should be.

Text can be a variety of different designs. These designs are called *fonts*. A font is the "design" of the text letter. Gothic, Swiss, script, and ISO are examples of fonts. See Figure 13-2.

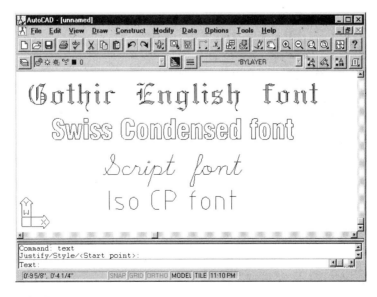

FIGURE 13-2 Text Fonts

TEXT, FONTS, AND STYLES   **13-3**

Each font can be stretched, compressed, obliqued, mirrored, or drawn in a vertical stack. The text font, the text height, and its modifications are saved and stored as a *style*. The text style, including the font and style modifications, are created in AutoCAD prior to placing the text. Once created, a style file can be used as many times in the same drawing as you wish. AutoCAD provides many text fonts for your use. Figure 13-3 shows some of the 85 fonts included with AutoCAD Release 13.

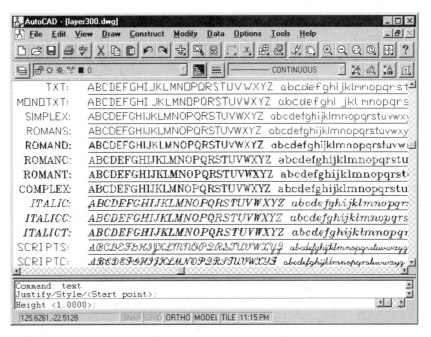

FIGURE 13-3 Some Fonts Supplied with Release 13 *(for complete font list see page 13-37)*

THE TEXT COMMAND AND ITS OPTIONS

When you enter the Text command, you see the following prompt:

Command: **Text**
Justify/Style/<Start point>:

The "start point" is the default. You used this in the previous exercise. Let's continue and look at the text options of Justify and Style.

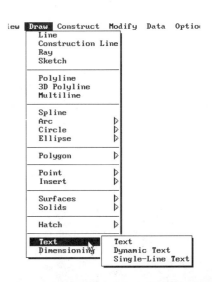

Placing Justified Text

The Justify option is used to specify how the text will be aligned. If you respond to the prompt with a **J** (Justify), AutoCAD will respond with a prompt for the alignment options.

Command: **Text**
Justify/Style/<Start point>: **J**
Align/Fit/Center/Middle/Right/TL/TC/TR/ML/MC/MR/BL/BC/BR:

Let's first look at the two letter abbreviated options. Don't worry about learning each of the abbreviated options. Let's stop and look at how text alignment options work and you will see that it's easy to remember.

Figure 13-4 shows the alignment modes for text. Notice that you can align text by both the vertical and horizontal alignments.

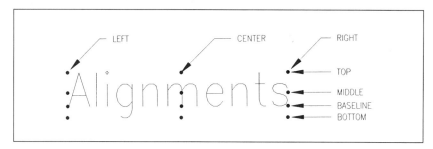

FIGURE 13-4 Text Alignments

Let's use single letter abbreviations to designate each alignment. Refer to Table 13-1.

ALIGNMENT	ABBREVIATION
Top	T
Middle	M
Bottom	B
Left	L
Center	C
Right	R

TABLE 13-1

You will notice in the last command prompt, there was a series of two letter options. To designate an alignment, simply use two letters; one to describe the vertical alignment, and one to designate the horizontal alignment.

If you know the type of alignment you desire, it is not necessary to enter the "J" option; just enter the two letters that describe the alignment you desire.

TUTORIAL

Let's place some simple text in a drawing. Start AutoCAD and start a new drawing named "TEXT". Next, type the Text command at the command prompt.

We will use the default style that is already contained in the drawing. When you enter the Text command, you will see the command prompt:

    Command: **Text**
    Justify/Style/<Start point>: (*pick*)

At this point, move your cursor into the drawing area and place a point on the screen. The command prompt will continue and AutoCAD will ask you for the text height:

    Height <*default*>: (*pick two points*)

Let's just "show" AutoCAD how high the text should be. Move the crosshair up from the point you entered. The point will "rubber band" to the crosshair. Enter a point approximately ¼ inch (actual distance) above the first point you entered. AutoCAD will continue and prompt you for the:

    Rotation angle <*default*>: **0**

Enter zero (0) in response to the prompt. Now it's time to type the text. Type your name. If you make a mistake, use the backspace key to erase back and start again. Notice how the text shows up on the command line, but not on the screen. When you are finished typing your name, press [ENTER]. You should now see your name on the screen!

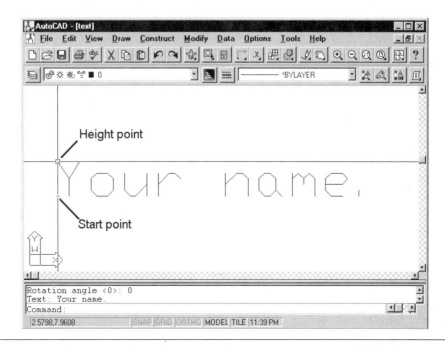

Additional Alignment Options

Several alignment modes listed under the Justify option are shown as full words. They are Align, Fit, Center, Middle, and Right. Let's take a look at each type of alignment.

Placing Aligned Text. Align allows you to select two points that the text will fit between. AutoCAD will adjust the text height so the baseline of the text fits perfectly between the two points. Note that the two points may be placed at any angle in relation to each other.

Command: **Text**
Justify/Style/<Start point>: **A**
First text line point:
Second text line point:
Text:

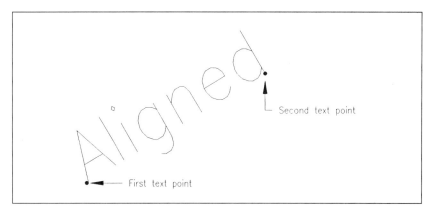

FIGURE 13-5 Aligned Text

Placing Centered Text. The Center option is used to center the baseline of the text on the specified point.

Command: **Text**
Justify/Style/<Start point>: **C**
Center point:
Height <*default*>:
Rotation angle <*default*>:
Text:

FIGURE 13-6 Centered Text

TEXT, FONTS, AND STYLES 13-7

Fitting Text in a Specified Distance. The Fit option is similar to the Aligned option. You are prompted for two points to place the text between. With Fit, however, you are prompted for a text height. AutoCAD calculates only the text width and adjusts the width to fit perfectly between the two entered points.

Command: **Text**
Justify/Style/<Start point>: **F**
First text line point:
Second text line point:
Height <*default*>:
Text:

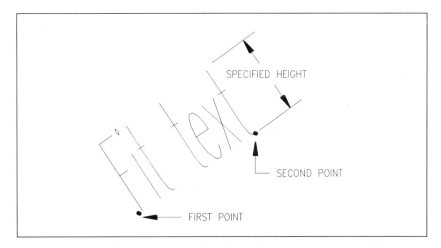

FIGURE 13-7 Fit Text Option

Placing Middle Aligned Text. Middle is used as a shortcut when centering text both horizontally and vertically. It achieves the same result as the MC option.

Command: **Text**
Justify/Style/<Start point>: **M**
Middle point:
Rotation angle <*default*>:
Text:

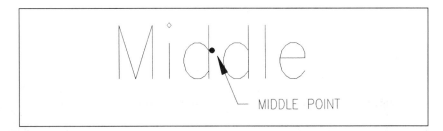

FIGURE 13-8 Middle Text Alignment

Placing Right Aligned Text. Using right justified text is similar to using "Start point", except the text *ends* at the reference point instead of beginning at it.

Command: **Text**
Justify/Style/<Start point>: **R**
End point:
Height <*default*>:
Rotation angle <*default*>:
Text:

FIGURE 13-9 Right-Justified Text

Selecting Different Text Styles

The Style option is used to change between defined text styles. Note that you must first create a text style before you can use it. After the style is changed, the text command is repeated. Refer to the following command sequence that changes from the default style named "Standard" to a user-built style named "Bigtxt".

Command: **Text**
Justify/Style/<Start point>: **S**
Style name (or ?) <Standard>: **BIGTXT**
Justify/Style/<Start point>:

Note that the Style option under the Text command and the Style command are not the same. The style option selects a style, while the style command creates the style.

Rotating Text

Text can be placed at any angle in your drawing. As you may have noticed, the previous command sequences included the prompt line:

Rotation angle <*default*>:

TEXT, FONTS, AND STYLES **13-9**

You can specify the angle at which the text is drawn by designating the angle at this point in any text prompt. Note that the zero-angle direction set with the Units command affects which direction is zero. Unless changed, the default zero-angle in AutoCAD is to the right. Thus, a text angle of zero would place text that runs from the left to the right. Figure 13-10 shows text placed at several angles.

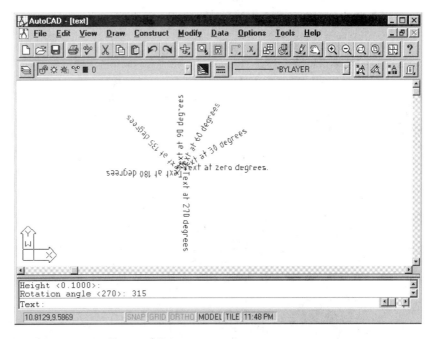

FIGURE 13-10 Rotated Text

Once the text angle is set, that angle remains the default angle until changed again.

PLACING MULTIPLE TEXT LINES

Many times you may wish to place several lines of text.

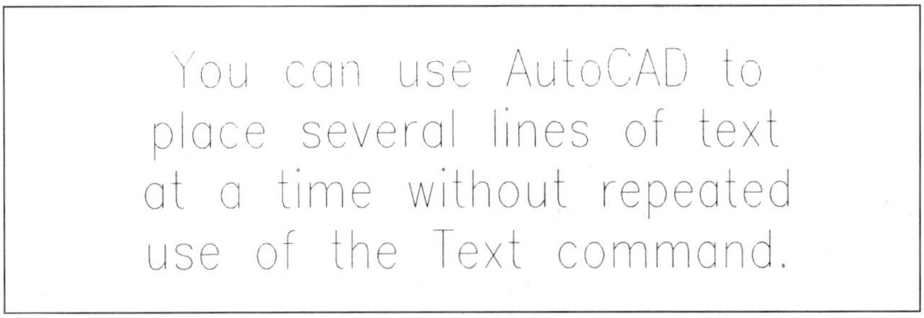

FIGURE 13-11 Placing Multiple Text Lines

To draw multiple text lines, place the first line of text, using the size and alignment you want for all the lines. Next, repeat the Text command. When AutoCAD prompts:

Justify/Style/<Start point>:

reply to the prompt by pressing [ENTER]. AutoCAD will step down one line space and prompt for the text. The next line of text will be sized and aligned the same as the first line.

DRAWING TEXT DYNAMICALLY

Text can be placed into your drawings "dynamically". As you have already noticed, the text you enter is not displayed in your drawing until you press [ENTER] after typing the string on the command line. Using dynamic text allows you to see the text in the drawing as you enter it. You can also place several strings of text at different positions in the drawing without exiting and reentering the text command again.

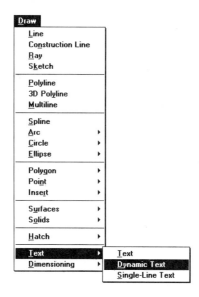

Dynamic text is placed with the Dtext command.

Command: **Dtext**
Justify/Style/<Start point>:
Height <*default*>:

Notice how the command prompts are the same as the regular text command.

Placing Dynamic Text

When you type the first string of dynamic text, you will notice that the text is displayed in the drawing as you are typing it. After you have placed a text string, you move the cursor to a new location, click, and start a new text string. To exit the Dtext operation, press [ENTER] twice.

When you use dynamic text, you will notice some differences.

- As you place dynamic text, a text box is displayed at the text location. The size of the text box represents the height and width of the current text style.
- When you first enter Dtext, the last text string entered will be highlighted. If you press [ENTER] instead of entering the text position with the cursor, the text box will be placed on the next line below the last text as though you had not previously exited Dtext. This is useful if you wish to go back and continue placing multiple text lines.
- When placing multiple lines of text, you can use the backspace key to backspace through all the text placed in that operation (even on previous lines).
- If you use text alignment codes other than "starting point", the text is not aligned properly until you finish the command.
- If you cancel the Dtext command at any time during the operation, all text placed during that operation will be canceled, not just the current line.
- All menu and tablet commands and options are "locked out". Only keyboard entry is permitted when placing dynamic text.
- When you use special text codes, the text codes are initially displayed instead of the effect. For example, if you enter:

    %%UTHIS IS UNDERLINED%%U

    the text codes will be shown in the drawing as you are entering the text. When the Dtext command is completed, the text will be redrawn showing the effect.

    THIS TEXT IS UNDERLINED

CREATING PARAGRAPH TEXT — THE MTEXT COMMAND

The Text and Dtext commands create lines of text. The Mtext command, in contrast, creates paragraphs of text. It fits the text into an invisible boundary (defined by the user) that is not printed or plotted. (Mtext is short for multiline text.)

To write the text, you need to use a text editor; by default, AutoCAD uses its internal editor. However, you can specify another text editor with the system variable MTextEd.

The Mtext command has the following options:

Command: **mtext**
Attach/Rotation/Style/Height/Direction/<Insertion point>: *(pick)*
Attach/Rotation/Style/Height/Direction/Width/2Points/<Other corner>: *(pick)*
   (Text editor appears.)

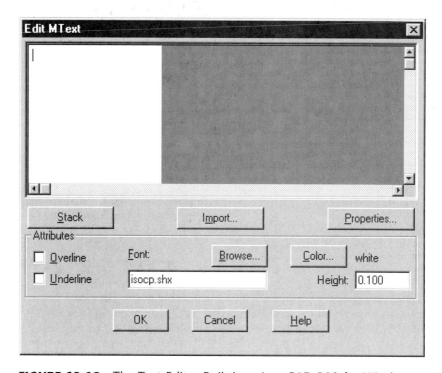

FIGURE 13-12   The Text Editor Built into AutoCAD R13 for Windows

The default option prompts you to pick two points on the screen. The two points define a rectangle that will contain the paragraph text. Then AutoCAD loads the text editor. Enter the text, then click OK. AutoCAD places the text in the rectangle.

The other options allow you to control the boundary box, as follows:

Command: **Mtext**
Attach/Rotation/Style/Height/Direction/<Insertion point>: **a**
TL/TC/TR/ML/MC/MR/BL/BC/BR: **tl**

The Attach option determines how the paragraph text is aligned to the boundary box. You probably recognize the justification options from the Text and DText commands. For example, the TL (short for top left) option attaches the text to the top left of the bounding box.

Attach/Rotation/Style/Height/Direction/<Insertion point>: **r**
Rotation angle <0>: **45**

The Rotation option determines the angle of the text boundary.

Attach/Rotation/Style/Height/Direction/<Insertion point>: **s**
Style name (or ?) <STANDARD>: **?**
Text style(s) to list <*>:
Text styles:
Style name: STANDARD   Font files: txt
  Height: 0.0000   Width factor: 1.0000   Obliquing angle: 0
  Generation: Normal
Current text style: STANDARD
Style name (or ?) <STANDARD>:

The Style option lets you select a predefined text style for the paragraph text.

Attach/Rotation/Style/Height/Direction/<Insertion point>: **h**
Height <0.2000>: **.3**

The Height option lets you specify the height of uppercase characters.

Attach/Rotation/Style/Height/Direction/<Insertion point>: **d**
Horizontal/Vertical: **v**

The Direction option draws the text horizontally or vertically.

Using the MText Dialog Box

The Edit MText dialog box is unique to the Windows versions of AutoCAD Release 13; it is not available in the DOS version. The dialog box lets you import text, set the style of the text, and stack fractions.

The Stack button lets you "stack" fractions. Instead of using the side-by-side format, like 11/32 inch, MText places the 11 over the 32 (see Figure 13-13).

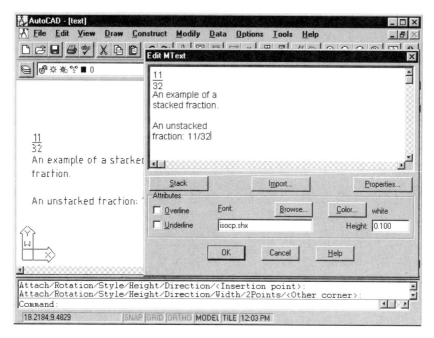

FIGURE 13-13 Example of Stacked and Unstacked Fractions

To create a stacked fraction, type the numbers you want stacked, such as "11/32". Use the cursor to highlight the five characters, then click on the Stack button. AutoCAD stacks the numerator over the denominator.

The Import button lets you bring text into the drawing from a file stored on disk. The file must be in plain ASCII format (also known as DOS text or Text document); you cannot import text saved in a word processing format, such as Write WRI, Word DOC, or WordPerfect files. MEdit's Import is limited to files 4KB in size.

To import a text file, click on the Import button. AutoCAD displays the Import Text File dialog box. Select the name of the file to import, then click on the Open button. AutoCAD places the text in the MText boundary. See Figure 13-14.

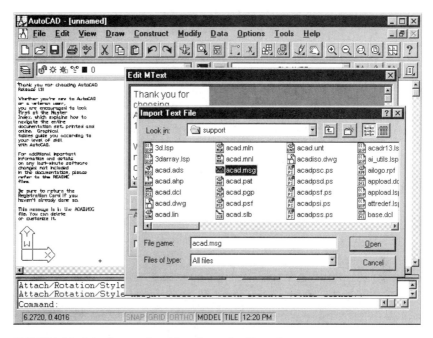

FIGURE 13-14 Importing Text into the Drawing

The Properties button lets you specify some of the properties of the text and the bounding box. For the text itself, you can select the text style, height, and direction. To select a text style, you have to use the DdStyle command to create the styles before using the MText command. For Direction, you have just two options: (1) left to right and (2) top to bottom for Chinese and Japanese text.

The Object part of the MText Properties dialog box lets you set some parameters for the bounding box, such as where the attachment point is, the width, and rotation angle. For example, if you set the rotation angle to 17 degrees, the entire block of text is rotated by that amount. Changing the width causes AutoCAD to reflow the text.

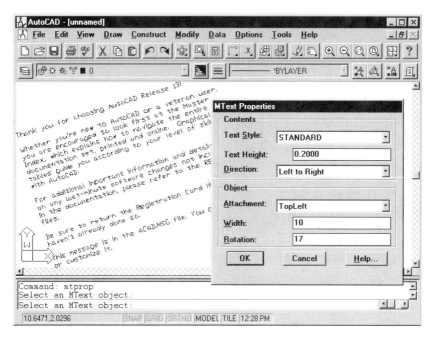

FIGURE 13-15 Setting Properties of MText

The Attributes area of the Edit MText dialog box lets you select a text font, specify the color of the text, text height, and whether the text is underlined or overlined.

> **NOTE:** Use the Leader command to place a leader and annotation in the drawing. Use the \P code to create line breaks in the text. When done, the text is an MText object.

CHANGING MTEXT

AutoCAD provides several ways to change paragraph text. You can use the MtProp, the DdModify, or the DdEdit commands. Each command has its strengths and weaknesses:

MtProp Command lets you change the properties of paragraph text as defined by the MText command, as follows:

Command: **Mtprop**
Select and MText object: *(Select.)*
Select and MText object: *(Press ESC.)*

The MtProp command displays the "MText Properties" dialog box, as shown below.

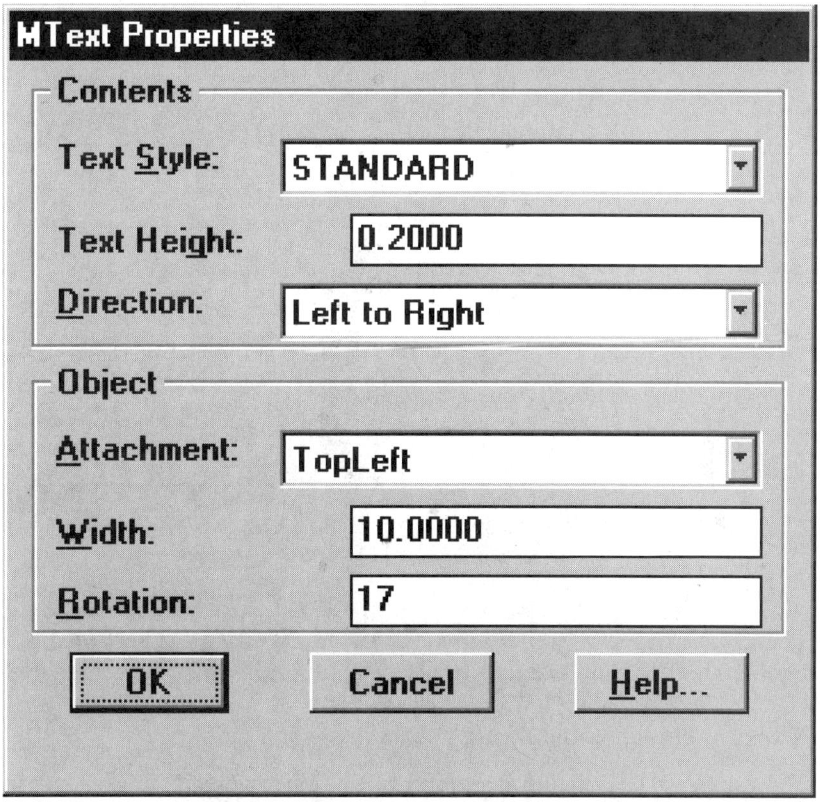

FIGURE 13-16 MText Properties Dialog Box

The dialog box lets you change the style, text height, and direction of the text (called "Contents" by dialog box), as well as the attachment, width, and rotation of the bounding box (called "Object" by the dialog box). These options have the same meaning as for the MText command.

After clicking on the OK button, the MtProp command prompts you to select another MText object. To cancel the command, press [ESC].

DdModify Command is the most powerful way to edit paragraph text, as follows:

    Command: **ddmodify**
    Select object to modify: *(Select.)*

After you select the Mtext object, the DdModify command displays the "Modify MText" dialog box, as shown in Figure 13-17. (Note that the DdModify command displays a different dialog box, depending on the object you select.)

FIGURE 13-17 Modify MText Dialog Box

The dialog box lets you change the color, layer, and insertion point of the MText. Because you cannot assign a linetype, change the linetype scale, or change the thickness of MText, those buttons are grayed out. You can change the elevation of the MText by changing the Z value of the insertion point.

The Contents box displays the text of the MText object. (Note that the \P characters indicate line breaks.) You cannot change the text in the Contents box; instead, click on the Edit Contents button, which causes AutoCAD to load the Edit MText dialog box.

When you click on the Edit Properties button, AutoCAD brings up the *MText Properties* dialog box (see page 13–17). When you click on the OK button, you are returned to the *Modify MText* dialog box.

DdEdit command loads the Edit MText dialog box, allowing you to make changes to the text, as follows:

Command: **ddedit**
Select a TEXT or ATTDEF object/Undo: *(select)*

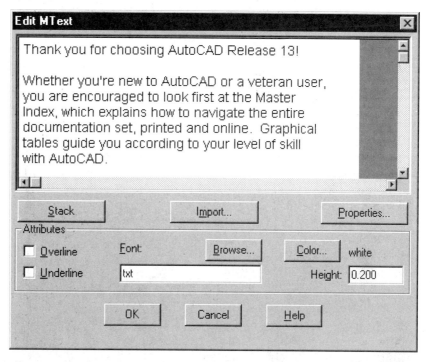

FIGURE 13-18 The DdEdit Command Uses the Text Editor when You Select an MText Object

Make your changes, and click the OK button. AutoCAD returns on the DdEdit command:

Select a TEXT or ATTDEF object/Undo: *(Press ESC.)*

Press ESC to exit the DdEdit command.

TEXT STYLES

A style is a collection of modifiers that are applied to the font of your choice.

For instance, you may want your text to be slanted, or perhaps expanded so that it is longer than normal.

The style that you choose is stored under a name of your choice and can be applied to the font of your choice.

A text style is composed of the following information:

1. A style name. This name may be up to 31 characters in length.
2. A font file that is associated with the style. (You must have some pattern to modify.)
3. A fixed height. This height will be the text height. If you want to specify the height each time you place text, enter a **0** for this value.
4. A width factor. This factor is a numerical representation of the width. A width of 1 is the standard width factor. A decimal value creates text that is "narrower". For example, a width factor of .5 produces text one half the width of text created with a factor of 1.
5. An obliquing angle. This angle determines the slant. A positive angle is a forward slant; a negative angle is a backward slant.
6. A draw backwards indicator.
7. A draw upside down indicator.
8. An orientation indicator to determine whether the text will be vertical (stacked), or horizontal.

After you have created a style, you can refer to it whenever you want that particular set of parameters. You do not have to go through a series of questions each time you want to define the "look" of your text!

Creating a Text Style

The Style command is used to create a style file. Note that the Style command is not the same as the Style option under the Text command. The Style command is used to *create* a text style while the Style option under the Text command is used to select a previously created text style for use.

To create a text style, type the Style command. The following is displayed.

    Text style name (or?) <standard>:
    Font file <*default*>:
    Height <*default*>:
    Width factor <*default*>:
    Obliquing angle <*default*>:
    Backwards? <Y/N>:
    Upside-down? <Y/N>:
    Vertical? <Y/N>:

Let's look at each of the settings.

Text Font File. There are many font files available for AutoCAD. The program comes with an excellent selection. Specialty fonts for map making (draws symbols instead of letters), cursive writing, and other applications are included with AutoCAD. Fonts with an SHX extension are AutoCAD's own vector format. Fonts with the PFA and PFB extension are PostScript fonts, while fonts with the TTF extension are TrueType fonts.

The font file should be appropriate for the application. Most mechanical applications use the ANSI type lettering. The Romans (roman, single stroke) font closely approximates this style. Architectural drawings typically use "hand-lettered" fonts. Engineering applications often use the Romans font since it closely resembles a traditional font constructed with a lettering template.

Text Height. The text height represents the height, in scale units, that the text will be drawn. For example, if the drawing is plotted at ⅛"=1' 0" and the text height is 12, the plotted text will be ⅛" high on the paper. Because of this, you must first determine the scale of the drawing (see Chapter 8 on setting up a drawing) before you can set the final plotted text height.

> **NOTE:** A standard architect's or engineer's drawing board scale can be used to "see" the size of the plotted text. Place the scale on a piece of paper, using the numerical scale that represents the final plotted scale and mark the height of the text.

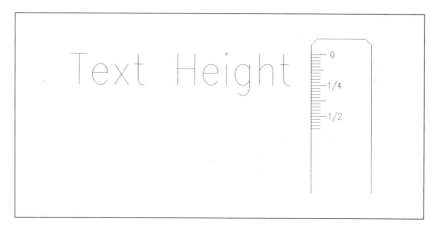

FIGURE 13-19 Measuring Scaled Text

Text Width Factor. The width factor determines how wide the text will be drawn. A width factor of one can be thought of as a "standard" width. A decimal value will draw text that is narrower, resulting in condensed text. Values greater than one will create expanded text. Figure 13-20 illustrates the same text font with different scale factors applied.

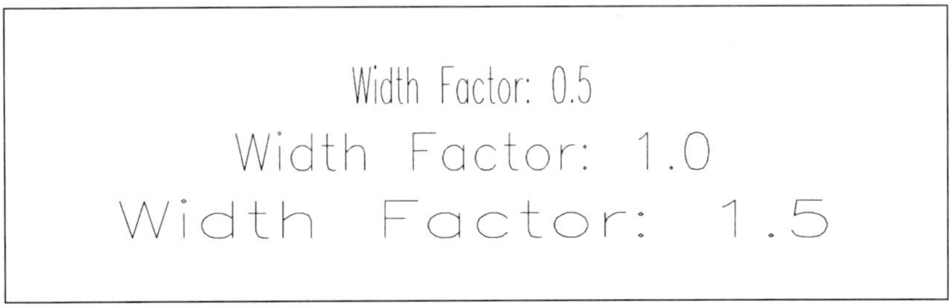

```
Width Factor: 0.5
Width Factor: 1.0
Width Factor: 1.5
```

FIGURE 13-20   Text Width Factors

Text Obliquing Angle.   A slant can be applied to a font with the obliquing angle setting. A zero obliquing angle draws text that is "straight up". Positive angles draw text that slants forward, while negative angles draw backward slants.

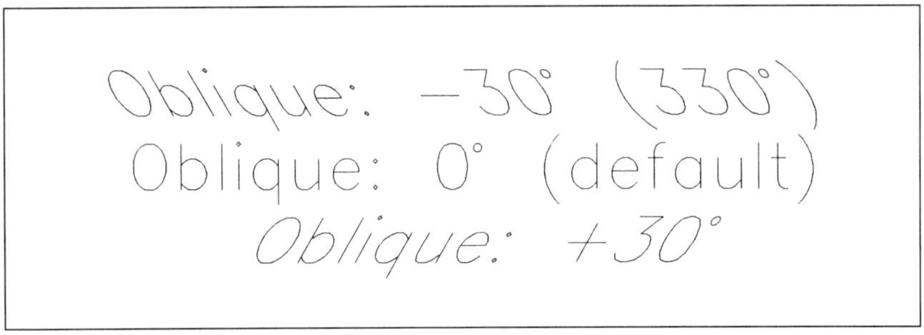

```
Oblique: −30° (330°)
Oblique: 0° (default)
Oblique: +30°
```

FIGURE 13-21   Text Obliquing Angles

Backwards Text.   Text can also be drawn backwards. Backwards text is useful if you want to plot your drawing on the back side of film media.

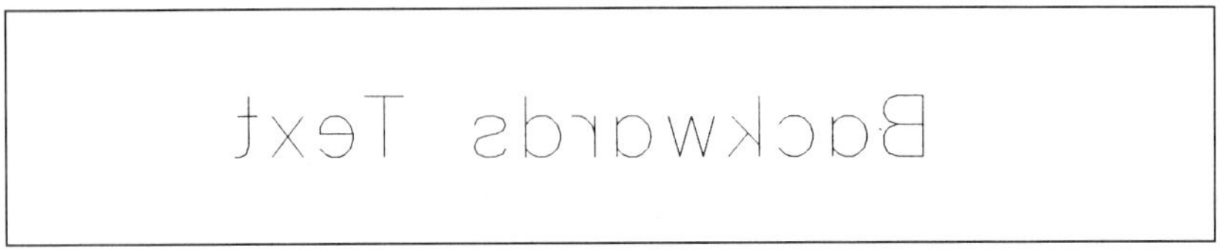

```
Backwards Text
```
(shown mirrored)

FIGURE 13-22   Backwards Text

Upside-Down Text.   If you want the text drawn upside down, reply **YES** to the "Upside down:" prompt. Setting the value to NO will draw normal, right side up text.

FIGURE 13-23 Upside-Down Text

Vertical Text. Your text can be oriented so it draws vertically. This is not the same as text that is *rotated* 90 degrees. Vertically oriented text is drawn so that each letter within the text string is vertical and the text string itself is vertical. PFA, PFB, and TTF fonts cannot be drawn vertically.

> ***NOTE:*** Obliquing, underscoring, and overscoring should not be used with vertical text, since the result will not be correct.

FIGURE 13-24 Vertical Line

Text Styles by Dialog Box

The c4 version of Release 13 adds a new command to AutoCAD: DdStyle. This is the dialog box version of the Style command and is not available in earlier versions of Release 13. The dialog box lets you create new styles and apply effects, just like the Style command.

The DdStyle command brings up the Text Style dialog box shown in Figure 13-25.

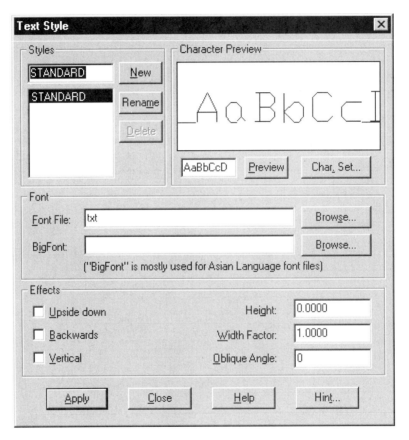

FIGURE 13-25   Text Style Dialog Box

To create a new text style:
1. Click on the New button.
2. AutoCAD automatically gives the style the name of STYLE1.
3. To change the name, type a different name and click on the Rename button.
4. To select the font file, click on the Browse button next to Font File.
5. AutoCAD displays the Select Font File dialog box. Font files are stored in the \Acad13\Com\Fonts subdirectory. You may have to navigate to this subdirectory to find the font files, which end in SHX (AutoCAD fonts), PFB (PostScript fonts), or TTF (TrueType fonts).

TEXT, FONTS, AND STYLES **13-25**

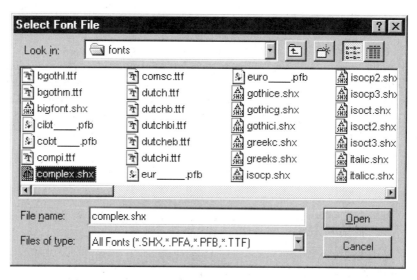

FIGURE 13-26 Select Font File Dialog Box

6. Select a font file, then click on the OK button.
7. AutoCAD loads the file into the drawing and returns to the Text Style dialog box.
8. If you wish, apply effects to the font, such as Upsidedown and Obliquing Angle. As you select effect options, the effect is immediately displayed in the Character Preview window.
9. To see all 255 characters in the font file, click on the Char.Set button. AutoCAD shows you all the characters; the ? means the character is not defined and will display as a question mark in the drawing.

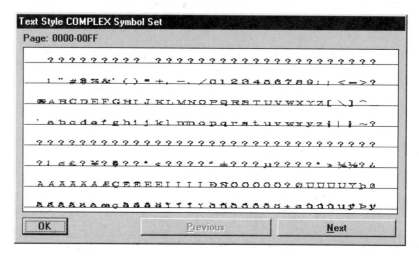

FIGURE 13-27 Text Style Character Set Dialog Box

10. To apply the changes in style to the current text style, click on the Apply button.
11. Click on the Close button to dismiss the dialog box.

Using Text Styles

The concept of fonts and styles is sometimes confusing to learn in the beginning. Remember that a style is a modification that is applied to a particular font (such as height, width factor, and slant factor). You must first create a style in order to use a font. The first step is to choose the particular font that you would like to use in your drawing, then determine the modifications you would like to apply to it by using the Style command parameters.

You may have different styles applied to the same font. Just store them in a different style file with the name of your choice.

> **NOTE:** "Fancy" styles (those created with multistroke fonts) are slow to regenerate and redraw. To speed your drawing operations, you can avoid such fonts, place the text in your drawing last, or use the Qtext command (covered later in this chapter) to make redraws and regenerations faster.
>
> Multistroke fonts are identified as "duplex" or "complex". AutoCAD names fonts such as "Roman C" with the "D" and "C" denoting *duplex* and *complex*.

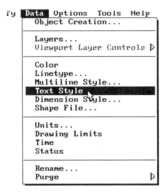

SPECIAL TEXT CONSIDERATIONS

AutoCAD allows for special text operations. You can use text "codes" to create text that is underlined, place degree and plus/minus symbols, and other special text notations. For the text and Dtext commands, this is accomplished by using a code consisting of a pair of percent characters (%%). For the Mtext command, the code consists of a backslash character (\).

Text and Dtext Codes

The following table shows the notations and their functions:

%%o	Toggle overscore mode on and off
%%u	Toggle underscore mode on and off
%%d	"Degrees" symbol
%%p	"Plus/minus" tolerance symbol
%%c	"Circle diameter" dimensioning symbol
%%%	"Percent" sign (think about it)
%%xxx	Draw a special character designated by "xxx"

Let's look at an example. The following text string:

> If the piece is fired at 400%%d F for %%utwenty%%u hours, it will achieve %%p95%%% strength.

would be drawn as:

> If the piece is fired at 400°F for twenty hours, it will achieve ±95% strength.

Notice that the underscore must be toggled on and off by typing %%U to start the underscore, and again to end the underscoring.

You may also "overlap" the symbols by using, for example, the degrees symbol between the underscore symbols.

The special characters refer to the ASCII (American Standard Code for Information Interchange) character set. This character set uses a number code for different symbols. Table 13-2 shows the ASCII character set. To use one of the characters, enter two percent signs (%%), followed by the ASCII character code. For example, to place a tilde (~) in your text, enter %%126.

MTEXT CODE CHARACTERS

You must use a set of text codes in paragraph text created with the MText command, codes that are different from the codes used for the Text and Dtext commands. Using these codes allows you to format text with a text editor or word processor external to AutoCAD. (You may even be able to write a set of macros that convert your word processor's codes into those recognized by AutoCAD.)

MText codes consist of a backslash symbol, followed by a character. (MText is case-sensitive, which means you must use an uppercase P and not a lowercase p in \P.) Table 13-3 shows the codes and their meanings.

32		space	64	@		96	'	left apost.
33	!		65	A		97	a	
34	"	doublequote	66	B		98	b	
35	#		67	C		99	c	
36	$		68	D		100	d	
37	%		69	E		101	e	
38	&		70	F		102	f	
39	'	apostrophe	71	G		103	g	
40	(		72	H		104	h	
41	)		73	I		105	i	
42	*		74	J		106	j	
43	+		75	K		107	k	
44	,	comma	76	L		108	l	
45	-	hyphen	77	M		109	m	
46	.	period	78	N		110	n	
47	/		79	O		111	o	
48	0		80	P		112	p	
49	1		81	Q		113	q	
50	2		82	R		114	r	
51	3		83	S		115	s	
52	4		84	T		116	t	
53	5		85	U		117	u	
54	6		86	V		118	v	
55	7		87	W		119	w	
56	8		88	X		120	x	
57	9		89	Y		121	y	
58	:	colon	90	Z		122	z	
59	;	semicolon	91	[		123	{	
60	<		92	\	backslash	124	|	vert. bar
61	=		93	]		125	}	
62	>		94	^	caret	126	~	tilde
63	?		95	_	underscore			

FIGURE 13-2 ASCII Character Set. *Courtesy of Autodesk, Inc.*

Code	Meaning
\P	New line
\S	Stacked fractions
^	Removes the previous character
\L	Start underline
\l	Stop underline
\O	Start overscore
\o	Stop overscore
\W	Width of the font
\Q	Obliquing
\F	Change font
\H	Height of font
\An	Alignment value
	\A0; = bottom alignment
	\A1; = center alignment
	\A2; = top alignment
\Tn	Tracking (spacing between characters)
	\T3; = results in wide spacing
\~	Prevent breaks (forces words to stay together)
\Cn	Color change
	\C1; red
	\C2; yellow
	\C3; green
	\C4; cyan
	\C5; blue
	\C6; magenta
	\C7; white
{	Opening brace to separate a function so that the entire text isn't affected by a change.
\}	Closing brace, used with \{
\\	Literal backslash (used when you need to include a backslash in the text)

TABLE 13-3

Let's look at some examples. Normally, MText automatically wraps text to fit a rectangle. If you want to force a line break in the text, you use the \P code. For example, AutoCAD reads the following line:

This is the first line of text\P and the second line.

and places it in the drawing as:

This is the first line of text
and the next.

The opposite is to force text to remain together, which you do as follows:

This\~is\~text\~that\~is\~forced\~to\~remain\~together.

To underline text, use the \L and \l codes, as follows:

This is \Lunderlined\l text.

Use the curly brackets, { and }, to specify a change in text for a specific part. For example, to write the following warning in white (color #7) and red (color #1) text:

{\C7;Do not} {\C1;change} {\C7;this drawing.}

Notice how the semicolon is required when the MText code consists of more than a single character:

\P   does not require the semicolon
\C7;   requires the semicolon

REDRAWING AND REGENERATING TEXT FASTER

Text strings redraw, regenerate, and plot slowly. As your drawings become more complex, the time required to handle text can slow down the drawing process. AutoCAD provides the Qtext (quick text) command to speed handling of text.

Qtext can be turned on or off by use of the Qtext command, followed by the Regen command.

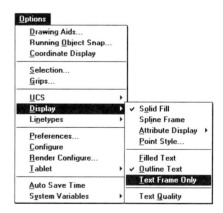

    Command: **Qtext**
    ON/OFF <*default*>:
    Command: **Regen**

Qtext uses rectangular boxes to represent the height and length of the text.

TEXT, FONTS, AND STYLES 13-31

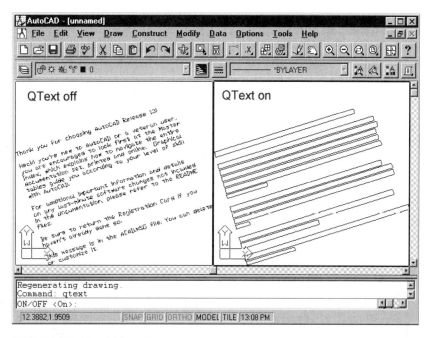

FIGURE 13-28 Using Qtext

The text boxes handle more quickly when you perform redraws, regens, and plots. The change to text boxes will not take place until the next regeneration.

Effects of Qtext on Plotting

The drawing will plot either text or quick text boxes. The display setting at the time of the plot determines the type of plot.

If you wish to plot the actual text, turn off the quick text, then regenerate the drawing with the Regen command before plotting.

Setting Qtext with a Dialog Box

You can use the *Drawing Aids* dialog box to set quick text. This dialog box is accessed by entering Ddrmodes at the command line or by selecting the "Options/Drawing Modes..." pull-down menu.

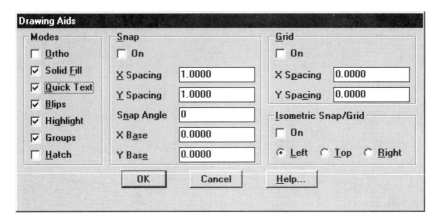

FIGURE 13-29   Drawing Aids Dialog Box

Click on the "Quick text" check box in the "Modes" section of the dialogue box. An "X" in the box means the Quick text option is on.

> **NOTE:**   When a drawing contains a large amount of text, several techniques can be used to minimize the time required to display and plot the text.
>
> 1. Place the text in the drawing last. This eliminates having to regenerate text while you are constructing the drawing elements.
>
> 2. Create a special layer for text. Freeze the layer when you are not using the text (see Chapter 14).
>
> 3. Use Qtext if you want the text placements shown, but still need fast regenerations.
>
> 4. Plot check drawings with the Qtext option on. The text will plot as rectangles, greatly reducing plot time.
>
> 5. Whenever possible, avoid multistroke, "fancy" fonts. These fonts regenerate and plot more slowly than single stroke fonts. If you wish to use fancy fonts, the quick text option reduces the regeneration time.

EXERCISES

1. We want to place the text "PART A" so that the text will be centered both vertically and horizontally on a selected point. We might choose this method of placement if we wanted to place text in the center of a space. Let's use the text command to place the text. Refer to the following command sequence and Figure 13-30.

   Command: **Text**
   Justify/Style/<Start point>: **MC**
   Middle point: *(Select point "A".)*
   Height <default>: *(Move up to show height.)*
   Rotation Angle <default>: **0**
   Text: **PART A**

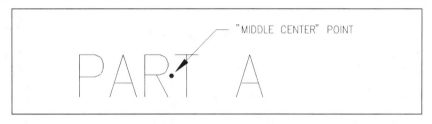

FIGURE 13-30   Middle-Center Alignment

Notice how we simply entered "MC" instead of selecting the Justify option. It is not necessary to select Justify unless you want to see the options. We entered "MC" because we wanted to place the text vertically in the "Middle" and horizontally in the "Center" (see Figure 13-4 and Table 13-1).

For the height, we moved up and entered another point to "show" AutoCAD the height. We could have entered a value. The height of text is in *scale* units, not actual size.

2. Figure 13-31 shows several text strings. The placement point is marked with a solid dot. Use the Text command to place the text as shown. If you need, refer back to Figure 13-4 and Table 13-1 to identify the alignment mode abbreviations.

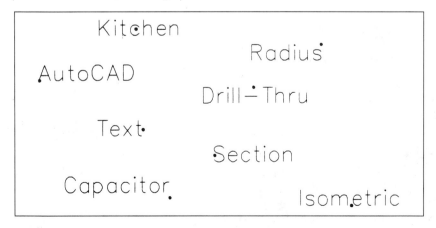

FIGURE 13-31   Text Alignment Exercise

3. Using the text angle prompt, place the following line of text at angles of 0, 45, 90, 135, and 180 degrees.

   THIS IS ANGLED TEXT

   Use the same text starting point for each line.

4. Let's place some multiple lines of text. We will use the following lines of text:

   THIS IS THE FIRST LINE OF TEXT
   THIS IS THE SECOND & CENTERED LINE OF TEXT

   Let's look at the command sequence used to do this.

   Command: **Text**
   Justify/Style/<Start point>: **C**
   Center point: *(Select point "1".)*
   Height <default>: *(Move up and place a point to define the height.)*
   Rotation angle <default>: **0**
   Text: **THIS IS THE FIRST LINE OF TEXT**

   Command: (ENTER *To repeat the text command.*)

   TEXT Justify/Style/<Start point>: (ENTER)
   Text: **THIS IS THE SECOND & CENTERED LINE OF TEXT**

   ```
   THIS IS THE FIRST LINE OF TEXT
   THIS IS THE SECOND & CENTERED LINE OF TEXT
   ```

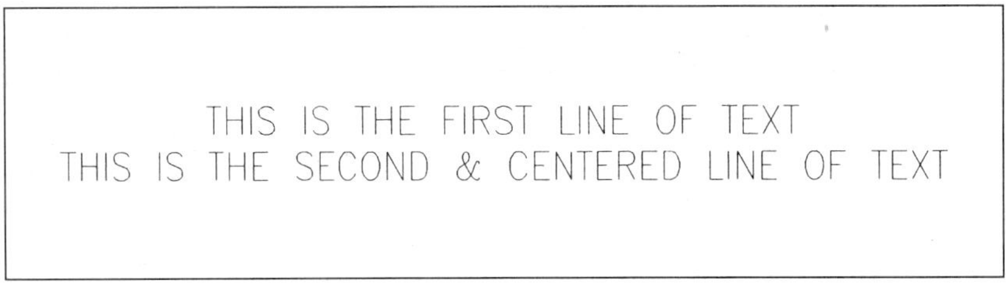

   FIGURE 13-32 Multiple Text Lines

   You can repeat this procedure for as many lines of text as you want.

5. Use the Dtext command to place a line of text. Press ENTER and place a second line of text.

   Use the backspace key to "erase" the text on the second line, then continue to use the backspace key to erase text on the first line.

   Place two more lines of text, then press ENTER twice to end the Dtext command. Press ENTER to repeat the Dtext command. Press ENTER again and notice how the text box is placed on the next line below the last text string. Enter another line of text.

   When you reach the end of the line, move the crosshairs and enter a new text location. Type a new line of text at this location.

6. Use text codes to construct the following text string in your drawing.

   The story entitled <u>MY LIFE</u> is ±50% true.

7. Start a new drawing and use the Text command to place several lines of text on the screen. Now turn on the Qtext command.

   Command: **Qtext**
   ON/OFF <OFF>: **ON**

   Next use the Regen command to regenerate the drawing. You should see the text boxes.

8. The electronic work disk contains a text file named "TEXTIMP.DOC". Use MText command's import function to import the text into your drawing.

9. If you have a word processor capable of creating an ASCII file (or know how to use Edlin in DOS), create a text file that contains your name, address, city, and state. Import the text file into your drawing with the MText command.

10. Use AutoCAD's text fonts to create styles you like and design your own business card.

11. Using the same font, create five different styles that represent very different text appearances, such as wide text, slanted, etc.

12. Use a combination of text and drawing commands to design a logo for a graphic design office.

CHAPTER REVIEW

1. How can you request a listing of stored style files?

2. What is a font?

3. What text characteristics can be altered by means of the Style command?

4. There are several methods of placing text. List them with a brief explanation of each.

5. When an underscore or overscore is added to text, is an entire string altered, or can segments be treated separately? Explain.

6. When a text height is asked for, is a numerical entry required? Explain.

7. List the alignment modes under the text command.

8. Can text be rotated at any angle?

9. What is justified text?

10. How can the text height be altered each time you enter text without redefining the style?

11. How can you compress or expand text?

12. What format must text be before importing into AutoCAD?

TEXT, FONTS, AND STYLES 13-37

cibt.pfb	ABC abc 123 !@#
cobt.pfb	ABC abc 123 !@#
eur.pfb	ABC abc 123 !@#
euro.pfb	ABC abc 123 !@#
par.pfb	ABC abc 123 !@#
rom.pfb	ABC abc 123 !@#
romb.pfb	ABC abc 123 !@#
romi.pfb	ABC abc 123 !@#
sas.pfb	ABC abc 123 !@#
sasb.pfb	ABC abc 123 !@#
sasbo.pfb	ABC abc 123 !@#
saso.pfb	ABC abc 123 !@#
te.pfb	ABC ABC 123 !@#
teb.pfb	ABC ABC 123 !@#
tel.pfb	ABC ABC 123 !@#

POSTSCRIPT FONTS

complex.shx	ABC abc 123 !@#
gothice.shx	ABC abc 123 !@#
gothicg.shx	ABC abc 123 !@#
gothici.shx	ABC abc 123 !@#
greekc.shx	??? ??? 123 !@#
greeks.shx	??? ??? 123 !@#
isocp.shx	ABC abc 123 !@#
isocp2.shx	ABC abc 123 !@#
isocp3.shx	ABC abc 123 !@#
isoct.shx	ABC abc 123 !@#
isoct2.shx	ABC abc 123 !@#
isoct3.shx	ABC abc 123 !@#
italic.shx	ABC abc 123 !@#
italicc.shx	ABC abc 123 !@#
italict.shx	ABC abc 123 !@#
monotxt.shx	ABC abc 123 !@#
romanc.shx	ABC abc 123 !@#
romand.shx	ABC abc 123 !@#
romans.shx	ABC abc 123 !@#
romant.shx	ABC abc 123 !@#
scriptc.shx	ABC abc 123 !@#
scripts.shx	ABC abc 123 !@#
simplex.shx	ABC abc 123 !@#
syastro.shx	☉♀♁ ✳'' 123 !@#
symap.shx	⌑△ ✝♘ 123 !@#
symath.shx	ℵ' ←↓∂ 123 !@#
symeteo.shx	↭ |\ 123 !@#
symusic.shx	♪♫ ♪♫ 123 !@#
txt.shx	ABC abc 123 !@#

AUTOCAD SHX FONTS

bgothl.ttf	ABC ABC 123 !@#
bgothm.ttf	ABC ABC 123 !@#
compi.ttf	±°' ©®° ○○▫ □
comsc.ttf	ABC abc 123 !@#
dutch.ttf	ABC abc 123 !@#
dutchb.ttf	ABC abc 123 !@#
dutchbi.ttf	ABC abc 123 !@#
dutcheb.ttf	ABC abc 123 !@#
dutchi.ttf	ABC abc 123 !@#
monos.ttf	ABC abc 123 !@#
monosb.ttf	ABC abc 123 !@#
monosbi.ttf	ABC abc 123 !@#
monosi.ttf	ABC abc 123 !@#
swiss.ttf	ABC abc 123 !@#
swissb.ttf	ABC abc 123 !@#
swissbi.ttf	ABC abc 123 !@#
swissbo.ttf	ABC abc 123 !@#
swissc.ttf	ABC abc 123 !@#
swisscb.ttf	ABC abc 123 !@#
swisscbi.ttf	ABC abc 123 !@#
swisscbo.ttf	ABC abc 123 !@#
swissci.ttf	ABC abc 123 !@#
swissck.ttf	ABC abc 123 !@#
swisscki.ttf	ABC abc 123 !@#
swisscl.ttf	ABC abc 123 !@#
swisscli.ttf	ABC abc 123 !@#
swisse.ttf	ABC abc 123 !@#
swisseb.ttf	ABC abc 123 !@#
swissek.ttf	ABC abc 123 !@#
swissi.ttf	ABC abc 123 !@#
swissli.ttf	ABC abc 123 !@#
stylu.ttf	ABC abc 123 !@#
swissk.ttf	ABC abc 123 !@#
swisski.ttf	ABC abc 123 !@#
swissko.ttf	ABC abc 123 !@#
swissl.ttf	ABC abc 123 !@#
swissli.ttf	ABC abc 123 !@#
umath.ttf	ABΨ αβψ + − × /
vinet.ttf	ABC abc 123 !@#

TRUETYPE FONTS

CHAPTER 14

LAYERS

Professional CAD drawings are constructed on different layers that can be turned on and off or changed. Chapter 14 covers the methodology of using layers in your drawings. After completing this chapter, you will be able to:

- Use the concept of CAD layering.
- Apply the methodology of using layers in your drawings.
- Perform layering with proficiency.

LAYERS

Traditional drafting techniques often include a method of drawing called overlay drafting. This consists of sheets of drafting media that are overlaid so the drawing below shows through the top sheet. Items can be placed on the top sheet that line up with the drawing below. Both sheets can be printed together, resulting in a print that shows the work on both sheets.

The first sheet is typically referred to as the *base drawing*. Each additional sheet can be used to place different items.

As an example, if you draw a set of floor plans, you must prepare a separate drawing for the dimensioned floor plan, the plumbing plan, the electrical plan, and so forth. Since most of your drawing time is spent redrawing the floor plan, you spend a great deal of time doing repetitive tasks. The floor plan can be thought of as the base drawing, with each discipline such as electrical and plumbing being placed on overlay sheets.

AutoCAD provides capabilities that eliminate the repetition of redrawing the base drawing. You can use drawing *layers* to place different parts of your drawings.

You may think of layers as transparent sheets of glass that are stacked on top of each other. You may draw on each layer and see through all the layers so that all the work appears as though it were on one drawing.

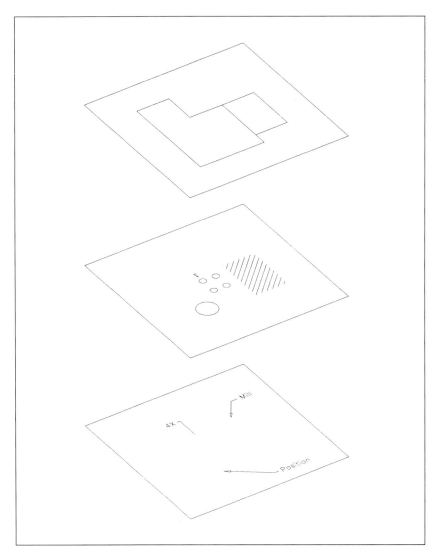

FIGURE 14-1 Transparent Drawing Layers

AutoCAD goes one step further. You may turn on or off each layer so that it is either visible or invisible!

Using the Layer Command

The Layer command is used to manipulate the layers in your drawing. Let's look at each option of the Layer command to learn how to use them.

Listing Layer Information. You may view a listing of the named layers and obtain a status report of each by entering a **?** in response to the prompt:

Command: **Layer**
?/Make/Set/New/ON/OFF/Color/Ltype/Freeze/Thaw/LOck/Unlock: **?**
Layer name(s) for listing <*>:

You may, at this point, enter any layer name to obtain a listing of the status. If you would like a listing of all the named layers, enter the wild card character "*". A typical listing would appear in the format shown in Figure 14-2.

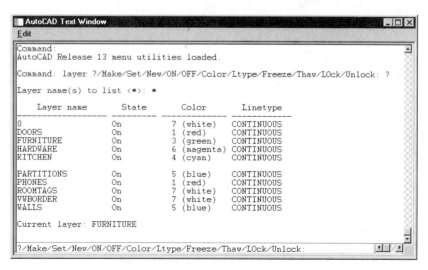

FIGURE 14-2 Sample Layer Listing

Notice that the current layer is shown at the bottom.

Naming a Layer. Each layer may be named. You determine the name to be used. To establish the layer name, respond to the prompt with **New** or **N**.

    Command: **Layer**
    ?/Make/Set/New/ON/OFF/Color/Ltype/Freeze/Thaw/LOck/Unlock: **N**
    Layer name(s) <default>: *(name)*

> **NOTE:** Choose names for layers that describe the items drawn on that layer. For example, use DIM as the layer name to contain dimensions. Also, try to standardize your layer names for drawings that are always similar.

You have now set a new layer with the specified name. You may name several layers at once. To do so, separate each name with a comma. Each name may contain up to 31 letters, but *no* spaces. If you enter a space, AutoCAD will treat it the same as an Enter and terminate the sequence. You may use letter characters, the dollar sign ($), underlines, and hyphens. Lowercase characters are converted to uppercase.

Making a Layer. The Make layer option allows you to make a new layer and simultaneously move to that layer. The new layer begins with a continuous linetype and a color assignment of 7 (white).

If the named layer already exists, AutoCAD makes it the current layer. If the named layer does not exist, it is created first, then becomes the current layer.

The option is selected while in the Layer command.

    Command: **Layer**
    ?/Make/Set/New/ON/OFF/Color/Ltype/Freeze/Thaw: **M**
    New current layer <0>: *(Layer name.)*
    ?/Make/Set/New/ON/OFF/Color/Ltype/Freeze/Thaw/LOck/Unlock: [ENTER]

Setting the Current Drawing Layer. You may only draw on the *current* layer. Even if other layers are on, the current layer is the only one that entities can be added to. The current layer name is displayed at the top of the screen. There can be only one current layer at a time.

There is no relationship to layers that are turned on and the current layer. If the current layer is turned off, your work can not be seen. Therefore, if you add a line, it will be placed on the drawing, but you will not be able to view it.

    Command: **Layer**
    ?/Make/Set/New/ON/OFF/Color/Ltype/Freeze/Thaw/LOck/Unlock: **S**
    New current layer <current layer>: *(Layer name.)*
    ?/Make/Set/New/ON/OFF/Color/Ltype/Freeze/Thaw/LOck/Unlock: [ENTER]

Turning Layers On and Off. The visibility of each layer is determined by whether the layer is currently on or off.

If a layer is off, it is invisible. It is still a part of the drawing, you just can not see it. To turn a layer on or off, respond to the prompt with either **on** or **off**. For example:

Command: **Layer**
?/Make/Set/New/ON/OFF/Color/Ltype/Freeze/Thaw/Unlock/LOck: **ON**
Layer name(s) to turn On: *(Name.)*

> **NOTE:** You may turn on or off several layers at a time by separating the names with commas, such as: floor, roof, walls.

Layer Colors. Each layer is associated with a color. The initial (default) color number is seven (white). AutoCAD sets the first seven color numbers and colors as follows:

1 — Red   5 — Blue
2 — Yellow   6 — Magenta
3 — Green   7 — White
4 — Cyan

If you have a color monitor, the entities drawn on a layer will appear in the color assigned to that layer. If your monitor is monochrome, the colors still exist, but, of course, you will see them as shades of gray.

> **NOTE:** Use colors to "code" different parts of your drawings. This will make recognition of the different parts easier.

You may set the color for each layer by responding to the prompt with C.

Command: **Layer**
?/Make/Set/New/ON/OFF/Color/Ltype/Freeze/Thaw/LOck/Unlock: **C**
Color: *(Select a color by name or number.)*
Layer name(s) for color n <current>:

Layers and Linetypes. Each layer can contain a linetype. Every entity that is placed on that layer will be shown on the drawing screen in that linetype.

Setting the *layer* linetype sets the default linetype that is used for each entity that is placed on that layer. You may also use the Change command to change one or more entities to different linetypes on any layer, regardless of the layer linetype setting.

Command: **Layer**
?/Make/Set/New/ON/OFF/Color/Ltype/Freeze/Thaw/LOck/Unlock: **LT**
Linetype (or ?) <current linetype>:
Layer name(s) for linetype *(selected linetype)* <default>:
?/Make/Set/New/ON/OFF/Color/Ltype/Freeze/Thaw/LOck/Unlock: [ENTER]

Freezing Layers. The Freeze option lets you turn off layers and eliminate them from subsequent regenerations. (See Regen command.) This can save time when performing operations that force regenerations. The current layer cannot be frozen.

Command: **Layer**
?/Make/Set/New/ON/OFF/Color/Ltype/Freeze/Thaw/LOck/Unlock: **F**
Layer name(s) to Freeze:

> **NOTE:** The Freeze option operates differently from the OFF option. When you turn a layer OFF, it is not displayed, but is regenerated by AutoCAD. A frozen layer is neither displayed nor regenerated. For faster regens, use the Freeze option.

Thawing Layers. The Thaw option is used to undo the effect of the freeze command.

Command: **Layer**
?/Make/Set/New/ON/OFF/Color/Ltype/Freeze/Thaw/LOck/Unlock: **T**
Layer name(s) to Thaw:

> **NOTE:** Use AutoCAD's layering ability to create a "base" drawing. Place specific parts of your drawing on other named layers. Freeze layers which you will not come back to. This will speed the drawing process, especially for complicated or "hatched" areas.

Locking and Unlocking Layers. Layers can be *locked* so that they can be viewed but not edited. When a layer is locked, it can not be either edited or made current. You can, however, perform some operations such as object snap, color and linetype changes, and freezing locked layers.

Command: **Layer**
?/Make/Set/New/ON/OFF/Color/Ltype/Freeze/Thaw/LOck/Unlock: **L**
Layer name(s) to Lock:
?/Make/Set/New/ON/OFF/Color/Ltype/Freeze/Thaw/LOck/Unlock:

> **NOTE:** Lock the base drawing layer(s) when other CAD drafters will be performing specific discipline work on other layers in the drawing.

Controlling Layers with a Dialog Box

Controlling layers is simplified by use of the *Modify Layer* dialog box. There are four ways of displaying the dialog box.

- Use the Ddlmodes command from the command line.
- Use the transparent command 'Ddlmodes during a current command.
- Select "Layers..." from the "Data" pull-down menu.
- Select Layers button from the Object Properties toolbar.

The following illustration shows the Layer Control dialog box.

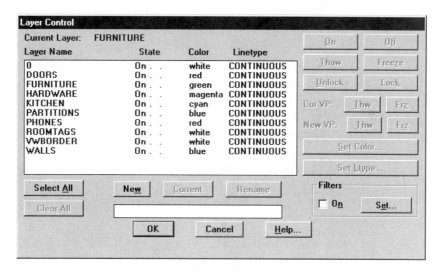

FIGURE 14-3 Layer Control Dialog Box

The dialog box can be used to perform all the operations possible with the Layer command plus a few more. Let's look at the methods used to manipulate layers with the Modify Layer dialog box.

Turning Layers On or Off. To turn one or more layers on or off, first highlight the target layer(s) by clicking on the layer name, then click either the On or Off button. The "State" column will immediately reflect the change. If a layer is on, the word "On" appears. If off, a period (.) appears in the column.

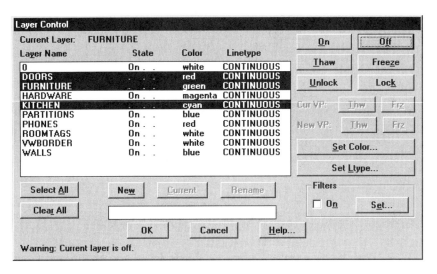

FIGURE 14-4 Turning Layers On and Off

Freezing and Thawing Layers. To freeze or thaw one or more layers, highlight the target layer(s), then click on the Freeze or Thaw buttons. A frozen layer is indicated by the letter "F" in the State column. A non-frozen or thawed layer is indicated by a period (.) in the column.

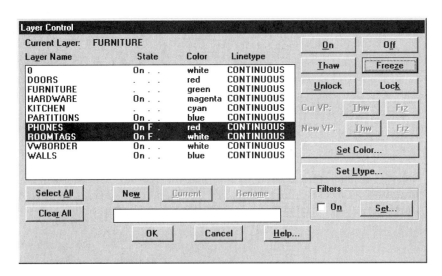

FIGURE 14-5 Freezing and Thawing Layers

Locking and Unlocking Layers. To lock or unlock layers, select the target layer(s), then click on the Lock or Unlock buttons. The letter "L" listed in the State column denotes a locked layer. A layer that is not locked shows a period (.) in this position.

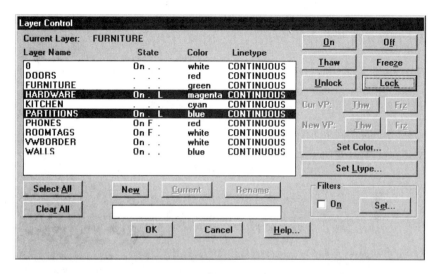

FIGURE 14-6 Locking and Unlocking Layers

Setting the Layer Color. To set the layer color, first select the layer(s) to be modified, then click on the "Set Color..." button. A Select Color sub-dialog box will be displayed as shown in Figure 14-7.

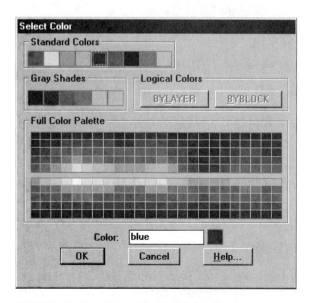

FIGURE 14-7 Select Color Dialog Box

To set the color, select a color from the color display chart. Note that this chart has the capability to display 255 colors. Not all display systems have the ability to use this many colors. The chart will display all the colors available with your display, with the remainder being displayed as gray. The color edit box will display either the name or number of the selected color. The first seven colors are listed by name, with the remaining colors listed by number. The first nine colors are displayed above the color edit box.

You may also enter the color name or number manually from the keyboard in the color edit box. After you have selected the color, click on the OK box to return to the Modify Layer dialog box.

Setting the Layer Linetype. You can set the linetype for all the entities residing on a layer. First pick the layer(s) for which you wish to set a linetype, then click on the "Set Linetype..." button. A Select Linetype sub-dialog box is displayed as shown in Figure 14-8.

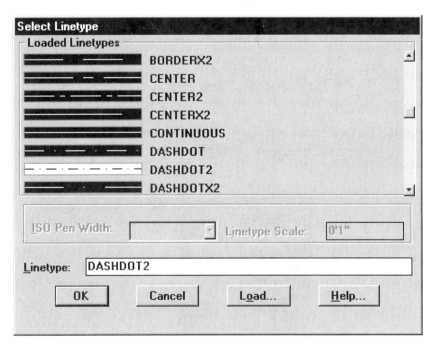

FIGURE 14-8 Select Linetype Dialog Box

The list box in the dialog box displays the linetypes that have been loaded. As you select each linetype, it is displayed at the top of the dialog box.

Filtering the Layer Listings. Some types of drawings may have many named layers. Manipulating through the layer listings can become tedious. AutoCAD allows you to "filter" the listings so that only certain types of listings are displayed.

For example, you can filter the listing so that only layers that are not frozen will be displayed. The filter capabilities can be used to display or suppress layers that are on or off, frozen or thawed, locked or unlocked, and whether the layer is frozen in the current viewport.

You can also filter by name, color, and linetype. With these filters, you can use the DOS wild-card characters of question mark (?) and asterisk (*) to designate the layers.

To set the filter type, select the "Set..." button in the "Filters" section of the dialog box. The following dialog box is displayed.

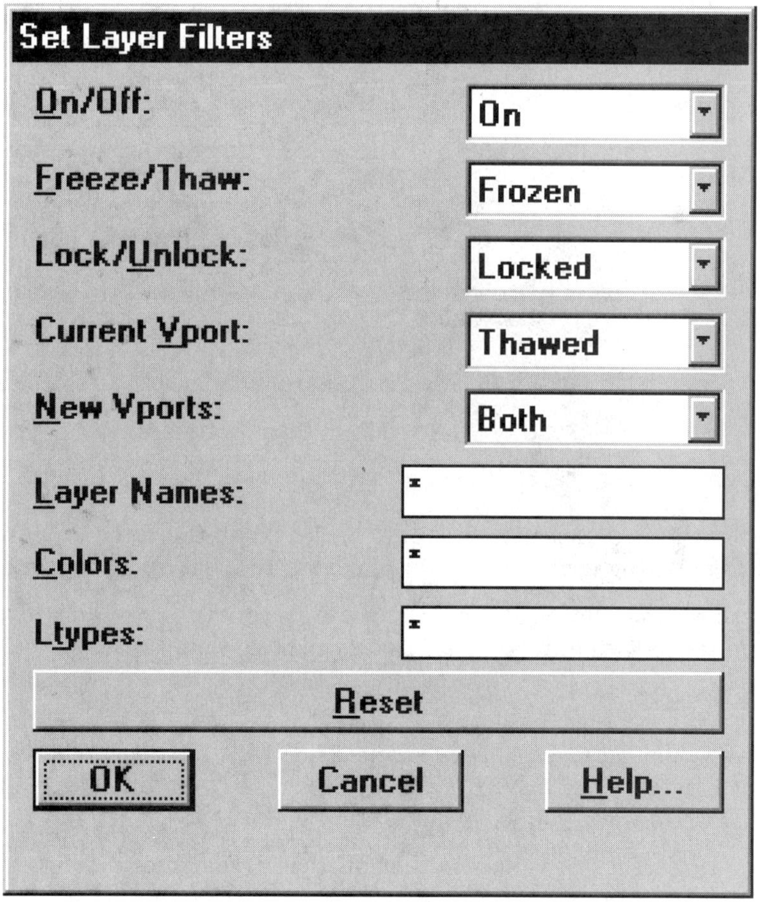

FIGURE 14-9 Layer Filter Dialog Box

The first five filters listed (see Figure 14-9) use popup boxes to select between the settings of either selection listed, or both selections to be listed.

For example, if you select the Freeze/Thaw popup box, the selections are Both, Frozen, and Thawed.

14-12 USING AUTOCAD

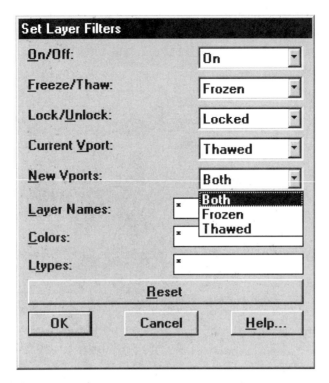

FIGURE 14-10 Filters Dropdown Box

The last three selections require you to enter the filter type in the text box. The default listing is an asterisk (*), which means that all layer names will be listed.

Let's assume that you wanted to list all layers that started with the letters LEVEL and ended with any two characters. The filter would read:

LEVEL??

This would list layers that are named LEVEL21, LEVEL_A, and LEVEL2C. It would not list 1LEVEL, LEV21, or 3RD_LEVEL. If you do not remember how wild-card characters work, you may want to review the DOS section in this book.

You can reset the values so that all the layer names are listed by clicking on the "Reset" button.

Controlling Layers with the Toolbar

When the Object Properties toolbar is available, you can select the current layer without using the Layer command or bringing up the Layer Control dialog box. By clicking on the down arrow at the right end of the layer name on the toolbar, AutoCAD displays a dropdown box, as shown in Figure 14-11.

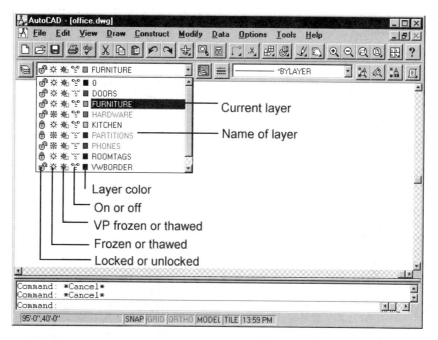

FIGURE 14-11 Layer Controls on the Toolbar

The dropdown box lists the names of all layers in the drawing, as well as a quintet of icons. Each icon has two states:

- Padlock open/closed: layer is unlocked (default) or locked.
- Sun or snowflake: layer is thawed (default) or frozen. Frozen layers are also shown by their name colored gray.
- Sun or snowflake on rectangle: layer is thawed (default) or frozen in VP (viewport) mode.
- Eyes open/closed: layer is on (default) or off.

The last icon is the small square showing the color assigned to the layer.

EXERCISES

1. Create a layer named "MYLAYER". Set "MYLAYER" as the current layer. Do you see the layer name in the toolbar at the top of the screen?
2. Set the layer color to green. Draw some objects on the layer.
3. Set the current layer to "0". Freeze the layer named "MYLAYER". Did the objects you drew disappear? Select the Layer dialog box from the pull-down menu and Thaw "MYLAYER". Did the objects reappear?

CHAPTER REVIEW

1. Why would a CAD drafter use layers?

2. What layer option would you use if you wanted to create a new layer and make that layer currently active?

3. How do you turn on a frozen layer?

4. What is the difference between turning layers off and freezing layers?

5. How would you obtain a listing of all the layers in your drawing?

6. Can you have objects of more than one color on the same layer? Explain.

7. What do the following layer symbols mean?
   - Snowflake:
   - Open lock:
   - Closed eyes:
   - Sun over square:
   - Colored square:

CHAPTER 15

INTRODUCTION TO DIMENSIONING

The ability to place dimensions into a drawing is one of the most essential elements of CAD drawing. In this chapter, the user will learn the basics of AutoCAD dimensioning. After completing this chapter, you will be able to:

- Utilize the basic elements of dimensioning.
- Identify the manner in which dimensions are placed in AutoCAD.
- Manipulate some of the special abilities of the dimensioning mode.

DIMENSIONING IN AUTOCAD

AutoCAD constructs dimensioning semi-automatically. That is, it can construct the dimensioning lines and measure the distances for you. All it needs is some basic information from you. This is a very powerful feature. If mastered, it can be a great time-saver and provide professional results.

DIMENSIONING COMPONENTS

Dimensioning is made up of several different parts. Before we try some dimensioning, you should know these parts. Let's look at each.

The Dimension Line

The dimension line is the line with the arrows or "ticks" at each end. The dimension text is placed either in this line, dividing it into two parts, or over the line.

FIGURE 15-1 Dimension Line

When you are using angular dimensioning, the dimension line is an arc instead of a straight line.

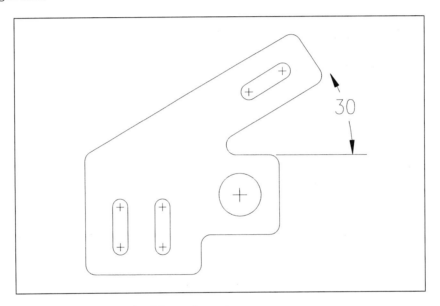

FIGURE 15-2 Angular Dimension

If you wish, the arrows at the end of the dimension line can be replaced with tick marks. Figure 15-3 shows dimension line with tick marks.

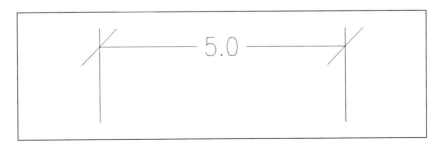

FIGURE 15-3 Dimension Line with Ticks

Extension Lines

The extension lines (sometimes called "witness lines") are the lines constructed perpendicular to the dimension line and extending to the point that is being dimensioned to.

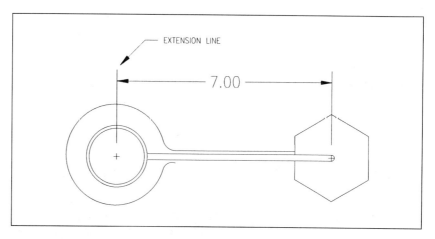

FIGURE 15-4 Extension Lines

Dimension Text

The dimension text is the text that appears at the dimension line. You may allow AutoCAD to measure the distance and enter the text, you may specify the desired text, or you may suppress the text entirely by entering a space in place of the text.

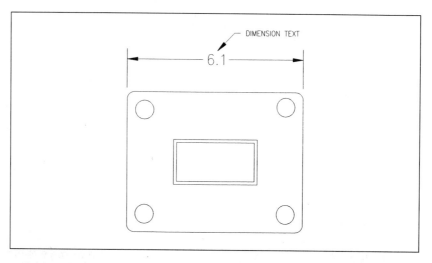

FIGURE 15-5 Dimension Text

The dimension text is drawn using the current text style and font. The format used is determined by the current units. That is, if the units are set to architectural, you will receive an output of feet and inches. The smallest fraction displayed will be as chosen by the Units command, and so forth. You may, of course, enter any text if you choose the manual method of inserting the dimension text.

> **NOTE:** You must use a horizontal text font when in the Dimension command. If the current text style is vertical, you must change it before entering the dimensions.

Dimension Tolerances

The dimension tolerances are the plus and minus amounts that are appended to the dimension text. These tolerances are added to the text that AutoCAD generates automatically. The plus and minus amounts are specified by you. They may be equal or unequal. If they are equal, they are drawn with a plus/minus symbol. If they are unequal, the tolerances are drawn one above the other. The following text shows examples of both equal and unequal tolerances.

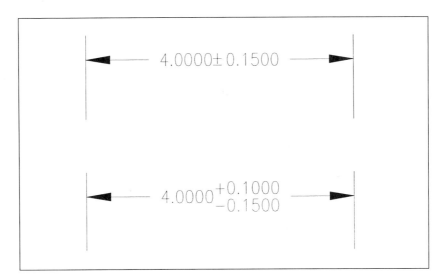

FIGURE 15-6 Dimension Tolerances

Dimension Limits

Instead of having the tolerance shown, you may choose to have them applied to the dimension. The example shown in Figure 15-7 is a measurement of 4.0000 units with a tolerance of ±0.15.

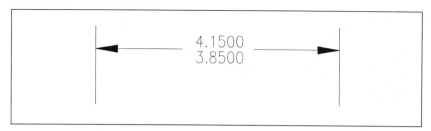

FIGURE 15-7 Dimension Limits

Alternate Dimension Units

Alternate units can be used to show two systems of measurement simultaneously. For example, you can show English and metric units on the same dimension line.

FIGURE 15-8 Alternate Dimension Values

Leader Lines

Leader lines are "arrowed lines" with text at the end. Leaders are often used to point out a specific part of a drawing to be noted.

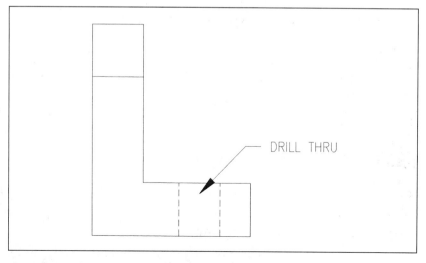

FIGURE 15-9 Dimension Leader

Center Marks and Center Lines

A center mark is a cross designating the center of a circle or arc. Center lines are lines that cross at the center of the circle or arc and intersect the circumference. Examples of each are shown in Figure 15-10.

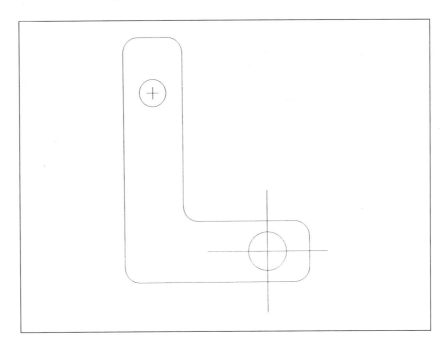

FIGURE 15-10 Center Marks and Center Lines

Changing the Look with Dimension Variables

Dimensioning variables determine the manner in which the dimension entities are drawn. Some variables are values and some are simply turned on and off. The variables may be changed by you to change the dimension "look" and function. Dimension variables are covered in Chapter 16.

ENTERING DIMENSIONING MODE

You begin dimensioning by selecting Draw/Dimensioning from the pull-down menu.

When we dimension, we first select the type of dimensioning we wish to perform, then use the associated commands to draw the dimensions.

Let's explore the different types of dimension commands.

DIMENSIONING COMMANDS

AutoCAD dimensioning commands can be grouped into four basic types of commands. These are:

>Dimension *drawing* commands
>
>Dimension *editing* commands
>
>Dimension *utility* commands
>
>Dimension *style* commands (covered in Chapter 16)

Let's now look at each of the types of dimensioning commands.

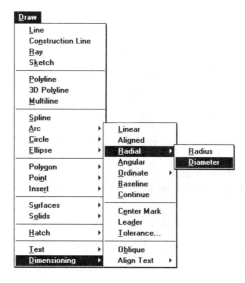

DIMENSION DRAWING COMMANDS

Dimension drawing commands are used to draw the different types of dimension lines. The following is a listing of dimension drawing commands.

>**Linear:** Draws a dimension with a linear dimension line.
>
>**Aligned:** Draws a dimension line parallel to an entity or to the dimension extension line origins you specify.
>
>**Rotated:** Constructs a dimension at a specified angle.
>
>**Radius:** Constructs a dimension line that shows the radius of an arc or circle.
>
>**Diameter:** Dimensions the diameter of a circle or arc.
>
>**Angular:** Dimensions the angle between two nonparallel lines. The dimension line is constructed as an arc, with the dimension value shown as the angle between the lines.

Ordinate: Dimensions the *X* or *Y* coordinate, referenced to a specified point, of an object.

Baseline: Continues a dimension line (similar to the Continue command) in reference to the first extension line origin of the first dimension line in the string. The dimension line is offset to avoid conflict with the previous dimension line.

Continue: Continues a dimension line string after you have placed the first dimension. The continuing string uses the first dimension to determine the correct positioning of the next dimension.

Center Mark: Places a center mark, or alternately, center lines, at the center point of a circle or arc.

Leader: Text attached to a line and arrowhead.

Tolerance: Displays a dialog box that lets you select a tolerancing symbol.

Let's look at each command and learn how to use each to place a dimension.

Placing a Linear Dimension Line

The Linear dimension command creates the most common types of dimension lines. The dimension line is drawn horizontally as shown in Figure 15-11.

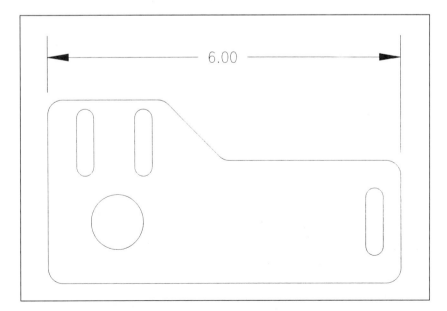

FIGURE 15-11 Horizontal Dimension Line

When placing a linear dimension, you can either specify an entity to dimension, or select the starting and ending points for the extension lines.

Let's place a horizontal dimension line. Draw a box as shown in Figure 15-12. Use the following command sequence and follow the instructions at the prompts.

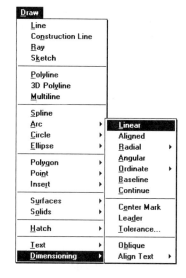

Command: **Dimlinear**
First extension line origin or RETURN to select: *(Press ENTER.)*
Select object to dimension: *(Select the top line of the box.)*
Dimension line location (Text/Angle/Horizontal/Vertical/Rotated): *(Move the cursor up to locate the dimension line.)*

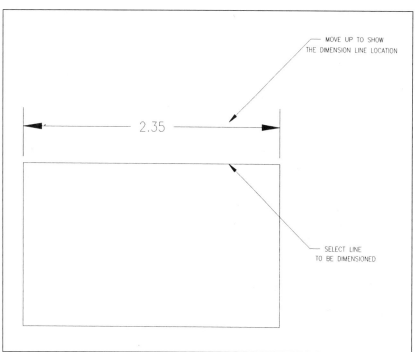

FIGURE 15-12 Placing a Horizontal Dimension Line

Note that you can also enter the extension line origin points. When AutoCAD prompts you for the first extension line origin, enter the actual point instead of pressing Enter. We will use this method in the next exercise.

> **NOTE:** Use object snap to set exact points for accurate dimensioning.

If you choose to select the object to be dimensioned by pressing [ENTER] instead of specifying the extension line points, you can save the steps necessary for entering each beginning point.

This method works on lines, arcs, or circles and may be used with the Dimlinear, DimAligned, and DimRotated dimension commands. The following illustration shows a sample command sequence and a typical pick point for dimensioning the three entities using this method.

Command: **Dimlinear**
First extension line origin or RETURN to select: [ENTER]
Select line, arc, or circle: *(Pick any part of the line, arc, or circle.)*
Dimension line location (Text/Angle): *(Select the dimension line location.)*

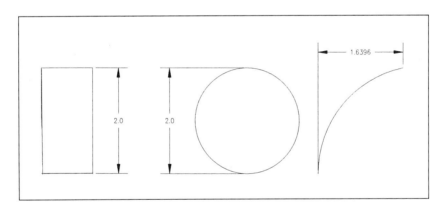

FIGURE 15-13   Selecting an Object to Dimension

Continuing the Dimension String

The Continue dimension command is used to place a dimension segment that follows the first dimension you place. The continued dimension is constructed by using the last extension line of the previous dimension as the first extension line point of the new dimension.

INTRODUCTION TO DIMENSIONING **15-11**

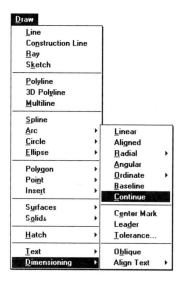

Construct the simple object shown in Figure 15-14. Place a horizontal dimension as shown.

Select DRAW Dimensioning from the menu bar, then click on Continue. AutoCAD prompts:

Command: **Dimcontinue**
Second extension line origin or RETURN to select:

Let's use object snap to capture the upper right intersection. Type **INT** from the keyboard and press [ENTER]. AutoCAD prompts you for the intersection point.

of *(Select the intersection.)*
(Press [ESC] to end the command.)

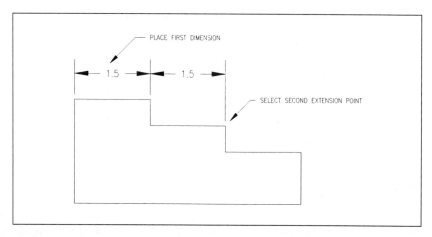

FIGURE 15-14 Continuing a Dimension String

> **NOTE:** There is no requirement for the dimension text to be a numerical value. You can type in any characters from the keyboard. For example, you may want to place a horizontal dimension and use the dimension text:
>
> FIVE EQUALLY SPACED NOTCHES

Placing a Vertical Dimension Line

The Dimlinear dimension command works equally as well for vertical dimensions as for horizontal dimensions, except that the dimension line is drawn vertically.

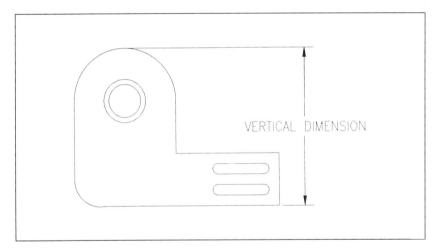

FIGURE 15-15 Vertical Dimension Line

To place vertical dimensions, either select an entity to dimension, or manually select the extension line origin points. Use Exercise 1, found at the end of the chapter, to place a vertical dimension line.

Dimensioning Angled Surfaces

Many times you will be required to place dimensions that are constructed at an angle. AutoCAD provides two commands for placing angled dimensions. Let's see how each one works.

Drawing Aligned Dimensions. The DimAligned dimension command draws the dimension line parallel to the extension line origin points.

INTRODUCTION TO DIMENSIONING   **15-13**

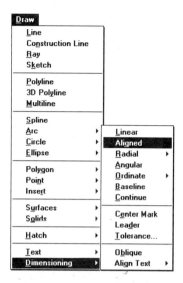

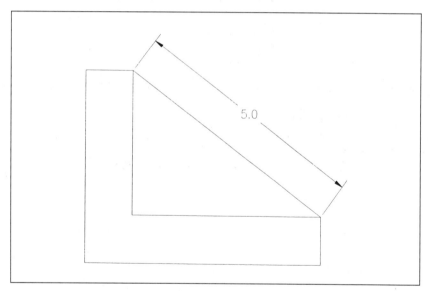

FIGURE 15-16 Aligned Dimension

To draw an aligned dimension, either select the entity that you want the dimension line to be parallel to, or manually select the extension line origins. The following exercise will guide you through the methodology of constructing an aligned dimension.

Constructing Rotated Dimensions. The Dimlinear dimension command functions the same as the aligned command, except that you must first specify the rotated option. This is especially useful in situations where the extension line origins do not accurately describe the desired dimension line angle. The illustration in Figure 15-17 shows such a situation.

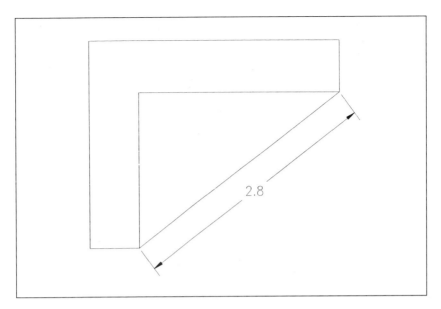

FIGURE 15-17 Rotated Dimension

When you select the Dimlinear dimension command, AutoCAD prompts:

Command: **Dimlinear**
First extension line origin or RETURN to select:
Second extension line origin:
Dimension line location (Text/Angle/Horizontal/Vertical/Rotated): **R**
Dimension line angle
Dimension line location:

You can tell AutoCAD the angle you want to use in two ways. The first is to enter the angle from the keyboard. The angle is specified in AutoCAD's standard angle notation, with the zero angle to the right (east) direction.

The second way is to use two points to "show" AutoCAD the angle. To do this, respond with two points when prompted for the dimension line angle. AutoCAD will measure the angle between the two points and use that angle to construct the rotated dimension.

Creating Baseline Dimensions

The Baseline dimension command creates continuous dimensions from the first extension line. The first extension line acts as a baseline from which the dimensions originate. AutoCAD offsets each new dimension line to avoid drawing on top of the first dimension line.

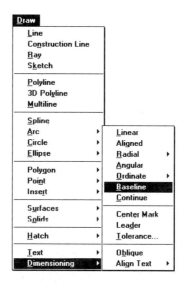

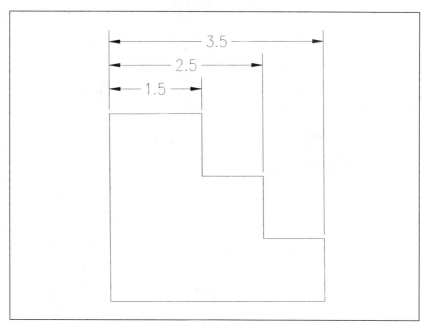

FIGURE 15-18 Baseline Dimension

To construct a baseline dimension, first construct an initial dimension line, then select Baseline from the menu. When AutoCAD prompts for the second extension line origin, select the next point. AutoCAD will construct the dimension line, using the first extension line origin of the first dimension line in the string as the first extension line origin, and the point you entered as the second extension line origin. AutoCAD also offsets the new dimension line so it is not constructed on top of the first.

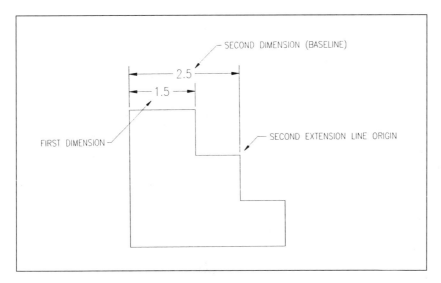

FIGURE 15-19 Constructing a Baseline Dimension

Dimensioning Angles

The DimAngular dimensioning command is used to dimension angles. There are four ways to specify an angular dimension:

- Two non-parallel lines
- An arc
- A circle and another point
- Three points

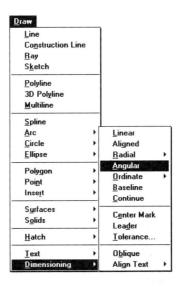

When you select Angular from the Draw Dimensioning menu, AutoCAD prompts:

Command: **Dimangular**
Select arc, circle, line, or RETURN:

Let's look at how to place an angular dimension in each circumstance.

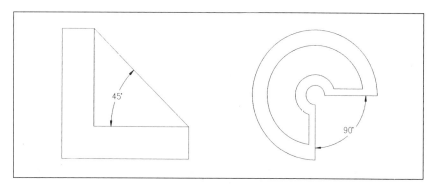

FIGURE 15-20 Angular Dimensions

Dimensioning Circles and Arcs

Diameter and radius dimensioning of circles and arcs is easy with AutoCAD. The Diameter and Radius commands are used to place these types of dimensions. Let's see how each works.

Dimensioning the Diameter of a Circle or Arc. The Diameter dimension command draws dimensions of the diameter of either a circle or an arc. An example of each is shown in Figure 15-21.

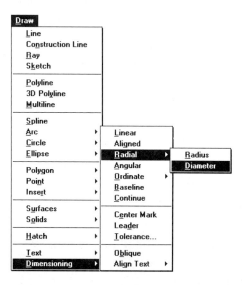

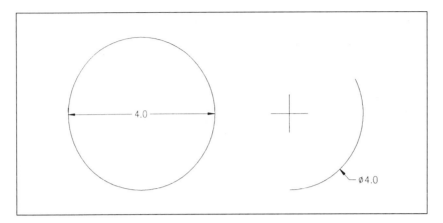

FIGURE 15-21 Diameter Dimensions

The dimension is placed according to the point you pick when selecting the arc or circle. The dimension line will intersect that point, passing through the center point of the arc or circle.

If the dimension line and text are too large to fit within the arc or circle, AutoCAD will place a leader line outside the circle or arc. The following illustration shows an example of such a leader line.

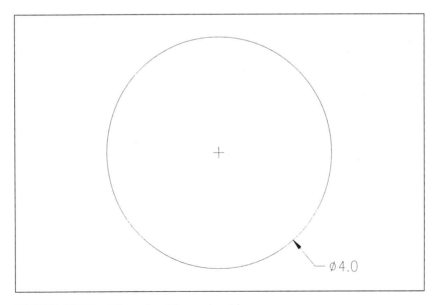

FIGURE 15-22 Diameter Dimension Line

The length of the leader line is placed by you. AutoCAD asks for the leader length with the following prompt:

Enter leader length for text:

Dimensioning the Radius of a Circle or Arc

The Dimradius dimension command draws the dimension of the radius from the center of a circle or arc.

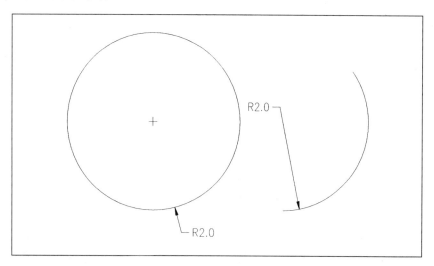

FIGURE 15-23 Radius Dimension

The point at which the radius dimension line intersects the circle or arc is determined by the location of the pick point used when selecting the object to be dimensioned.

Placing Center Marks

You may place a center mark at the center point of an arc or circle. Center marks are drawn as an intersection cross as shown in the following figure.

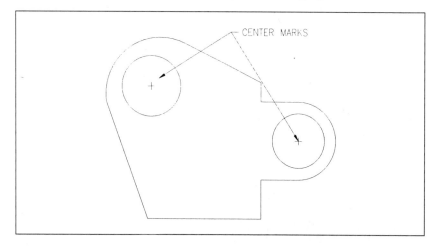

FIGURE 15-24 Center Marks

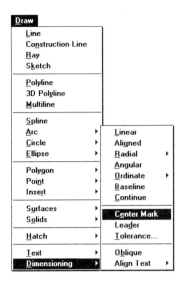

The center mark size is controlled by the Center Mark Size edit box and the Mark With Center Lines check box on the Extension Lines sub-dialog box.

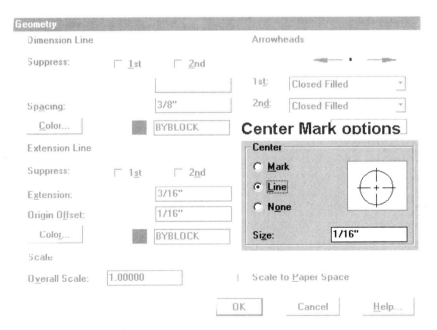

FIGURE 15-25 Dialog Box for Setting Extension Line Properties

To place a center mark, select Center from the dimension menu.

Command: **Dimcenter**
Select arc or circle:

Ordinate Dimensioning

The Ordinate dimensioning command places dimensions that are relative to a reference point.

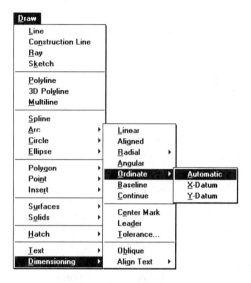

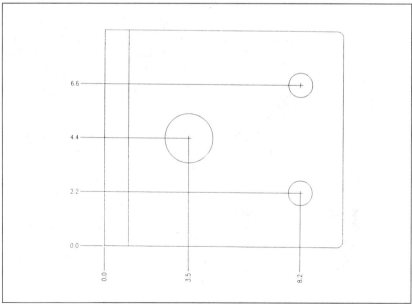

FIGURE 15-26 Ordinate Dimensioning

The reference point used is the current UCS origin. Values for ordinate dimensions are specified as either X or Y coordinates relative to the origin.

> **NOTE:** Always turn on the Ortho mode when placing ordinate dimensions.

The UCS origin is located, by default, in the lower left-hand corner of the screen (the *0,0* coordinate point). In most cases, you will want to relocate the UCS origin point so the ordinate dimensions will be referenced to a specific point. You can move the UCS origin by using the UCS command with the Origin option.

> Command: **UCS**
> Origin/Zaxis/3point/Entity/View/X/Y/Z/Prev/Restore/Save/Del/?/: **O**
> Origin point <0,0,0>: *(Select the new origin point.)*

Ordinate dimension lines are shown as simple leader lines. AutoCAD asks for a feature location. This is the location to which the dimension is measured. You can either stipulate whether the point is to be measured along the X-coordinate or the Y-coordinate. When you select Ordinate from the dimensioning menu, AutoCAD prompts:

> Command: **Dimordinate**
> Select feature: *(Enter the dimension location.)*
> Leader endpoint (Xdatum/Ydatum): *(Enter either X or Y.)*
> Leader endpoint: *(Enter the leader endpoint.)*

You can "shortcut" the process by entering a point on the drawing instead of specifying either a *X* or *Y* point. If you do this, AutoCAD will determine whether the measurement is along the X or Y direction by measuring between the two points.

DIMENSION EDITING COMMANDS

Dimension editing commands are used to alter a dimension after it has been placed in your drawing. AutoCAD's dimension editing commands are:

> **Dimedit:** Restores dimension text to its original position, changes existing dimension text, and forces extension lines to a specified angle.
>
> **DimTedit:** Changes placement and orientation of dimension text, and rotates dimension text.

Let's explore each function.

Restoring Dimension Text to Its Default Position

If the dimension has been moved to a new location, it can be restored to its default position by using the Dimedit command.

Command: **Dimedit**
Dimension edit (Home/New/Rotate/Oblique) <Home>: **H**
Select objects: *(Select the dimension.)*

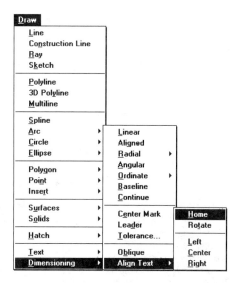

Changing Dimension Text

In the normal course of dimensioning, some dimension values or dimension text are changed. Instead of erasing the dimension and replacing it with a new dimension containing the desired value or text, the Dimedit command can be used.

Command: **Dimedit**
Dimension edit (Home/New/Rotate/Oblique) <Home>: **N**
Enter new dimension text: *(Enter new text.)*
Select objects: *(Select the desired dimension.)*

If you press [ENTER] when prompted for the new dimension text, AutoCAD measures the actual distance described by the dimension and uses that value as the dimension text.

Obliquing Dimension Extension Lines

In some situations, the standard orientation of dimension extension lines can conflict with other dimensions. The conflict can be remedied by obliquing the extension lines. Figure 15-27 shows an object with oblique extension lines.

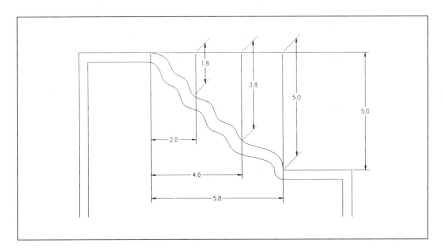

FIGURE 15-27 Oblique Dimensioning

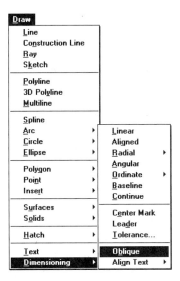

Extension lines can be obliqued by using the Dimedit dimensioning command.

Command: **Dimedit**
Dimension edit (Home/New/Rotate/Oblique): **o**
Select objects: *(pick)*
Enter obliquing angle:

Note that the dimension must first exist before the Oblique command can be used.

Relocating the Dimension Text

Some situations require that you move the dimension text. For example, dimensions that cross sometimes have overlapping dimension text. You can move the dimension text by using the DimTedit (text edit) dimension command.

When you select DimTedit, AutoCAD displays the following prompt:

Command: **DimTedit**
Select dimension: *(Select a single dimension.)*
Enter text location (Left/Right/Home/Angle):

After selecting the dimension, moving the cursor dynamically moves the dimension text. Move the text to the desired location, click, and the text is repositioned. Let's look at how each of the options work.

Left. Selecting Left positions the dimension text to the left side of the dimension line.

FIGURE 15-28 Move Dimension Text Left

Right. The Right option works the same as the Left option, except that the dimension text is positioned to the right side of the dimension line.

FIGURE 15-29 Move Dimension Text Right

Home. If you select Home, the text is restored to its default position. The default position is determined by the dimension variable settings.

Angle. Selecting Angle allows you to set the angle for the dimension text. You may set the angle in two ways. Entering an actual angle from the keyboard will cause the text to rotate to that angle. The default angle, with text reading from left to right, is 0. You can also enter two points to show AutoCAD the angle you wish to rotate the text.

FIGURE 15-30 Changing Dimension Text Angle

Rotating Dimension Text

The Dimedit dimension command is used to rotate dimension text to a specified angle. The Rotate option of the Dimedit command functions similar to the Angle option of the Tedit command. The difference is that Dimedit is used to rotate the dimension text of several dimensions.

   Command: **Dimedit**
   Dimension edit (Home/New/Rotate/Oblique): **R**
   Enter new text angle:
   Select objects:

You can either enter the actual angle or show two points to specify the angle. The Select objects prompt allows you to select several dimensions in one object selection process.

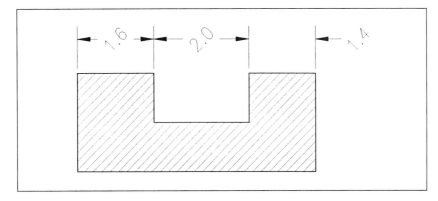

FIGURE 15-31 Dimensions with Text Rotated by 45 Degrees

> **NOTE:** Use Dimedit to change all the dimensions in your drawing by selecting the entire drawing with a window selection.

DIMENSION UTILITY COMMANDS

By typing "DIM" at the command prompt, the word "Command" changes to "DIM". This is the dimension mode. Type "Exit" to get out of the dimension mode.

The dimension utility commands are used within the dimension mode for special purposes. Let's look at each of the utility commands.

Displaying the Dimension Status

When you enter Status while in the dimension mode, AutoCAD displays the current dimension variable settings. If you are working with a single-screen system, the listing is displayed on a text page. Press ⌨ to return to the drawing screen. On two-screen systems, the listing is displayed on the text monitor.

Changing the Dimension Text Style

The dimension text style is drawn using the current text style. You can change the text style by using the Style command:

Command: **Style**
New text style <default>:

DEFINITION POINTS

When you create dimensions, AutoCAD places "definition points" in the drawing. These points are used as reference by AutoCAD in certain operations. The points are placed on a layer named "DEFPOINTS". If you do not wish to plot these points, freeze the layer named "DEFPOINTS".

ARROW BLOCKS

Many applications require specific icons at the end of the dimension line. AutoCAD provides arrows and tick marks for your use, but your application may specify other types of symbols (such as solid dots).

You can specify custom blocks to be used in place of the dimensioning arrows or ticks. The block reference must already exist in the drawing.

The following rules should be used when preparing a block for use as an "arrow".

1. The block should be prepared as the *right* arrow of a horizontal dimension line.
2. The insertion point (base point) should be placed at the point that would normally be the tip of the arrow.
3. The dimension line stops a distance from the tip of the arrow. Draw a small tail line to the left so it will connect to the dimension line.
4. In order for the arrow block to scale properly, draw the block exactly one drawing unit from the tip to the end of the tail line.

15-28 USING AUTOCAD

To select the block for use in dimensioning, select the Ddim command's Geometry sub-dialog box, then the Arrowheads 1st list box and select User Arrow, and enter the name of the block. When you place a dimension, the arrow block will be used in place of arrows or ticks.

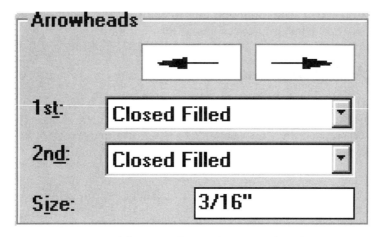

FIGURE 15-32 Dialog Box for Setting Arrowhead Properties

Separate Arrow Blocks

To create a separate arrow block at each end of a dimension line, enter a different block name for the 1st and 2nd User Arrow. Next, enter the names of the first and second arrow blocks in the edit boxes.

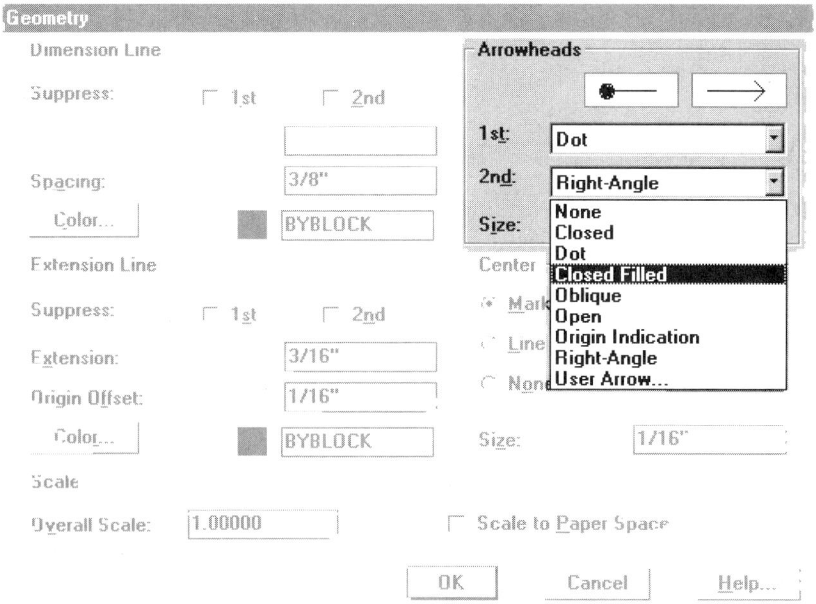

FIGURE 15-33 Creating Separate Arrowheads

EXERCISES

1. Draw the following objects and place the horizontal dimensions at the locations shown.

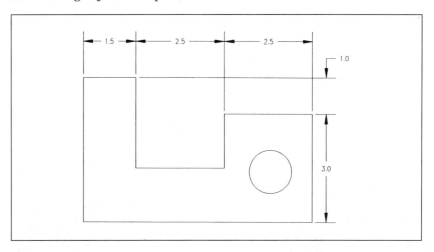

FIGURE 15-34

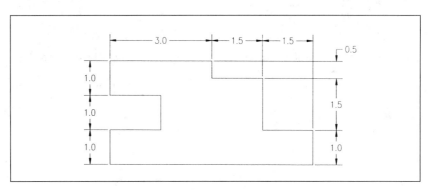

FIGURE 15-35

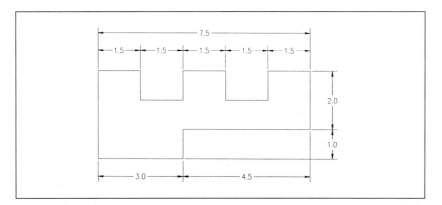

FIGURE 15-36

2. Draw the following objects and use the Dimlinear dimension command to place the vertical dimensions at the locations shown.

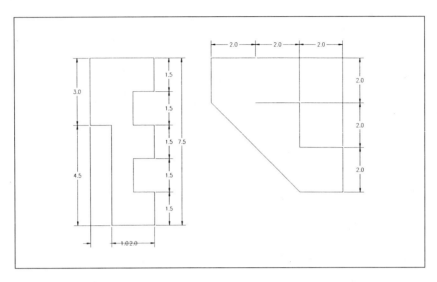

FIGURE 15-37

3. Construct the following aligned dimension drawing and place all the dimensions as shown.

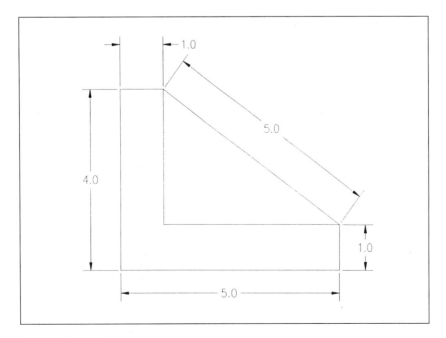

FIGURE 15-38

4. Construct the following drawing and place all the dimensions as shown.

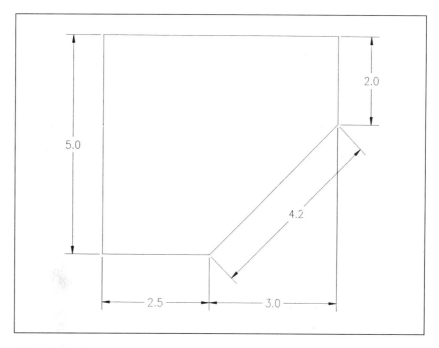

FIGURE 15-39

5. Construct the following object and place the baseline dimensions.

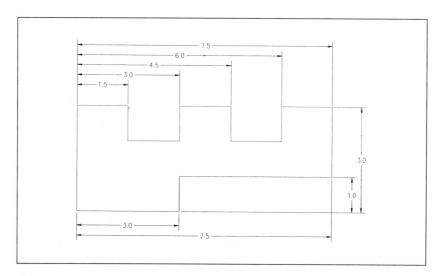

FIGURE 15-40

6. Let's place an angular dimension. Draw two lines as shown in Figure 15-41. Let's select Draw/Dimensioning/Angular from the menu bar. Use the following command sequence and Figure 15-41.

   Command: **Dimangular**
   Select arc, circle, line, or Return: *(Select "POINT 1".)*
   Second line: *(Select "POINT 2".)*
   Dimension arc line location (Text/Angle): *(Move the dimension arc into place and click.)*
   Dimension text <*default*>: [ENTER]

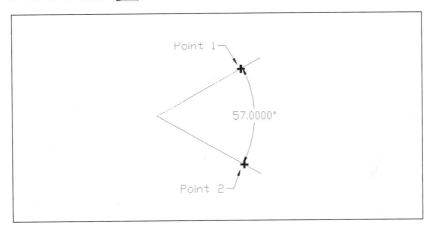

FIGURE 15-41 Placing an Angular Dimension

7. Draw four circles on the screen. From the screen root menu, select DRAW DIM:/Diameter. When AutoCAD prompts for the arc or circle, select any part on the first circle. Accept the default dimension. Notice how the dimension line is located at the point where you selected the circle.

8. Draw the following object and use radius dimensioning to place the dimensions shown.

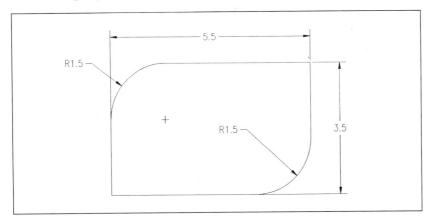

FIGURE 15-42

9. Draw a square approximately one-third the size of the screen.

   From the DRAW DIM screen menu, select "Linear:" from the menu.

   Command: **Dimlinear**
   First extension origin or RETURN to select:

   Use object snap intersection to capture the lower left corner of the box you drew. The command sequence will continue.

   Second extension line origin:

   Use object snap intersection again to capture the lower right corner. The command sequence will respond:

   Dimension line location (Text/Angle):

   Enter a point on the screen approximately ¼-inch actual distance from the bottom line of the square. You have just placed a dimension.

10. Repeat the same procedure for the upper line of the square. Instead of specifying the corners, however, just press [ENTER] when prompted and select the upper line to dimension. The command sequence will appear as follows:

    Command: **Dimlinear**
    First extension line origin or RETURN to select: ([ENTER])
    Select line, arc, or circle: *(Select the line.)*
    Dimension line location (Text/Angle/Horizontal/Vertical/Rotated): **T**
      (The EDIT MTEXT dialog box appears. Type in your name. The "< >" characters represent the default dimension measurements. Delete them if you do not want them to show. Click the OK button when finished.)
    Dimension line location (Text/Angle/Horizontal/Vertical/Rotated): *(Enter a point where you want the dimension line.)*

    Notice how AutoCAD measured the selected line length, placing extension lines in the proper positions. For the dimension text, you entered your name. Notice that you can accept the default measurement, enter your own numerical value, or enter a text string (such as your name).

11. Select Vertical from the menu and dimension one of the vertical sides of the box you drew. The vertical dimension works the same as the horizontal dimension.

12. Draw a triangle. Let's use the aligned option to dimension one of the angled sides. Refer to the following command sequence.

    Command: **Dimaligned**
    First extension line origin or RETURN to select: [ENTER]
    Select line, arc, or circle: *(Select an angled side.)*
    Dimension line location (Text/Angle/Horizontal/Vertical/Rotated): *(Select a point.)*

    AutoCAD will draw a dimension line parallel to the selected side.

13. Clear the screen and draw one horizontal dimension line. Accept the default dimension measurement. From the MOD DIM screen menu, select "DimEdit" and select "New". Delete the "< >" characters in the EDIT MTEXT editor and enter **250**. Click **OK**. AutoCAD will ask you to "select objects". Select the dimension text of the dimension line you drew. The text will update to "250".

14. Select "DimTedit". Select the dimension line. Then select "Angle". When prompted for the angle, enter **45** for 45 degrees. Did the text turn to a 45 degree angle?

15. Select "DimTedit" (text-edit). Select the dimension line. Now select "left" from the menu.

16. Select "DimTedit" again. When asked to "select dimension", select the dimension you drew. Now select "home" from the menu. The dimension text should return to its original position.

17. Use the dimension commands you have learned to draw and dimension the following objects, which are on the work disk.

FIGURE 15-43

INTRODUCTION TO DIMENSIONING **15-35**

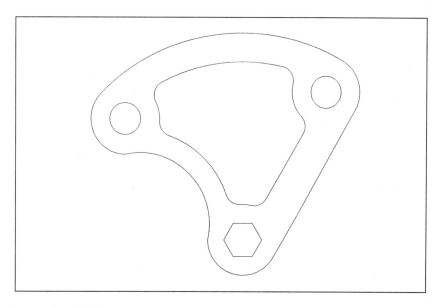

FIGURE 15-44

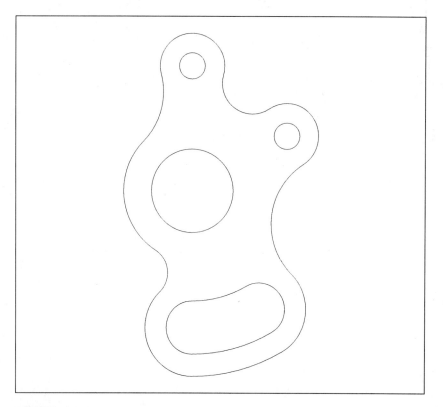

FIGURE 15-45

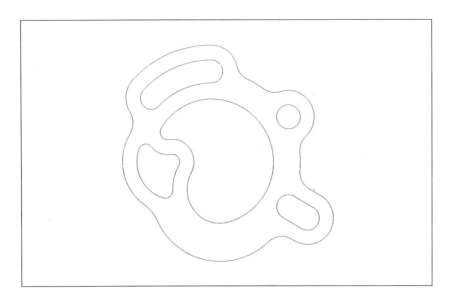

FIGURE 15-46

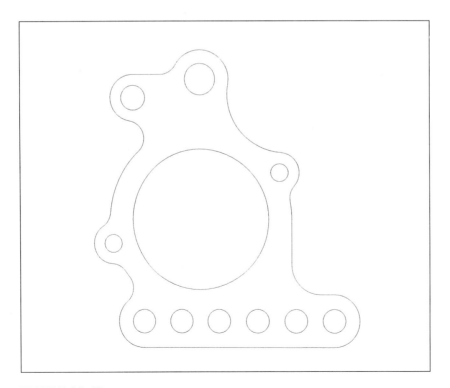

FIGURE 15-47

TUTORIAL

Start the drawing named "DIMEN" that is on the work disk. We will dimension this drawing, using several of the dimensioning commands you have learned. The finished drawing will look like Figure 15-48.

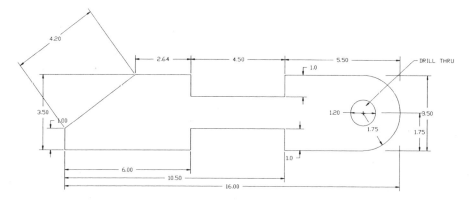

FIGURE 15-48 DIMEN.DWG Work Disk Drawing

Let's start with linear dimensioning.

Linear Dimensioning

From the menu bar, choose DRAW Dimensioning. You will notice that dimensioning has its own submenu.

From the dimension menu, choose Linear. Respond to the "First extension line origin" prompt by entering a point at the corner shown as "point 1" in Figure 15-49.

Respond to the "Second extension line origin" prompt by entering a point at the corner shown as "point 2" in Figure 15-49.

You should now be prompted with "Dimension line location". This is the distance above the object that the dimension line will be placed. Enter a point at the location of "point 3".

AutoCAD now has measured the distance and calculated it for you. It is displayed as "Dimension text value".

Isn't that easy? From now on, we will simply refer to the points labeled in the illustrations for each entry.

Let's continue this dimension line. From the Dimensioning menu, select "Continue".

Respond to "Second extension line origin" with point 4.

Repeat this procedure by selecting point 5 as the extension line origin. (Don't forget to press [ENTER] twice.)

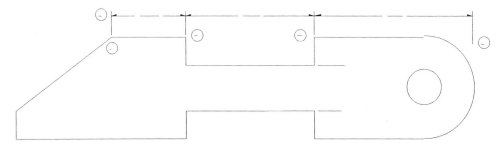

FIGURE 15-49 Linear Dimensioning (Continue)

You have now completed a horizontal and a continuing dimension line.

Baseline Dimensioning

Let's now do some baseline dimensioning.

This dimensioning method starts out the same way that continuing dimensioning does. From the Dimensioning menu, choose "Linear", then press [ENTER] so you can select an object to dimension. When prompted to select an entity, select the line defined by endpoints 6 and 7. Choose point 8 for the dimension line location.

Now, this is where we create the baseline dimensioning style. Choose "Baseline" from the Dimensioning menu and enter point **9** as the second extension line origin. Press [ENTER] twice.

Repeat the procedure by choosing "Baseline" again, entering point **10** as the second extension line origin.

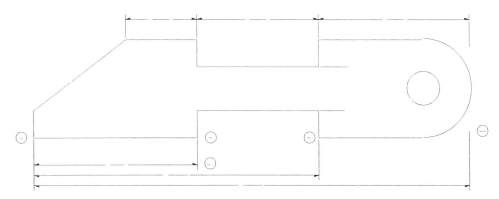

FIGURE 15-50 Linear Dimensioning (Baseline)

Notice that AutoCAD offset each dimension line the proper distance.

Vertical Dimensioning

Now let's draw the vertical dimensions.

Choose "Linear" from the Dimensioning menu. Choose point 11 as the first extension line origin, point 12 as the second extension origin and point 13 as the dimension line origin.

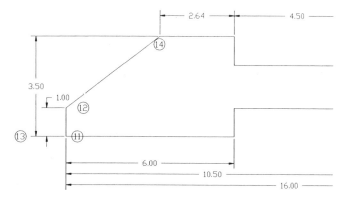

FIGURE 15-51 Linear Dimensioning (Vertical)

From the Dimensioning menu, choose "Baseline", and enter point 14 as the second extension line origin.

Repeat the procedure for the opposite end, using points 15 through 18 as the entered points. You can do this one on your own.

If you mess up, use the Undo command to eliminate the last dimension, or cancel if the dimension line is not yet drawn.

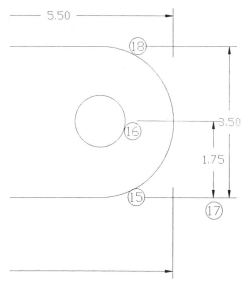

FIGURE 15-52 Completing the Vertical Dimension

Linear Dimensioning (Aligned)

Let's construct the dimension line which describes the angle at the left side of the part.

Choose "Aligned" from the Dimensioning menu, then press [ENTER]. Select the angled line on the object. When prompted for the dimension line location, select point 21.

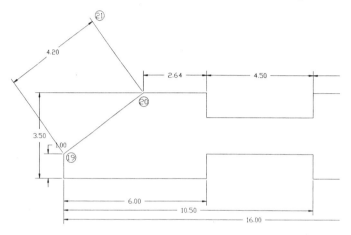

FIGURE 15-53 Linear Dimensioning (Aligned)

AutoCAD constructed the dimension line in alignment with the two extension line origin points.

Diameter Dimensioning

Select "Diameter" from the Radial submenu. You are prompted "Select arc or circle". Enter a point on the circle (point 22) to identify the object you want dimensioned. Press [ENTER] to accept AutoCAD's measurement. Show point 23 as the distance for the leader line.

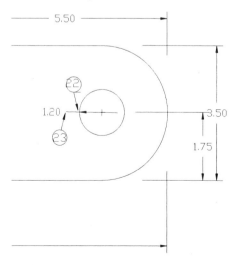

FIGURE 15-54 Diameter Dimensioning

Notice that AutoCAD used the point you entered on the circle as one of the endpoints of the dimension line. The other endpoint is positioned by drawing the dimension line from this point through the center of the circle to the opposite circumference. The extended text position is referenced beside the first point entered.

Radius Dimensioning

Now, let's dimension the arc which closes the right end of the object.

Select "Radius" from the Radial submenu. When prompted, choose the point on the arc which you want the dimension line to intersect (point 24). Move the crosshair to place the dimension leader.

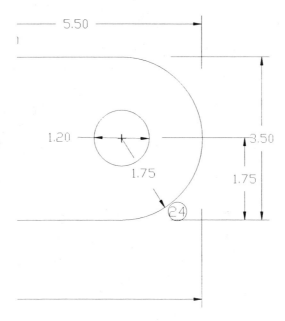

FIGURE 15-55 Radius Dimensioning

Leader Construction

Let's construct a leader line and some text. Select "Leader" from the Dimensioning menu.

Enter point **25** as the leader start and point **26** as the "to point". For the dimension text, let's put in some words (remember, we can use words instead of numerals for any dimension text). Press ENTER to accept the default action "ANNOTATION". Type in "DRILL THRU" and press ENTER twice.

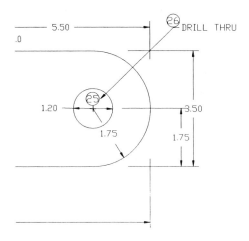

FIGURE 15-56 Constructing a Leader Line with Text

Your dimensioned drawing should look like the one in Figure 15-57.

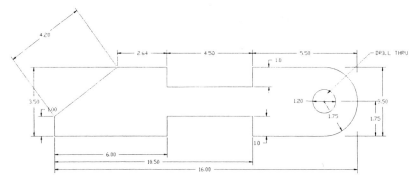

FIGURE 15-57 Completed Dimensioned Drawing

CHAPTER REVIEW

1. What is the dimension line?

2. What is the difference between the extension line and the witness line?

3. Draw an example of a baseline dimension.

4. Name the types of linear dimensions.

5. What does the DimDiameter dimension command do?

6. What command would you use to tilt the dimension text?

7. How would you apply new changes to an existing dimension?

8. What layer are the dimension definition points placed on?

9. What are "arrow blocks"?

CHAPTER 16

DIMENSION STYLES AND VARIABLES

AutoCAD is rich in dimensioning capabilities. To obtain professional results, you must understand how to set the many dimension variables that control the dimension. After completing this chapter, you will be able to:

- Manipulate different types of variables used to control the look and function of AutoCAD's dimensions.
- Control the dialog boxes to set the many variables.

DIMENSION STYLES AND VARIABLES

How dimensions look and act depends on the settings of *dimension variables*. The variables can control such characteristics as whether the dimension text is placed within or above the dimension line, whether arrows or ticks are used, and many other options.

After you set the variables for a dimension, you can store the settings in a file called a *style*. Storing dimension style files is similar to storing text style files. In both cases you can store many style files, then retrieve them for use.

In order to effectively use AutoCAD's dimensioning capabilities, you must become familiar with the variables available. The following sections show you what options are available and how to set and store them.

SETTING DIMENSION VARIABLES

AutoCAD's dimension variables are set through a dialog box. To display the dialog box, enter **Ddim** at the command line, or select Dimension Style from the Data menu.

The DDim command displays the Dimension Styles dialog box.

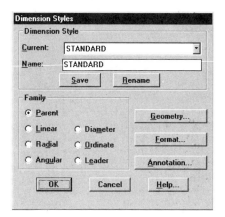 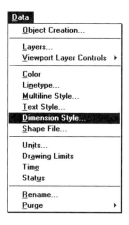

FIGURE 16-1 Dimension Styles Dialog Box

This dialog box controls almost all *dimension system variables* (called "dimvars," for short), which in turn affect the look of dimensions in your drawings. After you customize the look of the dimensions, you save that look by giving it a name, called a "dimension style" (or "dimstyle," for short). You can use many dimstyle names, which means you can create many customized sets of dimensions — one for each client and project, if need be.

There are two dimvars not set by DDim. These are DimSho (toggles whether dimensions are updated while dragged) and DimAso (determines whether dimensions are drawn as associative or non-associative). By default, both have a value of 1 (are turned on). You change the value of these two dimvars at the Command: prompt, as follows:

    Command: **dimsho**
    New value for DIMSHO <1>:
    Command: **dimaso**
    New value for DIMASO <1>:

The Dimension Styles dialog box contains three areas:

Dimension Style. You create, save, select, and rename dimension styles by name at the top of the dialog box. Create a dimstyle by entering the name next to "Name:", then click on the "Save" button. Select a dimstyle from the drop-down list box next to "Current:". Rename the current style by entering the new name in the "Name:" box, then clicking on the "Rename" button.

Family. In most cases, you change dimvars for every kind of dimension (called the "parent" dimstyle by AutoCAD). However, if you want the change in dimvars to affect only one kind of dimension, then click on the radio button next to Linear, Radial, Angular, Diameter, Ordinate, or Leader.

The lower right area of the Dimension Styles dialog box contains three buttons that lead to other dialog boxes. The "Geometry," "Format," and "Annotation" buttons lead to dialog boxes that display dimensioning parameters grouped logically together. Let's have a look at these now:

Dimension Line Geometry

The Geometry dialog box groups options for the display of dimension lines, extension lines, arrowheads, center marks, and dimension scaling.

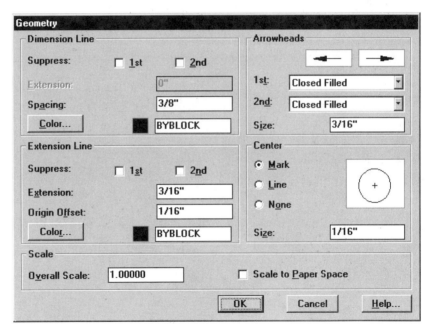

FIGURE 16-2 Geometry Dialog Box

Dimension Line. The "Dimension Line" area of the dialog box lets you suppress the first (the value is stored in system variable DimSd1) or second dimension line (stored in system variable DimSd2). The "Extension" option only applies when tick marks are used in place of arrowheads; this value is the distance that the tick mark extends beyond the extension line (stored in DimDle).

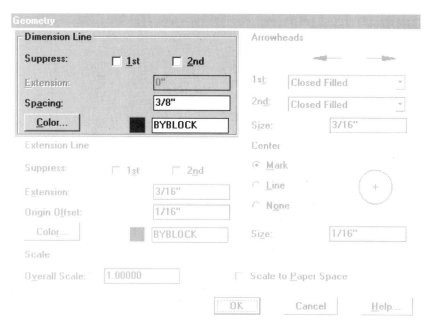

FIGURE 16-3   Dimension Line Area

The "Spacing" option determines the distance between dimension lines that are automatically stacked by the DimBaseline command (see Figure 16-4). The distance is stored in system variable DimDli.

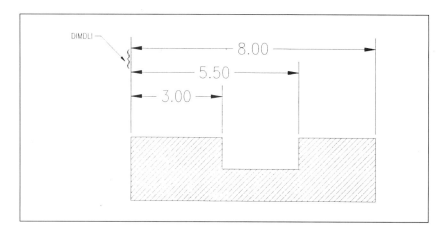

FIGURE 16-4   DimDli Controls the Distance Between Stacked Dimensions

To change the color of the dimension line from its default color of white, click on the "Color" button. This displays the standard Select Color dialog box, from which you can select the color, whose value is stored in DimClrd.

DIMENSION STYLES AND VARIABLES **16-5**

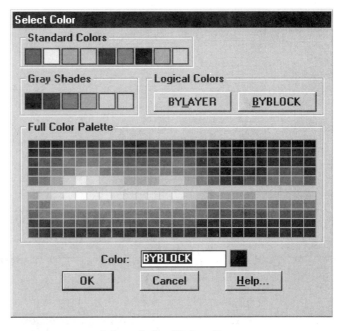

FIGURE 16-5 Select Color Dialog Box

Extension Line. As with dimension lines, you can opt to suppress the first (DimSe1) or second (DimSe2) or both extension lines.

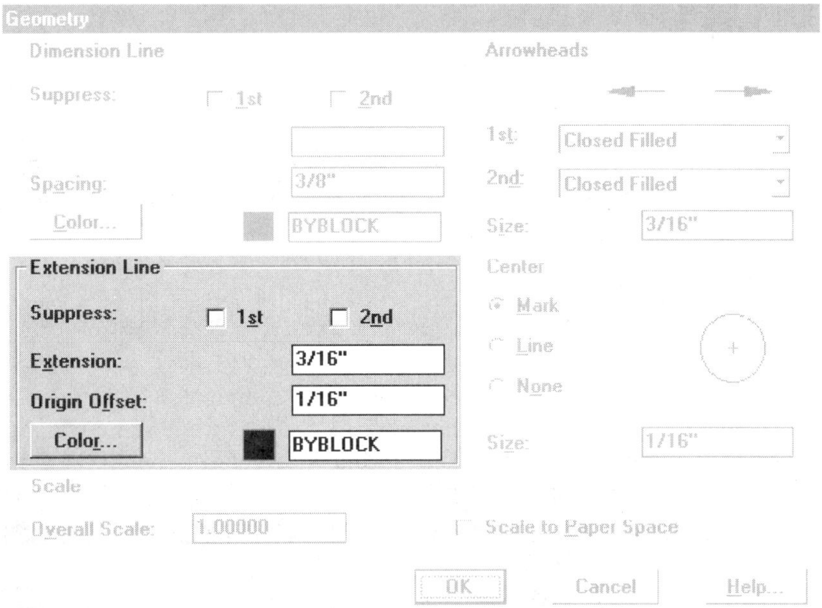

FIGURE 16-6 Extension Line Area

The "Extension" value is the distance the extension line protrudes beyond the dimension line (stored in DimExe) (see Figure 16-7).

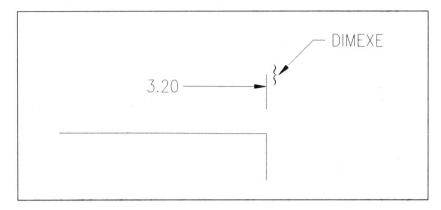

FIGURE 16-7 DimExe Controls the Distance That the Extension Line Extends Beyond the Dimension Line.

The "Origin Offset" value is the distance the extension line begins away from the object being dimensioned (DimExo) (see Figure 16-8).

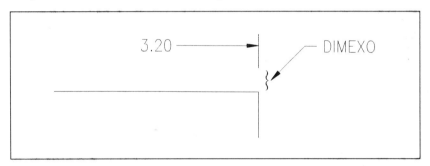

FIGURE 16-8 DimExo Controls the Distance Between the Pick Point and Start of the Extension Line.

You can select a different color for the extension line by clicking the "Color" button.

> **NOTE:** Selecting different colors for each component of a dimension is not as superfluous as it might appear at first. Since AutoCAD matches pen width to entity color, you can control the plotted width of dimension lines and extension lines through the use of color.

Arrowheads. You can select from eight different, predefined arrowhead styles, or create your own custom arrowhead. You have the option of specifying a different arrowhead at each end of the dimension line (DimBlk1 and DimBlk2).

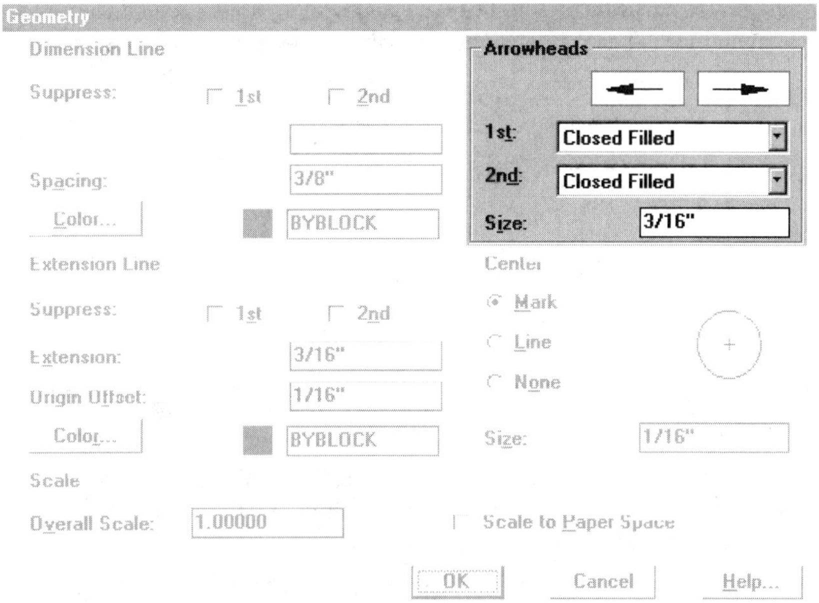

DIMENSION STYLES AND VARIABLES 16-7

FIGURE 16-9 Arrowheads Area

In addition to the standard arrowhead, AutoCAD includes the tick (Figure 16-10), the dot (Figure 16-11), open arrowhead, open dot, right-angle arrowhead, or no head at all.

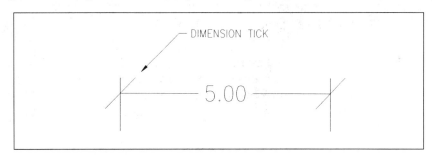

FIGURE 16-10 Dimension Tick

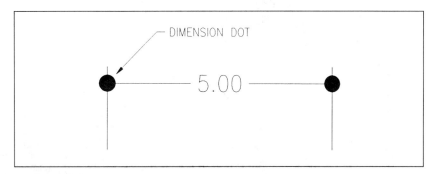

FIGURE 16-11 Dimension Dot

The custom arrowhead is defined by creating an object scaled to unit size, then saved as a named block. The name of the arrowhead block is stored in system variable DimBlk.

FIGURE 16-12 User-Defined Arrowhead

The size of the arrowhead is the distance from left end to right end; for custom arrowheads, AutoCAD scales the unit block to size (0.18 units, by default). The value is stored in DimAsz.

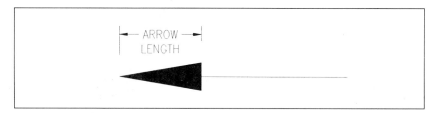

FIGURE 16-13 Length of the Arrowhead

Center. AutoCAD lets you mark the centers of circles and arcs with the DimCen command, which can place a center mark (see Figure 16-14), or a center mark with extending lines (see Figure 16-15), or no mark at all.

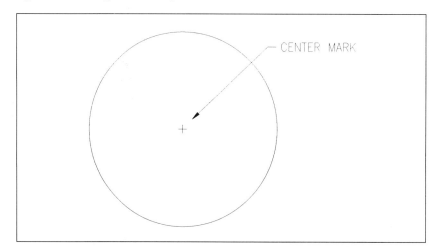

FIGURE 16-14 Center Mark

DIMENSION STYLES AND VARIABLES **16-9**

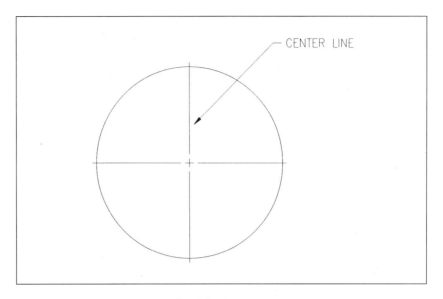

FIGURE 16-15 Center Mark with Lines

Click on the radio buttons next to "Mark," "Line," or "None" to change the look (stored in DimCen). To specify the size of the centermark, enter the value next to "Size."

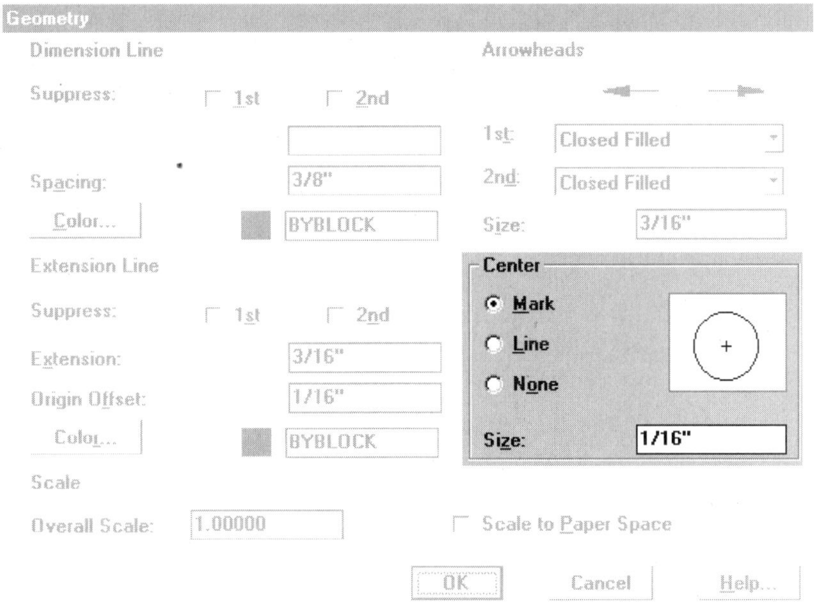

FIGURE 16-16 Center Line

Scale. This part of the dialog box controls the overall scale of dimensions when placed in the drawing (DimScale). It does not change the size of dimensions already in the drawing.

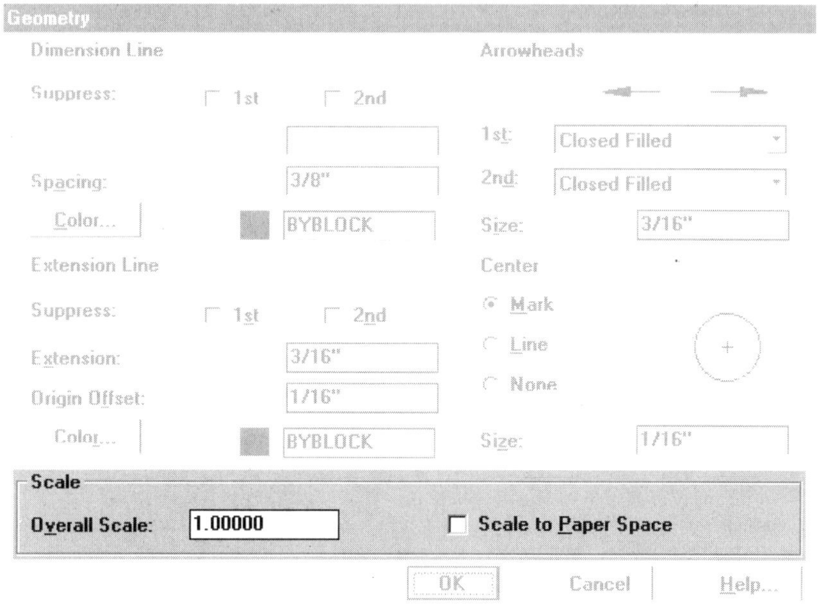

FIGURE 16-17 Scale Area

Entering a value of 2 doubles all aspects of a dimension. For example, the arrowheads and text are drawn twice as large. Clicking the check box next to "Scale to Paper Space" scales dimensions to a factor based on the scale between model and paper space.

Click on the "OK" or "Cancel" buttons to return to the Dimension Styles dialog box.

Dimension Format

From the parent Dimension Styles dialog box, click on the *Format* button to display the Format dialog box.

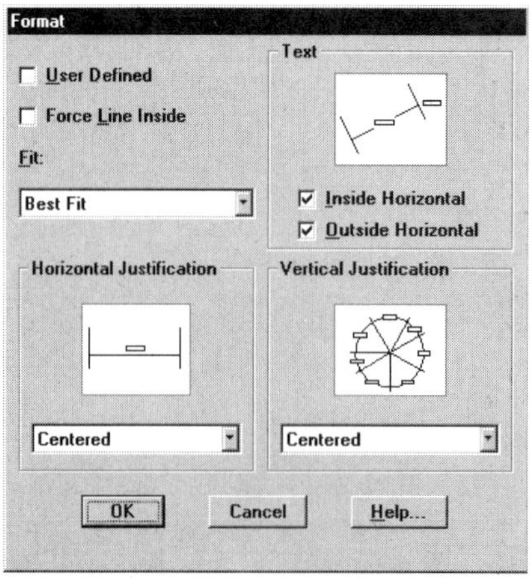

FIGURE 16-18 Format Dialog Box

User Defined. The "User Defined" check box is one of the most important. When checked (turned on), all dimension commands prompt you for the dimension text, allowing you to change the text, if necessary. By default, this value was turned on for earlier releases of AutoCAD; in Release 13, it is turned off, by default and stored in system variable DimUpt.

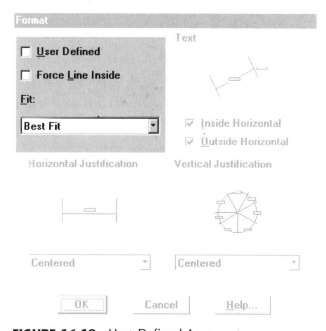

FIGURE 16-19 User-Defined Area

Force Line Inside. Normally, AutoCAD places the dimension lines, arrowheads, and text where there is room. Usually, this is inside the extension lines; if the extension lines are too close together, AutoCAD places those elements outside the extension lines.

Click on the check box next to "Force Line Inside" to force AutoCAD to always draw the dimension lines inside the extension lines (DimTofl); the arrowheads and text are placed outside, if there isn't enough room.

Fit. The dropdown list box next to "Fit" lets you control where the other elements are placed when the distance between extension lines is too narrow (DimFit):

> **Best Fit:** Dimension elements are placed where there is room to fit.
>
> **Text Only:** Dimension text is placed between extension lines, while arrowheads are placed outside — when space is lacking.
>
> **Text and Arrows:** Same as the "Force Line Inside" option.
>
> **Leader:** When there isn't enough room for the text, AutoCAD draws a leader line between the dimension line and the text.
>
> **Arrows Only:** When space is available for arrowheads only, places them between extension lines.

Horizontal Justification. The dropdown list box lets you select the horizontal placement of dimension text (DimJust):

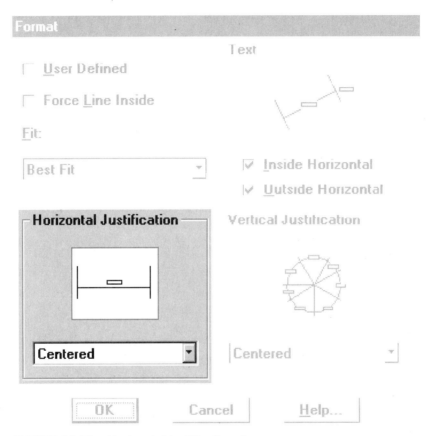

FIGURE 16-20 Horizontal Justification Area

Centered: Dimension text is centered between the extension lines and on the dimension line (the default).

1st Extension Line: Text is left justified against the first extension line (usually the left-hand extension line).

2nd Extension Line: Text is right justified against the second extension line.

Over 1st Extension Line: Text is positioned over the first extension line.

Over 2nd Extension Line: Text is positioned over the second extension line.

Vertical Justification. This dropdown list box lets you select the vertical placement of dimension text (DimTad):

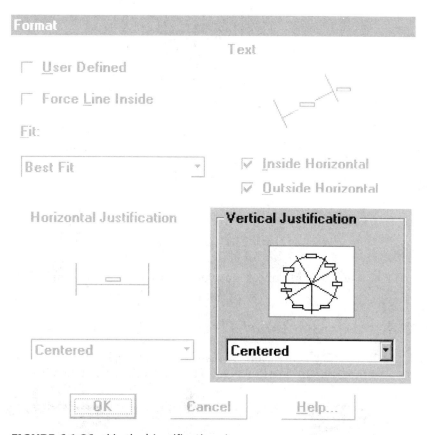

FIGURE 16-21 Vertical Justification Area

Centered: The dimension text is centered between the extension lines and on the dimension line (the default).

Above: Places text above the dimension line.

Outside: Places text "below" the dimension line (on the side furthest away from the dimension pick points).

JIS: Places text as per the Japanese Industrial Standards for dimensions.

16-14 USING AUTOCAD

Text. The "Text" area of the dialog box lets you force text outside the extension lines.

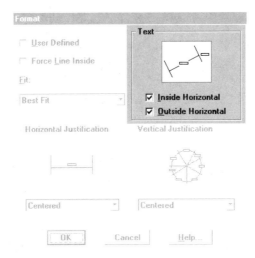

FIGURE 16-22 Text Area

Click on the "OK" or "Cancel" buttons to return to the Dimension Styles dialog box.

Dimension Annotation (Text)

From the parent Dimension Styles dialog box, click on the "Annotation" button to display the Annotation dialog box. This dialog box consists of five areas:

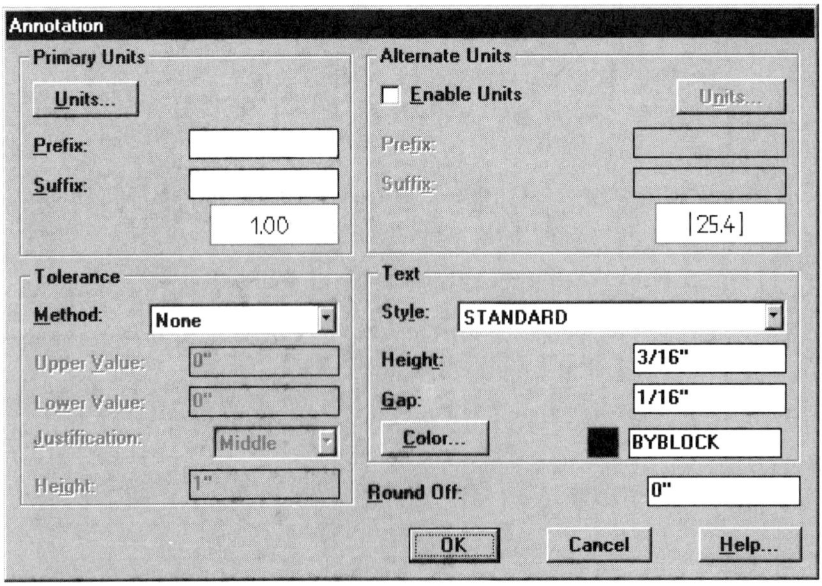

FIGURE 16-23 Annotation Dialog Box

Primary Units. Here you can specify the dimensioning units, and whether the dimension text will have a prefix or suffix.

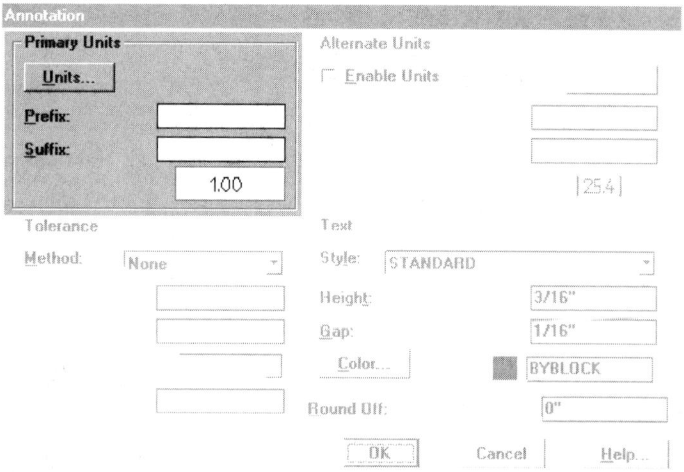

FIGURE 16-24 Primary Units Area

Enter any alpha-numeric value into the "Prefix" or "Suffix" text boxes. For example, prefix every dimension with the word "Verify," enter that in the Prefix box (see Figure 16-25).

FIGURE 16-25 Dimension Text with Prefix "VERIFY"

To suffix every dimension with "(TYPICAL)," enter that in the Suffix box (see Figure 16-26).

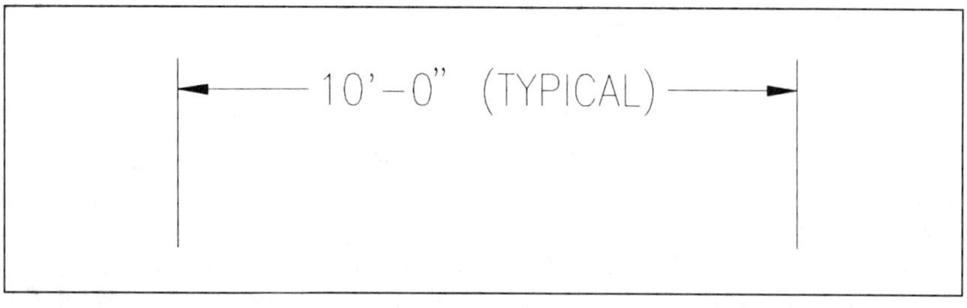

FIGURE 16-26 Dimension Text with Suffix "(TYPICAL)"

Since AutoCAD does not work with units (other than architectural), you may need to specify the units used by dimensions, such as mm. Click on the "Units" button to bring up the Primary Units dialog box.

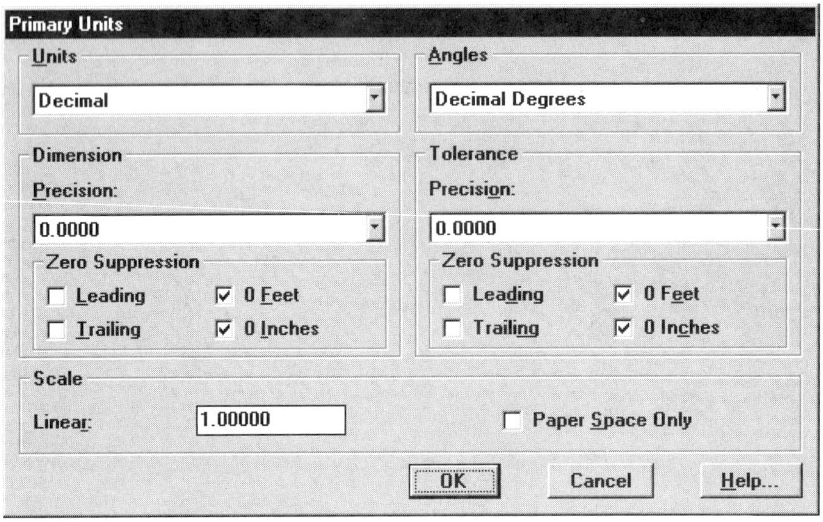

FIGURE 16-27 Primary Units Dialog Box

The dialog box has features similar to the Units dialog box displayed by the DdUnits command. Whereas the DdUnits command controls the units used by AutoCAD for everything in the drawing, this dialog box overrides the global units. The values you select here apply only to the dimension text.

> **Units.** You can select from decimal (the default), architectural, engineering, fractional, and scientific style of units (DimUnit).
>
> **Angles.** Select from decimal degrees (DDD.dddd, the default), Degrees/Minutes/Seconds (DD.MMSSdd), grads (DDg), radians (DDr), or surveyors units (NDDE). The value is stored in DimAUnit.
>
> **Dimension.** You can specify the precision of dimension text (DimTdec), and whether leading and trailing zeros, and the zero feet or inches are suppressed (DimZin).
>
> **Tolerance.** Specify the decimal places and zero suppression for tolerance text (DimTzin).
>
> **Scale.** This is repeated from the Geometry dialog box.

Click on the "OK" or "Cancel" buttons to return to the Annotation dialog box.

Alternate Units. AutoCAD allows you to place dimensions with double units: primary units, plus a second or alternate units. The alternate units are surrounded with square brackets. This is particularly useful for drawings that must show imperial and metric units.

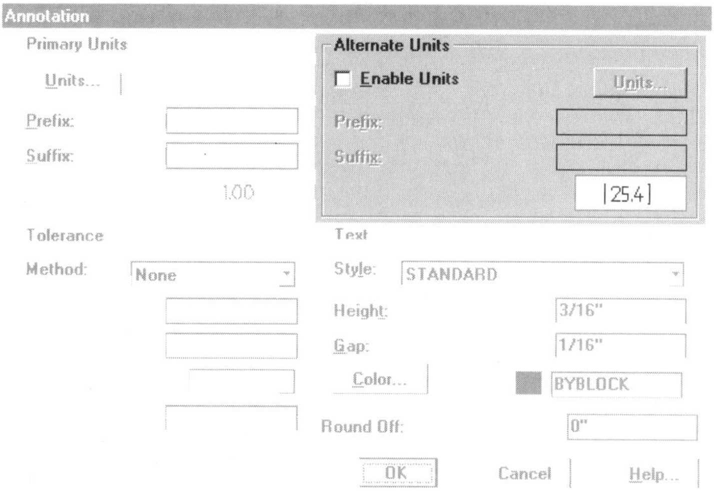

FIGURE 16-28 Alternate Units Area

When you select the check box next to "Enable Units" you can specify the parameters of the alternate units (DimAlt). The "Prefix" and "Suffix" text boxes have the same effect as in the "Primary Units" area. Clicking on the "Units" button displays the Alternate Units dialog box.

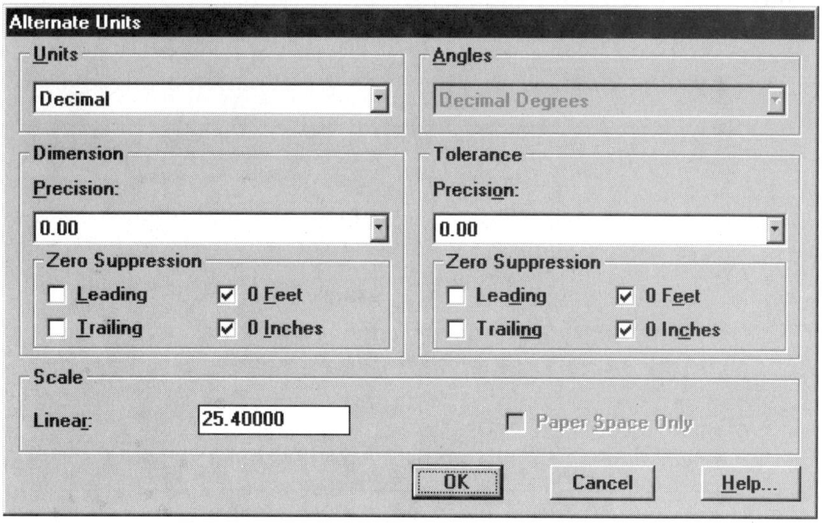

FIGURE 16-29 Alternate Units Dialog Box

This dialog box is exactly the same as the Primary Units dialog box. However, entering a scale factor (such as 25.4) makes AutoCAD calculate the alternate dimension text in the other units of measurement (mm, in this case).

Click on the "OK" or "Cancel" buttons to return to the Annotation dialog box.

Tolerance. The "Tolerance" portion of the dialog box lets you specify the look of tolerance notation in dimension text.

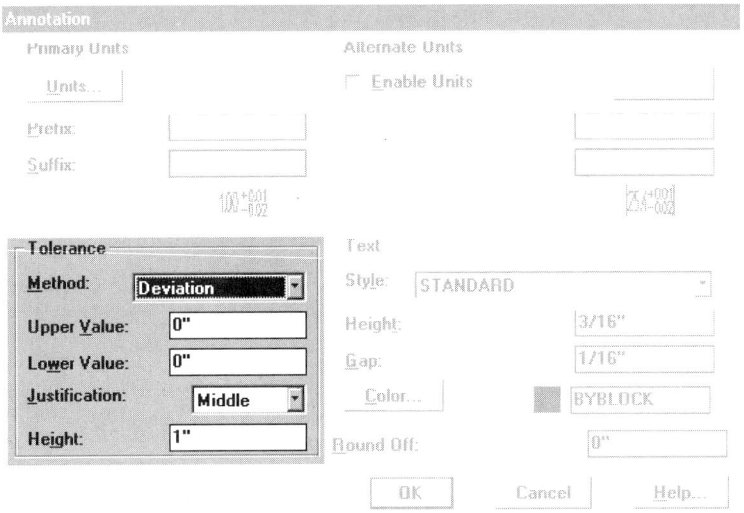

FIGURE 16-30 Tolerance Area

You can select from five different styles of tolerance text, as follows (controlled by DimTol and DimLim):

None.   The default: no tolerance text is placed.

Symmetrical.   A single plus/minus notation is added (see Figure 16-31).

FIGURE 16-31 Dimension Text with Tolerance

Deviation.   A pair of plus/minus notations are added (see Figure 16-32).

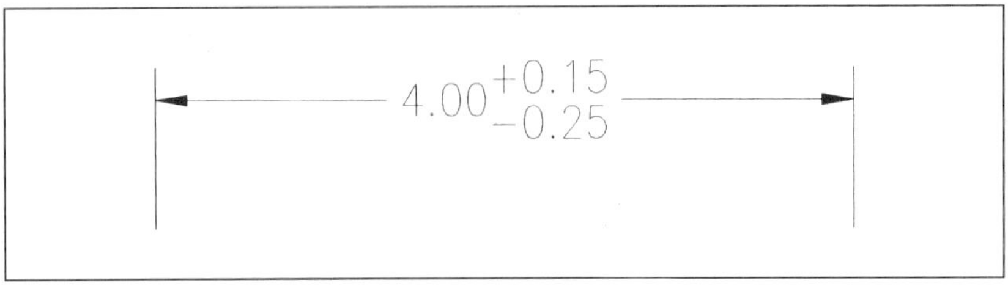

FIGURE 16-32 Dimension Text with Uneven Tolerance

Limits. A pair of dimension texts are placed (see Figure 16-33).

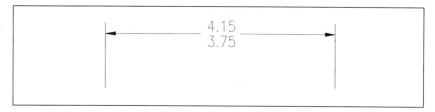

FIGURE 16-33 Stacked Dimension Text

Basic. A box is drawn around the dimension text (see Figure 16-34). Distance between the text and the box is stored in DimGap.

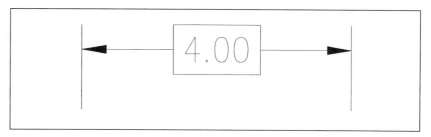

FIGURE 16-34 Boxed Dimension Text

In the remainder of the "Tolerance" area, you can specify the value of the upper (DimTp) and lower tolerance values (DimTm). The "Justification" options determine the relative placement of tolerance text (DimTolj): aligned to the top, middle, or bottom of the main dimension text. The "Height" scaling determines the size of the tolerance text relative to the main dimension text (DimTfac).

Text. The "Text" area lets you specify a style for the dimension text independent of other text used in the drawing.

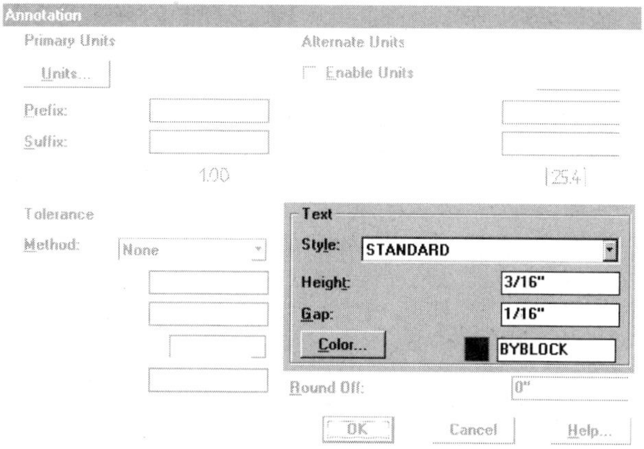

FIGURE 16-35 Text Area

16-20 USING AUTOCAD

You can select a text style name (DimTxsty), specify a fixed height (DimTxt), the gap between dimension line and text (DimGap; see Figure 16-34), and the color of the text (DimClrt). The size of the dimension text can be changed with the DimScale command.

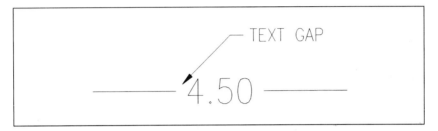

FIGURE 16-36 The Gap Between Dimension Line and Text

Round Off. This value applies a round-off to all dimension distances (not angles). For example, entering 1 rounds all dimensions to the nearest unit. Stored in DimRnd.

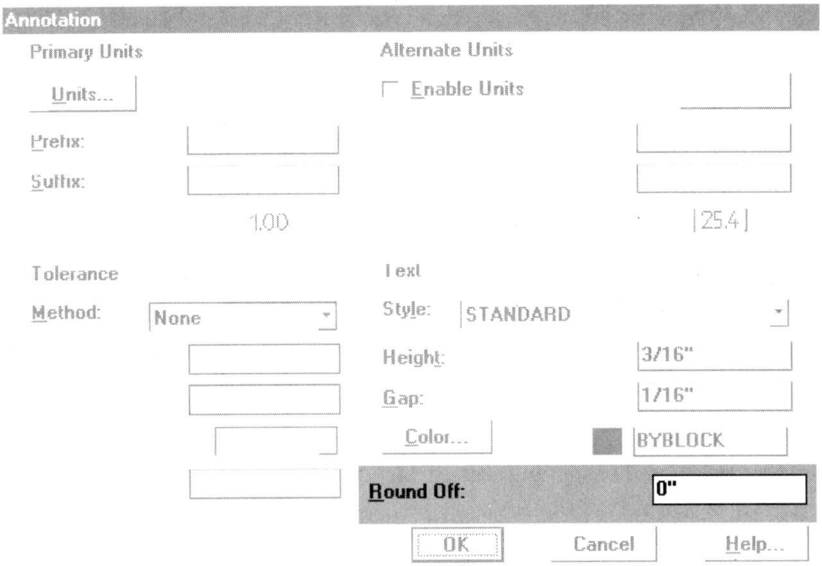

FIGURE 16-37 Round-Off Area

Click on the "OK" or "Cancel" buttons to return to the Dimension Styles dialog box. Remember to save your changes to the dimension variables!

CHAPTER 17

DIMENSIONING PRACTICES

Establishing dimensions practices ensures clear and concise graphic instruction. After completing this chapter, you will be able to:

- Perform standard dimensioning practices.
- Use dimensioning practices of different disciplines.

DIMENSIONING PRACTICES

The construction of an object can be pictorially shown by drawing methods. In order for the object to be constructed, it must also be described in terms of its dimensions. The combination of the pictorial representation and the dimensional information provides complete detail of the object that can be used for construction.

Proper dimensioning provides the necessary distances and notes to completely describe the object. The distances needed to draw the object are not necessarily those required for construction. Because of this, you must carefully select the dimensions you provide in your drawings. In order to properly dimension an object of any discipline, you should be familiar with the construction techniques used to build the object. It is helpful to mentally construct the object using the process that is common for building that object. Provide the dimensions that you would use when utilizing that method.

Placing Dimensional Information in a Drawing

There are two methods of placing dimensional information in a drawing: a dimension and a note. Dimensions are used to stipulate distances between two points. The extension lines of the dimension designate the points to which the dimension is applied. The dimension line shows the direction of the dimension, the arrows or ticks the extent of the dimension, and the dimension text conveys the actual distance in numerical terms.

Notes are used to indicate explanatory information that can not be properly conveyed with a dimension. Notes that are specific to a part of the object are indicated with a dimension leader.

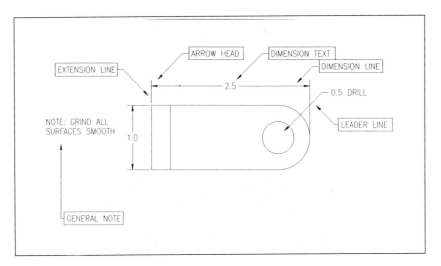

FIGURE 17-1 The Components of a Dimension

The arrow of the leader points to the part of the object to which the note applies. General notes are used to give information that is applied to the drawing as a whole. General notes are drawn without a leader.

Constructing Dimension Components

Many disciplines use drawing as a means of conveying information. The dimensioning practices of each discipline vary. The objective, however, is always the same: to provide clear, concise information that can be used to construct the object. Because the information must be conveyed in such a concise manner, the methodology of dimensioning must be carefully considered. Let's look at some of the principles of dimensioning.

Dimension Arrows. Dimension arrows are used to designate the extent of the dimension line. The length of the arrowhead will vary, depending on the scale of the drawing. Generally, arrowheads are 1/8 inch in length when used in small drawings, and 3/16 inch in larger drawings. Arrowheads that are too large or small are either distracting or difficult to read.

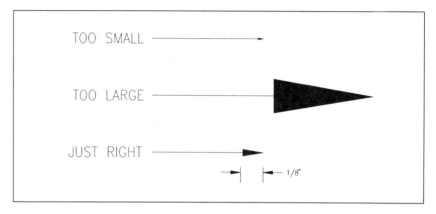

FIGURE 17-2 Dimension Arrowsize

Constructing Extension Lines. Extension lines designate the points to which the dimensions measure. The extension lines should not touch the points they reference. The normal offset from the reference points is 1/16 inch. Extension lines should extend approximately 1/8 inch beyond the dimension line.

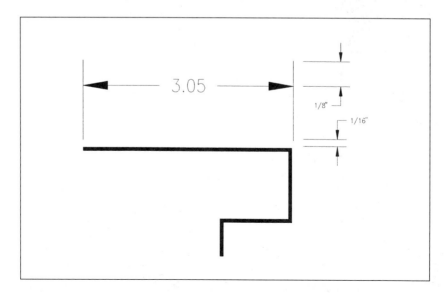

FIGURE 17-3 Extension Lines

Constructing Dimension Lines. Dimension extension lines and their dimension lines should be drawn outside the object whenever possible. Dimensions should only be drawn within the object when no other option is available.

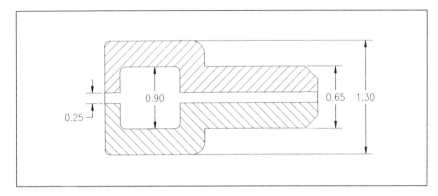

FIGURE 17-4 Dimension Lines

Constructing Dimension Text. Dimension text is used to provide the actual distance described by the dimension components. Text used in dimensions is generally ⅛ inch for small drawings and ⁵⁄₃₂ inch for larger drawings. The location of dimension text is dependent on the discipline. In architectural and structural drawings, the dimension is placed on top of the dimension line. In mechanical drawings, the dimension is usually placed within the dimension line.

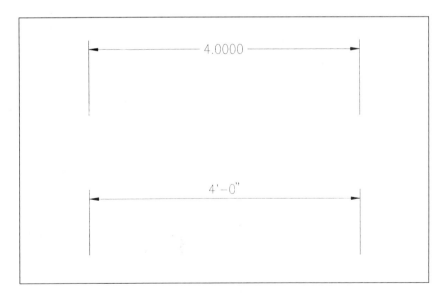

FIGURE 17-5 Dimension Text Practices

When stipulating distances, the text designating feet and inches is separated by a dash, such as 5'-4". If there are no inches, a zero is used; 6'-0". If dimensions are stipulated strictly in inches, the inch mark should be used to avoid confusion.

Fractions. Fractions are given either as common fractions such as ½, ¾, etc., or as decimal fractions such as 0.50 and 0.75. Normally, inches and common fractions are stipulated without a dash between them. In CAD, however, many text fonts do not have "stacked" fractions. When fractions must be constructed from standard numerical text, you should use a dash to avoid confusion, such as 3-1/2.

Dimensioning 3D Objects

Three-dimensional objects are generally dimensioned by stipulated rules. Let's look at some 3D objects and the accepted methodology of dimensioning each.

Dimensioning Wedges. Wedges are dimensioned in two views. The three distances that describe the length, width, and height are dimensioned in the two views.

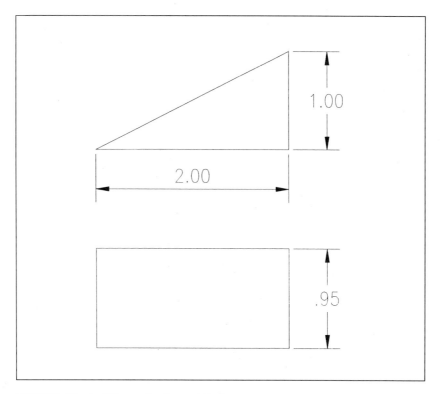

FIGURE 17-6 Dimensioning a Wedge

Dimensioning Cylinders. Cylinders are dimensioned for diameter and height. The diameter is typically dimensioned in the non-circular view. If a drill-thru is dimensioned, it is described by a diameter leader.

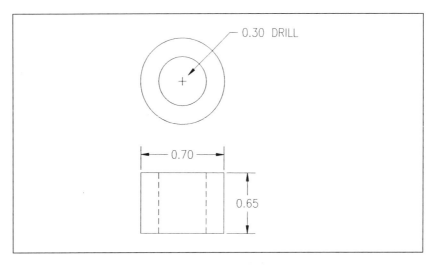

FIGURE 17-7 Dimensioning a Cylinder

Dimensioning Cones. Cones are dimensioned at the diameter and the height. Some conical shapes require two diameter dimensions.

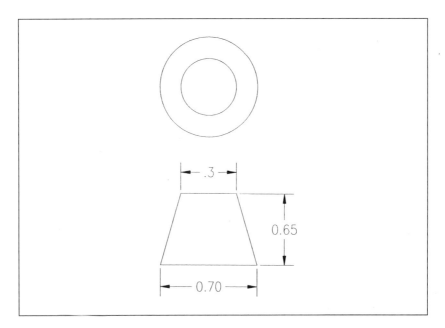

FIGURE 17-8 Dimensioning a Cone

Dimensioning Pyramids. Pyramids are dimensioned in a manner similar to cones.

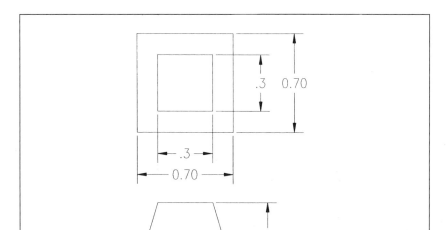

FIGURE 17-9 Dimensioning a Pyramid

Dimensioning Arcs and Curves. Arcs are dimensioned as a radius. The dimension line should be placed at an angle, avoiding placement as either horizontal or vertical. The dimension text designating the radius value should be followed by the letter R, designating that it is a radius dimension.

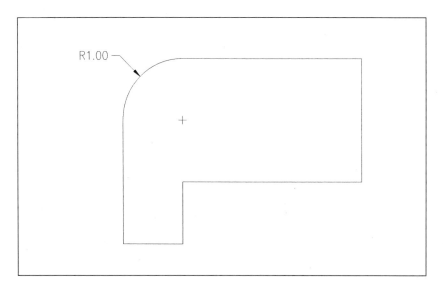

FIGURE 17-10 Dimensioning an Arc

An object constructed of several arcs is dimensioned by locating the center of the arcs with horizontal or vertical dimensions, and showing the radii with radius dimensioning.

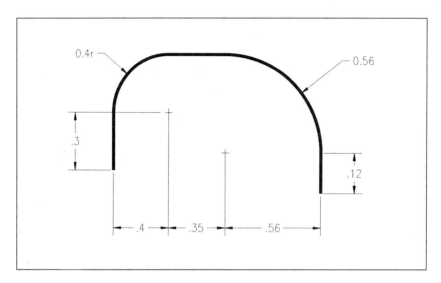

FIGURE 17-11 Dimensioning Multiple Arcs

Irregular curves are dimensioned with offset dimensions as shown in the following illustration.

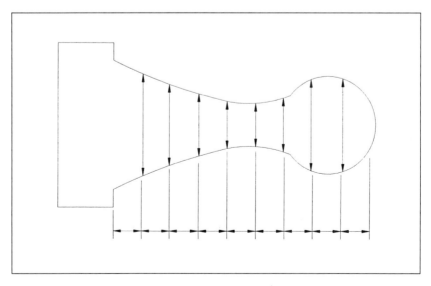

FIGURE 17-12 Dimensioning Irregular Curves

Dimensioning Mechanical Components

The following sections illustrate the methodology of dimensioning mechanical components. Note that many techniques vary due to individual interpretation.

Dimensioning Chamfers. A *chamfer* is an angled surface applied to an edge. Chamfers of 45 degrees are dimensioned by a leader, with the leader text designating the angle and one (or two) linear distances.

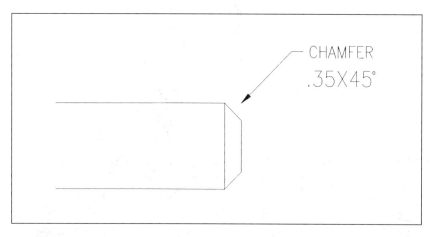

FIGURE 17-13 Dimensioning a Chamfer

If the chamfer is not 45 degrees, the part is described by dimensions showing the angle and the linear distances.

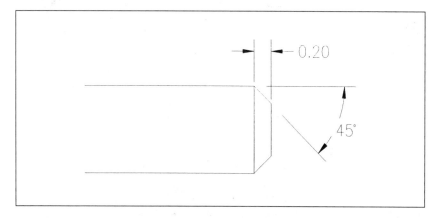

FIGURE 17-14 Dimensioning a Chamfer by Angle

Enlarging Parts for Dimensioning. A portion of a part may be too small to properly dimension. To properly show the dimensions, a segment may be enlarged.

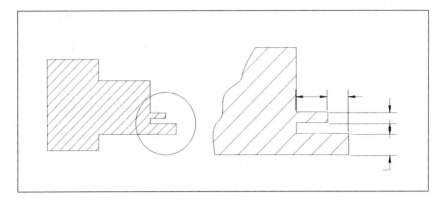

FIGURE 17-15 Enlarging a Detail for Dimensioning

Dimensioning Holes. Holes can be drilled, reamed, bored, punched, or cored. It is preferable to dimension the hole by note, giving the diameter, operation, and (if there is more than one hole) the number. The operation is used to describe such techniques as counterbored, reamed, and countersunk.

Standards dictate that drill sizes be designated as decimal fractions.

Whenever possible, point the dimension leader to the hole in the circular view.

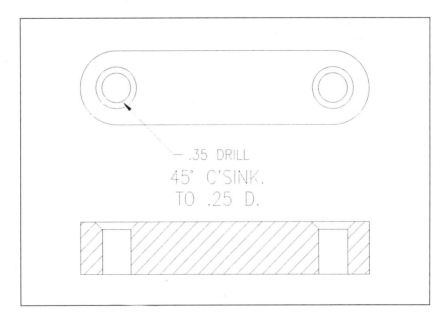

FIGURE 17-16 Dimensioning a Hole

Holes that are made up of several diameters can be dimensioned in their section.

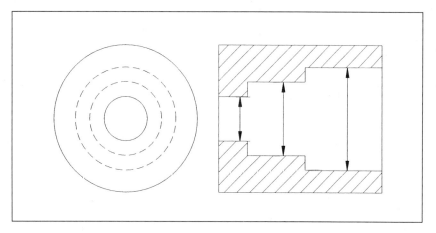

FIGURE 17-17 Dimensioning Holes of Several Diameters

Dimensioning Tapers. A taper can be described as the surface of a cone frustum. Tapers are dimensioned by giving the diameters of both ends and the rate of taper, given as the distance of taper per foot.

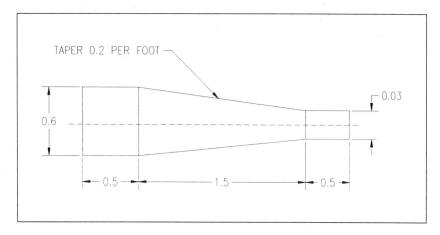

FIGURE 17-18 Dimensioning a Taper

EXERCISES

Dimension the following drawings.

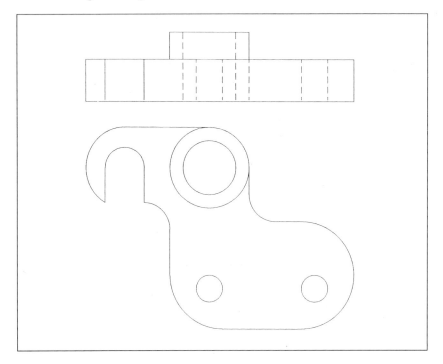

FIGURE 17-19

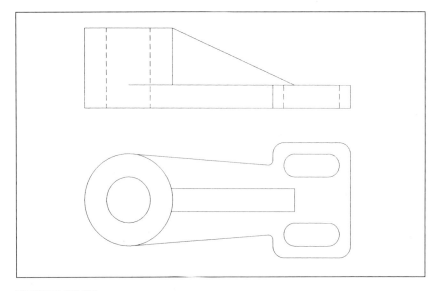

FIGURE 17-20

DIMENSIONING PRACTICES 17-13

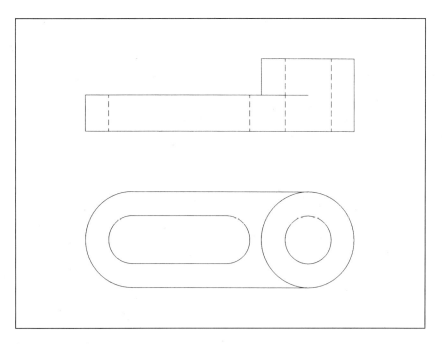

FIGURE 17-21

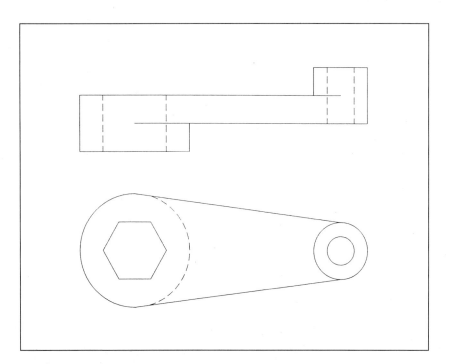

FIGURE 17-22

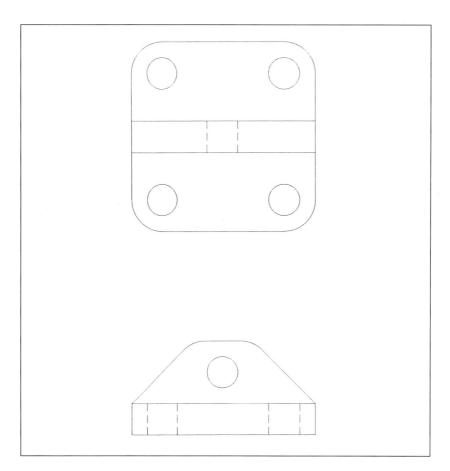

FIGURE 17-23

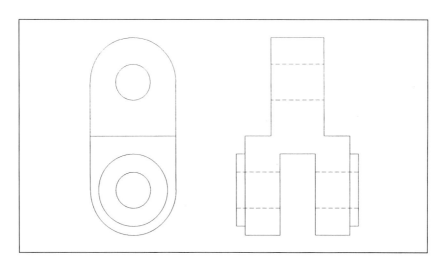

FIGURE 17-24

DIMENSIONING PRACTICES 17-15

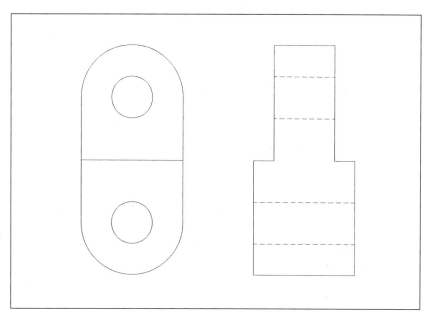

FIGURE 17-25

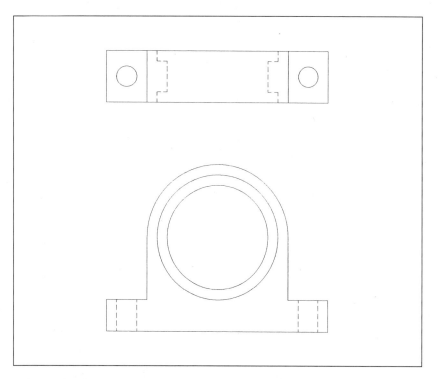

FIGURE 17-26

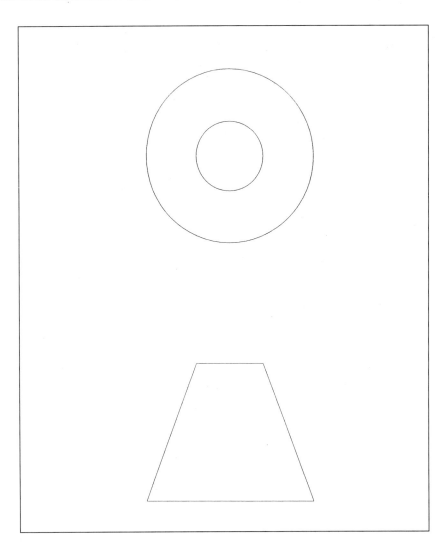

FIGURE 17-27

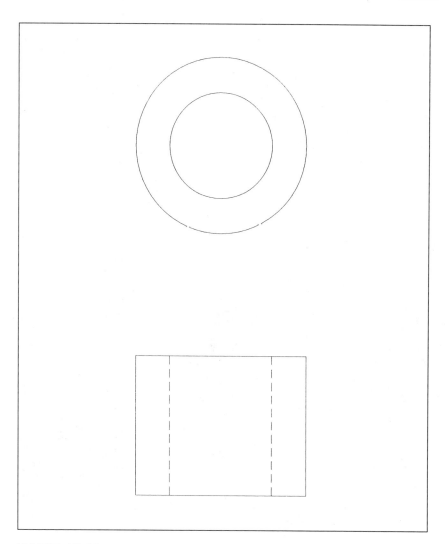

FIGURE 17-28

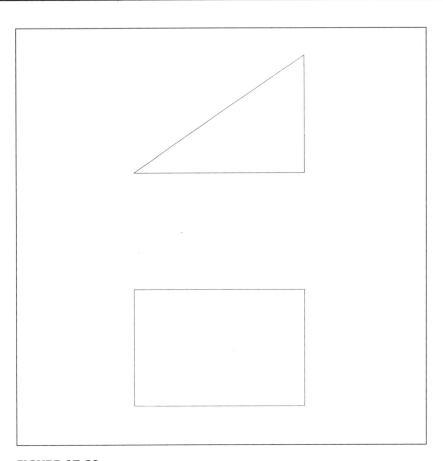

FIGURE 17-29

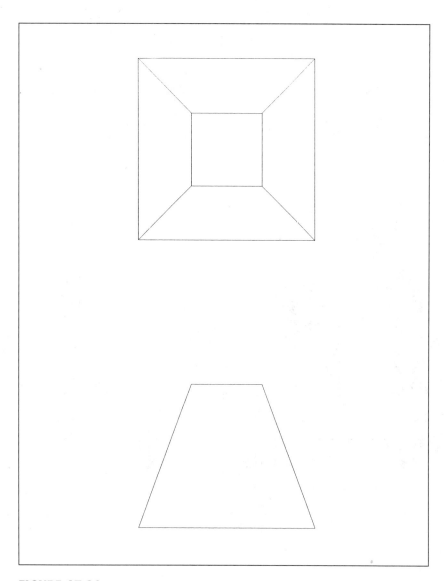

FIGURE 17-30

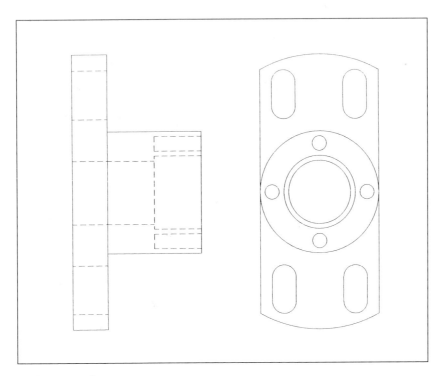

FIGURE 17-31

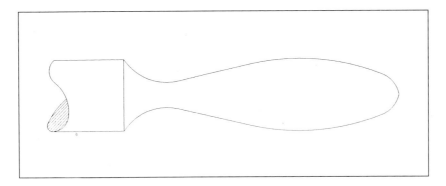

FIGURE 17-32

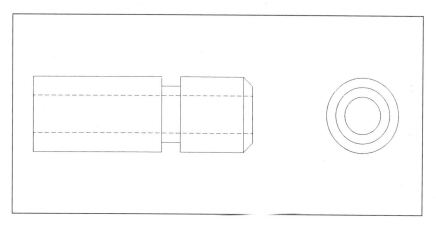

FIGURE 17-33

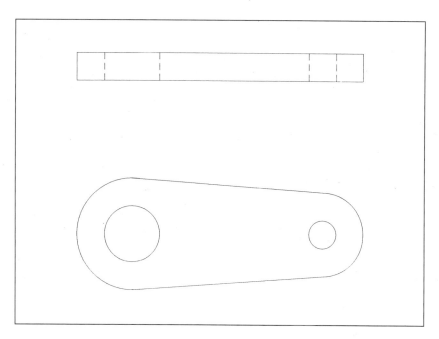

FIGURE 17-34

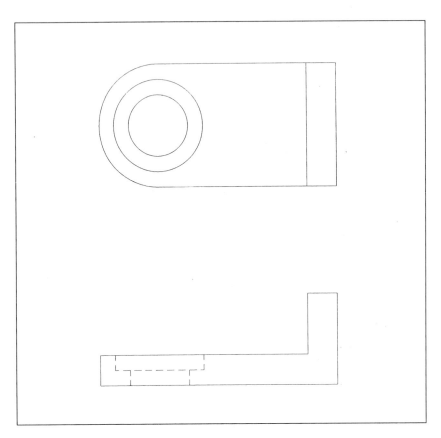

FIGURE 17-35

CHAPTER 18

PLOTTING YOUR WORK

The intended product of most CAD drawing products is to produce a hard copy of the work. In this chapter, you will learn how to plot your drawings. After completing this chapter, you will be able to:

- Demonstrate the advantages and disadvantages of different types of plots.
- Initiate a plot.
- Use AutoCAD's plot commands.
- Comprehend the relationship of plot scales and limits.
- Arrange a drawing for a plot of a specified scale and paper space.

PLOTTING OVERVIEW

You have learned so far how to build and edit drawings using AutoCAD commands, but wouldn't you really like to see them on paper? After all, the end product is usually a finished drawing on paper.

AutoCAD allows you to plot your work on a plotter or printer (referred to as a *printer plot*). There are several other types of output devices. Before you can plot a drawing, the particular plotter and/or printer must be installed by using the Config command.

Printer Plots

Printer plots are easy (and cheap), but the line quality is usually not very good. Printer plots are excellent for check plots of your work. Some printer plotters have the ability to plot up to C-size drawings, while others can plot in color.

Plotter Plots

Plots produced from a pen plotter are very high quality. The plotter uses a technical or marker pen to actually "draw" your work.

Plotters use different sizes of paper. You will learn about this later in this chapter. The scale at which you plot is determined by your requirements and the paper size.

PLOTTING THE WORK

There are two basic ways to plot a drawing:

- By using AutoCAD's Plot dialog box
- From the command line

The Cmddia system variable determines whether the Plot command displays dialog boxes or command line prompts. If Cmddia is set to 1, dialog boxes are displayed. If set to 0, command line prompts are used. Let's see how AutoCAD's plot dialog boxes are used to plot a drawing.

Using the Plot Command Dialog Boxes. To plot a drawing, enter the Plot command at the command line. When you use the Plot command (and Cmddia is set to 1), the following dialog box is displayed.

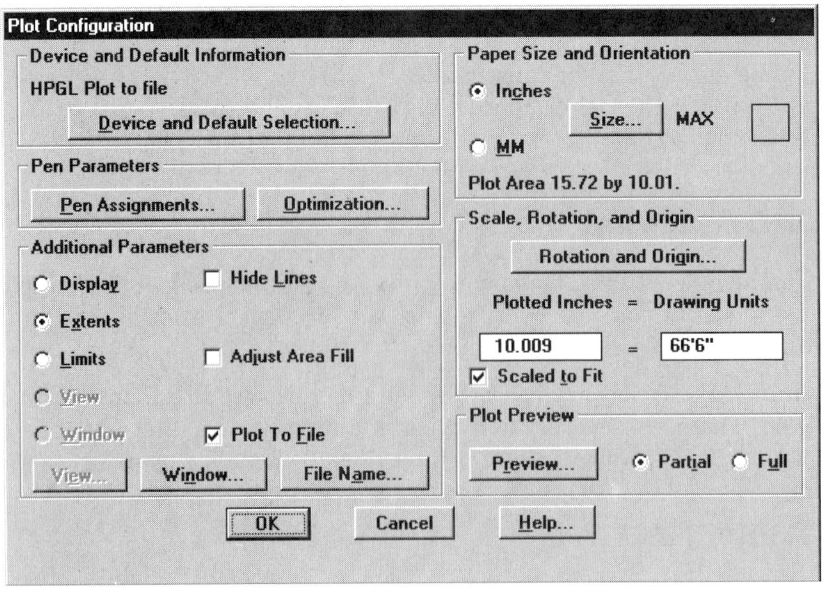

FIGURE 18-1 Plot Dialog Box

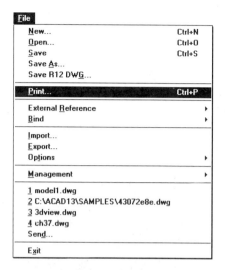

How to Plot. In order to plot a drawing, you must first determine several aspects concerning the plot. Let's take these in recommended order.

Determine the Proper Plotting Equipment. AutoCAD can store several settings for different plot devices. For example, you may have two different types of plotters available to you. The "Device and Default Information" section of the Plot dialog box displays the currently selected devices.

FIGURE 18-2 Device and Default Selection Dialog Box

To change the device, select the "Device and Default Selection..." button. AutoCAD displays a listing of the devices you have installed. To select a different plotter, click on the name of the device in the box, then select "OK".

Select Pen Assignments.   Many plotters are capable of plotting with several pens. The pens can be different widths or colors. The pens are stored in a pen carousel or in a pen corral. Other plotters simulate a multipen device by stopping and prompting the user to install a different pen.

AutoCAD uses entity colors to assign pens. For example, the user can specify that all entities in the drawing that are red will be plotted with a certain pen, while those that are blue are plotted with a different pen.

The pen used for each color is specified in the Pen Assignments dialog box. This dialog box is accessed by selecting "Pen Assignments.." from the Plot dialog box. The following sub-dialog box is displayed.

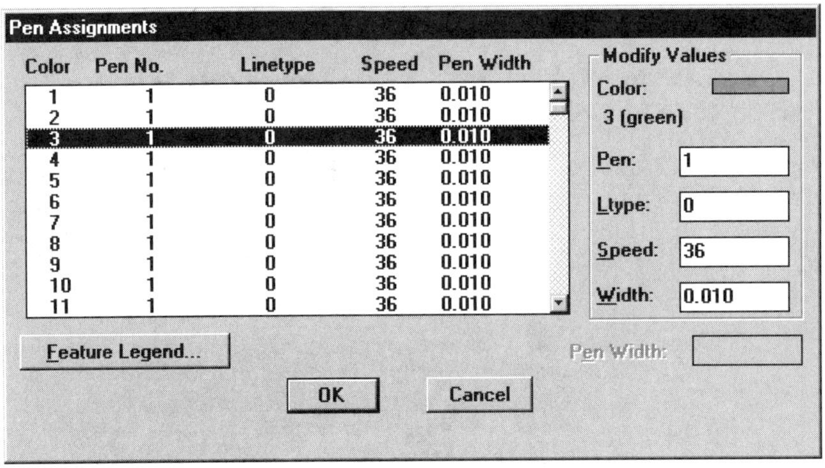

FIGURE 18-3   Pen Assignments Dialog Box

You can change the pen assignment, the linetype, the speed that the pen moves, and the width of the pen.

To make a change, first select the color number. For example, color number 1 represents red. The following listing shows the numbers AutoCAD uses to represent each color.

COLOR NUMBER	COLOR
1	RED
2	YELLOW
3	GREEN
4	CYAN
5	BLUE
6	MAGENTA
7	WHITE

Colors above 8 are listed by number only.

When you select a listing, the values in the "Modify Values" section of the dialog box lists the values set for that color. To change a value, click in the edit box and enter the new value.

Let's continue and look a bit closer at each of the items you can change.

Pen.   The pen number corresponds to the pen in a numbered carousel in the plotter. On single pen plotters that emulate a multipen plotter, AutoCAD will prompt for the pen by number.

Different pens can be used to plot drawings with different pen widths or colors. Using varying pen widths is common in complex drawings that are visually clearer when plotted in different line widths.

Some non-pen plotters (such as electrostatic) use the pen information to assign different line widths and/or colors.

Experienced CAD operators plan the drawing for the different pen widths as they construct the drawing. They draw objects in specified colors that correspond to pens they will use to plot the drawing. For example, all objects that will be plotted with a "wide" pen may be drawn in red. Entity colors can be assigned "BYLAYER" (see the Layer command), or by using the Color command to set all subsequent entity colors. The Change command can be used to change an entity color after it is constructed.

Linetype.   The linetype setting stipulates the type of line that is drawn for entities drawn in the color listed. You can see the linetypes available by selecting the "Feature Legend..." button.

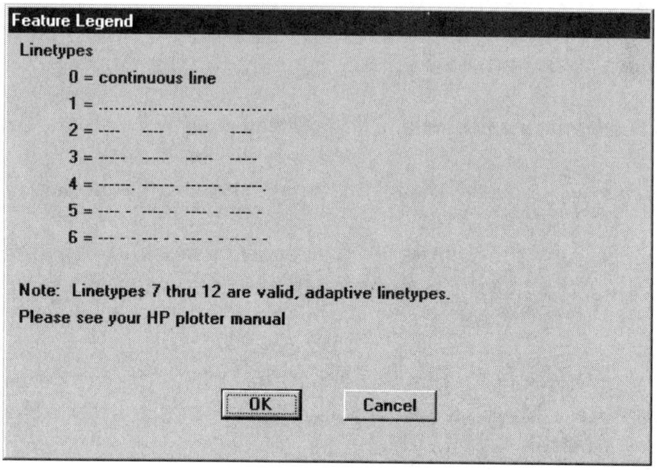

FIGURE 18-4   Feature Legend Dialog Box

> **NOTE:**   You should not mix "onscreen linetypes" and plotter linetypes. Either use onscreen linetypes or set the linetype by color. Using both will yield unsuitable linetypes.

Speed. The Speed setting controls the speed at which the pen moves. This is useful if some of your pens skip when moved too quickly. Note that the speed of the pen movement is limited by the speed at which the plotter is capable of moving the pen.

Width. The width setting specifies the actual linewidth drawn by the pen. The width should be set accurately since AutoCAD uses this information to calculate how much the pen is moved laterally when drawing in solids, polylines, and trace lines.

> **NOTE:** Most pens list the width somewhere on the pen. In most cases, these widths are in millimeters (for example, 0.5 mm). If you are drawing in inch units, be sure to convert the pen width to inches.

Select the Part of the Drawing to Plot. You can specify the portion of the drawing that you wish to plot. The choices are contained in the "Additional Parameters" box of the dialog box. Let's look at the choices.

Display. Clicking in the Display check box will result in a plot of the current screen view. This is a "what you see is what you get" plot.

Extents. The Extents option plots the portion of the drawing that contains any entities, eliminating "blank" areas. The plot would be the same as performing a Zoom Extents and then plotting that display.

If you choose the Extents option to plot a drawing that is in a perspective view, and the camera position is in the drawing extents, the following message is displayed.

    PLOT Extents incalculable, using display

The plot process will continue, but using the Display option instead of the Extents.

Limits. Plotting the Limits of a drawing uses the drawing limits as the border definition of the plot. This option will plot all areas of the drawing bounded by the limits, including "blank" areas.

View. The View option plots a named view. The drawing that you stipulated to plot must contain a stored view. If you wish to plot a view, select the "View..." button. AutoCAD prompts you for the name of the view.

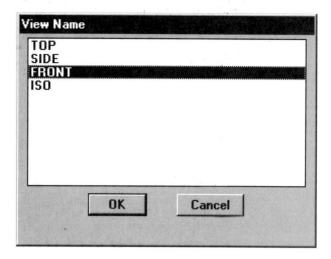

FIGURE 18-5 View Name Dialog Box

All the named views existing in the drawing are displayed in the sub-dialog box.

Window. Window allows you to plot any portion of your drawing by identifying it with a window. AutoCAD displays a Window Selection dialog box.

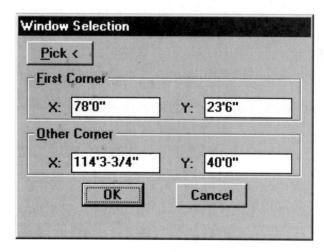

FIGURE 18-6 Window Selection Dialog Box

You can specify the drawing in two ways. The first is to use the text boxes to specify the *X* and *Y* coordinates of two corners of a window bounding the part of the drawing to be plotted.

> **NOTE:** Use AutoCAD's ID command to determine exact coordinates for precise window placement when accuracy is of great importance.

The second way to specify a plotting window is to "show" the window on the drawing screen. Select "Pick" from the Window Selection sub-dialog box. The drawing screen will be displayed. Enter a window on the screen in the same manner as an object selection window. When the second window corner is selected, the sub-dialog box will be redisplayed. The actual coordinates of the window you picked will be displayed in the text boxes.

Select Other Parameters. After selecting the portion of the drawing to plot, choose any other parameters that will affect the plot. Let's look at the choices. All selections are found in the "Additional Parameters" section of the dialog box.

Hide Lines. Selecting "Hide Lines" causes AutoCAD to process the drawing and remove hidden lines when plotting a 3D drawing. Removing hidden lines from complex drawings can take a period of time and will slow down the plotting process.

> **NOTE:** Using the Hide command to remove hidden lines on the drawing screen does not carry over to the plotting routine. The hidden lines must be removed at the time of the plot even if the display shows the hidden lines removed.

The Hideplot option of the Mview command does not affect plots made from model space. You must select the Hide Lines check box if you want the plot to be drawn with hidden lines removed.

Adjust Area Fill. Selecting "Adjust Area Fill" causes AutoCAD to move the pen inward ½ pen width when plotting solids and filled areas. This results in very accurate fill areas. In most applications, this is not necessary. In some applications requiring extreme precision, such as printed circuit board design, this option should be used.

Plot to File. AutoCAD can write your plot to a file. The plot file can then be used to plot the drawing without AutoCAD actually being present.

This ability is very useful if you wish to continue drawing on one computer while plotting from another. In most cases, you will need a third-party program that allows you to plot a plot file without AutoCAD.

To create a plot file, select the "Plot To File" check box. Next, select the "File Name..." button. A sub-dialog box will be displayed.

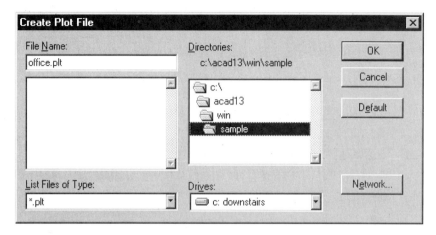

FIGURE 18-7 Create Plot File Dialog Box

The sub-dialog box contains list boxes for the directory path (on the left), and the plot files (on the right). To enter a new plot file name, click in the "File" text box and enter the plot name.

The plot file name should be entered without a file extension. AutoCAD will append a file extension of .plt for plot files.

A plotter file can be used with a plot file utility program supplied by a third party to plot the file, or you can import the plot file into another program.

In some cases, you can send a .plt file to the "plotter" with the DOS copy command. For example, some PostScript laser printers will allow you to send a .plt file if you are configured for PostScript devices, and parallel (as opposed to a COM port) without a third-party plot utility program.

For example, if the plot file is named "WIDGET1.PLT", you can enter the following to send the plot file to the printer:

COPY WIDGET1.PLT LPT1 /B

This would send the file to the number one printer port (LPT1).

Select Paper Size and Orientation. CAD drawings are plotted on many different sizes of paper or film. AutoCAD allows you to select all the sizes available for your plotter, plus sizes that you can specify.

The plot size is determined by the Paper Size sub-dialog box. To access this dialog box, select "Size..." from the Plot dialog box. AutoCAD displays the sub-dialog box in Figure 18-8.

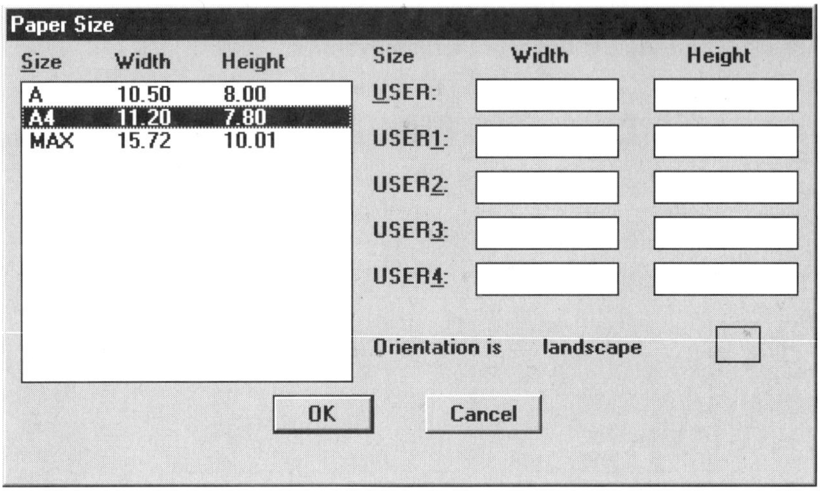

FIGURE 18-8 Paper Size Dialog Box

The list box displays the sizes available. Highlight the size you want by clicking on the entry. If you want to enter a custom size, click inside a text box next to one of the "USER" listings and enter the width and height of the custom plot size.

The orientation is listed below the user entries. The orientation listed is the normal direction of the plot for the plotter selected. You can rotate the plot if you want the orientation changed. Plot rotation is covered next in this chapter.

Select the Scale, Rotation, and Plot Origin. You can rotate the plotted drawing in 90-degree increments. This has the effect of "turning" the drawing on the paper. To rotate the plot, select the "Rotation and Origin..." button.

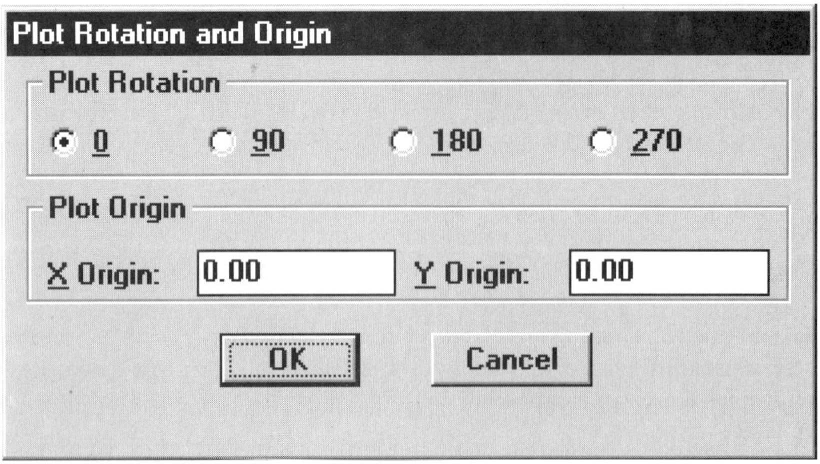

FIGURE 18-9 Plot Rotation and Origin Dialog Box

The degree of plot rotation is selected with the radio buttons at each rotation listing. The following illustration shows the effect of each rotation.

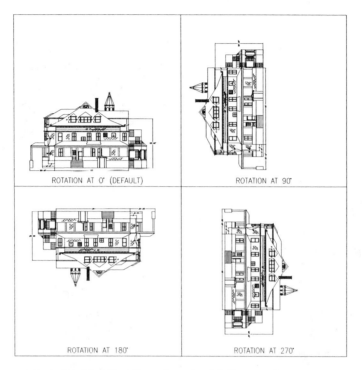

FIGURE 18-10 Plot Rotations by 90-Degree Increments

The plot origin is the distance that AutoCAD will offset the corner of the drawing from the origin corner of the paper. The origin point of the paper can be different for various devices. For most plotters, however, the origin point is the lower left corner.

For example, if you are using a pen plotter and plotting in inches, a plot origin of 6,4 would move the plot origin six inches to the right and four inches up the paper.

The distance that the plot is offset is set in the Plot Rotation and Origin sub-dialog box. Click in the text boxes and enter the X and Y distances to offset the drawing. The offset distance is calculated in either inches or millimeters, depending on the setting in the "Paper Size and Orientation" box in the Plot Configuration dialog box. The plot is *not* measured in drawing units.

Set the Plot Scale. The plot scale is determined by setting the number of drawing units to be plotted on a specified number of inches or millimeters on the paper. For example, a scale of 1" = 8'-0" means that 1" of paper contains 8'-0" of drawing. This is the same, of course, as the scale ⅛" = 1'-0". To set this scale, enter 1" under "Plotted Inches" and 8'0" under "Drawing Units".

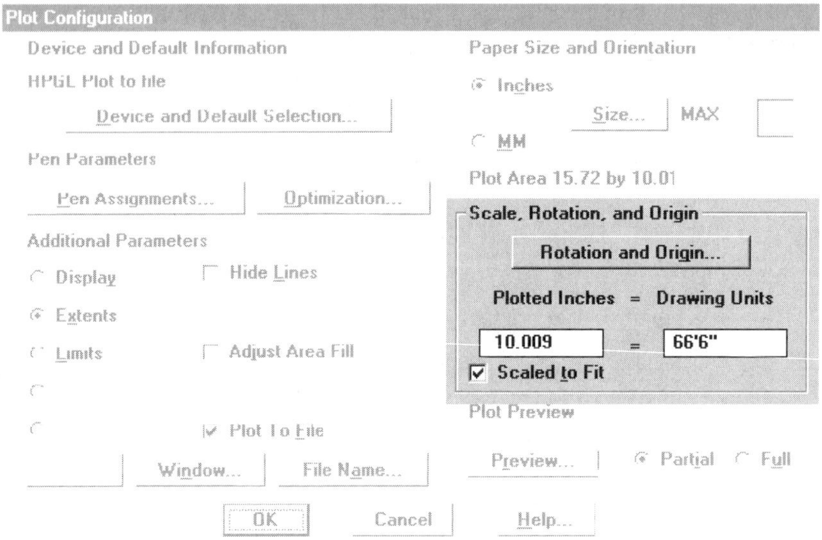

FIGURE 18-11 Setting the Plot Scale

You can also tell AutoCAD to plot the drawing to the maximum size allowed by the paper size you have selected. To do this, select the "Scaled to Fit" box.

Many CAD operators initially calculate the proper plot scale. As unusual as it might seem, proper scaling begins at the time you set the drawing limits. The limits, paper size, and the scale at which you intend to plot must all match for a properly scaled plot.

If you begin a drawing using traditional drafting techniques, you must know the size of paper on which you are drawing (of course), and the scale at which you are drawing. The same is true for CAD drafting. In fact, the limits determine the page size. But how do all these work together?

Let's first look at how the plotter translates the scale factor you give it. When you give the plot ratio, you are stipulating how many *units* will be contained in one inch on the paper that is being plotted on (or millimeters, if you choose that option). For example, if you intend to plot the drawing at ¼" = 1'-0", and you have stipulated that one unit equals one foot, you may also say that ¼" = 1 unit. (One unit = 1 foot, remember?) We may take this further and say that 1" = 4 units, or there are four units in one inch. Let's review this process:

    If   ¼" = 1'-0" & one unit = 1'
    then ¼" = 1 unit
    then 1" = 4 units

Let's take this one step further. If 1" = 4 units, then each inch on the paper will contain 4 units. If you intend to plot your drawing on paper that is 36" wide by 24" high, you may multiply the number of inches in each direction of the paper by the number of units in one inch.

  36" × 4 units/inch = 144 (X limit)
  24" × 4 units/inch = 96 (Y limit)

Therefore, assuming the following parameters:

  Intended scale: ¼" = 1'-0"
  1 unit = 1 foot
  Paper size = 36 × 24

the limits would be 0,0 and 144,96.

From this, we can derive a formula that will determine the proper limits when we have the other necessary information.

  (No. of units/inch) × (paper width) = X limit
  (No. of units/inch) × (paper height) = Y limit

When you are using architectural units, one unit = one inch.

  If ¼" = 1'-0" & 1 unit = 1"
  then ¼" = 12 units
  then 1" = 48 units

therefore,

  36" × 48 units/inch = 1728 (X limits)
  24" × 48 units/inch = 1152 (Y limits)

You could also say that 1728 inches is equal to 144 feet and 1152 inches is equal to 96 feet. Therefore, you could enter either 1728,1152 or 144',96' as the upper right limits.

As you can see, calculating limits requires some basic mathematics skills. In fact, any type of drafting requires these types of skills. With practice, the process becomes easier.

The appendix has a chart that makes things quite a bit easier. This chart lists plot scales, paper sizes, and limits. You can use the chart to set up a drawing for a specific plot scale and paper size.

Preview the Plot. You can preview the plot to see the area of the paper on which the plot will appear. Before doing this, you should first set the paper size, orientation, and scale. AutoCAD's plot preview allows two types of preview: partial and full. Let's look at each.

> ***NOTE:*** Use plot previews to discover plotting results before committing time, paper, and pens to plotting a drawing that may have been set up incorrectly for the plot.

Partial Preview. A partial preview shows the effective plotted area on the "page size" you have selected. On color displays, the plotted area is shown as a blue rectangle, and the paper area as a red rectangle. A partial preview does not show the drawing, but rather the position of the plot on the paper.

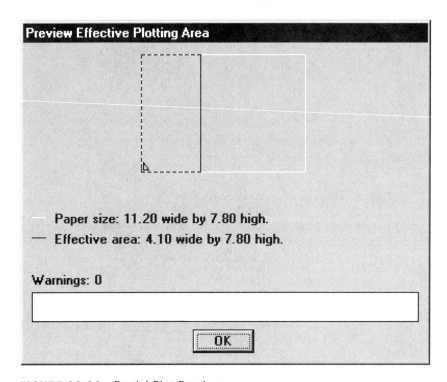

FIGURE 18-12 Partial Plot Preview

The plotted area also shows a rotation icon. This triangular icon represents the lower left corner of the drawing as it is positioned on the drawing screen. When the drawing is rotated to angle 0 on the paper, the icon appears at the lower left. At 90 degrees rotation, it is at the upper left. At 180, it appears at the upper right, and is positioned at the lower right at 270 degrees rotation.

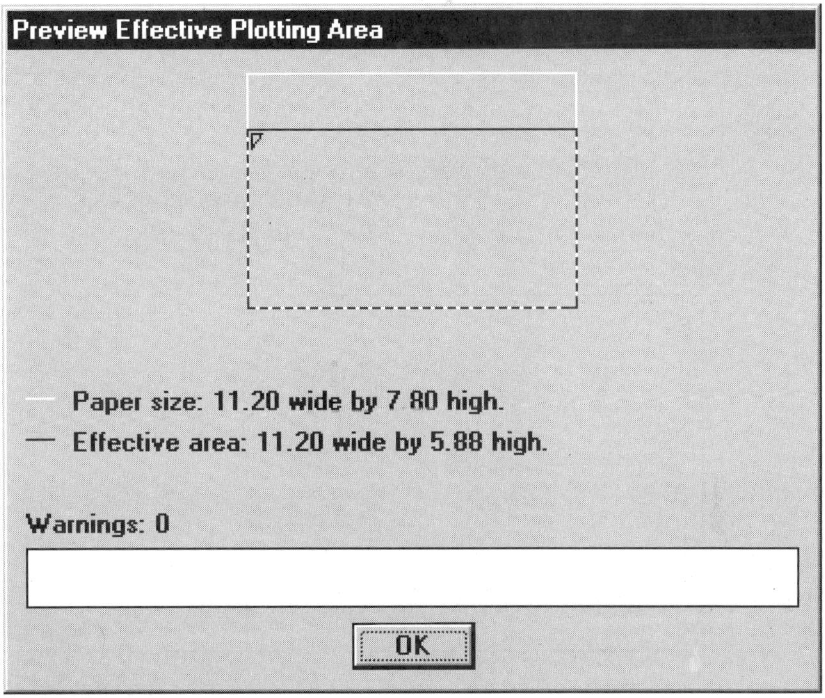

FIGURE 18-13 Rotation Icon

If the plotted area is entirely within the paper size, the plot area is slightly offset from the lower left corner so the plot area lines and the paper lines do not lie on top of each other. If the plotted area and the paper size are exactly the same, the lines are shown dashed, with the dashes alternating red and blue. If the plot origin is offset so that the plotted area extends outside the paper area, AutoCAD displays a green line at the clipped side and displays the following message:

Effective area clipped to display image

AutoCAD also uses the partial preview box to display warning messages such as "Plotting area exceeds maximum".

Full Preview. A *full preview* shows the drawing as it will appear as the final plot on paper. When you select full preview, the drawing is regenerated to display the plot. Since the regeneration takes a period of time, AutoCAD displays a meter showing the percent of regeneration completed in the lower right corner of the Plot dialog box.

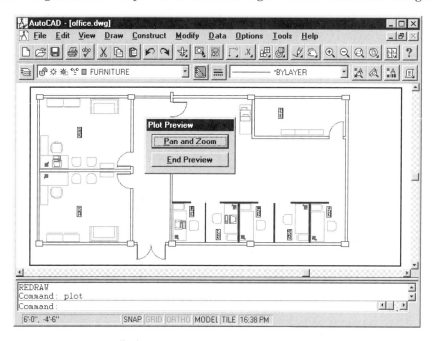

FIGURE 18-14 Full Plot Preview

When the full preview is displayed, AutoCAD places a small Plot Preview dialog box on the screen. If you wish, you can move this box by placing the arrow pointer on the "Plot Preview" bar at the top of the dialog box, and then clicking and holding as you move the box.

The dialog box has two choices: "Pan and Zoom" and "End Preview". The pan and zoom is used to examine parts of the drawing to be plotted and works in the same way as the dynamic Zoom command. When you select Pan and Zoom, a zoom box is displayed with an "X" in the center. Move the box to a new location and press [ENTER] to pan. Clicking the pick button on the input device will display an arrow at the right side of the box. If you move the arrow, the box will change size, allowing you to zoom the display size. You can use pan and zoom in the same operation by clicking the input button to change between pan and zoom mode.

After you have panned or zoomed, the dialog box displays a "Zoom Previous" button that will return the drawing to the previous zoom.

To end the preview, select the "End Preview" button. You will be returned to the Plot dialog box.

Finally, Plot the Drawing. When you have completed all the steps outlined, make sure the plotter is ready and select the "OK" button at the bottom of the Plot dialog box.

PLOTTING FROM THE COMMAND LINE

If the Cmddia variable is set to 0, the plotting dialog boxes will not be displayed. Instead, the plot information will be entered from the command line. Let's look at how AutoCAD prompts for the information. You may initiate a plot by entering:

AutoCAD prompts:

What to plot - Display, Extents, Limits, View, or Window <default>:

Choose the portion of the drawing to be plotted by entering the first letter of your choice and [ENTER].

AutoCAD then displays the default plot specifications. Figure 18-15 is an example of this listing (your listing may vary, depending on the equipment configuration, and if changes have been made to the defaults).

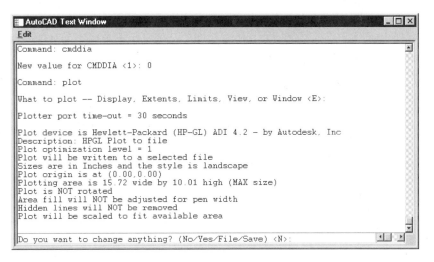

FIGURE 18-15 Sample of Default Plot Specifications

The default settings for these values were set when you configured AutoCAD for the plotter you are using.

If the settings are correct for your use, press [ENTER] to accept all the default values.

If you wish to change any part, enter a **Y**. AutoCAD will prompt:

Do you want to change plotters? <N>:

Responding **Yes** will display a listing of the installed plotters for choice. If you respond with **No**, AutoCAD will prompt for the time-out period, then display a chart displaying the layer colors, pen numbers, linetypes, and pen speeds on the screen. Figure 18-16 is an example of the chart. Note that your listing may be different, depending on the type of plotter you have configured.

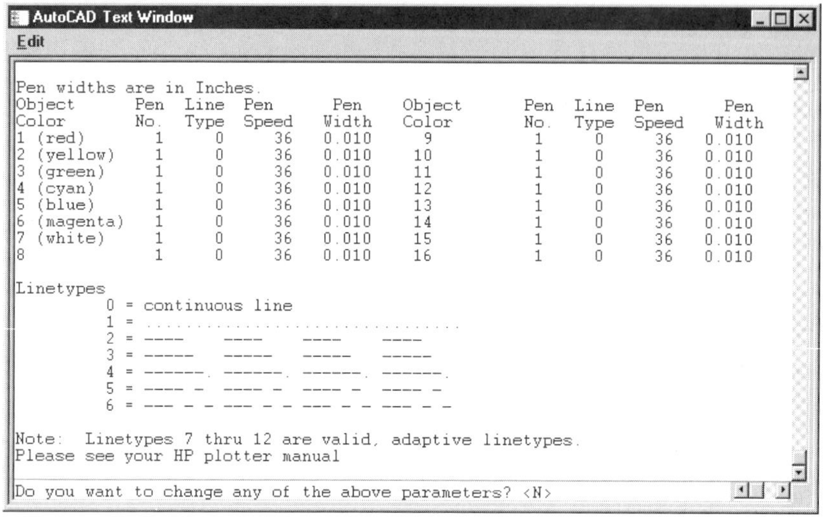

```
AutoCAD Text Window
Edit

Pen widths are in Inches.
Object      Pen  Line  Pen     Pen     Object     Pen  Line  Pen     Pen
Color       No.  Type  Speed   Width   Color      No.  Type  Speed   Width
1 (red)      1     0    36     0.010      9        1     0    36     0.010
2 (yellow)   1     0    36     0.010     10        1     0    36     0.010
3 (green)    1     0    36     0.010     11        1     0    36     0.010
4 (cyan)     1     0    36     0.010     12        1     0    36     0.010
5 (blue)     1     0    36     0.010     13        1     0    36     0.010
6 (magenta)  1     0    36     0.010     14        1     0    36     0.010
7 (white)    1     0    36     0.010     15        1     0    36     0.010
8            1     0    36     0.010     16        1     0    36     0.010

Linetypes
      0 = continuous line
      1 = ................................
      2 = ----    ----    ----    ----
      3 = -----   -----   -----   -----
      4 = ------. ------. ------. ------.
      5 = ---- -  ---- -  ---- -  ---- -
      6 = --- - - --- - - --- - - --- - -

Note:  Linetypes 7 thru 12 are valid, adaptive linetypes.
Please see your HP plotter manual.

Do you want to change any of the above parameters? <N>
```

FIGURE 18-16 Plotting Pen Chart Dialog Box

If you would like to change anything, respond with a **Y** to the prompt:

Do you want to change any of the above parameters? <N>

If everything is correct, press [ENTER] to accept the default of "N" (NO).

If you answer **Y** to the prompt, AutoCAD will prompt:

Enter values. blank=Next value, Cn=Colorn, S=Show current values, X=Exit

A listing will appear at the bottom of the screen, similar to the following:

Layer Color	Pen No.	Line Type	Pen Speed
1 (red)	1	0	16

Pen number <1>:

You may flip through the listing of values for each color number and enter a change when you wish. You will first see the color number, then the pen number for color number 1, then the linetype for color number 1, then the pen speed for the pen set for color number 1. After that, the same sequence is displayed for color number 2, and so forth. You can change all colors to the same value by preceding the value with an asterisk (*). AutoCAD uses the following entries to step through the choices:

Blank (spacebar): Keeps the current value and advances to the next parameter. After you have advanced to color 15, AutoCAD returns to color 1.

Cn (color number): Proceeds directly to the color number entered. For example, if you want to proceed directly to color 12 to make a change, enter **C12**. This eliminates flipping through each color number.

S (Show current values): Entering an **S** redisplays the table, showing the changed values. Use this after changing any parameter to verify the change.

X (Exit): This entry will exit you from this portion of the listing and keep any changes made. This is the way you will exit the repeating sequence of color/pen/linetype/pen speed choices.

AutoCAD continues:

Write the plot to a file <N>:

You are next prompted for the type of plotting units.

Size units (Inches or Millimeters) <default>:

You may choose to plot in inches or millimeters by entering either **I** or **M**.

Choosing either the "inches" or "millimeters" option will not change the units the drawing was drawn with.

Next you are prompted for the plot origin.

Plot origin in Inches <0.00,0.00>:

The normal plot origin (home position) for a pen plotter is the lower left corner. For a printer plotter, it is usually the upper left corner.

You may set a new plot origin; that is, a specified position from the "home" position. The units used are either inches or millimeters, depending on which was specified in the previous plotting units option. The format is the coordinate type of X,Y entry.

Next you will be prompted for the plotting size. AutoCAD has to know the plot size you are using. Different plotters have the capability of plotting on various size sheets of paper. The standard paper sizes are A, B, C, D, and E (architectural sizes). The following table lists some paper sizes and the plot size for each. Notice that the plot size is less than the paper size.

	Paper		Plot	
Size	Width	Height	Width	Height
A	11.00	8.50	10.50	8.00
B	17.00	11.00	16.00	10.00
C	24.00	18.00	21.00	16.00
D	36.00	24.00	33.00	21.00
E	48.00	36.00	43.00	33.00

There are two other settings: MAX and USER. MAX is the largest plot that your plotter can plot. USER will describe the size if you enter anything other than a standard plot size (A, B, C, D, or E) by stipulating a width, height description. Note that not all plotters will plot all the sizes listed. Your display will vary, depending on the configured plotter.

The prompt for plot size is:

> Enter the Size or Width, Height (in units) <*default*>:

AutoCAD continues by prompting for the plot rotation.

Rotating a plot is especially useful when you are using printer plotters and wish to take advantage of the paper width. The prompt for plot rotation is:

> Rotate plot clockwise 0/90/180/270 degrees <0>:

Next, set the pen width. When AutoCAD is using the Fill mode to color in solids and traces, it needs to know the width of the pen used in the plotter. This allows for an appropriate offset for each pass of the pen to obtain a properly filled solid area. Pens are often marked with the width of the "nib". These widths are usually in mm. If the units are set to inches, be sure to convert the pen width to mm. The prompt for pen width is:

> Pen width in <*units*>: <*default*>:

If you are plotting drawings that require more accuracy than normal (such as printed circuit board drawings), AutoCAD will adjust the boundaries around filled areas one half pen width. Printer plotters do not display this option. The next prompt AutoCAD displays controls this adjustment:

> Adjust area fill boundaries for pen width? <N>:

For most applications, the response should be **N**.

Next, determine whether AutoCAD will remove the hidden lines from the plot.

> Remove hidden lines? <N>:

Now set the scale. AutoCAD allows you to plot at a scale of your choice. The prompt from which the scale is set is:

> Specify scale by entering:
> Plotted units=Drawing units or Fit or ? <*default*>:

The Fit option will adjust the scale so that the drawing will fit the paper size. Thus, if you choose **F**, the drawing will not be plotted to any standard scale.

Now, after all that, you are ready to plot. The following prompt will appear:

> Effective plotting area: x wide by y high

Get your paper and pen or printer ready! AutoCAD continues:

> Position paper in plotter
> Press RETURN to continue or S to Stop for hardware setup.

The hardware setup option is required by some plotters as the time to perform certain preparation. If you are configured for a multipen plot, and your plotter is a single-pen plotter, AutoCAD will stop the plot at this point and ask for a specific pen. After everything is ready, press [ENTER] and watch the plotter plot your work.

EXERCISES

1. If you have a printer plotter (most dot-matrix printers will work as a printer plotter), use the Plot dialog box to verify the plotter is configured for use.

   Use the chart in the appendix to set the limits for an A-size drawing. Draw a simple drawing and save your work. Plot the drawing, using "Scaled to Fit" to scale the drawing.

2. Plot the drawing again, using the scale you configured from the chart in the appendix. After the drawing is plotted, use a draftsman's scale to check the scale of the drawing at a known dimension length. Did your drawing plot to scale?

3. Plot the drawing again, rotating the plot 90 degrees.

4. Enter the drawing you constructed. Select the Plot command. Plot the drawing, using a Window. Place a window around any part of the drawing. Set the plot units to Scaled to Fit. Plot the drawing.

5. Select the Plot command again. Set the plot for Limits. Plot to a file. Select Scaled to Fit. Set the plot file name to "MYPLOT". After the plot has been written to file, use the Shell command to display the DOS prompt. Type the following:

   COPY MYPLOT.PLT LPT1 /B

   (If your printer is connected to a second parallel port, you will need to specify LPT2.) You do not need to have AutoCAD running, or even installed on the computer, to printer plot a plot file.

CHAPTER REVIEW

1. What is the advantage of a printer plot?

2. What part of the drawing would be plotted if you plotted the extents of the drawing?

3. How would you plot only a portion of a drawing?

4. How would you rotate the plot on the paper?

5. How would you write a plot file to a disk for a printer plotter?

6. What DOS command would you use to print a plot file named "WIDGET3" to a printer connected to the second parallel port on the computer?

7. What is the *plot origin*?

8. How would you offset the plot origin?

9. What units does the plot origin use?

10. Why does AutoCAD need to know the plotter pen width?

11. What drawing limits would you set if you set up a drawing to be plotted in architectural units, with a scale of ¼ in. = 1 ft. -0 in., and on architectural C-sized paper?

CHAPTER 19

INQUIRY AND UTILITY COMMANDS

When drawing with CAD, the user must be capable of obtaining drawing information. This chapter covers the Inquiry and Utility commands available to achieve this. After completing this chapter, you will be able to:

- Utilize the ID, List, and Dblist commands to obtain information on points and entities in the drawings.
- Calculate areas within a drawing.
- Manage AutoCAD's ability to execute operating system commands while in the AutoCAD program.
- Use the time-keeping function that is built into AutoCAD.

INQUIRY AND UTILITY COMMANDS

AutoCAD's inquiry and utility commands allow you to obtain information and perform utility functions while working within your drawing. We will learn about the following commands:

   ID: Identify a point
   List: List information on one or more entities
   Dblist: List information on all the entities in the current drawing
   Dist: Compute distance between two points
   Area: Compute area of a closed polygon
   Files: On-line operating system capabilities
   Time: Time management facility

ID SCREEN COORDINATES

The ID command allows you to identify the coordinates of any point on the drawing. If you enter a location on the screen as the response to the prompt for a point, the coordinates of that location will be displayed on the command line.

The information displayed will contain the X-coordinate value, the Y-coordinate value, and the current Z-value elevation.

To execute the ID command, enter:

Command: **ID**
Point: *(Pick a point.)*
X= <X-coordinate> Y= <Y-coordinate> Z= <current elevation>

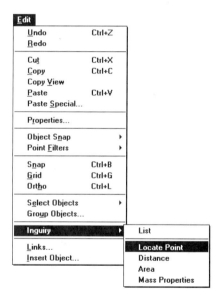

You may also use the ID command to show a known point on the screen. If you enter a set of coordinates in response to the prompt, a marker blip will be displayed at that point on the screen.

For example, if you enter the ID command and enter the absolute coordinate value of **8,5** in response to the "Point:" prompt, a blip mark will be displayed at the X,Y coordinates of 8,5. Note that Blipmode must be "on" before marker blips are displayed.

LISTING DRAWING INFORMATION

The LIST command is used to display the data stored by AutoCAD on any entity.

The format for List is:

Command: **List**
Select objects: *(Select.)*

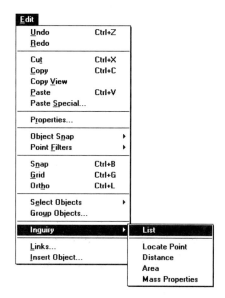

The information listed varies for each type of entity. For example, if you select a circle, you will obtain the layer the circle is drawn on, the radius, circumference, area, and center point of the circle.

If the object you select contains a lengthy amount of data, you can terminate the listing by using [ESC].

The following shows a sample listing of a circle:

```
CIRCLE     LAYER: 0
           Space: Model space
center point, X = 6.0264 Y= 4.8211 Z= 1.0239
radius 1.7577
Circumference = 11.0439,
Area = 9.7059
```

LISTING DRAWING DATABASE INFORMATION

Whereas the List command is used to obtain information on a single entity in a drawing, the Dblist command is used to display the data stored on *all* entities in a drawing.

This listing can be very long! AutoCAD pauses the listing. The listing will begin scrolling again if you strike any key. You may terminate the listing entirely by using [ESC].

COMPUTING DISTANCES

The Dist command computes the distance between two points, the angle created by the relative position of the points, and the difference of the X and Y values of the points. To invoke the Dist command, enter:

Command: **Dist**
First point: *(Pick first point.)*
Second point: *(Pick second point.)*

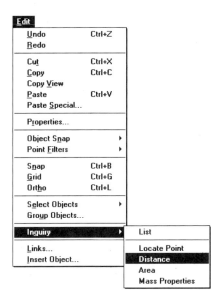

The following is a sample listing of distance information:

Distance=1.4971   Angle in X-Y Plane=45   Angle from X-Y Plane=0
Delta X=1.0612   Delta Y=1.0560   Delta Z=0.0000

The following is an explanation of the listing:

Distance: Distance between the two points in drawing units.

Angle in X-Y plane: Angle created by relative position of the points in the 2D X,Y plane.

Angle from X-Y plane: Angle "up" from the paper into the Z-plane (used in 3D construction).

Delta X: Change in X-coordinate values between the points.

Delta Y: Change in Y-coordinate values between the points.

Delta Z: Change in Z-coordinate values between the points.

You may also show a desired distance on the screen by specifying a relative coordinate as the response to the "Second point" prompt.

For example, if you enter the first point on the screen and respond with a relative coordinate of **@10,0** for the second point, AutoCAD will display a blip mark 10 drawing units to the right (positive 10 units X) of the first point.

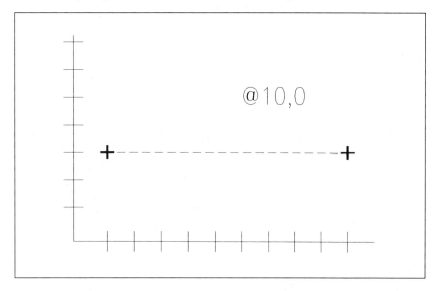

FIGURE 19-1 Using Distance with a Relative Coordinate

Note that Blipmode must be "on" before blip marks are displayed.

You may also use polar coordinates with the Dist command. For example, if you respond to the second point prompt with **@15<45**,

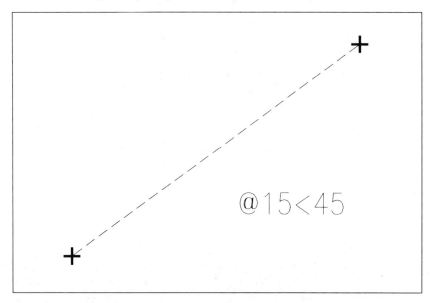

FIGURE 19-2 Using Distance with a Polar Coordinate

19-6 USING AUTOCAD

AutoCAD will display a marker blip at a distance of 15 drawing units and 45 degrees from the point you entered in response to the "First point" prompt.

> **NOTE:** Use the Dist command and relative coordinate responses to set reference markers on the screen. You may then line up the crosshairs on these markers. Remember, if you redraw the screen or use a command that forces a redraw (such as a zoom), the markers will be removed.

CALCULATING AREAS IN YOUR DRAWING

The Area command is used to calculate the area, in current drawing units, of a closed polygon.

The Area command is especially useful for rough computations of land bounded by property lines or calculating the areas of floor plans.

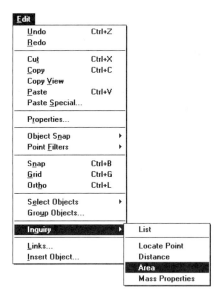

Methods of Calculating Areas

The Area command permits different methods of calculating areas. The following explanations explain how to use the options in the Area command.

Point Method. One method of calculating the area of an object is to simply enter points at the intersections of a polygon. The following example shows this method. Refer to Figure 19-3.

Command: **Area**
First point/Object/Add/Subtract: *(Select point "1".)*
Next point: *(Select point "2".)*
Next point: *(Select point "3".)*
Next point: *(Select point "4".)*
Next point: *(Select point "5".)*

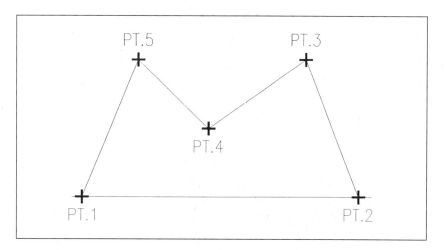

FIGURE 19-3 Using the Area Command

Notice that the "Next point" prompt is repeated until you press [ENTER] to tell AutoCAD that you have entered all the desired points. You do not have to "close" the polygon by entering the last point on top of the first point entered. AutoCAD will automatically close from the last point to the first point entered.

After you press [ENTER], you are shown the following information in the prompt area:

Area=<calculated area>
Perimeter=<calculated area>

Object Method. If you select the Object option under the Area command, you can compute the area of a circle or polyline. The following prompt appears:

Select objects:

AutoCAD calculates each in the following manner.

> **Circle:** Displays the area and the circumference.
>
> **Polyline:** If the polyline is closed, the area and perimeter are calculated. If the polyline is open, the area and perimeter are calculated as though a line were drawn between the endpoints to close the polyline.
>
> A wide polyline is calculated from the centerline of the polyline. An area for a polyline is valid only if the polyline can describe a closed area. In other words, you cannot compute the area of a polyline of a single segment.

Add Areas. The Add option is used to calculate running totals of calculated areas. To use the Add option, select Add before calculating the areas. The prompt will be reissued after each calculation and AutoCAD will display the running total.

Subtract. Subtract is used to subtract an area from the running total. Select Subtract before selecting the area(s) to be subtracted from the total.

Enter. Pressing only [ENTER] will exit the Area command.

FILE UTILITIES

At times, you may need to perform operating system functions while in your drawing. For example, you may want to list the directory of your disk. AutoCAD allows you to do this and other functions without having to exit your drawing by using the Files command.

The Files command allows you to:

1. List any specified files on any disk.
2. Delete files from any disk.
3. Rename files on any disk.
4. Copy files.
5. Unlock files.

To use the Files facility, enter:

Command: **Files**

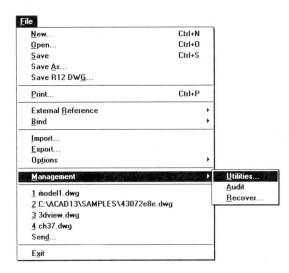

AutoCAD displays a File Utilities dialog box.

FIGURE 19-4 File Utilities Dialog Box

Selecting any of the buttons causes AutoCAD to display a sub-dialog box with a list box containing file names. The following figure shows the File List sub-dialog box.

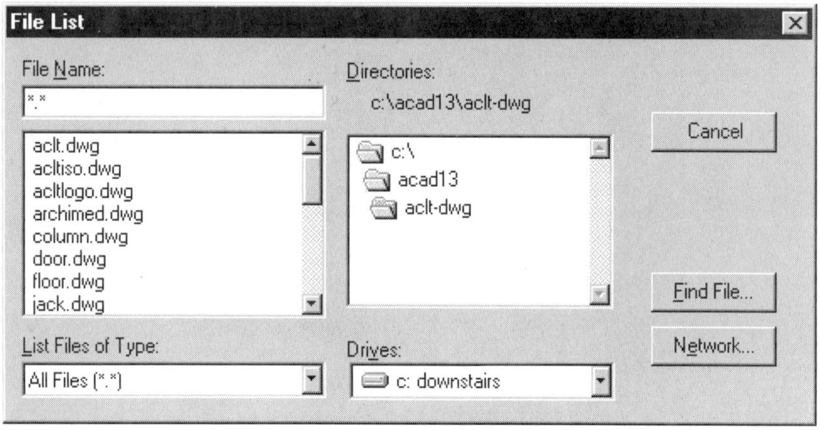

FIGURE 19-5 File List Dialog Box

Select the file name on which you wish to perform the operation.

Alternatively, you can use the Windows File Manager to perform all these and many more file-related functions. See Chapter 5.

AUTOCAD OPERATING SYSTEM COMMANDS

Certain operating system commands can be accessed straight from the command line in AutoCAD. These commands are contained in a file named "ACAD.PGP". When you issue a command, AutoCAD looks for the command, and if it cannot find a command by that name, it checks the "ACAD.PGP" file. Some of the commands included with AutoCAD include the following:

CATALOG (directory)
DIR (directory)
DEL (delete)
EDIT (Edit)
TYPE (file listing)

TIME COMMAND

The Time command uses the clock in your computer to keep track of time functions for each drawing. Some computers keep constant track of the date and time. This information must be entered on others at boot-up.

To obtain a listing of the time information for the current drawing, enter the Time command.

Command: **Time**

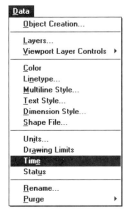

Figure 19-6 displays a text screen as shown. The explanation of each follows:

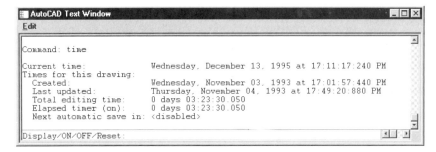

FIGURE 19-6 Time Command Text Screen

"Stopwatch" timer. You may turn this timer on or off. It is independent of the other functions.

Cumulative time spent on drawing. The Save command does not reset the time. If you exit the drawing by using Quit, the time does not count. (Printer and plotter time is not added.)

Last time the drawing was updated and saved using the Save or End command.

Date and time the current drawing was initially created. If the drawing was created using Wblock, the date of that execution is the creation time. If you edit a drawing created with a previous version of AutoCAD, the first edit time is used as the creation time.

The last line contains options for execution. The options are as follows:

Display: Redisplays the time functions with updated times.

On: Starts the "stopwatch" timer (the timer is initially on).

Off: Stops the "stopwatch" timer, freezing the display at the accumulated time.

Reset: Resets the "stopwatch" timer to zero.

If you do not wish to execute any of the options, enter a null response ([ENTER]) or [ESC] (cancel).

EXERCISES

1. Start a new drawing named "INQUIRE". You can use this drawing for all exercises in this chapter. Draw a line, using the absolute endpoints of 2,2 and 6,6.

   Select the ID command, then object snap ENDpoint. Select the first endpoint of the line with the object snap aperture. The command line should show the ID point as 2,2. Repeat the procedure with the other endpoint location of 6,6.

2. Draw a line, circle, and arc on the screen. Use the List command to select the circle. You should obtain a listing of the circle similar to the one shown previously. Select each of the entities and notice the information displayed for each entity. The List command can be useful if you need to obtain information on a single entity in your drawing.

3. Use the same drawing with the entities you used for the List exercise. Enter Dblist at the command line. Notice how the information scrolls on the screen. If you are using a single display system, you will need to use [F1] to switch back to the graphics display.

4. Use the Copy command to copy all the entities in the drawing three times. Use the Dblist command again and notice the length of the listing.

5. Select the Dist command. Using object snap endpoint, select two endpoints of one of the lines on the screen. Notice how the length of the line is displayed on the command line.

6. Draw two boxes on the screen. Use the Area command to figure the area of one of the boxes. Be sure to use object snap INTersection to precisely capture the intersections of the box. Repeat the same procedure on the second box. Add the areas together to obtain the area total.

19-12 USING AUTOCAD

7. Use the Area command again to obtain the area of the first box. This time, use the Add option to add the area of the second box. Did the two areas add up to the same area you computed before?

8. Start the drawing named "AREA" from the work disk. Compute the area of each object and write down the figures.

9. Enter **Dir** at the command line. AutoCAD will respond with:

   File specification:

   Reply with **C:** and press [ENTER]. You will see the directory listing of the root directory of the C-drive.

10. Enter **Type** at the command line. AutoCAD responds with:

    File to list:

    Enter **ACAD.MSG**. If the file named ACAD.MSG has not been deleted from the AutoCAD directory, you will see a listing of the contents of the file.

11. Use the Time command to check the current time in the drawing you are currently using.

12. Use the elapsed timer option in the Time command to time the period needed to draw a car.

CHAPTER REVIEW

1. If you wanted to identify the coordinates of a specific point in a drawing, what command would you use?

2. What is displayed when you use the List command?

3. What is the difference between the List and the Dblist command?

4. What are the six distances returned by the Dist command?

5. What values are displayed when a circle is selected under the Area command?

6. What option under Area would you use if you wanted to calculate the area of a circle?

7. What are the six functions of the Files command?

8. Why would you not want to delete a file with a .$AC file extension?

9. What operating system commands can be accessed from the AutoCAD command line?

10. How would you determine the last time a drawing was updated?

CHAPTER

20

INTERMEDIATE DRAW COMMANDS

After you have become proficient with the basic draw commands in AutoCAD, the next step is to master the intermediate commands. Learning the use of these commands will allow you to construct more complex drawings. After completing this chapter, you will be able to:

- Control commands for special objects such as ellipses and doughnuts.
- Construct new objects from existing ones with commands such as Offset and Minsert.
- Operate one of AutoCAD's most powerful features: symbol libraries. These are built with the Insert, Block, and Wblock commands.

DRAWING ELLIPSES

The Ellipse command is used to construct ellipses by specifying the axis endpoints, the axis center points, and the length of the axis. You may also place ellipses properly into any of the three planes of an isometric drawing.

Before constructing an ellipse, let's look at the parts of an ellipse. An ellipse is defined by a major axis and a minor axis. The major axis consists of a line drawn along the longest direction of the ellipse. The minor axis consists of a line drawn along the shortest direction of the ellipse. The major and minor axes intersect at the center point of the ellipse.

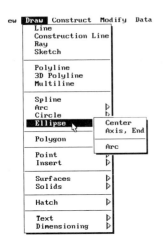
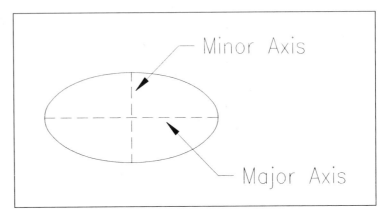

FIGURE 20-1 Ellipse Axes

You may specify these points in constructing an ellipse. Let's look at several ways to specify an ellipse.

Specifying an Ellipse by Axis and Eccentricity

In the first example, we will construct an ellipse by showing AutoCAD the length of one axis, then the length of the other. First, enter the Ellipse command:

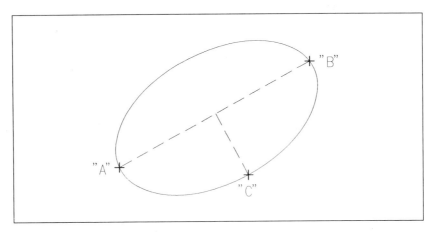

FIGURE 20-2 Ellipse by Axis and Eccentricity

Command: **Ellipse**
Arc/<Axis endpoint 1>/Center: *(Enter point "A".)*
Axis endpoint 2: *(Enter point "B".)*
<Other axis distance>/Rotation: *(Enter point "C".)*

You may also specify point "C" by using a numerical distance.

The first two points entered may be either the major or minor axis. The distance specified for point "C" will determine which is the major axis. (Remember, the major axis is the longer axis.)

Specifying an Ellipse by Axis and Rotation

If you respond to the prompt **<Other axis distance>/Rotation:** with **ROTATION** (or **R**), you will be allowed to rotate the ellipse around the axis first specified. The ellipse will be considered as a circle, with the previously specified axis acting as a diameter line. The rotation will take place in the Z-plane. That is, it will be as though you rotated the circle "into" the screen. The circle may be rotated into the Z-plane at any angle from 0 to 89.4 degrees. Specifying a zero angle, however, will result in a full circle being drawn.

You may also show AutoCAD the angle by drag specification. The angle is relative to the midpoint of the ellipse.

The following illustrations show ellipses of varying rotation angles. Points "A" and "B" designate the endpoints of the major axis.

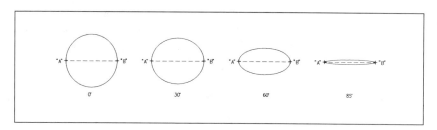

FIGURE 20-3 Ellipse by Axis and Rotation

Specifying an Ellipse by Center and Two Axes

An ellipse may also be defined by specifying its center point, the endpoint of one axis, and the length of the other axis.

Let's construct an ellipse by this method. Enter the Ellipse command:

Command: **Ellipse**
Arc/<Axis endpoint 1>/Center: **C**
Center of ellipse: *(Enter point "A".)*
Axis endpoint: *(Enter point "B".)*
<Other axis distance>/Rotation: *(Enter point "C".)*

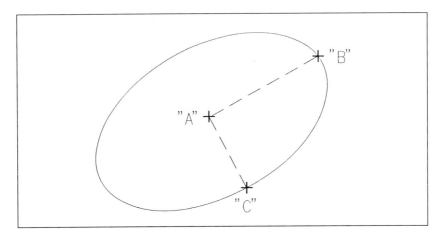

FIGURE 20-4 Ellipse by Center and Two Axes

Notice that the location of point "B" determines the angle of the ellipse. The prompt for "Axis endpoint" may be a numerical distance. This distance will be the distance from the center point, and perpendicular to the first axis specified by points "A" and "B". Note that a numerical distance represents one-half of the axis defined.

Figure 20-5 shows the results of constructing ellipses by the center and two axes method. Note that the angle and the major axis is determined by the placement and distance of the points.

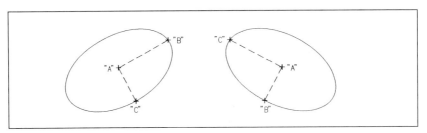

FIGURE 20-5 Specify Ellipse Axis Angle

You may also choose the "Rotation" option when constructing an ellipse using the center method. Figure 20-6 shows the effects of this type of construction.

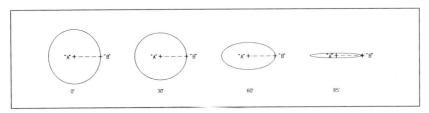

FIGURE 20-6 Using Rotation Option with Ellipse

CONSTRUCTING ISOMETRIC CIRCLES AND ELLIPSES

The Ellipse command allows you to correctly place circles in isometric drawings. You must be in "ISO" mode (use the Snap command's Style option to set Isometric mode) and execute the Ellipse command, the following prompt is issued:

    Command: **Ellipse**
    <Axis endpoint 1>/Center/Isocircle:

Responding with "Isocircle" or **I** from the keyboard allows you to draw a circle in the current isometric plane. Let's look at an example of how isocircles work. First, use AutoCAD's Isometric Snap mode to construct a cube as shown in the following illustration. Then, enter the Ellipse command:

    Command: **Ellipse**
    Arc/<Axis endpoint 1>/Center/Isocircle: **I**
    Center of circle: *(Enter point "A".)*
    <Circle radius>/Diameter: *(Enter point "B".)*

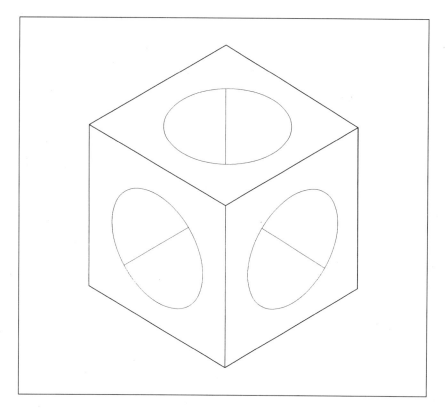

FIGURE 20-7 Drawing Isometric Circles with the Ellipse Command in Iso Mode.

You must be in the correct isometric plane to properly place an isocircle. Use `CTRL` `E` to toggle between the planes. (You may use this toggle while in the command without problem.) Chapter 21 covers isometric drawing in detail.

CONSTRUCTING AN ELLIPTICAL ARC

To construct an elliptical arc, use the Arc option of the Ellipse command, placing the axes as before. However, you also specify the starting and ending angles of the arc, as follows:

Command: **Ellipse**
Arc/Center/<Axis endpoint 1>: **a**
<Axis endpoint>/Center: (*pick*)
<Other axis distance>/Rotation: (*pick*)
Parameter/<start angle>: (*pick*)
Parameter/Included/<end angle>: (*pick*)

In addition to picking the start and end angle, there are two additional options for constructing an elliptical arc, Parameter and Included.

The Parameter option lets you specify the arc's angle by sweeping in real time, while the Included option lets you specify the included angle for the elliptical arc.

TWO TYPES OF ELLIPSES

AutoCAD Release 13 can draw two kinds of ellipses: (1) the mathematically accurate ellipse made from the Ellipse entity new to Release 13; and (2) the less accurate ellipse made from a connected series of polyline arcs, which approximate an elliptical shape. The second type of ellipse is the kind drawn by AutoCAD Release 12 and earlier.

You cannot draw elliptical arcs with the polyline ellipse, although you can use the Trim command to "cut" the ellipse into an arc. The only time you would use the second type of ellipse is if you need to exchange drawings with someone using an earlier version of AutoCAD (via the SaveAsR12 command).

The system variable PEllipse determines which style of ellipse is drawn:

Pellipse	Meaning
0	True ellipse is drawn (default)
1	Ellipse is made up of polyline arcs

You change the value of PEllipse as follows:

Command: **Pellipse**
New value for PELLIPSE <0>: **1**

Properties of Ellipses

If you wish to edit the ellipse, use the Trim, Extend, Offset, and Break commands.

DRAWING SOLID-FILLED CIRCLES AND DOUGHNUTS

The Donut command constructs solid-filled circles and doughnuts. The Donut command can be executed by entering either Donut or Doughnut. The following explanation uses both forms of the command.

To construct a doughnut, you must tell AutoCAD both the inside and the outside diameters. If you wish to make a solid-filled circle, specify an inside diameter of zero.

Let's first make a regular doughnut. Enter the Donut command:

Command: **Donut**
Inside diameter <*default*>: **1.5**
Outside diameter <*default*>: **3**
Center of doughnut: *(Pick point "A".)*
Center of doughnut: [ENTER]

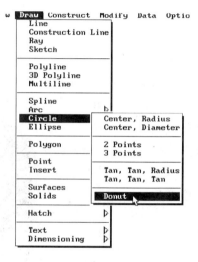

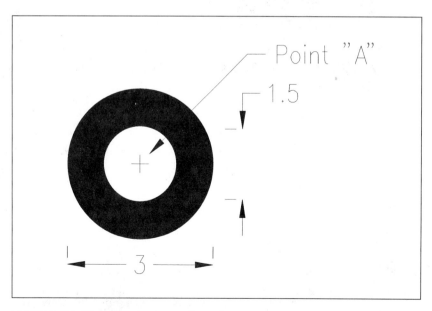

FIGURE 20-8 Constructing a Doughnut

Notice that the "Center of doughnut" prompt repeats to allow placement of several identical doughnuts. The default values are saved from the last values used.

Entering a value of zero (0) for the inside diameter results in a solid-filled circle. Let's construct a solid-filled circle using the Donut command.

Command: **Donut**
Inside diameter <default>: **0**
Outside diameter <diameter>: **3**
Center of doughnut: *(Pick point "A".)*
Center of doughnut: [ENTER]

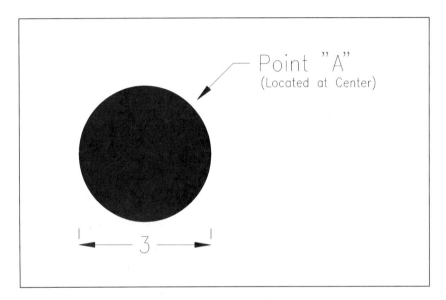

FIGURE 20-9   Constructing a Solid-Filled Circle

Doughnuts are constructed as closed polylines using wide arc segments. To edit a doughnut, use the Pedit (polyline editing) command or the edit functions.

The display of the solid fill in the doughnut is determined by the current setting of the fill mode. If fill is on, the doughnut will be solid-filled. If fill is off, only the outline of the doughnut will be displayed and plotted.

OFFSETTING ENTITIES

The Offset command allows you to construct a parallel copy to an entity or to construct a larger or smaller image of the entity through a point.

Constructing Parallel Offsets

Let's first look at how to construct a parallel entity. First, construct a vertical line on the screen similar to the one shown in the following illustration. Now, enter the Offset command:

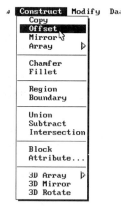

Command: **Offset**
Offset distance or Through <default>: **2**
Select object to offset: *(Select the line.)*
Side to offset? *(Pick point "A".)*
Select object to offset: [ENTER]

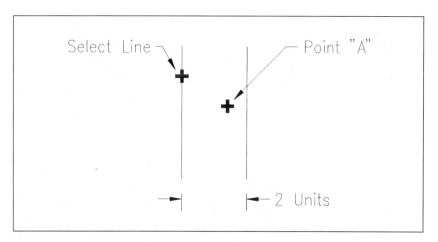

FIGURE 20-10 Offsetting a Line

The default value shown in the brackets will be either a numerical value or "Through," whichever was last chosen.

You may choose to show AutoCAD two relative points to designate the offset distance, instead of entering a numerical value.

The "Select object to offset:" prompt repeats to allow you to offset as many copies as you wish. Enter a null response (press [ENTER]) to terminate the command.

The direction of the offset must be parallel to the Z-axis of the current user coordinate system (UCS systems are explained in Chapter 32). If it is not, AutoCAD will display the message:

That entity is not parallel with the UCS.

The "Select object" prompt will then be repeated.

Constructing "Through" Offsets

The Through option allows construction of the image through a point. Let's create an offset to a circle. Draw a circle as in the following illustration. Enter the Offset command.

Command: **Offset**
Offset distance or Through <*default*>: **T**
Select object to offset: *(Pick the circle.)*
Through point: *(Pick point "A".)*
Select object to offset: [ENTER]

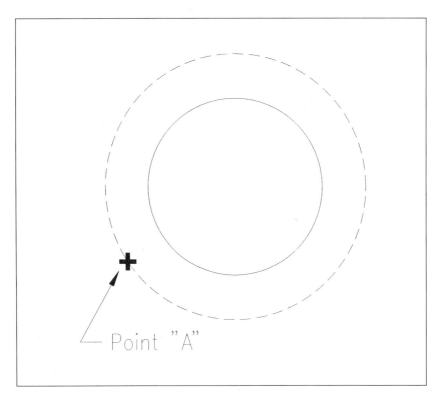

FIGURE 20-11 Using a Through Point with the Offset Command

> **NOTE:** You may choose only one object at a time to offset. If you choose to offset a complex object, it must be one entity. Constructing objects with polylines will allow you to do this.

CHAMFERING LINES AND POLYLINES

The Chamfer command is used to trim segments from the ends of two lines or polylines and draw a straight line or polyline segment between them.

The amount to be trimmed may be specified by entering either a numerical value or by showing AutoCAD the distance using two points on the screen.

The distance to be trimmed from each segment may be different or the same. The two entities do not have to intersect, but should be capable of intersecting within the limits. If the limits are off, the segments should be capable of intersecting at some point.

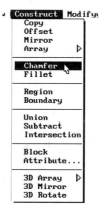

The following example shows how the Chamfer command works. The following drawing will be used:

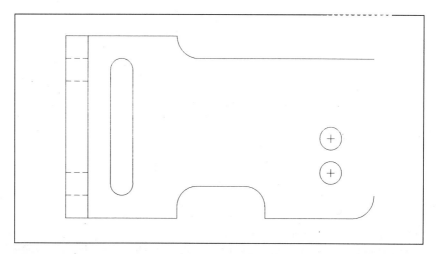

FIGURE 20-12 Chamfer Example Drawing

20-12 USING AUTOCAD

We will first set the Chamfer defaults:

Command: **Chamfer**
Polyline/Distance/<*Select first line*>: **D**
Enter first chamfer distance <0.0>: **2**
Enter second chamfer distance <*default*>: **4**

You have now set the default chamfer distances. When you next enter the Chamfer command, the first line you choose will be trimmed by two units and the second by four units.

Reenter the command:

Command: **Chamfer**
Polyline/Distance/<*Select first line*>: *(Select point 1.)*
Select second line: *(Select second line.)*

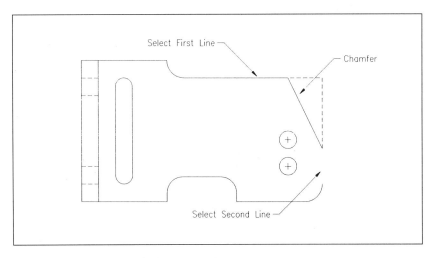

FIGURE 20-13   Chamfering a Corner

The default settings will remain until you reset them or you enter a new drawing. The initial setting is determined by the prototype drawing.

DRAWING POLYGONS

The Polygon command allows you to construct regular polygons (a closed object with all edges of equal length) with a specified number of sides. You may choose any number of sides from 3 to 1,024.

There are three methods of constructing polygons: inscribed, circumscribed, and by specifying one edge of the polygon. Let's look at each method of constructing polygons.

INTERMEDIATE DRAW COMMANDS **20-13**

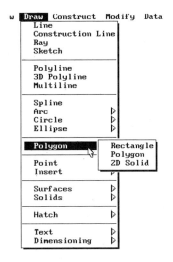

Inscribed Polygons

Inscribed polygons are constructed inside a circle of a specified radius (the circle is not drawn). The center point of the circle is first specified, followed by any point on the circumference of the circle. A vertex point of the polygon will start on the point chosen on the circumference, establishing the angle of the polygon. Figure 20-14 shows a polygon constructed using the circumscribed method.

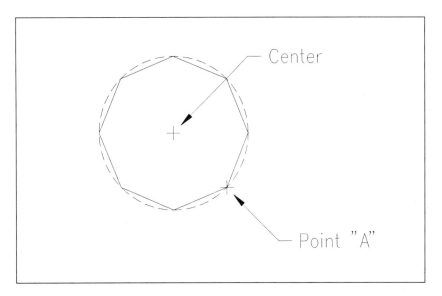

FIGURE 20-14 Inscribed Polygon

Circumscribed Polygons

Circumscribed polygons are constructed outside a circle of a specified radius. The center point of the circle is first specified, followed by any point on the circumference of the circle. The midpoint of one edge of the polygon will be placed on the point specified on the circumference, establishing the angle of the polygon. The following illustration shows a polygon constructed using the circumscribed method.

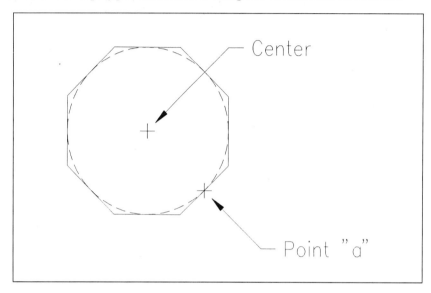

FIGURE 20-15 Circumscribed Polygon

Edge Method of Constructing Polygons

You may also construct a polygon by specifying the length of one edge. Simply enter the endpoints of the edge. The angle of the two points specifies the angle of the polygon. The following illustration shows construction of a polygon using the "edge" method.

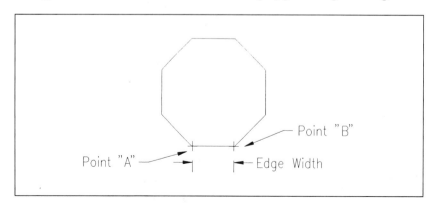

FIGURE 20-16 Edge Method of Constructing a Polygon

Constructing Polygons

To construct a polygon using either the inscribed or circumscribed methods, enter the Polygon command:

Command: **Polygon**
Number of sides: *(Enter number of sides.)*
Edge/<Center of polygon>: *(Pick center point.)*
Inscribed in Circle/Circumscribed about circle (I/C) <I>: *(Enter I or C.)*
Radius of circle: *(Pick.)*

To construct a polygon using the edge method, respond to the "Edge/<Center of polygon>" prompt with **E** or choose "EDGE" from the screen menu.

Command: **Polygon**
Number of sides: *(Enter number of sides.)*
Edge/<Center of polygon>: **E**
First endpoint of edge: *(Enter first edge point.)*
Second endpoint of edge: *(Enter second edge point.)*

Polygons are constructed of closed polylines and may be edited using Pedit or edit commands.

BLOCKS AND INSERTS

One of the most valuable parts of CAD drafting is the ability to library and reuse parts of drawings. AutoCAD provides commands to store such symbol details and drawings. These commands are Block, Wblock, and Insert.

The Block and WBlock commands are used to create and store the symbols and the Insert command is used to place them in a drawing. Let's look at how to use the Block and Insert commands.

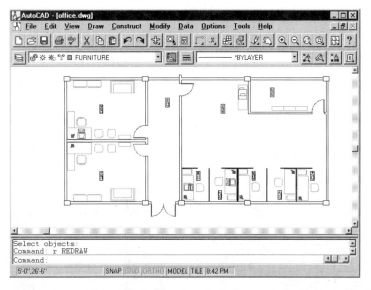

FIGURE 20-17 This Office Drawing Is Constructed of Many Blocks

Combining Entities into a Block

A block is a group of entities that have been identified by a name. This grouping can be placed into a drawing. Blocks may be placed any place in the drawing, scaled, and rotated to your specifications.

A block is considered as a single entity. Because of this, you may move or erase a block as though it is a single entity.

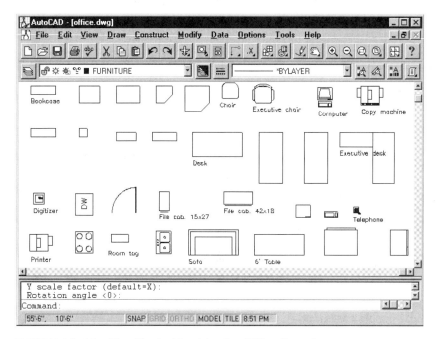

FIGURE 20-18 The Blocks Used in the Office Drawing

To define a block, identify the entities on the current drawing that you wish to capture.

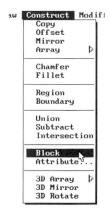

Command: **Block**
Block name (or ?): **NAME**
Insertion base point: [ENTER]
Select objects: *(Select.)*

The block name does not need a drive specifier since the block reference is stored with the drawing.

The insertion base point is the reference point for the block. The block will be placed into the drawing in reference to this point. To specify the base point, enter the point on the screen in the desired location.

Use the object selection process to identify the objects that make up the block.

After you capture the block, AutoCAD will erase the entities that make up the block from the screen. If you wish to keep the entities, use the Oops command to restore them to the drawing.

Blocks and Layers. When inserted, a block retains its original layer definitions. That is, if an entity was originally located on a layer named "PCBOARD", it will be on that named layer when inserted. The exception is an entity which was originally on Layer 0. Entities on Layer 0 in the block are assigned to the layer on which the block is inserted.

Nested Blocks. Blocks may be nested into each other. That is, you may place one block into another. Consider the following example. Each object is the same block and was inserted into another block. If you wish, you may "Block" the entire part (made up of other blocks) and have a new block which contains the original and the nested blocks together.

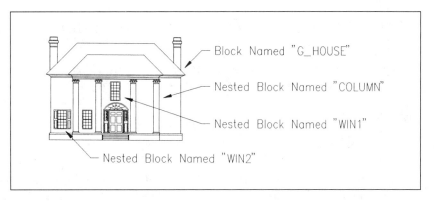

FIGURE 20-19 Nested Blocks

Listing the Defined Blocks. If you respond to the prompt with a **?**, a listing of all the defined blocks which are associated with the drawing are displayed.

    Command: **Block**
    Block name (or ?): **?**

Creating a Drawing File from a Block

The Wblock command can be used to capture a portion or all of a drawing and write it to the disk as a separate drawing.

To execute the WBlock command, enter:

    Command: **Wblock**
    File name: *(Name)*
    Block name: *(Name)*
    Insertion base point: *(Select.)*
    Select Objects: *(Select.)*

Notice that you must first have a named block. Let's suppose that you want to save a part of a mechanical drawing and create a separate drawing on your disk that consists of those entities.

First you must designate the part of the drawing that you wish to capture. Designate this part as a block using the Block command. Let's use an example of a block named "PART-A".

Next, use the WBlock command to write the block named "PART-A" to the disk and call the drawing "GEAR".

> Command: **Wblock**
> File name: **GEAR**
> Block name: **PART-A**
> Insertion base point: *(Select.)*
> Select Objects: *(Select.)*

The block is now recorded to disk and has a file name of "GEAR.DWG". Do not specify the file type (.dwg), since AutoCAD will attach the file extension for you.

This method of storage must be used if you wish to utilize the block in other drawings. A separate drawing may be placed in any drawing, whereas one created "on the fly" as a block may only be used in the drawing in which it was created.

After you have placed the block in the drawing, the block may be used repeatedly without having the original (separate) drawing present from which the block was extracted.

A separate drawing may be used as part of any other drawing by simply inserting it with the Insert command.

Inserting Blocks into Your Drawing

The Insert command is used to place blocks or other entire drawings into your drawings. After inserting, the block becomes a part of the drawing.

To place a block or other drawing into your drawing, enter:

> Command: **Insert**
> Block name (or ?): *(block or drawing name)*
> Insertion point: *(Select.)*
> X scale factor <<1>/Corner/XYZ: *(Select.)*
> Y scale factor (default=X): *(Select.)*
> Rotation angle: [ENTER]

Let's look at the options presented by the prompts.

Insertion Point. The insertion point is the reference point for the block. When you identified the base point for the block, you chose the reference point from which the block will be inserted. The block is inserted into the drawing so that the insert point and the base point of the block are the same.

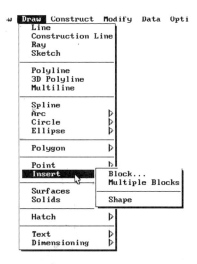

Scale Factors. AutoCAD automatically scales the inserted block to fit, regardless of the new drawing's limits. There are times, however, when you may want the block to be inserted at a different scale. The scale factor is a multiplier by which the block is scaled. The X and Y scales may be the same or different. The following examples show the same drawing inserted at different X and Y scales.

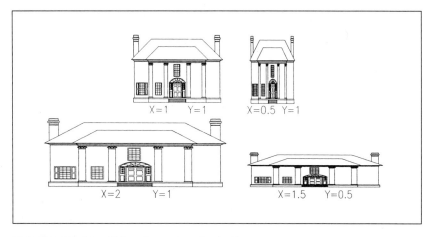

FIGURE 20-20 Using Insertion Scale Factors

Negative Scale Factors. If you use negative scale factors, the drawing will be mirrored around the axis to which the negative factor is applied. The following example shows illustrations of negative scale factor combinations for each axis:

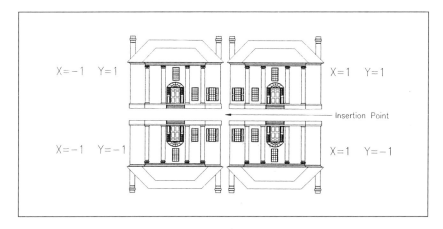

FIGURE 20-21 Negative Insertion Scale Factors

Note that the Mirror command is a more efficient manner of producing images of blocks and other entities than using negative scale factors. The Mirror command is covered in Chapter 22.

Corner Specification of Scale. You may specify the X and Y scales by responding to the scale prompt with the corners of a window. The X and Y dimensions of the window become the X and Y scales that are applied to the insert. If you enter a point in response to the prompt, AutoCAD will assume that you are showing the scale with a window. You may also enter a **C** and AutoCAD will prompt you with:

   Corner point:

You can then enter the first corner point of the window.

This method of scaling is very useful if you have a space of a certain size in which you wish to fit a block.

Rotation Angle. The angle of the inserted block is specified in the current AutoCAD angle format. This angle is in reference to the original orientation of the drawn block. If you wish, you may enter a point showing the desired angle. This point will be entered immediately after you have been prompted for the insertion point. If you move the crosshairs, the point will rubber band between the previously set insertion point and the crosshair. Move the cursor until the rubber-banded line shows the desired angle and enter a point. The distance between the two points is irrelevant. The angle measured by the rubber-banded line between the insertion point and the angle point determines the angle of insertion.

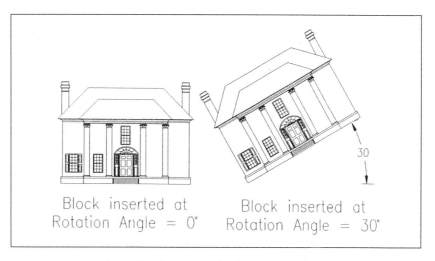

FIGURE 20-22 Insertion Rotation Angle

Preset Block Values. The rotation and scaling of a block are normally defined after the block's insertion point is specified. When using AutoCAD as a design tool, it is often desirable to "drag" the block into place at the desired scale and rotation to see the final results. By default, AutoCAD normally places the block into the drawing at a scale of 1 and a rotation of 0. Sometimes, it is helpful to drag the block at a size and/or rotation that is predetermined. To do this, first specify the block name you wish to insert. When AutoCAD prompts for the insertion point, select one of the following items to preset from the menu. Necessary information that is not preset will be prompted for after the insertion point is selected.

The following is an explanation of each menu item.

Scale: Applies an overall (x,y,z) scale factor to the named block. The block is displayed at the specified scale as it is "dragged" into position.

Xscale: Sets only the X scale factor of the block.

Yscale: Sets only the Y scale factor of the block.

Zscale: Sets only the Z scale factor of the block.

Rotate: Presets the rotation angle of the block. The block is displayed at the specified rotation angle as it is "dragged" into position.

PScale: Sets a temporary x,y,z scale. The block is displayed at the specified scale until the insertion point is selected, then AutoCAD prompts for the actual scale.

PXscale: Same as Pscale, but only sets the X scale factor.

PYscale: Same as Pscale, but only sets the Y scale factor.

PZscale: Same as Pscale, but only sets the Z scale factor.

PRotate: Similar to rotate. The rotation angle set is used temporarily to drag the object into the drawing area. After the insertion point is selected, AutoCAD prompts for the actual rotation angle.

Whole Drawings as Inserts. You may insert an entire drawing into another drawing by using the Insert command. When you are prompted for the block name, specify the desired drawing name, including the drive specifier.

The drawing will be inserted using the 0,0 point of the original drawing as the insertion base point.

If you wish to have greater control over the insertion of the drawing, use the Base command. The Base command is executed in the drawing to be inserted before you insert it into another drawing. In the drawing to be inserted, designate the Base command by entering:

Command: **Base**
Base point <default>: *(Select the point you desire for the insertion base point.)*

Be sure to End the drawing to save the base point location. The point you stipulate will become the reference when you insert this drawing into another.

Redefining Inserts. If you have used many inserts of the same block in a drawing, all the block duplications can be changed by "redefining" only one of the blocks. This is an especially powerful feature. Imagine being able to change 100 drawing parts in a single operation!

There are two methods of redefining a block. If you inserted an entire drawing as a block, you may edit the original drawing. This alone will not redefine the block. You must then (while in the drawing in which the block was inserted) reissue the Block command. When you are prompted for the block name, use the

Block name=file name

form of identification. This will force the regeneration of all instances of the inserted block and will result in your changes being incorporated in all of them!

If you defined the block "on the run" (that is, you defined a portion of your current drawing as a block), you may still redefine the block, but must use a different method.

Insert the block into the drawing using the *name method. Make the desired changes. Then re-block the edited block using the same block name. You will then be informed by AutoCAD that you already have an existing block by this name and will be asked whether or not you wish to redefine it. Respond to this prompt with Yes (**Y**) and press [ENTER]. All instances of the block will be redefined (updated).

Another method of updating blocks is to use the "external reference" of specifying an insert. This method allows you to insert a separate drawing in a procedure similar to a standard insert. Instead of storing the inserted drawing in the database of the new drawing, however, an external reference to the original drawing is stored. If the original inserted drawing is changed, the changes will be loaded into the "second" drawing when it is next loaded.

External references are covered in Chapter 27.

Advantages of Blocks. Using blocks in your drawings has several distinct advantages.

> **Libraries:**   Entire libraries of blocks can be built which can be used over and over again for repetitive details.
>
> **Time savings:**   Using blocks and nested blocks is an excellent method of building larger drawings from "pieces." (Nested blocks are blocks that are placed within each other.)
>
> **Space savings:**   Several repetitive blocks require less space than copies of the same entities. AutoCAD must only store information on one set of entities instead of several sets. Each instance of the block can be referred to as one entity (a Block reference). The larger the block, the greater the space savings. This can be very valuable if there are many occurrences of the block.

Attributes.   An attribute is a text record that is stored with a block. The text can be set to be displayed or invisible. Attributes are "attached" to a block. These attributes can be loaded into database or spreadsheet programs or printed as a listing. This is useful in facilities management where there are multiple occurrences of items such as desks which may be stored as blocks and have attributes such as a person's name or telephone number associated with them.

Attributes are covered in Chapter 26.

Inserting Blocks with a Dialog Box

You can use a dialog box to insert blocks and drawing files into your drawing. To use a dialog box, use the Ddinsert command.

> Command: **Ddinsert**

AutoCAD displays the Insert dialog box.

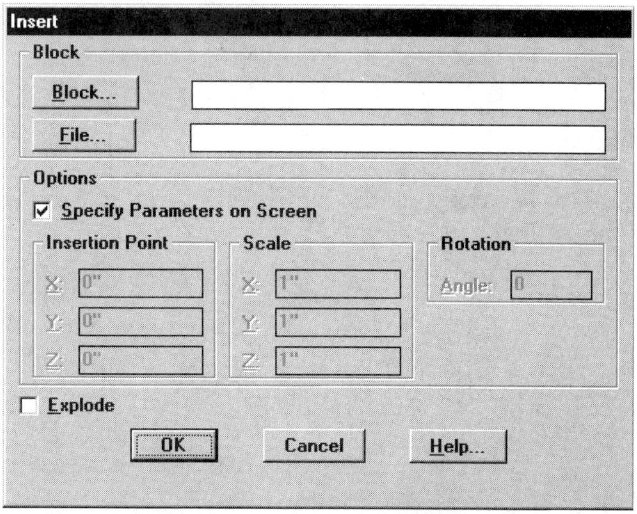

FIGURE 20-23   Insert Dialog Box

Inserting Blocks. A predefined block can be inserted in two ways. The first is click in the text box next to the "Block..." button, then enter a block name.

The second way to insert a block is to click on the "Block..." button to display a sub-dialog box.

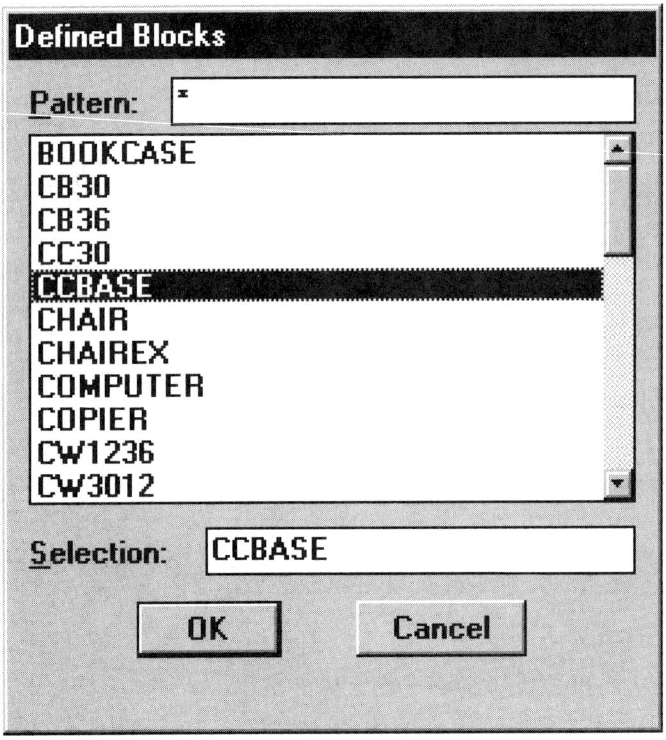

FIGURE 20-24 Defined Blocks Dialog Box

The sub-dialog box contains the names of all the predefined blocks in the drawing. Select a block, then click on the "OK" button.

Inserting Drawing Files. The methods of inserting entire drawing files into your drawing are similar to inserting blocks. You can either enter a drawing name in the text box next to the "File..." button, or click on the "File..." button to display the Select Drawing File sub-dialog box.

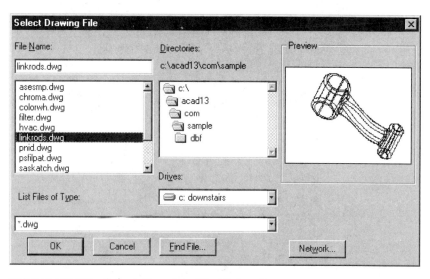

FIGURE 20-25 Select Drawing File Dialog Box

You can select a drawing file from the list box displayed in the sub-dialog box.

Setting the Scale and Rotation. The scale factor and rotation can be "preset", or entered at the time of insertion in the traditional manner. To preset the scale factors and rotation angle, deselect the "Specify Parameters on Screen" check box. When the box is selected, the factors will be selected when the block is inserted and the scale factor text boxes will be "grayed out".

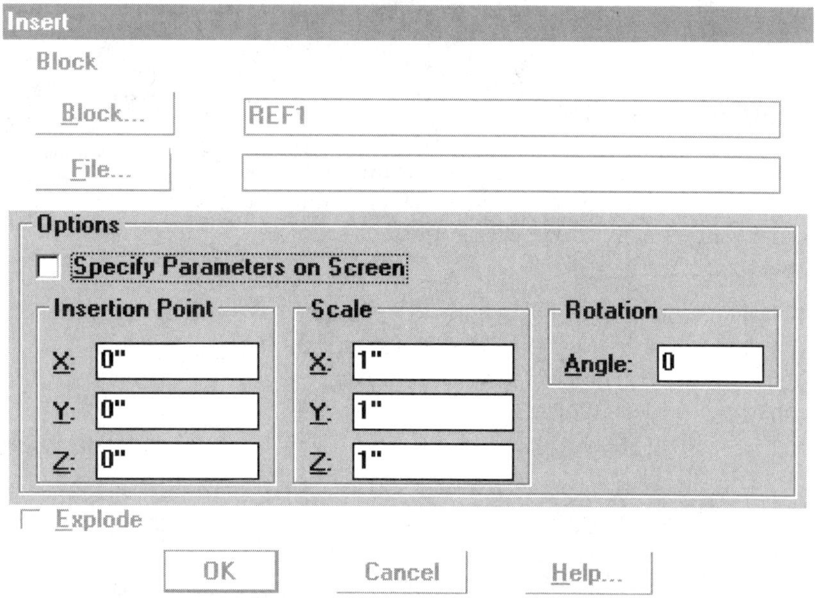

FIGURE 20-26 Insertion Point, Scale, and Rotation Parameters

20-26 USING AUTOCAD

Exploding the Inserted Item. You can designate that the block or drawing file be automatically exploded. To do this, select the "Explode" check box. Note that you can only specify a single scale factor for an exploded block.

FIGURE 20-27 Explode Check Box

Multiple Insertions

The Minsert (multiple insert) command is actually a single command that combines the Insert and Rectangular Array operations.

The sequence starts by issuing prompts in the same manner as the Insert command. You are then prompted for information to construct a rectangular Array. Let's explore the procedure step-by-step.

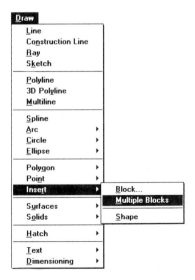

Minsert Operations. The Minsert command produces entire arrays that have many of the same properties as blocks, with some exceptions. The following qualities apply to Minserts:

1. The entire array reacts to editing commands as if it were one block. You may not edit each individual item. If you select any one object to move or copy, for instance, the entire array will be affected.

2. You may not use the "*" method of inserting blocks with individual entities.

3. You cannot "explode" the block into individual entities. (The Explode command is covered in Chapter 22.)

4. If the initial block is inserted with a rotation, the entire array will be rotated around the insertion point of the initial block. This creates an array in which the original object appears to have been inserted at a standard zero angle, then the entire array rotated. The following illustration shows the block inserted at a 30 degree angle using the Minsert command, with four rows and six columns.

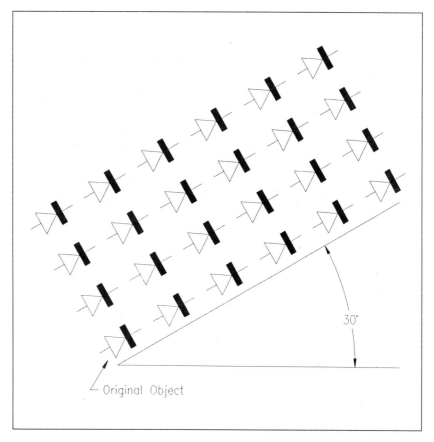

FIGURE 20-28 MInsert at 30 Degrees

Using Minsert. If you wish to create a true horizontal and vertical array in which the objects are themselves rotated, block the original object in its desired rotation, or use the Insert and Array commands separately.

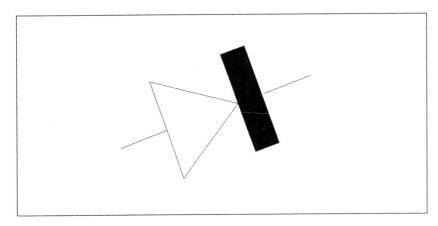

FIGURE 20-29 Original Diode Drawing Rotated

The Minsert command can be used to save steps if you intend to array an inserted block by combining two command functions into one command.

Create arrays using the Minsert command if you may need to edit the array as a whole later. For example, a seating arrangement consisting of several rows of chairs may need to be moved around in a space for design purposes. Creating the arrangement by using the Minsert command would allow you to move and rotate the seating as a whole.

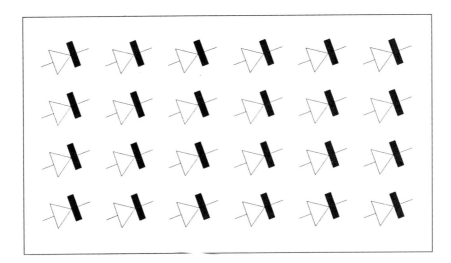

FIGURE 20-30 Minserted Diodes

INTERMEDIATE DRAW COMMANDS **20-29**

EXERCISES

1. Let's use the Ellipse command to draw a can. Start by entering the Ellipse command. Use the following illustration to draw the ellipse. Don't worry about the size.

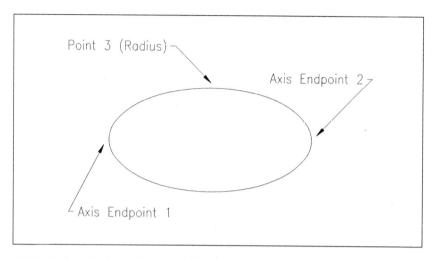

FIGURE 20-31 Drawing the Ellipse

Next, use the Copy command to copy the ellipse up to create the top of the can. Use [F8] to turn on Ortho mode so the copy will align perfectly with the original ellipse.

Now use the Line command to draw lines between the two ellipses as shown in the following illustration. Use object snap QUADrant to capture the outer quadrant of the ellipses.

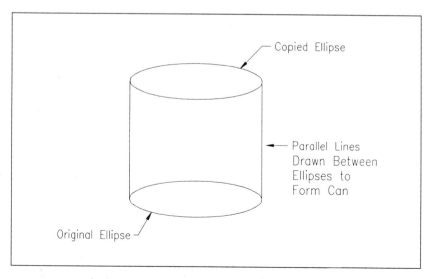

FIGURE 20-32 Completing the Can

2. Draw the bicycle as shown in Figure 20-33. Use polylines for the frame, lines for the spokes, and doughnuts for the tires.

FIGURE 20-33 Bicycle

3. Let's use the Offset command to help draw a site plan. Start the drawing named "OFFSET" from the work disk. This is a drawing of a city scape. The edges of the streets are drawn with a different type of line called a polyline (polylines are covered in Chapter 24). Polylines can be joined so that each segment is "glued" to the other segments.

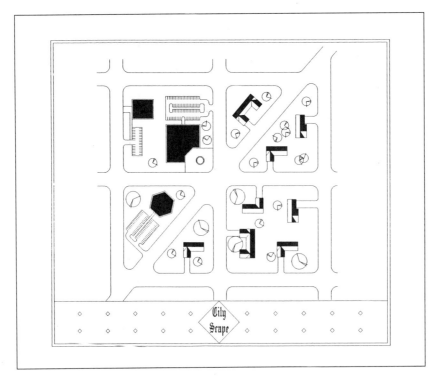

FIGURE 20-34 OFFSET.DWG Work Disk Drawing

Use the Offset command to offset a curb thickness (curbs are typically 6" thick). You can do this by using Offset/Through. Start by selecting the curb to be offset, then entering a relative coordinate to offset the 6". You may want to review relative coordinates in Chapter 6.

4. Draw the object as shown on the left in the following illustration. Use the Chamfer command to edit the object so that it looks like the object on the right. Hint: You can set distances in AutoCAD by entering two points.

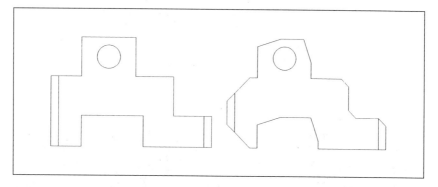

FIGURE 20-35 Chamfer Exercise

5. Use the Polygon command to draw the following object.

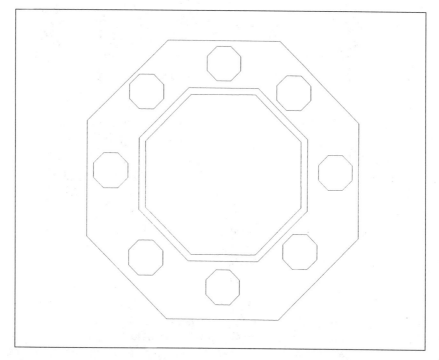

FIGURE 20-36 Polygon Exercise

6. Let's capture a block from a drawing. Start the drawing from the work disk named "OFFICE". The following illustration shows the drawing. Execute the Block command and use the following command sequence. Refer to Figure 20-37 for the points to enter.

Command: **Block**
Block name (or?): **Sdesk**
Insertion base point: *(Select point "1".)*
Select objects: **W** *(Place a window around the desk.)*

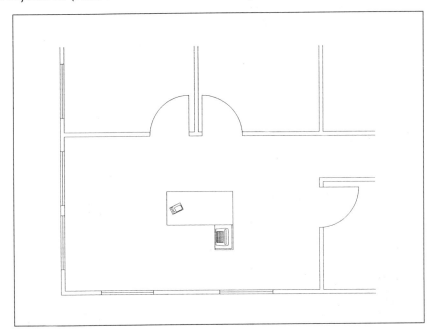

FIGURE 20-37 OFFICE.DWG Work Disk Drawing

The block is now stored with the drawing and may be used as many times as you desire. Keep the drawing open or use End if you want to exit the exercise drawing now. We will use this drawing again for the Insert command which follows later in this chapter.

7. Use the work disk drawing named "OFFICE". This is the same drawing you captured a block from when we learned about the Block command earlier in this chapter. Use the Insert command to place the block "SDESK" into the drawing.

8. Use the work disk drawing named "INSERT". This is the site plan you used in the Copy exercise in Chapter 10. The landscape items have already been drawn and blocked and are contained as resident blocks in the drawing.

Use Insert/? to display the names of the blocks, then use Insert to insert the landscaping to create a design of your own.

9. We will draw a box and perform a multiple insert containing four rows and six columns. We will use the "window" method of showing AutoCAD the column and row spacing. As you proceed, notice the similarities to the Insert and Rectangular Array commands.

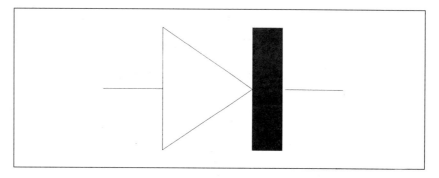

FIGURE 20-38 Diode Drawing

Construct a drawing as shown in the illustration above. Block the rectangle and name it "DIODE1". Use the lower end of the diode as the insertion point.

Now use the Minsert command:

Command: **Minsert**
Block name (or ?) <name>: **diode1**
Insertion point: [ENTER]
X scale factor <1>/Corner/XYZ: [ENTER]
Y scale factor *(default=X)*: [ENTER]
Rotation angle <0>: [ENTER]

You have now placed the block "DIODE1" in the drawing. (Notice, however, that you can not see it.) AutoCAD then continues and prompts:

Number of rows (---) <1>: **4**
Number of columns (|||): **6**
Unit cell or distance between rows (---): *(Enter point "A".)*
Other corner: *(Enter point "B".)*

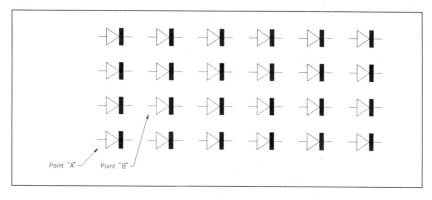

FIGURE 20-39 Minserted Diodes

(Notice the notations in the rows and columns prompts that make it easy to remember which way rows and columns operate.)

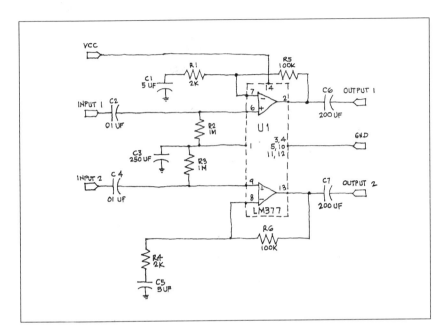

FIGURE 20-40 Create a Symbol Library and Use It to Draw the Stereo Amplifier

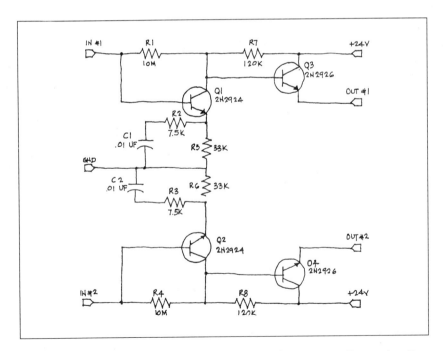

FIGURE 20-41 Create a Symbol Library and Use It to Draw the Pre-amplifier

CHAPTER REVIEW

1. What are the two axes of an ellipse?

2. How would you draw an isometric circle?

3. What two types of entity arc ellipses constructed from?

4. What type of entities are constructed with the Donut command?

5. How do you control whether the solid-filled areas of doughnuts are displayed and plotted?

6. How many entities can be offset at one time?

7. What does the Through option of the Offset command perform?

8. Draw an example of using the Offset option of Through.

9. What does the Chamfer command perform?

10. What is the procedure for setting chamfer distances?

11. The Polygon command constructs polygons of a specified number of sides. What is the minimum number of sides?

    The maximum?

12. What is a Block?

13. What is the insertion base point of a block?

14. When placed into a drawing, how does a block handle its layer definitions?

15. What is a nested block?

16. How would you create a separate drawing file from an existing block?

17. How do you place a block into a drawing?

18. Can you place a block into another drawing?

19. When you place a block into the drawing, what is the insertion point?

20. Can you place one AutoCAD drawing into another AutoCAD drawing?

21. Name two advantages of using blocks.

22. What commands are combined to create the Minsert command?

CHAPTER

21

ISOMETRIC DRAWINGS

One of the standards in the drawing profession is the isometric drawing. This chapter covers the methodology of constructing isometric drawings with AutoCAD. After completing this chapter, you will be able to:

- Distinguish the three basic types of pictorial drawings.
- Manipulate the aspects of isometric drawing.
- Use and understand the isometric drawing commands in AutoCAD. You will achieve this through example and by tutorial.

ISOMETRIC DRAWINGS

AutoCAD provides a mode for drawing isometric drawings. This mode uses three drawing planes which you utilize to construct your drawings. The Snap and Grid commands are used to aid in the drawing process.

There are three basic types of engineering pictorial drawings: axonometric, oblique, and perspective. Isometric drawings are one type of axonometric drawings. Figure 21-1 shows examples of each type of pictorial drawing.

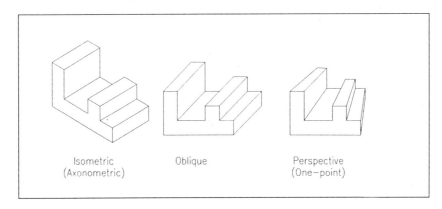

FIGURE 21-1 Engineering Pictorial Drawings

Axonometrics are mostly used for engineering drawings. Each face of the object is shown in true length, resulting in the ability to measure each length.

Oblique drawings are primarily used as quick design drawings, since the front face is shown in plane, and the sides are drawn back at an angle. This makes an oblique drawing easy to construct from an elevation view of the front face.

Perspectives are the most realistic types of drawings. Perspective drawings show the object as it would appear to the eye from a specified location and distance.

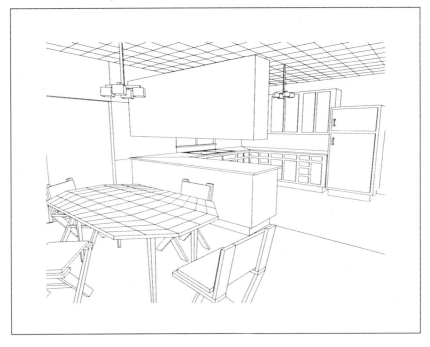

FIGURE 21-2 Perspective Drawing

This involves the use of vanishing points. (See Figure 21-2.) Although perspectives are the most realistic, most of the lengths in the drawing are not true lengths and cannot be accurately measured.

PRINCIPLES OF ISOMETRICS

As mentioned earlier, isometric drawings are a type of axonometric drawing. In isometric, the lines that are used to construct a simple box are drawn at 30 degrees from the horizontal. Figure 21-3 shows a box drawn in isometric.

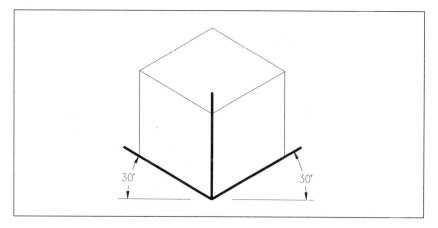

FIGURE 21-3 Box in Isometric

From this, we can derive an isometric axis as shown in Figure 21-4.

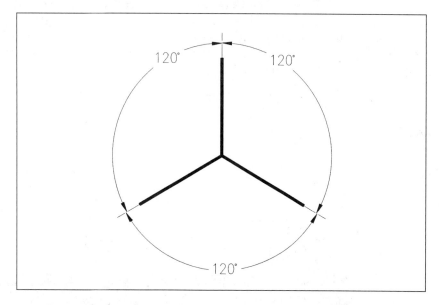

FIGURE 21-4 Isometric Axes

If we look at each of the isometric axes, we can identify a plane lying between each axis. These are referred to as isometric planes. There is a top, left, and right plane as shown in Figure 21-5.

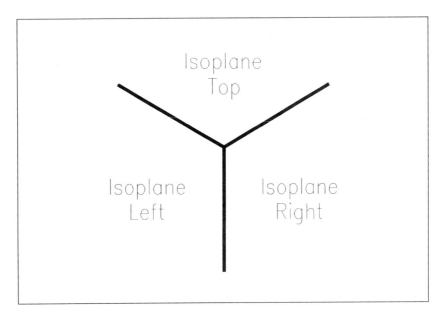

FIGURE 21-5 Isometric Planes

As you draw in AutoCAD, you will be drawing in one of these three planes. Any lines that are parallel to one of the isometric axes are called isometric lines. Some objects are made up of lines that are not parallel to an axis. For example, a sloping surface will require lines to describe the slopes that are not parallel to any axis. These lines are called nonisometric lines. The endpoints of these lines, however, are derived from points determined by isometric lines.

Let's continue to learn how to create a drawing with AutoCAD's isometric capabilities.

ENTERING ISOMETRIC MODE

The isometric mode is actually a Snap style. To enter isometric mode, use the Snap command:

Command: **Snap**
Snap spacing or ON/OFF / Aspect/ Rotate/Style <default>: **S**
Standard/Isometric: **I**
Vertical spacing <default>: (Select.)

You will notice that the crosshairs are displayed at angles. The particular angle depends on the current axis.

You may also use the pull-down menus to set Isometric mode. Select the Options pull-down menu. A Drawing Aids dialog box will be displayed on the screen as shown in Figure 21-6.

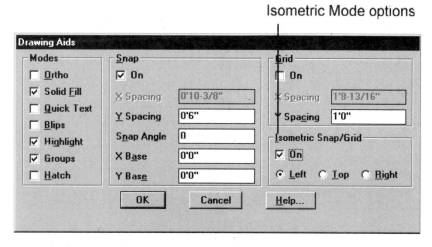

FIGURE 21-6 Drawing Aids Dialog Box

Select the box next to Isometric to set the Isometric mode. You can also select the snap spacing and the isoplane setting from the dialog box.

SWITCHING THE ISOPLANE

In order to draw in each axis, you must change between the axes. This is accomplished by using either the Isoplane command or by using [CTRL] [E] or function key [F5] as a toggle. Let's look at how to use each.

The Isoplane command is used to toggle between the three isometric planes.

Command: **Isoplane**
Left/Top/Right/<Toggle>:

The following explains the options for the Isoplane command:

Left: Selects the left plane and uses axes defined by 90° and 150°.

Top: Selects the top plane and uses axes defined by 30° and 150°.

Right: Selects the right plane and uses axes defined by 90° and 30°.

Enter: Entering an Enter will toggle the current plane. The planes are displayed in a rotating fashion.

You may switch between the isoplanes transparently by using [CTRL] [E] or [F5]. Each time you press [CTRL] [E], the isoplane changes between the top, left, and right in a repeating order.

Figure 21-7 shows the crosshair configuration for each of the isoplane settings.

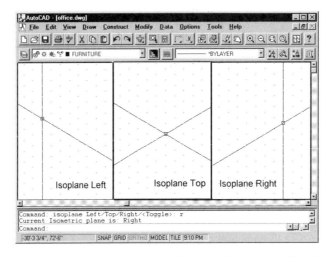

FIGURE 21-7 The Three Isoplanes: Left, Top, and Right

If you display a grid while in Isometric mode, the grid display will reflect the axis markings.

DRAWING IN ISOMETRIC

After you have set up the isometric mode, you may use AutoCAD's commands to construct and edit the drawing. An isometric drawing is constructed by first boxing in the general shape of the object. The boxing-in process uses all isometric lines, creating the intersection points for nonisometric. Figure 21-8 shows the sequence for drawing an isometric object.

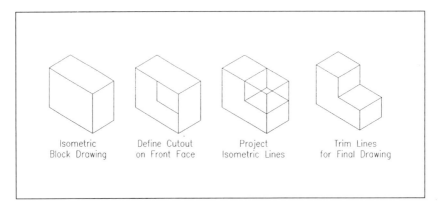

FIGURE 21-8 Isometric Drawing Sequence

ISOMETRIC CIRCLES

Isometric circles are constructed with the Isocircle option under the Ellipse command.

Command: **Ellipse**
<Axis endpoint 1>/Center/Isocircle: **I**
Center of circle:
<Circle radius>/Diameter:

The isocircle will be drawn correctly for the current isoplane.

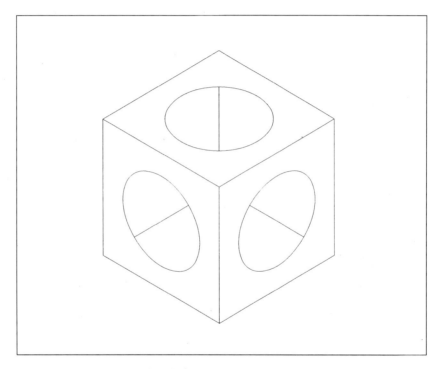

FIGURE 21-9 Isometric Circles

ISOMETRIC TEXT

Text can be aligned with the isometric planes by creating text styles that align with the axes. This is achieved by creating three styles that have text rotation angles of 90 degrees, 30 degrees, and –30 degrees. Figure 21-10 shows the use of three text styles used with the text rotation angles of 90 degrees, 30 degrees, and –30 degrees.

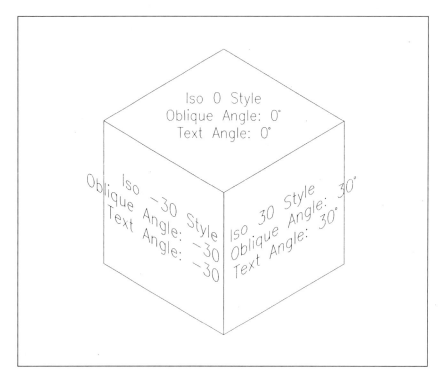

FIGURE 21-10 Isometric Text Styles

ISOMETRIC DIMENSIONING

Dimensioning an isometric object requires the dimension line, extension line, and text to sit properly in the isometric plane. To dimension an isometric drawing, the following steps are recommended.

1. Create three text styles of the desired text font and height. Set the text angle to 0 degrees, 30 degrees, and −30 degrees. (You can use the same text styles for adding notes to the drawing.) You could name the style ISO, ISO30, and ISO-30 to denote the text rotation angles of 0 degrees, 30 degrees, and −30 degrees, respectively.
2. Set the text style to the style that uses the proper text angle. (See Figure 21-11.)
3. Use either the aligned or vertical dimension style to place the dimension. The type selected depends on the plane in which the dimension is placed. (See Figure 21-11.)
4. Use the Oblique dimension option to change the extension lines to the proper angle. (See Figure 21-11.)
5. Use the Trotate dimension option to rotate the dimension text to the proper angle. (See Figure 21-11.)

Figure 21-11 shows the settings for each dimension placement.

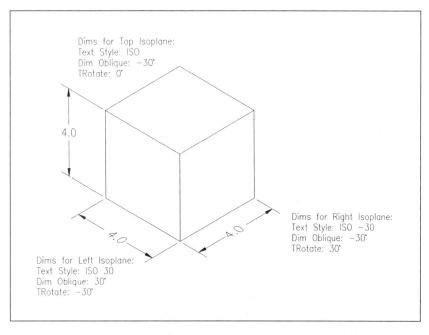

FIGURE 21-11 Isometric Dimensioning

EXERCISES

1. Enter the isometric mode by selecting "Drawing Aids" from the "Options" pull-down menu. In the resulting dialog box, enter **.25** in the "Y spacing" text box. Click on the check box next to "On" in the Isometric Snap/Grid section. Click "OK" to exit the dialog box.

   Move the crosshairs into the screen area. Use [CTRL] [E] to switch the crosshairs between the three isoplanes.

   Start the Ellipse command. Choose the Isocircle option, then enter a point at the center of the screen. Move the crosshairs away from the center point and watch the isocircle dynamically drag into form. Use [CTRL] [E] to switch between the isoplanes. Note how the isocircle changes form for each isoplane.

2. Draw the following isometric objects and dimension each. After you have drawn the objects, you may want to use the Color command to set different colors for each face and use the Solid command to make the faces solid.

   You can make the isometric drawing process easier by setting the snap spacing to be an increment of the dimensions of the object. Also, using the Ortho mode will assist in drawing isometric lines perfectly along the isometric axes.

21-10 USING AUTOCAD

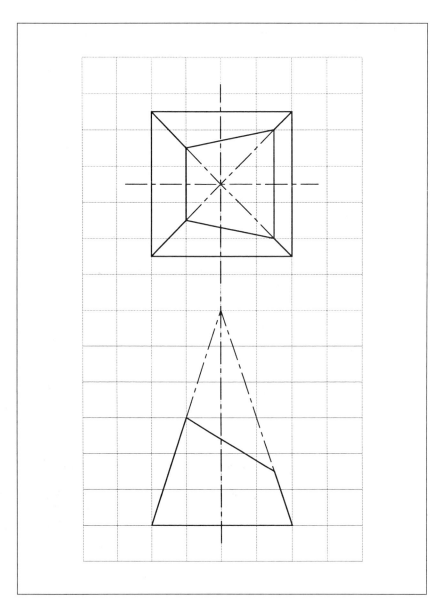

FIGURE 21-12

ISOMETRIC DRAWINGS **21-11**

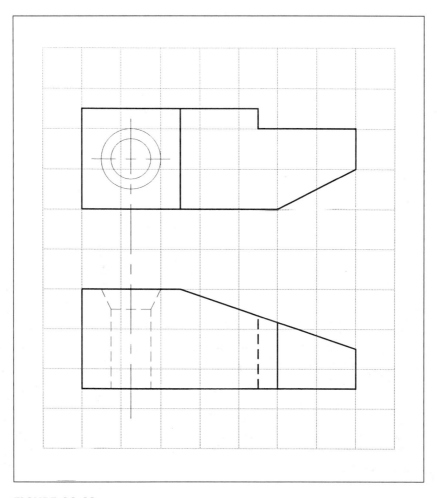

FIGURE 21-13

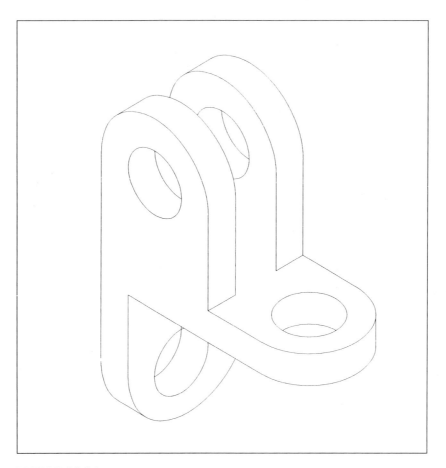

FIGURE 21-14

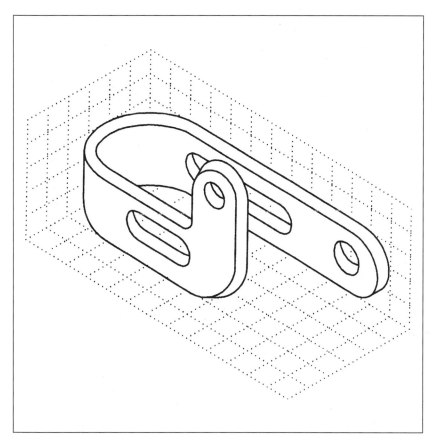

FIGURE 21-15

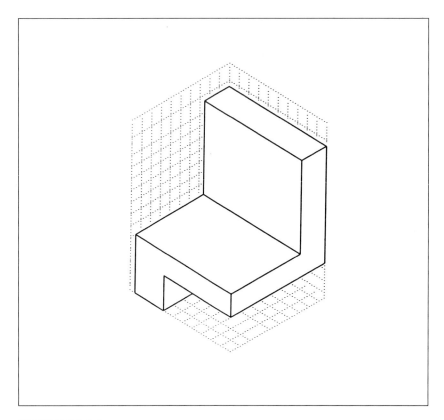

FIGURE 21-16

CHAPTER REVIEW

1. What are the three types of engineering pictorial drawings?

2. Which type is the most realistic in appearance?

3. What type of drawing is an isometric?

4. How many axes are used in isometric drawing?

5. How many degrees above horizontal are the isometric axes?

6. What are the three isometric planes?

7. How do you change between isometric planes in AutoCAD?

8. What command is used to enter isometric mode?

9. How would you set isometric mode from a pull-down menu?

10. How would you draw a circle in isometric mode?

11. What command do you use to set up isometric text?

12. What dimension commands are used to convert standard dimension components to isometric dimension components?

CHAPTER

22

INTERMEDIATE EDIT COMMANDS

After you have become proficient with the basic edit functions in AutoCAD, you can learn more powerful edit commands. This chapter covers the intermediate edit commands. After completing this chapter, you will be able to:

- Change the properties of an existing object or objects.
- Produce multiple copies of an object or objects with the array functions.
- Edit existing objects by rotating, stretching, scaling, etc.
- Undo operations just performed with AutoCAD.

CHANGING ENTITY PROPERTIES

The Change command is used to modify existing entities.

>   Command: **Change**
>   Select objects: *(Select.)*
>   Properties/<Change point>: *(Select.)*

Enter the option desired. Notice the capitalized letters that denote the abbreviation that may be entered. If you are using the screen menu, it is not necessary to enter the "P" when choosing to change entity properties. Just choose the desired property from the menu. Let's look at the execution of each sub-option.

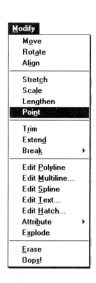

Changing Properties

The Change command allows changes in common entity properties. Each entity (line, circle, arc, etc.) has an associated layer on which it resides, a color, and a linetype. Let's go through the options in the Change command under the Properties sub-option.

    Command: **Change**
    Select objects: *(Select entities to change.)*
    Properties/<Change point>: **P**
    Change what property (Color/Elev/LAyer/LType/LtScale/Thickness).

Following is an explanation of each option.

    Color: Changes the color for the selected entity. You will be prompted:

        New Color <current color>:

You may enter a color number, or choose a color name from the screen menu. If you have chosen entities of more than one color, the default will be:

        New color <varies>:

If you wish the entity to be the color that is assigned to the layer on which it resides, enter "BYLAYER" in response to the prompt. If the object will be placed into a block and you wish it to inherit the color of the block, enter "BYBLOCK" in response to the prompt.

    Elev: Changes the elevation. You are prompted:

        New elevation <current>:

    LAyer: Changes the layer on which the entity resides. The prompt is:

        New layer <current>:

    LType: Changes the linetype of the chosen entity or entities. The prompt is:

        New linetype <current>:

You may only change the linetype of a line, arc, circle, or polyline. All other entities are drawn in the continuous linetype.

    LtScale: Changes the linetype scale. You are prompted:

        New linetype scale <current>:

    Thickness: Changes the 3D thickness of the chosen entity or entities. The prompt is:

        New thickness <current>:

Changing Properties with a Dialog Box

If you wish to use a dialog box to change entity properties, type the DdChprop command.

AutoCAD will prompt:

    Select objects:

Select the objects you wish to change and the following dialog box will be displayed.

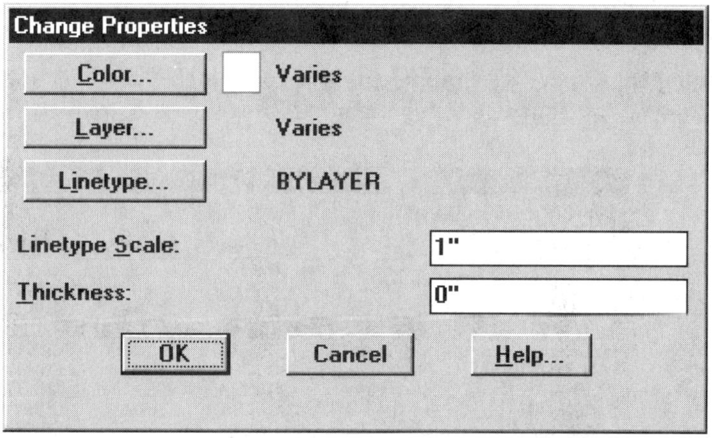

FIGURE 22-1 Change Properties Dialog Box

Let's look at each part of the dialog box.

Changing the Color. You can change the selected entities' color by selecting the "Color..." button. The Select Color sub-dialog box is displayed.

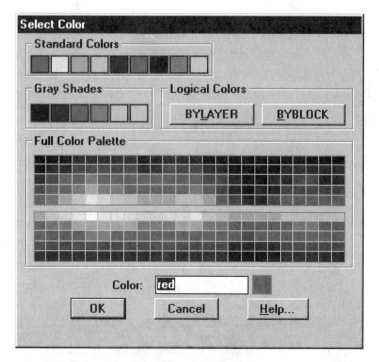

FIGURE 22-2 Select Color Dialog Box

Select the desired color from the palette.

You can also set the color of the selected entities to be set to the layer color. Selecting the "BYLAYER" check box will set the selected entity color to the layer color.

Changing the Layer. To change the layer on which the selected entities reside, select the "Layer..." button. The following sub-dialog box is displayed.

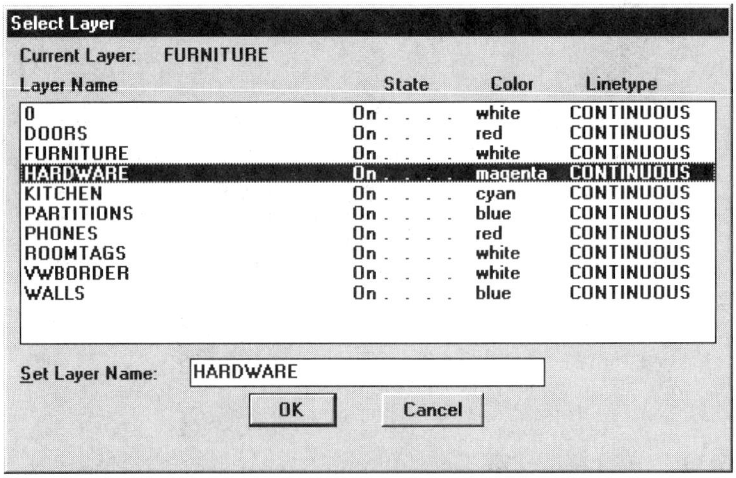

FIGURE 22-3 Select Layer Dialog Box

To change the entities to a new layer, select the layer name from the listing in the sub-dialog box, then click on the OK button.

Changing the Linetype. The linetype of the selected entities can be changed by selecting the "Linetype..." button. The following sub-dialog box is displayed.

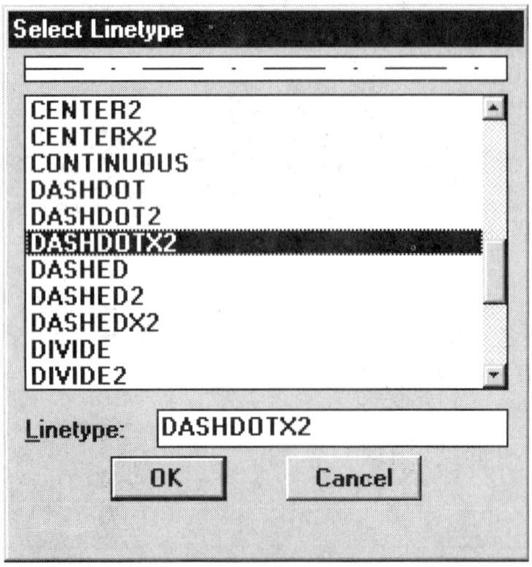

FIGURE 22-4 Select Linetype Dialog Box

The sub-dialog box displays the linetypes that have been loaded. Select the linetype that you desire and click on the OK button. If the linetype you want is not displayed, it must first be loaded with the Linetype command.

Changing the Entity Thickness. The thickness of the selected entities can be changed in the "Thickness:" text box. Thickness settings are mostly used in 3D drawings. To set the thickness, click in the box and enter the numerical value for the thickness.

The dialog box does not let you change the elevation. Use the Change command, instead.

Changing Entity Points

You can change the properties of existing lines, circles, text, attribute definitions, and inserted blocks by responding to the Change Point prompt with a new point. The effect is different on each entity.

The following describes the change for each:

> **Line:** The nearest endpoint of the line is changed to the change point. The line length will be modified if necessary.

NOTE: Use the Change command to "grow" lines longer.

The following illustration shows an example of line modification using the Change command.

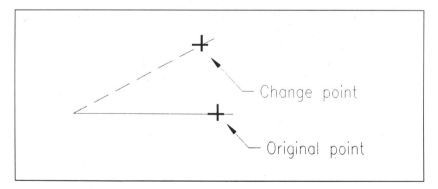

FIGURE 22-5 Using the Change Command

> **Circle:** The change point becomes a point on the circumference of the circle. The radius is changed accordingly. Thus, the distance from the original center of the circle to the change point becomes the new radius.

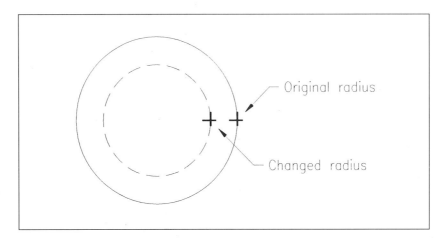

FIGURE 22-6 Change Point on the Circumference of a Circle

Text: You may show a new text location by entering the change point at the desired point. You are then prompted for a new text style, height, rotation angle and text string.

You may leave the text string unchanged by pressing [ENTER] when prompted for the "New text".

The height and angle of the text may be chosen by showing points relative to the change point or by entering numerical values.

If you only wish to change the text and not the location, press [ENTER] in response to the Change Point prompt. You may then proceed to change either the text or the height or rotation angle. By using the Enter key to accept the parts of the text you do not wish to change, you may make only the desired changes.

Attribute Definition: Changes may be made in attribute definitions.

Blocks: When the Change command is used to change a block, the change point becomes the new insertion point. You are then prompted for a new rotation angle. The angle may be shown by entering a point at the desired angle from the change point.

If you do not want to relocate the insertion point, press [ENTER] when prompted for the change point. The angle is then shown from the original insertion point (you may, of course, enter a numerical value for the angle).

> **NOTE:** Be sure that a block made up of several separate entities is really a block. If you select several separate entities in the object selection process and then enter a change point, you will scramble your drawing. Each entity will change separately.

CHANGING PROPERTIES WITH THE CHPROP COMMAND

If you wish to change properties (such as color, layer, etc.) of a drawing, the Chprop command is a more efficient method than Change.

You could, of course, use the Change command to facilitate this. The Chprop command, however, functions in the same manner if you desire to only have the option to change entity properties.

The following properties may be changed with the Chprop command:

Color of entities.
Layer of entities.
Linetype of entities.
Linetype scale.
Thickness of entities.

The Chprop command dialog is as follows:

Command: **Chprop**
Select objects:
Change what property (Color/LAyer/LType/LtScale/ Thickness):

Notice that the options provided are similar to those under the "Properties" option of the Change command, with the exception of the Elev (elevation) option.

> **NOTE:** The Chprop command is a more efficient method of changing object properties than the multi-optioned Change command.

ARRAYING OBJECTS IN THE DRAWING

There are times that multiple copies of an object or objects are desirable. Consider the number of seats in a movie theater. Or the number of parking space lines in a shopping center parking lot. If you were using traditional drafting techniques, you would have to draw each one separately. AutoCAD uses the Array command to draw repeated objects.

The Array command is used to make multiple copies of one or more objects in rectangular or circular patterns. After you have arrayed the object, each one may be edited separately.

Constructing Rectangular Arrays

Let's first look at rectangular arrays. Rectangular arrays copy an object in a rectangular pattern that is made up of rows and columns.

The following example shows a rectangular array and identifies the parts that are defined as rows and columns:

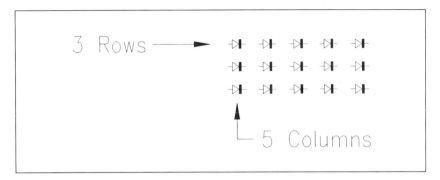

FIGURE 22-7 Rectangular Array

The first item of a rectangular array must occupy the most lower and left position. This is the object that will be identified as the selected object in the standard object selection process (notice how this process keeps showing up?).

Before invoking the Array command, this object must already be in existence. After you have this object in place, enter:

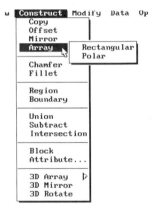

Command: **Array**
Select objects:
Rectangular or Polar array (R/P): **R**
Number of rows (---):
Number of columns (|||):
Unit cell distance between rows (---):
Unit cell distance between columns (|||):

The number of objects in the rows and columns should include the identified item(s).

The unit cell distances are "center to center". Many new CAD users make the mistake of thinking the unit cell distance is the distance *between* the items. It is not. The unit cell distance may be thought of as the distance from the center of one item to the center of the next item.

The unit cell distance is defined in the current drawing units. The distance between the items in the columns can be different from the distance between the items in the rows.

> **NOTE:** Set your first point on the selected object so you may "see" the distances that the objects will be spaced.

Rotated Rectangular Arrays

A rotated rectangular array may be constructed by changing the Snap rotation angle. For example, if the snap angle is set to 30 degrees, the rectangular array will be rotated as a whole to a 30-degree angle.

Constructing Polar Arrays

Polar arrays are used to array objects in a circular pattern.

To construct a polar array, you must define the angle between the items (again, from center to center, not actually between) and either the number of items or degrees to fill.

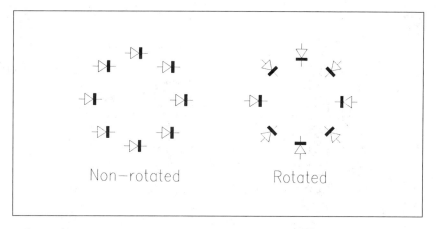

FIGURE 22-8 Non-Rotated and Rotated Polar Arrays

MIRRORING OBJECTS

The Mirror command is used to make a mirror image of objects. You may choose to retain or delete the original objects.

To use the Mirror command, enter:

    Command: **Mirror**
    Select objects: *(Select.)*
    First point of mirror line: *(Select.)*
    Second point: *(Select.)*
    Delete old objects? <N>: **Y** (or **N**)

The mirror line is the line from which the objects will be mirrored.

Mirrored Text

You may mirror objects with text (and attribute-associated text) without producing "backward" text in the image. This is achieved using the Mirrtext option under the Setvar (set variable) command. Enter **Setvar**:

Command: **Setvar**
Variable name or ?: **Mirrtext**
New value for MIRRTEXT <default>: **0**

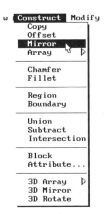

Setting the variable to zero produces non-mirrored text, while a variable setting of one produces a true mirror image of text.

The following illustration shows examples of text mirrored with the Mirrtext variable set to each variable.

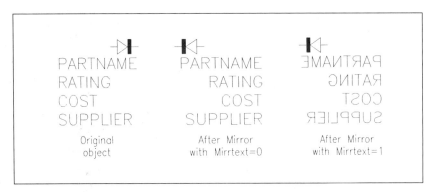

FIGURE 22-9 Using the MIRRTEXT System Variable

The mirror command may be used to draw only half or one fourth of an object and construct the other parts by mirroring them.

DIVIDING AN ENTITY

The Divide command divides an entity into an equal number of parts and places either a specified block or a point entity at the division points on the entity.

You may divide a line, arc, circle, or polyline. The crosshair pick selection process of object selection must be used to select the entity. You may not use Window, Crossing, or Last. Choosing an entity other than a line, arc, circle, or polyline will result in the message:

> Cannot divide that entity.

The current point setting will determine the type of point used at the divide points.

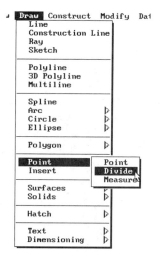

Using Blocks to Divide

You may also place blocks of your own definition at the division points. The block reference must already exist in the drawing. If you are unsure whether a block reference exists, use the ? option from the Block command. Let's look at an example using a block for division markers.

> **NOTE:** You may use the Divide command to divide an object into an equal number of parts, then snap to these points using the Node object snap.
>
> Create custom blocks, used with the Divide command to place desired symbols at intervals along an entity.

USING THE MEASURE COMMAND

The Measure command is similar to the Divide command, except that you may choose the length of segment that the points are spaced.

The rules for execution are the same as for the Divide command. You may only measure a line, arc, circle, or polyline. You must use only the single point method of object selection. The current point setting is the one used to place the point entities. If you use a block, the block reference must already exist in the drawing.

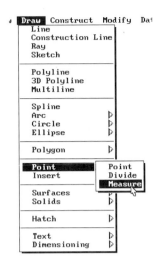

EXPLODING BLOCKS

The Explode command breaks down the entities that make up a block. The block is replaced with the individual entities from which it was constructed. This command is similar to the *block method of inserting blocks as individual entities, except that it is used after the block is inserted.

To explode a block into individual entities, enter the Explode command:

Command: **Explode**
Select objects: *(Select.)*

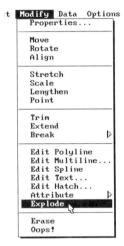

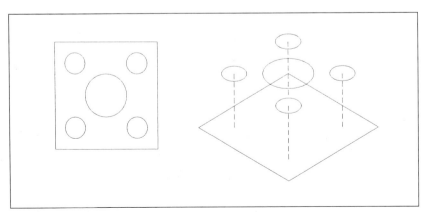

FIGURE 22-10 Using the Explode Command

Explode breaks down the block one level at a time.

The Explode command breaks blocks into the simple entities that comprise the object. You can use Explode on polylines to break them down into simple lines and arcs. 3D polygon meshes are replaced with 3D faces and polyface meshes with 3D faces, lines, and points.

Nested blocks must be exploded after the initial block is exploded. As you can see, some objects must be "exploded" several times to reduce them to the graphic primitives of lines, arcs, and circles.

The width and tangent information of polylines are discarded, and the resulting lines and arcs follow the center line of the old polyline. If the exploded polyline has segments of non-zero width or has been curve fitted, the following message appears:

> Exploding this polyline has lost (width/tangent) information.
> The UNDO command will restore it.

The new lines and arcs are placed on the same layer as the polyline and will inherit the same color.

Block attributes are deleted, but the attribute definitions from which the attributes were created will be redisplayed. The attribute values and changes made by using Attedit are discarded.

Blocks inserted with the Minsert command may not be exploded.

If you explode a dimension, the dimension entities are placed on layer "0", with the color and linetype of each "BYBLOCK".

> **NOTE:** Explode standardized inserted objects into individual entities so they may be edited to suit the specific purposes for that drawing.

TRIMMING ENTITIES

The Trim command allows you to trim objects in a drawing by defining other objects as cutting edges, then specifying the part of the object to be "cut" from between them.

To use Trim, you must first define the object(s) to be used as the cutting edges, then the portion of the desired object to be removed. The cutting edges are defined by using the standard object selection process. The portion of the object to be trimmed must be specified by pointing to that part. Other types of object selection are not valid for this choice. Cutting edges may be lines, arcs, circles, and/or polylines. If you use a polyline with a non-zero width as a cutting edge, the center line of the polyline will be used as the point to trim to.

Combinations of circles, arcs, and lines may be used as cutting edges. The following illustration shows how the Trim command can be used in a more complex manner.

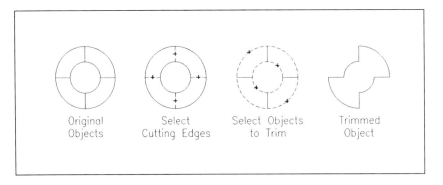

FIGURE 22-11 Use of the Trim Command on a Complex Object

Trimming Polylines

Polylines are trimmed at the intersection of the center line of the polyline and the cutting edge (polylines are covered in Chapter 24). The trim is performed with a square edge. Therefore, if the cutting edge intersects a polyline of non-zero width at an angle, the square-edged end may protrude beyond the cutting edge. The following illustration shows an example of such a situation.

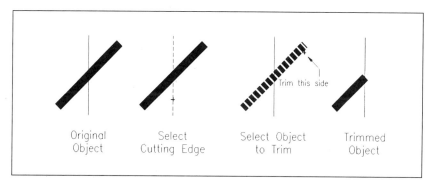

FIGURE 22-12 Trimming a Wide Polyline

If you select entities that cannot serve as cutting edges, AutoCAD displays the message:

No edges selected.

If an entity is chosen that cannot be trimmed, AutoCAD displays:

Cannot TRIM this entity.

If the entity to be trimmed does not intersect a cutting edge, the following message is displayed:

Entity does not intersect an edge.

Trimming Circles

In order to trim a circle, it must either intersect two cutting edges, or the same cutting edge twice (such as a line drawn through two points on the circumference). If the circle intersects only one cutting edge, AutoCAD displays the following message:

Circle must intersect twice.

EXTENDING OBJECTS

The Extend command allows you to extend objects in a drawing to meet a boundary object. The Extend command functions very much like the Trim command.

22-16 USING AUTOCAD

To use Extend, you must first specify the object(s) to be used as the boundary objects (the point to which a selected object will be extended), then the object to be extended.

You may use any type of object selection to choose the boundary objects. The selection of the object to be extended, however, must be performed by pointing to the end (or close to the end) of the part to be extended. This is the process by which you tell AutoCAD which end of the object to extend. Let's look at an example.

If several boundary edges are selected, the object to be extended will be lengthened to meet the first boundary object encountered. If none of the selected boundary objects can be met, AutoCAD displays the message:

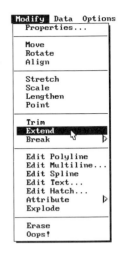

   No edges selected.

If an object that cannot be lengthened is chosen, the following message is displayed:

   Cannot EXTEND this entity.

Using Extend with Polylines

Polylines are extended in much the same manner as lines and arcs. Polylines of non-zero width are extended until the center line meets the boundary object. Objects extended to a polyline used as a boundary are extended to the center line of the polyline.

Polylines of non-zero width are extended until the center line intersects the boundary. Therefore, if the wide polyline and the boundary intersect at an angle, a portion of the square end of the polyline may protrude over the boundary.

Extending tapered polylines results in a continuation of the existing taper. If the additional length causes the polyline to taper past the "zero point" and become a negative taper, the polyline stops the taper at zero. Examples of using the Extend command on tapered polylines follow.

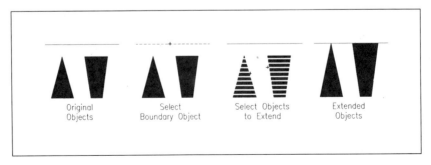

FIGURE 22-13 Extending Tapered Polylines

Only open polylines may be extended. If you attempt to extend a closed polyline, the following message will appear:

Cannot extend a closed polyline.

LENGTHENING LINES

The Lengthen command is a faster version of the Trim and Extend commands. Lengthen changes the length of open objects, such as lines, polylines, arcs, and splines; closed objects, such as circles and doughnuts, cannot be lengthened. Lengthen can change the length by making open objects shorter or longer. Lengthen is faster than Trim and Extend because you do not need to trim or extend to an existing object: you just point on the screen for the change to occur. Or, you can specify a percentage change numerically.

The Lengthen command has the following options:

Command: **Lengthen**
DElta/Percent/Total/DYnamic/<Select object>:

The default option, <Select object>, lets you pick an object before applying one of the four lengthen options. AutoCAD reports its length, as follows:

Command: **Lengthen**
DElta/Percent/Total/DYnamic/<Select object>: *(Pick.)*
Current length: 2.9486
DElta/Percent/Total/DYnamic/<Select object>: *(Pick.)*

You can then use one of the following options:

Delta Option. The DElta option changes the length by the distance from the object's endpoint to the screen pick point, as follows:

Command: **Lengthen**
DElta/Percent/Total/DYnamic/<Select object>: **de**
Angle/<Enter delta length>: **a**
Enter delta angle:
<Select object to change>/Undo: *(Select.)*
<Select object to change>/Undo: *(Press* [ESC]*.)*

The Angle option changes the angle of a selected arc. The default response, <Enter delta length>, prompts you to select an object, changes its length, then prompts you for another object until you press [ESC].

Percent Option. The Percent option changes the length by the percentage of the object. For example, entering **25** shortens a line to 25 of its original length, while entering **200** doubles its length, as follows:

> Command: **Lengthen**
> DElta/Percent/Total/DYnamic/<Select object>: **p**
> Enter percent length: **25**
> <Select object to change>/Undo: *(Select.)*
> <Select object to change>/Undo: *(Press* [ESC] *.)*

Total Option. The Total option changes the length of a line to an absolute length. For example, a value of 5 changes the line to a length of 5.0000 units, no matter its existing length.

> Command: **Lengthen**
> DElta/Percent/Total/DYnamic/<Select object>: **t**
> Angle/<Enter total length>: **a**
> Enter total angle:
> <Select object to change>/Undo: *(Select.)*
> <Select object to change>/Undo: *(Press* [ESC] *.)*

The Angle option changes the length of an arc to an absolute circumference.

Dynamic Option. The DYnamic option lets you visually change the length of open objects, as follows:

> Command: **Lengthen**
> DElta/Percent/Total/DYnamic/<Select object>: **dy**
> Specify new end point.
> <Select object to change>/Undo: *(Select.)*
> <Select object to change>/Undo: *(Press* [ESC] *.)*

After you select the object, the length of the object changes as you move the cursor about. You can use object snap modes to make the dynamic lengthening more accurate.

ROTATING OBJECTS

The Rotate command allows you to rotate an entity or group of entities around a chosen base point. The entities are not required to be part of a block.

You may choose the angle at which the chosen object rotates by specifying the angle to rotate from its existing angle, by dragging the angle, or by choosing the angle of rotation from a reference angle. Let's look at examples of each type of angle rotation.

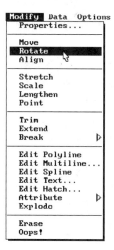

Rotating by Specifying Angle. An object or a group of objects can be rotated by specifying an angle. You may specify a simple angle, or you may want to change one angle to another. For example, if an object is currently oriented at 58 degrees and you wish to rotate the object to 26 degrees, you could rotate the object the difference of –32 degrees. Let's use the Rotate command to rotate an object by specifying an angle.

Entering a positive angle will result in a counterclockwise rotation of the object; a negative angle will result in a clockwise rotation.

Rotating from a Reference Angle

It is sometimes necessary to obtain a particular rotation angle. The Rotate command allows you to achieve this by using the Reference option.

In order to use the Reference option, you must first know the existing rotation angle of the object. Let's look at an example of rotating an object in reference to its existing angle.

> **NOTE:** The List command is helpful in determining the existing angle of an entity when using the Reference option.
>
> You may also rotate an object a specified number of degrees from its existing angle by entering the "Reference angle" as zero, regardless of the actual angle.

Rotating an Object by Dragging

If you wish to drag-rotate an object, just move the cursor in response to the "<Rotation angle>/Reference:" prompt. If the updating screen coordinates are set for relative distance and angle, you may read the angle from the screen. Pressing [ENTER] or the pick button on the input device will fix the object at the location currently shown on the screen.

You may rotate part of an object by choosing only the parts to rotate in response to the "Select objects:" prompt.

SCALING OBJECTS

The Scale command allows you to change the scale of an entity or entities. The X and Y scales of the object are changed equally. The entities are not required to be part of a block.

A base point is specified on the object. This point remains stationary, while the object is scaled from that point.

You may change the scale by a numerical factor, by dragging, or by referencing a known length, then entering a new length. Let's look at each method.

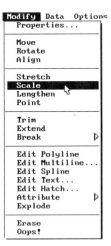

Changing Scale by Numerical Factor

You may change the scale of an object by entering a numerical factor which serves as a "multiplier." Entering a positive number will have the effect of multiplying the size of the object by that number. For example, a factor of two will result in an object twice the original size.

INTERMEDIATE EDIT COMMANDS **22-21**

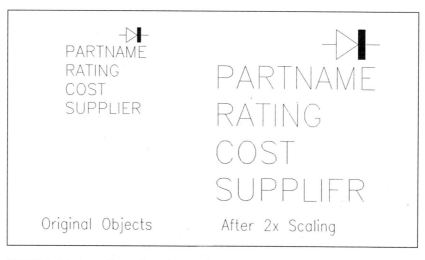

FIGURE 22-14 Changing the Scale Factor

Entering a decimal factor will result in an object smaller than the original. A factor of .25 will result in an object 25 percent of the original size.

Changing Scale by Reference

You may change scale by specifying the known length of an entity, then choosing the desired length of that object. Let's look at an example of this method.

STRETCHING OBJECTS

The Stretch command permits you to move selected objects while allowing their connections to other objects in the drawing to remain unchanged. As you will see, the Stretch command can be one of the most useful edit commands in AutoCAD.

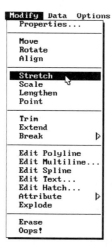

There are several rules associated with the Stretch command that must be understood in order to execute it properly. Let's look at some of these rules.

Lines, arcs, solids, traces, and polylines may be stretched.

You may choose the objects to be edited by any combination of object selection options, but one of these options must be a window option such as "Window" or "Crossing."

Any objects that are entirely within a window will be moved in the same manner as the Move command.

If a line, arc, or polyline segment is chosen by using the "Crossing" method, the endpoints that are contained in the window will be moved, but those outside the window will remain "fixed."

When stretching traces and solids, vertices that are outside of the Crossing window will remain fixed, while those inside will be moved.

> **NOTE:** Turning Ortho mode on allows you to move the window along the wall perfectly.
>
> You may enter a relative coordinate in response to the "New point:" prompt to move the object(s) an exact distance.

Stretch Rules

Let's look at some other rules that apply to the Stretch command.

- Arcs are stretched similar to lines, except that the arc's center, start, and endpoints are adjusted so the distance from the midpoint of the chord to the arc is constant.
- Polylines are handled by their individual segments. The polyline width, tangent, and curve fitting are not affected.
- Some entities are just "moved" or left alone. The decision depends on the "definition point." If the definition point lies inside the window, the entity is moved. If it occurs outside the window, it is not affected. The definition point of certain entities is as follows:

    Point: Center of the point
    Circle: Center point of the circle
    Block: Insertion point
    Text: Left most point of the text line

- If more than one window specification is used in the object selection process, the last window used will be the one used for the Stretch.
- If you do not use a window selection, AutoCAD displays the message:

    You must select a window to stretch.

UNDOING DRAWING STEPS

The Undo command is used to undo several command moves in a single operation. You may also identify blocks of commands for reference later. To execute the Undo command, enter:

Command: **Undo**
Auto/Back/Control/End/Group/Mark/<number>:

Responding to the prompt with a number will cause AutoCAD to undo the specified number of commands. This has the same effect as entering the U command the same number of times, except it is done in one operation, thus causing only one regeneration.

```
Edit
  Undo          Ctrl+Z
  Redo

  Cut           Ctrl+X
  Copy          Ctrl+C
  Copy View
  Paste         Ctrl+V
  Paste Special...

  Properties...

  Object Snap         ▶
  Point Filters       ▶

  Snap          Ctrl+B
  Grid          Ctrl+G
  Ortho         Ctrl+L

  Select Objects      ▶
  Group Objects...

  Inquiry             ▶

  Links...
  Insert Object...
```

Let's look at the options for the Undo command.

Mark: Think of your drawing as a list of functions that is added to the drawing file, one at a time. If this list were on paper, you could place a mark at a certain point for reference. The Mark option allows you to do this. At a later time, you can use the Back option to undo everything back to the mark.

Back: The Back option will cause AutoCAD to undo all operations back to the preceding mark. Each operation will be listed and the message "Mark encountered" will be displayed. When the next marker back is encountered, it is removed. The next Undo will then undo all operations back to the previous marker, if any. If there is no preceding mark, AutoCAD displays the message:

This will undo everything. OK? <<Y>>

Entering **Y** will undo every operation since last entering the drawing editor. Answering **N** will cause the Back option to be ignored.

Multiple undos are stopped by a mark. Entering a number in response to the undo prompt that is greater than the number of operations since the last mark will have no greater effect than the Back option. The undo will still stop at the mark.

Group/End: The Group and End options cause all operations from the time Group is entered and the time End is entered to be treated as a single command. This means that all operations between the Group and End entries will be undone in a single step. For example, consider the following string of operations:

```
Circle
Arc
GROUP
    Line
    Line
    Fillet
END
Arc
U
U
U
```

The first U will remove the arc. The second U will remove all the operations between the Group/End entries (Line, Line, Fillet). The third U will remove the Arc. Only the first circle will remain.

If you enter the Group option while in a current group, the current group will be ended and a new one will be started.

If a Group has been started, but not ended, and the U command is entered, it will undo one operation at a time, but cannot back up past the point where Group was entered. To continue back, the current group must be ended, even if nothing remains in the current group.

When using the Group option, Undo will not display each operation. The message "GROUP" will be displayed.

Auto: The Auto option causes macro commands from menus to be treated as single commands. Some menus combine several operations such as inserting a window in a wall of a floor plan by Breaking the wall and Inserting the window. If Auto is on, the entire command string will be undone as if it were a single command. Auto prompts with: "ON/OFF".

Control: The Control option limits the Undo commands or disables them entirely. This is sometimes necessary because of the large amount of disk space required if the Undo edit is extensive.

The Control option prompts:

All/None/One/ <<All>>:

Following is an explanation of the sub-options:

All: Enables all Undo commands and features.

None: Disables all Undo commands and features. Using None removes all markers, groups, and other stored information. If Undo is entered when the None sub-option is on, AutoCAD will display the Control prompt:

   All/None/One <<All>>:

You may then enter the level of Undo performance that you desire.

One: Allows the Undo commands to function for one undo at a time. All markers, groups, and other stored information are removed. This frees disk space, making it a preferred setting for floppy-disk based systems. The Group, Mark, and Auto options are not available. The following prompt is displayed when Undo is executed:

   Control/<<1>>:

General Notes

Undo has no effect on the following:

ABOUT	FILES	PLOT	RESUME
AREA	FILMROLL	PSOUT	SAVE
ATTEXT	GRAPHSCR	QSAVE	SAVEAS
COMPILE	HELP	QUIT	SHADE
CONFIG	HIDE	RECOVER	STATUS
DBLIST	ID	REDRAW	TEXTSCR
DELAY	LIST	REDRAWALL	U
DIST	MSLIDE	REGEN	UNDO
DXFOUT	NEW	REGENALL	
END	OPEN	REINIT	

Plotting clears the Undo information from the drawing.

The Undo command should not be confused with the Undo option contained in some commands such as the Line command. Using the Undo option from the Line command will remove one line segment at a time. If you use the Undo command, however, all line segments in the sequence will be undone in a single operation. For example, if you draw a box containing four lines, then exit the Line command and execute the Undo1 command, all four lines will be undone. If this does not produce the intended results, the Redo command will restore the lines.

EDITING WITH GRIPS

In Chapter 10, you learned how to use the object selection process. This process is used to select objects for editing. AutoCAD provides a shortcut process to object selection and editing. You can set AutoCAD to display grips on the entities displayed on the screen. Grips are small rectangles that are placed on objects and can be used as *grip points* when editing. The grip boxes are generally placed at the positions available with object snap. The following illustration shows AutoCAD drawing objects with their grip locations.

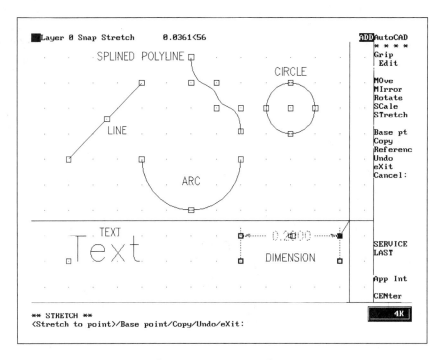

FIGURE 22-15   Grips Selection Points on Objects

Enabling Grip Editing

Grip editing is enabled or disabled by using the Ddgrips command to display the Grips dialog box.

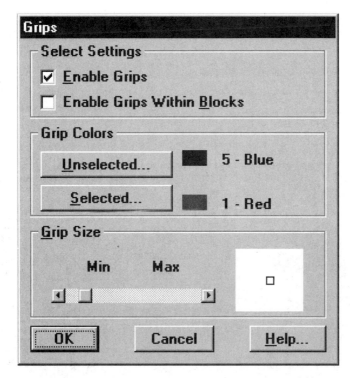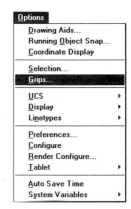

FIGURE 22-16 Grips Dialog Box

To enable grip editing, select the "Enable Grips" check box. Normally, a grip box is displayed at the insert point of an inserted block. If you wish to display grips on the entities within blocks, select the "Enable Grips Within Blocks" check box.

Setting Grip Colors. As a visual aid, grip boxes can be assigned a color. If a grip box is selected, the interior of the box is filled with a solid color. Both the grip box and solid fill can be assigned a color through the dialog box.

To set the colors, choose either the "Unselected" (grip box), or the "Selected" (solid-fill) buttons. AutoCAD will display the Select Color sub-dialog box.

22-28 USING AUTOCAD

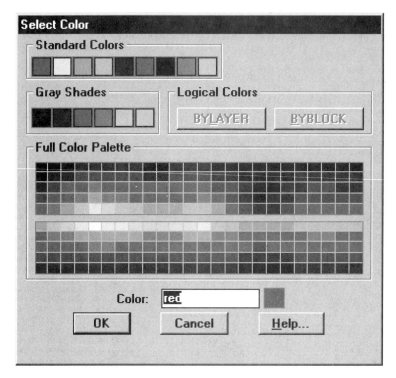

FIGURE 22-17 Select Color Dialog Box

Select the desired color from the displayed palette, then click on the "OK" button to set the color.

Setting the Grip Size. The size of the grip boxes is set with the slider bar in the dialog box.

FIGURE 22-18 Setting the Size of Grips

The panel to the right of the slider bar shows the current grip box size. As you move the slide bar, the grip box changes scale to indicate the actual grip box size. The grip box size can be set from 1 to 255 screen pixels.

Using Grips for Editing

When grip editing is enabled, a grip selection box is displayed at the intersection of the crosshairs.

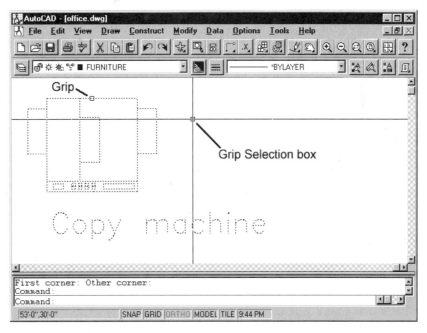

FIGURE 22-19 Grip Selection Box

To select an object for editing, place the selection box at any point on an object in the drawing and click. The object will be highlighted and grip boxes will be placed on the object. You can continue to select as many objects as you wish by selecting several times.

After you select the object(s) to be edited, you can select any of the displayed grip points as a base point for editing. As you move the selection over the grip points, the crosshairs will snap to the grip box in a similar manner as when using object snap. To select a grip box, move to it and click. The grip box will change colors to denote selection.

To select several grip boxes, hold down the shift key on the keyboard when selecting the grips.

Pressing [ESC] twice will clear all selected objects and their grips. Using [ESC] once will clear the selected objects, but not the grips. This facilitates using a grip on a non-selected object as a base grip.

Grip Mode. When you select a single grip, AutoCAD displays edit command options on the command line. The selections rotate as you press the Spacebar. The following command line listings show the options available as you "page" through with the Spacebar.

STRETCH
<Stretch to point>/Base point/Copy/Undo/eXit:

ROTATE
<Rotation angle>/Base point/Copy/Undo/Reference/eXit:

SCALE
<Scale factor>/Base point/Copy/Undo/Reference/eXit:

MIRROR
<Second point>/Base point/Copy/Undo/eXit:

Press the Spacebar until the desired edit command is listed, then proceed with the command. The selected grip point is used as a base point for the edit operations. You can alternately select one of the options listed on the command line. The commands and their options are covered later in this section.

When selecting several grip points by holding down the Spacebar, the base point grip is selected last. The base grip can be either a selected or nonselected grip. It is selected, however, after all the target grips are selected and the [SHIFT] key is no longer held. It is only after the base grip is selected that the command options are displayed.

Grip Editing Commands

The grip editing commands of Stretch, Move, Rotate, Scale, and Mirror are displayed on the command line when you enter grip mode. You can rotate through the edit command options by pressing the Spacebar. Let's look at each choice.

Stretch Mode. Stretch mode functions similarly to the Stretch command, except that AutoCAD uses the selected grip points to determine the stretch results.

STRETCH
<Stretch to point>/Base point/Copy/Undo/eXit:

If you select the midpoint grip of a line or arc or the center of a circle, the object is moved, but not stretched. Figure 22-20 shows an example of using the Stretch mode.

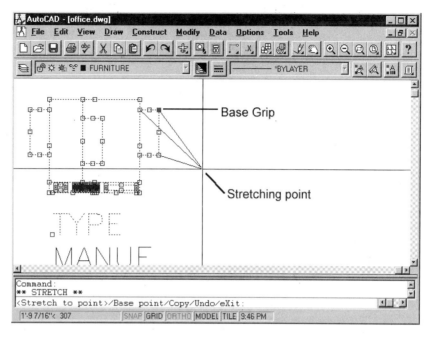

FIGURE 22-20 Stretching with Grips

Options:

Base point: Entering **B** at the command line allows you to use a base point other than a grip location.

Copy: Entering **C** (for Copy) makes copies from the stretch entities, leaving the original entities intact.

Undo: Performs an undo of the operation.

eXit: Exits grip editing.

Move Mode. If you press the Spacebar until the Move mode is displayed, you can use grip editing to move the selected objects.

```
**MOVE**
<Move to point>/Base point/Copy/Undo/eXit:
```

The distance the object(s) is moved is determined by the distance from the base grip to a point entered in response to AutoCAD's "Move to point" prompt. You can alternately enter either an absolute or relative coordinate to specify the distance and direction of the move.

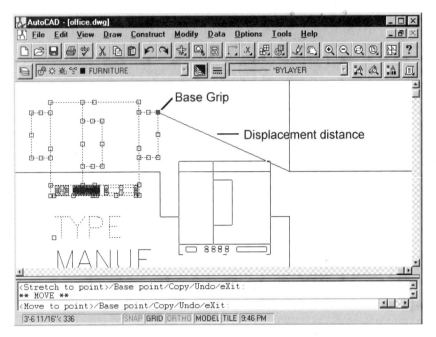

FIGURE 22-21 Moving with Grips

If you hold down the [SHIFT] after selecting the base grip and before placing the point designating the move distance and direction, AutoCAD places a copy of the object(s) at the new location, leaving the original object(s) unchanged. You can use this technique to place multiple copies. This functions similarly to the Copy option (explanation follows).

Options:

> **Base point:** Entering **B** at the command line allows you to use a base point other than a grip location.
>
> **Copy:** Entering **C** (for Copy) makes copies of the entities, leaving the original entities intact.
>
> **Undo:** Performs an undo of the operation.
>
> **eXit:** Exits grip editing.

Rotate Mode. The Rotate option allows you to rotate the selected objects around the base point.

　　ROTATE
　　<Rotation angle>/Base point/Copy/Undo/Reference/eXit:

Unless you use the "Base point" option to position the base point at a location other than a grip, the rotation will occur around the base grip.

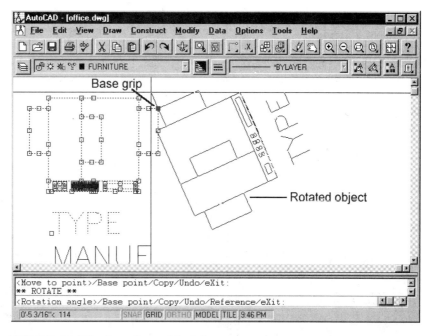

FIGURE 22-22 Rotating with Grips

When rotating, you can dynamically set the rotation angle by moving the crosshair, or you can specify a rotation in degrees.

Options:

Base point: Entering **B** at the command line allows you to use a base point other than a grip location.

Copy: Entering **C** (for Copy) makes copies of the rotated entities, leaving the original entities intact.

Undo: Performs an undo of the operation.

Reference: Allows you to set a reference angle. When you enter **R**, AutoCAD prompts for the "Reference angle". This is the angle at which the entity is currently rotated. AutoCAD next prompts for the "New angle". The new angle is the actual angle you want the entity to be.

eXit: Exits grip editing.

Scale Mode. If you wish to scale the selected objects, cycle through the mode list until the Scale mode is listed on the command line.

```
**SCALE**
<Scale factor>/Base point/Copy/Undo/Reference/eXit:
```

The scale mode is used to rescale the selected object(s). The base grip serves as the base point for the scaling operation. You can dynamically scale the objects by moving the crosshairs away from the base grip. You can also scale the selected objects by entering a scale factor. A scale factor greater than one scales the object by that factor. For example, a scale factor of two results in an object twice the size. Entering a decimal scale factor results in an object that is smaller than the original. For example, entering a scale factor of **0.5** scales the object to one-half size of the original.

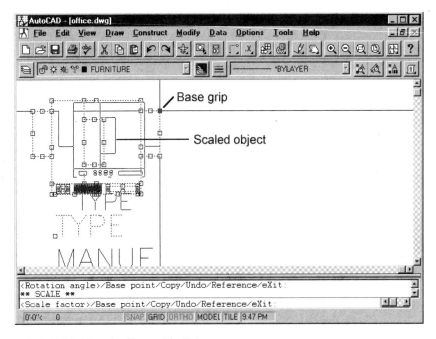

FIGURE 22-23 Scaling with Grips

Options:

Base point: Entering **B** at the command line allows you to use a base point other than a grip location.

Copy: Entering **C** (for Copy) makes copies of the scaled entities, leaving the original entities intact.

Undo: Performs an undo of the operation.

Reference: Allows you to set a reference scale. Let's look at an example. Let's suppose you have an object is 6 units in length and you wish to scale it to 24 units in length. Select the Reference option by entering **R** at the command line. AutoCAD prompts:

Reference length <1.0000>:

Enter **6** and press [ENTER]. AutoCAD then prompts for the new length.

<New length>/**B**ase point/**C**opy/**U**ndo/**R**eference/e**X**it:

Enter **24** and press [ENTER]. AutoCAD calculates the scale factor and applies the scale.

eXit: Exits grip editing.

Mirror Mode.   Mirror mode is used to mirror the selected object(s) in a similar manner to the Mirror command.

MIRROR
<Second point>/**B**ase point/**C**opy/**U**ndo/e**X**it:

When using Mirror mode, the first point of the mirror line is stipulated as the point of the base grip. The second point of the mirror line is entered at any point on the drawing.

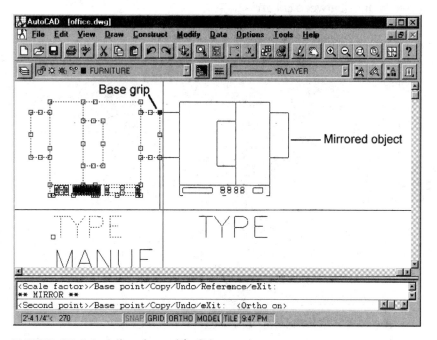

FIGURE 22-24   Mirroring with Grips

Options:

Base point: Entering **B** at the command line allows you to use a base point other than a grip location.

Copy: Entering **C** (for Copy) makes copies of the mirrored entities, leaving the original entities intact.

Undo: Performs an undo of the operation.

eXit: Exits grip editing.

EXERCISES

1. Let's try a rectangular array. Start a new drawing called "ARRAY". Set limits of 0,0 and 12,8. As an aid, display a grid with a value of one.

   Draw a circle with the center point located at 1,1 and a radius of one.

   Execute the Array command:

   Command: **Array**
   Select objects: **L**
   Rectangular or Polar array (R/P): **R**
   Number of rows (---): **3**
   Number of columns (|||): **5**
   Unit cell or distance between rows (---): **2**
   Unit cell distance between columns (|||): **2**

   Your array should look like the following illustration:

   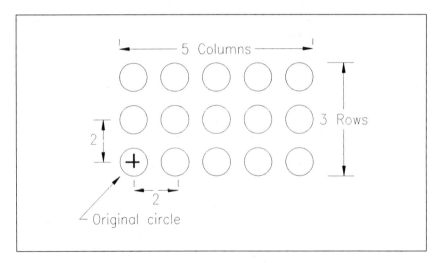

   FIGURE 22-25 Example of Using the Array Command

You may also use a window to specify the distances between the items. Simply enter a point when prompted for the first unit cell distance and enter a second point to define a window. The X and Y distances will be measured by AutoCAD and used as the values. The following illustration shows this method of distance definition:

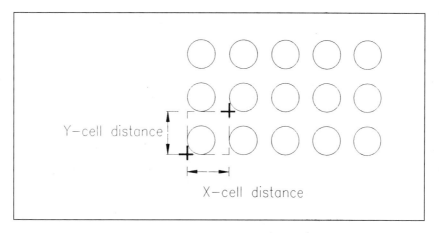

FIGURE 22-26 Using a Window for Array Spacing

2. Let's start another drawing called "PARRAY" (for polar array). Set limits of 0,0 and 12,8. Place a square with sides that are one unit each.

   Execute the Array command:

   Command: **Array**
   Select objects: *(Select.)*
   Rectangular or Polar array(R/P): **P**
   Center point of array: *(Select.)*
   Number of items: **7**
   Angle to fill (+=ccw, -=cw) <360>: **270**
   Rotate objects as they are copied? *<default>*: **N**

   Your polar array should look like the one in the following illustration:

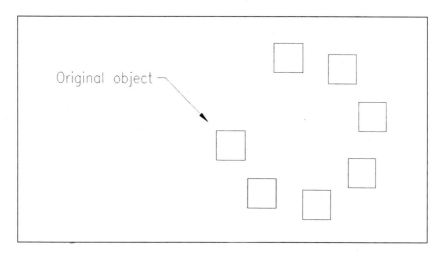

FIGURE 22-27 Polar Array Exercise

The item will be arrayed around the center point you chose.

A positive angle will cause the array to build in a counterclockwise direction; while a negative angle will cause the array to build in a clockwise direction from the selected object.

If the selected object is a circle or a block, you will be asked whether you wish the block to be rotated about the center point:

Rotate objects as they are copied <*default*>? (**Y** or **N**)

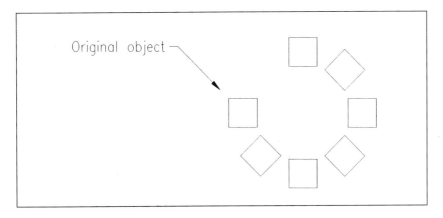

FIGURE 22-28 Rotated Polar Array

Respond with a **Y** or **N** (yes or no) to tell AutoCAD which you require. If the square you used in your circular array was a block and you responded with **Y**, the block would be rotated around its base point. Other objects are rotated about specific points. The following table lists these points.

Point:	Insertion point
Circle, Arc	Center point
Block, Shape	Insertion base point
Text	Start point
Line, Trace	Nearest endpoint

TABLE 22-1

3. The following example shows how the Mirror command can be a great time-saver. Start the work disk drawing named "MIRROR1". Let's suppose that you are designing a house and you want to reverse the layout of the bathroom. The following drawing shows the room to be reversed.

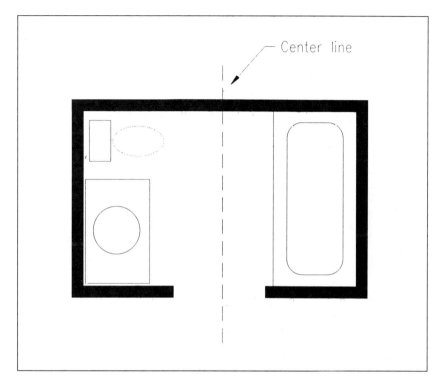

FIGURE 22-29 MIRROR1.DWG Work Disk Drawing

Let's reverse the room:

Command: **Mirror**
Select objects: *(Select the fixtures.)*
First point of mirror line: *(Enter point 1.)*
Second point: *(Enter point 2.)*
Delete old objects? <N>: **Y**

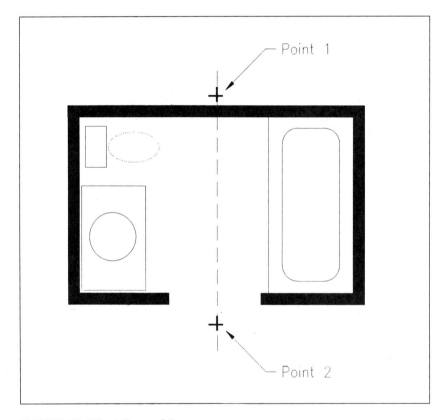

FIGURE 22-30 Mirrored Room

The mirror line may be placed at any angle in respect to the selected object.

4. Let's look at an example using a Pdmode of 34. First, draw a circle on the screen. Now, enter the Divide command:

Command: **Divide**
Select object to divide: *(Select the circle.)*
<Number of segments>/Block: **8**

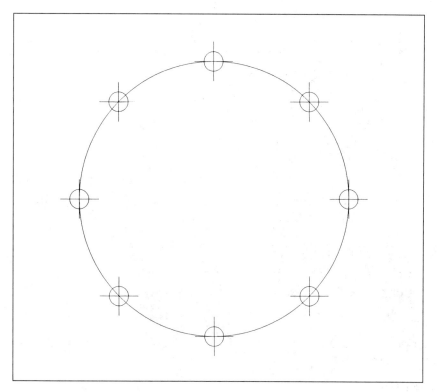

FIGURE 22-31 Using the Divide Command

The circle should now be divided into eight equal segments as shown in the preceding illustration.

You may choose the number of divisions to be between 2 and 32,767. The entity will be equally divided into the specified number of parts and a point entity will be placed at the division points.

5. First, draw a symbol similar to the one in the following illustration.

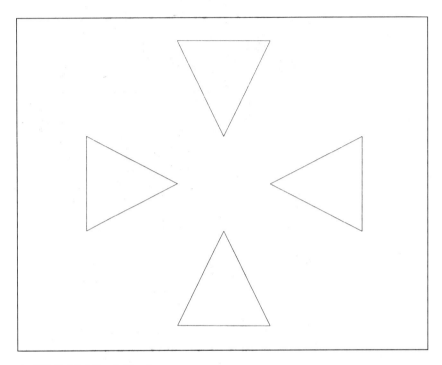

FIGURE 22-32 A Block

Now, block the symbol and name it "SYMBOL1". Draw another circle on the screen. Enter the Divide command.

Command: **Divide**
Select object to divide: *(Select the circle.)*
<Number of segments>/Block: **B**
Block name to insert: **SYMBOL1**
Align block with object? <Y>: **Y**
Number of segments: **8**

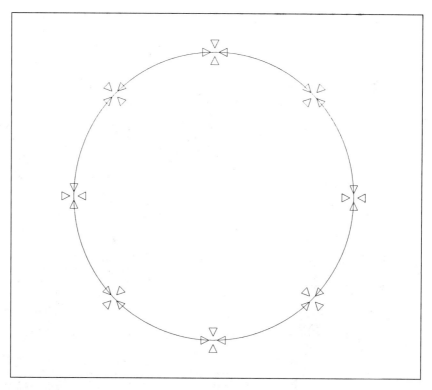

FIGURE 22-33 Dividing with the Rotated Block

If you respond to the "Align block with object?" prompt with **Y**, the Divide command will rotate the block around its insertion point so that its horizontal lines are tangent to the object that is divided.

The following illustration shows the same procedure, except the block is not rotated:

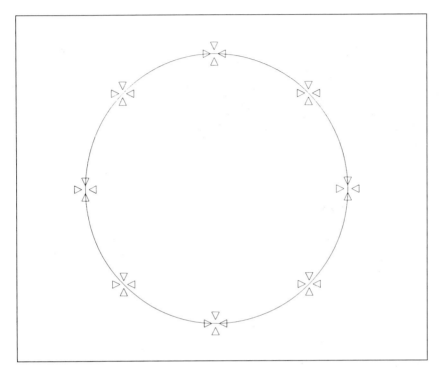

FIGURE 22-34 Dividing with the Non-rotated Block

6. Let's perform a measure on an entity. Draw a horizontal line on the screen, 6 units in length. (A point setting of 34 is used to make the points visible.) Enter the Measure command:

Command: **Measure**
Select object to measure: (*Select the line.*)
<Segment length>/Block: **1**

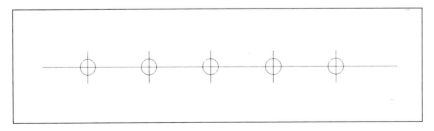

FIGURE 22-35 Using the Measure Command

You may also show AutoCAD the segment distance by placing two points on the screen.

The measurements of lines and arcs start at the endpoint closest to the point used to select the entity. The measurement of a circle starts at the angle of the center set by the current snap rotation. Measurements of polylines start at the first vertex drawn.

7. Let's use the Trim command to trim an intersection. Draw four intersecting lines as shown in the following illustration. Now, enter the Trim command:

Command: **Trim**
Select cutting edge(s)...
Select objects: **W** *(Window the four lines.)*
Select objects to trim: *(Select points A,B,C, and D.)*

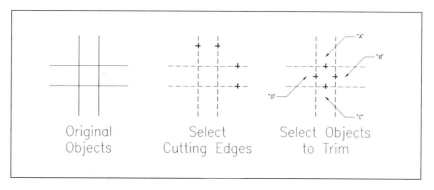

FIGURE 22-36 Using the Trim Command

Notice that the "Select objects to trim" prompt repeats to allow multiple trimming. Entering a null response will return you to the command line.

The intersection will be trimmed and should appear as shown in the following illustration.

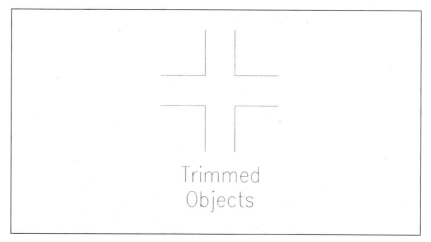

FIGURE 22-37 Objects After Trimming

8. Draw two vertical parallel lines and one horizontal line as shown in the following illustration. We will now extend the horizontal line to the vertical lines. Enter the Extend command.

Command: **Extend**
Select boundary edge(s)...
Select objects: *(Select the vertical lines and* ENTER*)*
Select object to extend: *(Select.)*

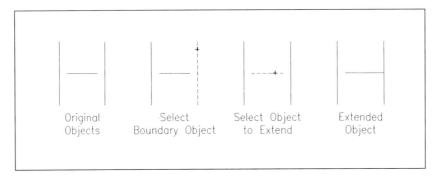

FIGURE 22-38 Using the Extend Command

If you wish to extend the other end (remember, you chose both vertical lines as boundaries), simply respond to the repeating "Select object to extend" prompt by selecting a point at the other end of the horizontal line. Pressing ENTER (null response) will return you to the command line.

9. Draw the object as shown in the following illustration. Enter the Rotate command:

Command: **Rotate**
Select objects: **W** *(Window the entire object.)*
Base point: *(Enter point "A".)*
<Rotation angle>/Reference: **45**

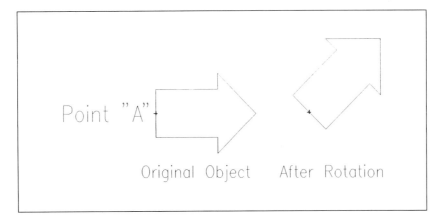

FIGURE 22-39 Rotating an Object

10. Draw an object similar to the one shown in the following illustration. Enter the Rotate command.

Command: **Rotate**
Select objects: **W** *(Window the object.)*
Base point: *(Enter point "A".)*
<Rotation angle>/Reference: **R**
Reference angle <*default*>: **0**
New angle: **-45**

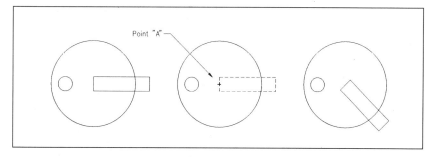

FIGURE 22-40 Rotating in Reference to a New Angle

11. Draw the object as in the following illustration. Enter the Scale command.

Command: **Scale**
Select objects: *(Select object.)*
Base point: *(Select bottom left corner.)*
<Scale factor>/Reference: **R**
Reference length <*default*>: **2**
New length: **4**

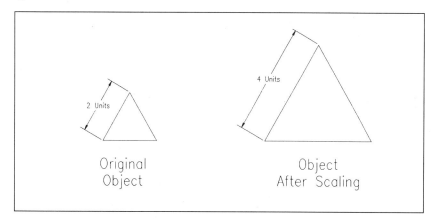

FIGURE 22-41 Change the Scale by Reference

Notice that AutoCAD rescales all the entities that make up the object, not just the length of the known entity.

12. Before looking at more rules, let's use the Stretch command to move the location of a window in a wall. The following illustration shows a wall containing a window that we wish to move to the right. This is a work disk drawing. Start the drawing named "STRETCH1" and use the following command sequence. Let's enter the Stretch command and move the window.

Command: **Stretch**
Select objects to stretch by window...
Select objects: **C** (AutoCAD enters this for you automatically if
  selected from the menu.)
First corner: *(Enter point "A".)*
Other corner: *(Enter point "B".)* [ENTER]
Base point: *(Enter point "C".)*
New point: *(Enter point "D".)*

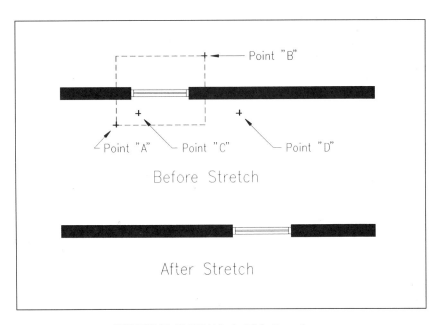

FIGURE 22-42 STRETCH1.DWG Work Disk Drawing

13. Start the drawing from the work disk named "STRETCH2". This is a pencil as shown in the following illustration.

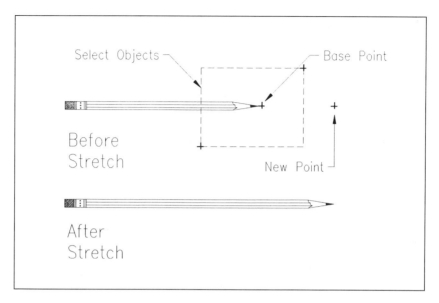

FIGURE 22-43 STRETCH2.DWG Work Disk Drawing

Use the Stretch command to make the pencil longer and shorter.

CHAPTER REVIEW

1. What properties of an object can be altered with the Change command?

2. What is meant when you change an entity's color to "BYLAYER"?

3. Why would you use the Chprop command instead of the Change command?

4. What are the two types of arrays?

5. Using simple circles, draw an example of an array using five columns and three rows.

6. Using four-sided boxes, draw an example of a polar array with six objects that are rotated.

7. What does the Mirror command perform?

8. What variable controls whether text is mirrored?

9. What is the *mirror line*?

10. Use point type 34 to show a horizontal line divided with the Divide command into four segments.

11. Show a circle divided into six segments with the Divide command, using a small square as the block used with Divide. Show the same situation again, using the rotated block option.

12. What entities can you use with the Measure command?

13. What is the result when you explode a block?

14. What happens when you explode a polyline?

15. What is a trim *cutting edge*?

16. At what point is a polyline trimmed?

17. What are the requirements for trimming a circle?

18. What is the object called that is selected with the Extend command and acts as a borderline for the extended object?

19. What are the three ways to rotate an object?

20. Can an object be scaled differently in the X- and Y-axes with the Scale command?

21. What value would you enter to scale an object to one-fourth of its original size?

22. What entities can be stretched?

23. When using the Stretch command, if you select objects with three window operations, which operation is used to determine the stretch?

24. What does the Mark option of the Undo command do?

25. What effect does plotting from the drawing have on the Undo command?

CHAPTER 23

INTERMEDIATE OPERATIONS

Intermediate use of AutoCAD requires the CAD operator to possess knowledge of several operations. After completing this chapter, you will be able to:

- Set and control the color of entities independent of the layer color settings.
- Store and retrieve drawing views and pictorial images.
- Save and view slide images of a drawing screen.
- Remove unwanted layers, blocks, styles, and other named objects from a drawing.
- Rename objects in the drawing.
- Import and export PostScript image files.

SETTING THE CURRENT COLOR

The Color command is used to set the color type for all subsequently drawn entities.

This differs from the Color option in the Layer command, which sets the color for entities drawn on that layer. The use of the Color command allows you to set colors that are contrary to the current color setting for the layer. Thus, it is possible that a layer will contain entities of different colors, regardless of the color set for that layer using the Layer/Color menu picks.

To set a new color for subsequently drawn entities, enter the Color command.

Command: **Color**
New object color <*current*>:

23-2 USING AUTOCAD

You can set a new color by one of two methods. You can specify a number from 1 to 255 or enter a standard color name such as "BLUE". All new entities will be displayed in the designated color until a new color is selected by the Color command.

You may also enter **BYLAYER** or **BYBLOCK** in response to the prompt.

Entering **BYLAYER** will cause all subsequently drawn entities to inherit the layer's color, thus relinquishing control of entity colors to each layer's color setting.

Entering **BYBLOCK** causes all subsequent entities to be drawn in white until they are blocked. When the block is inserted, the entities will inherit the color of the block insertion.

To select a color by dialog box, use the DdColor command or click on the Color control button on the toolbar.

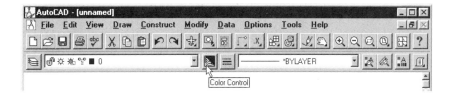

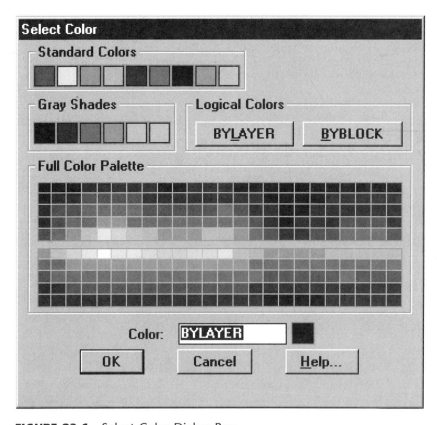

FIGURE 23-1 Select Color Dialog Box

STORING AND DISPLAYING DRAWING VIEWS

Views are stored zooms that are identified by name. You may recall a view at any time in the drawing. This has the same effect as zooming to that location.

To invoke the View command, enter:

Command: **View**
?/Delete/Restore/Save/Window:
View name to save

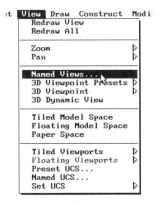

The following is an explanation of the options to the View command.

?: Entering a **?** will result in a listing of the previously named views. AutoCAD prompts:

View(s) to list <*>:

You can enter either a list of view names, separated by commas, or use wildcard characters to list views. If you respond with an Enter, all the saved views will be listed.

D (Delete): Deletes a stored view.

R (Restore): Restores a saved view to the screen. You will be prompted for the name of the previously saved view.

S (Save): Selecting the "S" option saves the current screen view. You will be prompted for a name of the saved view. If you respond with a name of a view that is already saved, the present view will replace it.

W (Window): The window option works in the same manner as the Save option, except that you may show a window which describes the view. You do not have to be in the zoom that will be saved.

View names may be up to 31 characters in length but no spaces may be used. Some examples of named views would be "OFFICE_1", "BEDROOM", and " LIVING_ROOM".

> **NOTE:** Views are an invaluable aid in saving "zoom time" on complex drawings.

You may specify a view to be plotted from the Plot command (see Chapter 18).

STORING AND DISPLAYING DRAWING SLIDES

There may be times when you want to show several views of a drawing or several drawings to someone. You could, of course, take the time to load each one, but this would be very laborious. A much faster way to load pictures is the Vslide command.

The slide commands can capture a view of a drawing like a camera shot, then show it later. Think of it as making a slide with a camera, then showing the shot with a projector.

The slide commands are Mslide and Vslide. Let's look at how to use each.

Making a Slide

The Mslide (make slide) command is used to make a slide file. First, display the drawing or any part of the drawing you wish to make a slide of on the screen. Then enter:

Command: **Mslide**

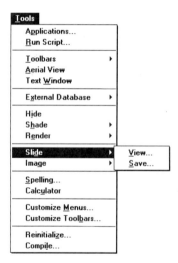

AutoCAD displays the Create Slide File dialog box. Enter the name you want the slide file to have. AutoCAD will add a file extension of .sld to the slide name.

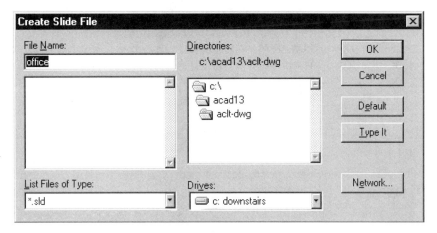

FIGURE 23-2 Create Slide File Dialog Box

Portions of the drawing not currently displayed are not captured. The Mslide command operates as a "what you see is what you get" operation.

Viewing a Slide

The Vslide command is used to view a slide. The following format is used to view a single slide.

    Command: **Vslide**

AutoCAD displays the Select Slide File dialog box.

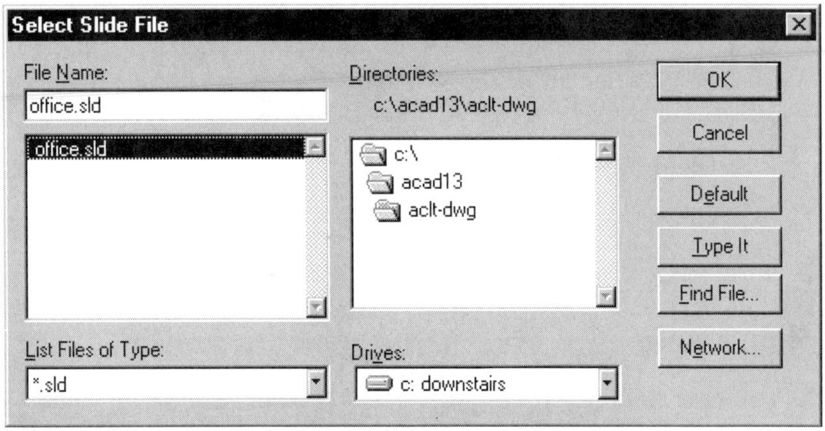

FIGURE 23-3 Select Slide File Dialog Box

If you wish to view a slide that is part of a slide library, select the "Type it" button and enter the slide library name, followed by the slide file name. (See the next section for an explanation of slide libraries.) For example, if you wanted to view a slide named "sofa" from the "furnlib" library, the following format would be used.

Command: **Vslide**
Slide file name: **furnlib(sofa)**

It is not necessary to include the file extension. You may view the slide at any time and from any drawing. The current screen display is overwritten by the slide file.

The current drawing is not affected by the Vslide command. To return to the original drawing, issue a Redraw command.

A slide may not be edited or zoomed. It is a fixed snapshot of the drawing.

Slide Libraries

Slide libraries are a collection of slides stored in a special library format. Slide libraries are excellent for storing slides for use with icon menus.

To build a slide library, you must first prepare a text file (ASCII format) that contains the names of the slide files you wish to assemble into a library. List one slide file per line in the document. Next, shell out to DOS and use the Slidelib utility program included in the AutoCAD\Com\Support subdirectory. The correct usage for the Slidelib program is as follows.

C>**slidelib** *(library name)* <*(file list name)*

Let's look at an example.

We want to prepare a slide library that contains names of furniture. First, we will prepare an ASCII text file that contains a list of the slide names, one to a line. Next, we use the Slidelib utility to assemble the slides into a library. If the text file is named "FLIST" (for furniture list) and we wish the library name to be "FURNLIB" (for furniture library), we would enter the following:

C>**slidelib furnlib <flist**

Once a library is built, you cannot change a slide. If you wish to add or change a slide, you must rebuild the library.

Slide Shows

You may write a self-running script that presents a series of slides on the screen.

To do so, utilize the Script command. You can eliminate the required loading time by entering an "*" before the slide name, which makes AutoCAD "preload" the next slide while the current one is being displayed. The following is an example of a script for a self-running slide show:

VSLIDE SLIDE_A   (begin slide show)
VSLIDE *SLIDE_B   (preload slide B)
DELAY 5000   (insert a delay)
VSLIDE   (display slide B)
VSLIDE *SLIDE_C   (preload slide C)
DELAY 1000   (insert a delay)
VSLIDE   (display slide C)
DELAY   (insert a delay)
RSCRIPT   (recycle show)

Refer to the Script command for more information.

Purging Objects from a Drawing

AutoCAD stores named objects (blocks, layers, linetypes, shapes, and styles) along with the drawing. When the drawing is loaded using the Open command, the drawing editor determines whether each named object is referenced by other objects in the drawing. You may use the Purge command to delete any unused named objects.

The Purge command can eliminate unused blocks, text styles, dimension styles, etc. This eliminates space-consuming parts of a drawing and can simplify drawing management. For example, you can purge unused layers so the dialog box does not contain useless layers to filter through.

Command: **Purge**
Purge unused Blocks/Dimstyles/LAyers/LTypes/SHapes/STyles/APpids/Mlinestyles/All:

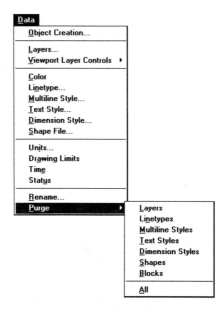

You may use the capitalized letters to select any of the options or "A" to purge all the objects.

AutoCAD will prompt you with the name of each unused object and will ask if that object should be purged.

The Purge command cannot be used to purge layer "0", the Continuous linetype, or the Standard text style.

RENAMING PARTS OF YOUR DRAWING

The Rename command is used to rename certain parts of a drawing. These parts are as follows:

Blocks	Text styles
Dimension styles	Named user coordinate systems
Layers	Named views
Linetypes	Named viewports

To rename a part of the drawing, enter the Rename command.

Command: **Rename**
Block/Dimstyle/LAyer/LType/Style/Ucs/VIew/VPort:
Old (object) name:
New (object) name:

Select the option that describes what you want to change. Then enter the existing (old) name in response to the prompt. You are then prompted for the new name. Names can be up to 31 characters in length.

There are items that cannot be renamed. These are as follows.

Layer "0"
Linetype named "CONTINUOUS"
Names of shapes

Using a Dialog Box to Rename

You may use the Rename dialog box to rename drawing objects. The dialog box is displayed either by entering Ddrename from the keyboard or by selecting "Rename Dialog" from the Data/Rename menu.

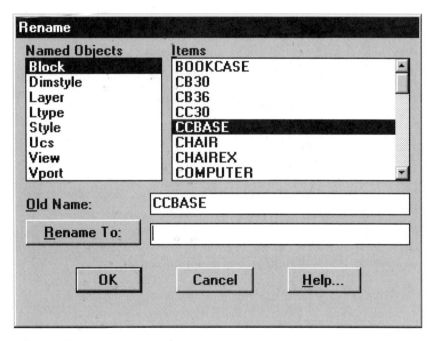

FIGURE 23-4 Rename Dialog Box

PRODUCING AND USING POSTSCRIPT IMAGES

Images using the PostScript drawing format can be imported or exported within AutoCAD. PostScript images are universally used by desktop publishing packages and other types of drawing programs. The use of PostScript images allows a direct interface with these types of programs and makes AutoCAD an invaluable illustrating program for these programs.

Exporting a PostScript Image

You can export a drawing in PostScript format by using the Psout command. When you issue the Psout command, the following dialog box is displayed.

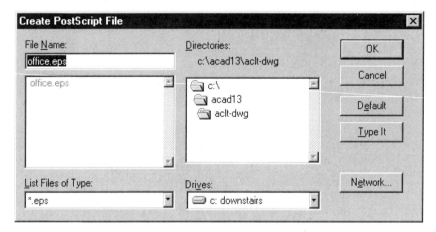

FIGURE 23-5 Create PostScript File Dialog Box

The following series of prompts is displayed. Let's look at how each prompt controls the PostScript file.

Determining What Part of the Drawing to Plot. You may plot any part of the drawing to a PostScript file. The following prompt is displayed.

What to plot — Display, Extents, Limits, View or Window <D>:

Enter the part of the drawing to export to the PostScript file. The choices are similar to the plot routine where you select the part of the drawing you want to plot.

Adding a Screen Preview. You are next prompted for the screen preview image.

Include a screen preview image in the file? (None/ESPI/TIFF) <None>:

A screen preview image can be used by an external program to preview the image before placement. If you want to preview the image, select one of the image format options: ESPI or TIFF.

> **NOTE:** Check your desktop publishing or drawing program documentation to verify the use of either format with your program.

Setting the Preview Resolution. If you select either the ESPI or TIFF format, AutoCAD prompts for the pixel resolution to use.

    Screen preview image size (128x128 is standard)? (128/256/512) <128>:

> **NOTE:** Although higher resolution screen previews are clearer, they require larger file sizes and "drag" more slowly in your desktop publishing or drawing program. The lowest acceptable resolution is recommended.

After you enter the screen preview image size, AutoCAD displays the following message:

    Effective plotting area: ww by hh high.

where *ww* is the width and *hh* is the height in the current size units of either inches or millimeters.

Setting the Output Units. The output units of either inches or millimeters can be selected. AutoCAD prompts:

    Size units (Inches or Millimeters) <current>:

> **NOTE:** Set the size units to the same setting as the external program to which you will import the PostScript file.

Setting the Output File Scale. The scale for the output file determines the final size of the drawing in the same manner as the plot scale function in the Plot command.

The scale is important if you will be printing the PostScript file to paper (see also the next section on output size). AutoCAD prompts:

    Specify scale by entering:
    Output *units*=Drawing units or Fit or ? <default>:

where *units* are either inches or millimeters (previously selected).

Setting the Output Size. The output size is the size to which the plot will be drawn. A listing similar to the following is displayed.

```
Specify scale by entering:
Output Inches=Drawing Units or Fit or ? <Fit>:

Standard values for output size
Size    Width       Height
A        8.00       10.50
B       10.00       16.00
C       16.00       21.00
D       21.00       33.00
E       33.00       43.00
F       28.00       40.00
G       11.00       90.00
H       28.00      143.00
J       34.00      176.00
K       40.00      143.00
A4       7.80       11.20
A3      10.70       15.60
A2      15.60       22.40
A1      22.40       32.20
A0      32.20       45.90
USER     7.50       10.50

Enter the Size or Width,Height (in Inches) <USER>:
Effective plotting area:  7.50 wide by 3.41 high
2600 objects
Command:
```

FIGURE 23-6 Output Size Listing

AutoCAD will write the drawing to a file. Exported PostScript files contain a DOS file extension of .EPS (Encapsulated PostScript File).

Importing a PostScript Image

PostScript images can be imported into AutoCAD with the Psin (PostScript in) command. When you enter the Psin command, AutoCAD displays the following dialog box.

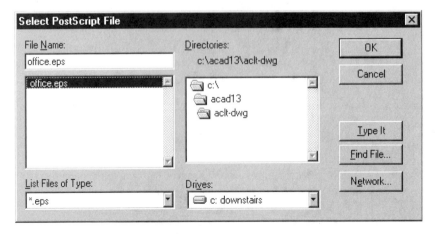

FIGURE 23-7 Select PostScript File Dialog Box

INTERMEDIATE OPERATIONS **23-13**

The PostScript file is imported into the drawing as a block. The block can be exploded and edited with AutoCAD edit commands.

When you select a PostScript file to import, depending on the setting of the Psdrag command, AutoCAD displays a box representing the size of the file. The box contains the name of the PostScript file. AutoCAD then prompts:

```
Insertion point <0,0,0>:
Scale factor:
```

The insertion point and scale factor function in the same manner as when inserting a block. You may also visually place the file by moving the cursor around the screen, moving the representative box into position.

The scale can also be visually and dynamically set by moving the cursor away from the box.

Setting the Psdrag. Psdrag controls how the PostScript image is displayed when it is imported into your drawing. The drag setting is controlled by the Psdrag command.

```
Command: Psdrag
PSIN drag mode <0>:
```

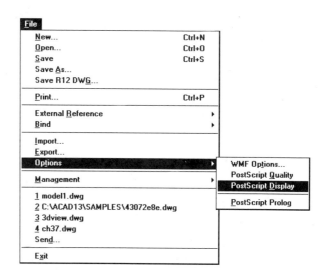

If Psdrag is set to 0, the PostScript image is displayed as a box when imported. If set to 1, the actual image is displayed.

> **NOTE:** Setting Psdrag to 1 will cause the image to handle slowly. If your system works slowly, set Psdrag to 0.

Setting PostScript Input Quality. The quality of a PostScript image during insertion is determined by the Psquality command.

Command: **Psquality**
New value for PSQUALITY <current>:

If Psquality is set to 0, the PostScript image is displayed only as a box with the file name enclosed within. The box is displayed even if the Psdrag setting is set to 1.

Any positive value sets the number of pixels per drawing unit. For example, a setting of 50 displays 50 pixels per drawing unit.

A negative value will display the drawing outline, but will not include the fills. The absolute value of the negative number controls the quality.

Displaying a PostScript Fill

AutoCAD allows you to fill a closed polyline boundary with any of several PostScript fill patterns. The fill patterns will appear on the screen when you perform a PsOut and PsIn cycle, and the fill patterns will print when you output the drawing to a PostScript printer.

To place a PostScript fill, use the Psfill command.

Command: **Psfill**

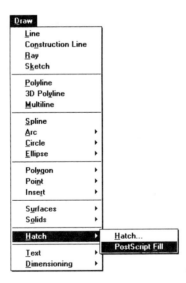

AutoCAD prompts you to select the polyline border, then asks for the PostScript fill pattern.

Select polyline:
PostScript fill pattern (.=none) <.>/?:

To see the selections, enter a question mark (?). AutoCAD displays the following listing of patterns:

Grayscale RGBcolor Allogo Lineargray Radialgray Square Waffle Zigzag Stars Brick Specks

then repeats the prompt.

PostScript fill pattern (.=none) <.>/?:

Figure 23-8 shows the available PostScript fills.

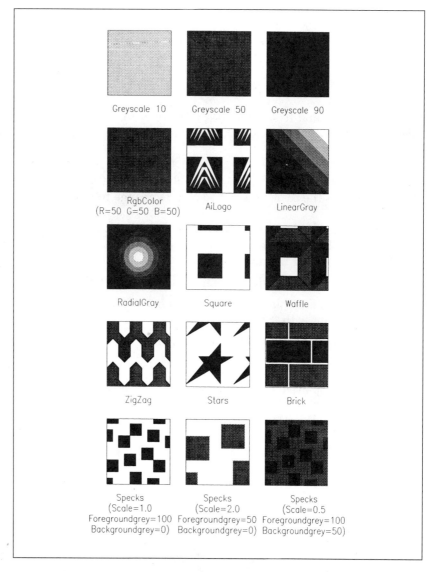

FIGURE 23-8   PostScript Fill Patterns Supplied with Release 13

Enter the name of the desired PostScript fill. AutoCAD continues.

Scale <*default*>:

The scale factor works in a similar manner to a hatch pattern scale.

The remaining prompts are dependent on the pattern you select. When selecting gray scales, a value of 0 is white and 100 is black. Values between 0 and 100 represent the percent of black area of the gray scale.

EXERCISES

1. Set the current color to yellow and draw three circles. Next, set the current color to red. Draw three boxes. Did the objects appear in the correct color?

2. Use the Layer command to set the *layer* color to cyan. Draw some lines. Did the circles and boxes you drew change color? Notice that the Color command overrides the layer color setting.

3. Use the Change command and select all the objects on the screen. Select Properties, then Color. Select blue as the color. Did all the objects change to blue?

4. Start either an existing drawing you have completed or one of AutoCAD's sample drawings (you can use the Files command to display the available drawing files). Use a zoom window to select a part of the drawing.

   Use the Mslide command to capture a slide named "SLIDE1".

   Now use Zoom All to redisplay the entire drawing. Next, issue the Vslide command and respond to the prompt with "SLIDE1".

   Note that you can view a slide from any drawing, not just the one from which it was captured.

5. While the slide is displayed, issue a Redraw. Did the slide go away?

6. Use Vslide to display the slide again. Select the Line command and try to draw a line on the slide. What happens? You cannot edit or zoom a slide. (Be sure to exit the drawing with Quit if you happen to edit the existing drawing and do not wish to save the edits.)

CHAPTER REVIEW

1. When can you display a slide?

2. How do you "capture" and store a view?

3. Can a view be edited?

4. How do you make a slide?

5. Can you edit a slide?

6. What is the file extension of a slide file?

7. From where can you save a slide?

8. What can be purged with the Purge command?

9. When can you use the Purge command?

10. What items cannot be purged?

11. What file extension is used with a PostScript file?

12. Name some uses for PostScript export files.

CHAPTER 24

ADVANCED OPERATIONS

Once you have become proficient with the draw, edit, and display commands of AutoCAD, you may want to use some of AutoCAD's advanced features. This chapter covers some of the advanced operations in AutoCAD. After completing this chapter, you will be able to:

- Utilize the construction, editing, and special features of polylines.
- Create multilines.
- Use different types of spline curves.
- Produce special output files from AutoCAD.
- Understand the processes of recovering damaged AutoCAD drawing files.

DRAWING POLYLINES

Polylines are similar to trace lines, except that they can be used in many more ways.

A polyline can have a tapering width, curves, be made of arcs, be composed of different linetypes, and a variety of other configurations.

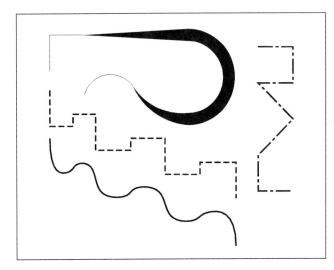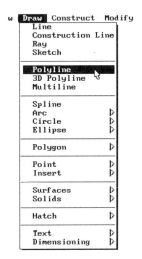

FIGURE 24-1 Examples of Polylines

Polylines have their own set of draw and edit commands. This chapter describes the draw commands first, then the edit commands.

Polylines are a bit more difficult to learn than most AutoCAD commands. You should practice each option of the Pline and Pedit commands until you become familiar with the characteristics of each.

USING THE PLINE COMMAND

To draw a polyline, the Pline (polyline) command is used:

Command: **Pline**
From point:

You will always start a polyline by entering a "from point". After this point is entered, the following prompt is issued:

Current line-width is <u>x</u>

The stated width is the width for all your polylines unless you make a change. This is followed by a rather long prompt:

Arc/Close/Halfwidth/Length/Undo/Width/<Endpoint of line>:

This prompt contains all the options from which you can branch from this point. The following list explains each of these commands:

A (Arc): Switches to arc mode.

C (Close): Works the same as the Close option for line. A polyline will be drawn back to the starting point and the command will be terminated. The last width entered is used as the width for the closing polyline.

H (Halfwidth): Functions like the Width option, except that you can "show" AutoCAD the width by moving the crosshairs from the "from point" to one edge of the desired polyline width. This represents one-half the total width.

L (Length): Draws a line segment at the same angle as the previous segment. You must specify the length.

U (Undo): Works the same as the Undo command for line entities. The last polyline segment entered will be "undone".

W (Width): Selects the width for the succeeding polylines. A zero width produces a polyline like the Line command. A width greater than zero produces a line of the specified width similar to Trace lines. Polylines have both a starting and an ending width. You are prompted for both widths when choosing the Width option:

Starting width <0.0000>:
Ending width <0.0000>:

When you specify a starting width, that value becomes the default for the ending width. Setting both the starting and ending widths the same produces polylines of continuous width. If you specify different values, the polyline will be tapered with end widths of the specified values.

DRAWING ARCS WITH POLYLINES

To draw polyline arcs, you must enter the polyline arc mode. This is done by choosing the Arc option.

Arc/Close/Halfwidth/Length/Undo/Width/<Endpoint of line>: **A**

AutoCAD then switches to the arc mode and displays the following (another long one) prompt:

Angle/CEnter/CLose/Direction/Halfwidth/Line/Radius/Second pt/Undo/Width/
  <Endpoint of arc>:

The default is "Endpoint of arc". Therefore, if you enter a point, it is used as the endpoint of the arc.

The Halfwidth, Undo, and Width options are the same as for polylines. The Width and Halfwidth options determine the width of the polyline used to draw the arc and the undo removes the most recent arc.

The arc starts at the previous point and is tangent to the previous polyline segment. If this is the first segment, the direction will be the same as the direction of the last entity drawn. The options allow you to modify the manner in which the arc is drawn. The following lists the options and their functions:

A (Angle):   Permits you to specify the included angle. The prompt is:

   Included angle:

Just like regular arcs, a polyline arc is drawn counterclockwise. If you desire a clockwise rotation, use a negative angle.

CE (Center):   Polyline arcs are normally drawn tangent to the previous polyline segment. The CE option allows you to specify the center point for the arc. The prompt is:

   Center point:

After the center point is entered, a second prompt is displayed:

   Angle/Length/<End point>:

The Angle option refers to the included angle and Length refers to the chord length.

Notice that the CEnter option must be specified by entering **CE** in order to distinguish it from the CLose option.

CL (Close):   Functions the same as the Close option for normal polylines, except that the close is performed using an arc.

D (Direction):   As previously discussed, the starting direction is determined by the last entity's direction. The Direction option allows you to specify a new starting direction for the arc. You are prompted:

   Direction from starting point:

You may then show AutoCAD the starting direction by entering a second point in the desired direction. The next prompt is:

   End point:

H (Halfwidth):   Same as straight line segments.

L (Line):   Exits Arc mode and returns to regular Polyline mode.

R (Radius): Allows you to specify the radius of the arc. After choosing the Radius option, you are prompted with:

   Radius:

Then with:

   Angle/Length/<<End point>>:

S (Second pt): Using the Second pt option allows you to switch to a three-point type of arc construction. You are prompted:

Second point:
End point:

U (Undo): Same as for straight line segments.

W (Width): Same as for straight line segments.

POLYLINE EDITING

The polyline editor is used to modify your polylines. The Pedit command is used to begin editing.

Command: **Pedit**
Select polyline:

The desired polyline may be selected by any of the standard object selection processes. If the selected entity is not a polyline, AutoCAD prompts you with:

Entity selected is not a polyline.
Do you want it to turn into one?

If you respond with a **Y**, the entity will be converted into a polyline.

After selecting a polyline, the following prompt is displayed:

Close/Join/Width/Edit vertex/Fit/Spline/Decurve/Ltype gen/Undo/eXit <X>:

NOTE: The Close option replaces the Open option, depending on whether the polyline is open or closed.

The following list describes each option and its function:

C (Close): Closes the polyline by connecting the last point with the first point of the polyline. This command performs the same function as the Close option in the Pline command, except it allows you to close the polyline after you have exited the Pline command.

O (Open): Opens a closed polyline by removing the segment created by the Close option. If the closing segment was not created by the Close option, the Open command will have no effect.

J (Join): The Join option is used to convert and connect non-polyline entities to a polyline. The string of connected entities becomes a polyline that is part of the original polyline.

When you enter the **J** option, AutoCAD prompts:

Select objects or Window or Last:

You may then select the objects that you wish to join. Several objects may be selected in a continuous string. AutoCAD will then determine any arc, line, or polyline that shares a common endpoint with the current polyline and merge them into that polyline.

In order to be successfully joined, the entities must have a perfect endpoint match (pretty close doesn't count). Use Fillet or Change and/or object snap to insure a perfect match.

The following example shows the use of the Join option:

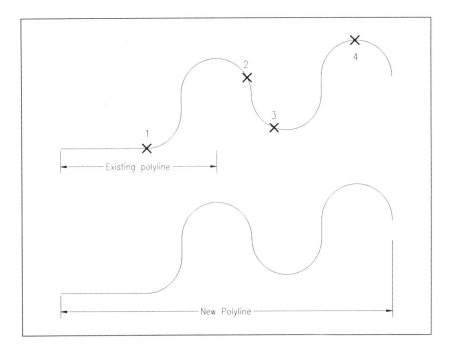

FIGURE 24-2 Using the Join Option

Command: **Pedit**
Select polyline: *(Enter point "1".)*
Close/Join/Width/Edit vertex/Fit curve/Decurve/eXit <X>: **J**
Select objects or Window or Last: *(Enter points 2, 3, and 4.)*

The objects picked by points 2, 3, and 4 will be converted into polylines and merged into the polyline selected by point 1.

W (Width): The "W" option allows you to choose a new line width for the entire polyline. When you choose the Width option, AutoCAD prompts:

Enter new width for all segments:

The new width may be entered by either a numerical value from the keyboard or by showing two points on the screen. The following example shows a polyline of varying widths and the same polyline after use of the Width option:

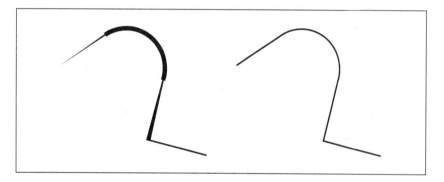

FIGURE 24-3 Using Pedit with Width Option

E (Edit vertex): The connecting points of the segments of a polyline are called a vertex. The "E" option allows you to edit a vertex and all the segments that follow it. Vertex editing is another mode, and is covered later in this chapter.

F (Fit curve): The fit curve option constructs smooth curves from the vertices in the polyline. Extra vertices are inserted by AutoCAD where necessary.

S (Spline curve): Creates a spline curve from the edited polyline. See the following section on spline curves.

D (Decurve): Negates the effects of the two curve options.

L (Linetype generation): Controls the generation of linetype through the vertices of the polyline. If the polyline linetype is set to other than continuous, the linetype dashes and dots are generated for each individual segment of the polyline. Depending on the linetype scale, some polyline segments will not show the linetype. If linetype generation is turned on, the polyline is considered as one segment when the linetype is applied, resulting in a uniform linetype pattern along the entire length of the polyline.

U (Undo): Reverses the last Pedit operation.

X (eXit): Exits the Pedit mode and returns to the command prompt.

CONSTRUCTING MULTIPLE PARALLEL LINES

The Mline command lets you draw up to 16 parallel lines at the same time. Each of the lines can have a different color, different linetype, and have a different offset distance. Unlike lines drawn parallel to each other with the Offset command, the objects drawn with the Mline command are treated as a single object.

The Mline command draws multilines; by default, a pair of parallel lines 1.0 units apart is drawn. The MlEdit command displays a dialog box that lets you perform some changes to the multiline. The MlStyle command lets you define custom multilines. The DdModify command only lets you change the multiline's layer and linetype scale. Other editing commands, such as Fillet, Chamfer, and Trim, cannot be used on multilines.

Using the Mline Command

You begin the Mline command by selecting "Draw/Multiline" from the popdown menu, or by entering **Mline** at the Command: prompt, as follows:

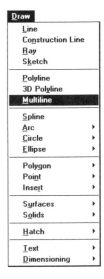

Command: **Mline**
Justification = Top, Scale = 1.00, Style = STANDARD
Justification/Scale/STyle/<From point>:

Drawing a multiline is just like drawing a line with the Line command. Keep picking points on the screen or supply coordinates, as follows:

Justification = Top, Scale = 1.00, Style = STANDARD
Justification/Scale/STyle/<From point>: *(Pick)*
<To point>: *(Pick)*
Undo/<To point>: *(Pick)*
Close/Undo/<To point>: *(Pick)*
Close/Undo/<To point>: *(Pick)*
Close/Undo/<To point>: *(Pick)*
Close/Undo/<To point>: *(Pick)*
Close/Undo/<To point>: **c**

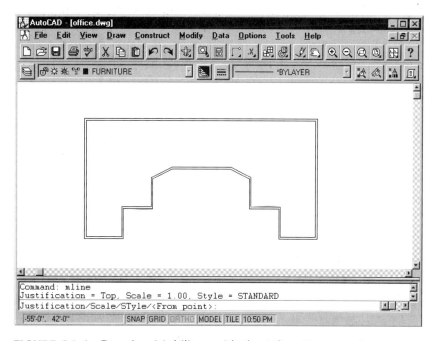

FIGURE 24-4 Drawing Multilines with the Mline Command

The Close option joins the start point with the last end point. The Undo option lets you back up, segment by segment.

The Mline command has three options and always reports the current status of each option: justification, scale, and style.

Justification Option. The three Justification options determine how the multiline is drawn relative to the cursor, as follows:

Justification = Top, Scale = 1.00, Style = STANDARD
Justification/Scale/STyle/<From point>: **j**
Top/Zero/Bottom <top>:

Option	Meaning
Top	Draws multiline below the cursor
Zero	Draws multiline centered on the cursor
Bottom	Draws multiline above the cursor

When a multiline is defined, you specify the offset of each parallel line. Thus, the Top Justification draws the multiline so that cursor picks the points that the parallel line with the greatest positive offset is drawn.

Scale Option. The Scale option draws the multiline larger or smaller than as defined:

Justification = Top, Scale = 1.00, Style = STANDARD
Justification/Scale/STyle/<From point>: **s**
Set Mline scale <1.00>:

Style Option. The STyle option lets you pick from a selection of predefined, named multiline styles. By default, AutoCAD comes with just one style, named Standard, which consists of a pair of lines 1 unit apart. You created custom multiline styles with the MlProp command.

Justification = Top, Scale = 1.00, Style = STANDARD
Justification/Scale/STyle/<From point>: **st**
Mstyle name (or ?): **?**
Loaded multiline styles:

Name	Description
STANDARD	

Mstyle name (or ?):   standard

As with the Text command, the ? option lists the names of available styles.

Editing Multilines with MlEdit

After the multiline is drawn, you edit it with the MlEdit command,

Command: **Mledit**

which displays the Multiline Edit Tools dialog box (see Figure 24-5).

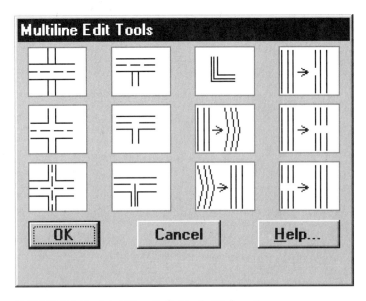

FIGURE 24-5   Multiline Edit Tools Dialog Box

As you click on the images, their meaning is displayed at the bottom of the dialog box. From left to right, top to bottom, the symbols in the Multiline Edit Tools dialog box have the following meaning:

Closed Cross	Closed Tee	Corner Joint	Cut Single
Open Cross	Open Tee	Add Vertex	Cut All
Merged Cross	Merged Tee	Delete Vertex	Weld All

The "Closed" options do not cut the multiline; the "Open" options cut multilines; the "Merged" options cut just the outermost lines of the multiline, while interior lines remain. You cannot use the Tee options on closed multilines.

Defining Multilines with MlStyle

You can define many different kinds of multilines, each with its own style name. You might want to create one multiline for exterior walls, another for interior lines, and yet another for walls to be demolished.

New multiline styles are created by defining the color, linetype, and offset for each line. You can specify whether the multiline be filled with a color, or be capped with a line or arc at its ends. Here's how:

Command: **Mlstyle**

displays the Multiline Styles dialog box (see Figure 24-6).

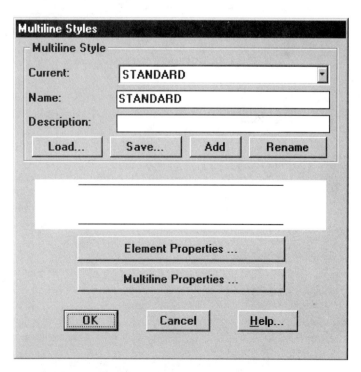

FIGURE 24-6 Multiline Styles Dialog Box

The top half of the dialog box lets you enter the name of a new style (maximum eight characters), along with a description up to 64 characters long. (After you are finishing defining the style, click the Save button to save the style to the Acad.Mln file.) Two buttons bring up sub-dialog boxes for defining the properties of individual elements and of the multiline as a whole.

Element Properties. Clicking on the "Element Properties" button brings up the sub-dialog box, shown below. Here you add and change elements from the multiline style by changing the offset distance, color, and linetype. Click the Add button to add a new element; click the Delete button to remove it. A multiline can have up to 16 elements. The only element available is the straight line; a multiline cannot have arcs or splines.

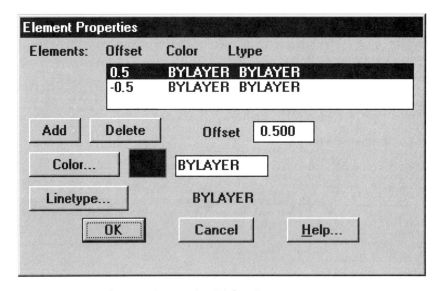

FIGURE 24-7 Element Properties Dialog Box

While the Multiline Styles dialog box displays the result of changes you make in the Elements Properties sub-dialog box, the changes are not made in real time. Instead, you need to exit the sub-dialog box periodically to check the effect of the changes you made.

Multiline Properties. Clicking on the "Multiline Properties" button brings up a second sub-dialog box, shown below.

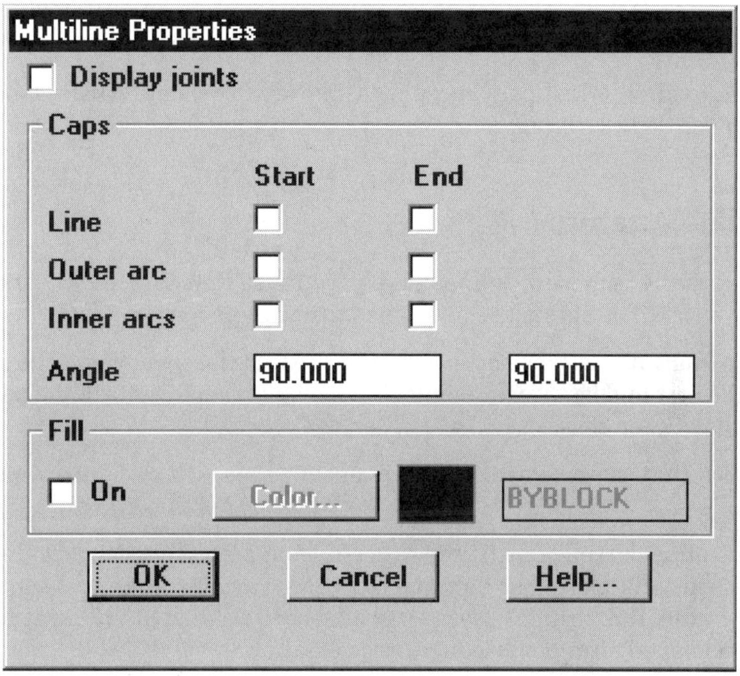

FIGURE 24-8 New Properties for a Multiline Style

Here you have a number of options over the multiline's inside and endpoints:

> **Display Joints:** when turned on, makes the multiline display a joint line at every vertex.
>
> **Caps:** Either or both ends of an open multiline can have a "cap" consisting of a straight line or an arc. The line cap is drawn at right angles (90 degrees, the default) or at any angle down to 10 degrees (or up to 170 degrees). You can have a different angle of line cap at both ends.

The alternative is to draw an arc as the cap. An "outer" cap draws an arc between the two outermost lines making up the multiline; an "inner" cap draws an arc between the innermost lines. Once again, you can have a different arc cap at each end of the multiline.

You can have both a straight line and arc cap at an end. For example, you can arc cap the innermost lines but straight line cap the outermost lines.

> **Fill:** You can specify that the multiline be filled with any of AutoCAD's 255 colors. You cannot specify which multiline elements form the boundary of the fill: the entire multiline is filled. The preview image in the main dialog box does not display the look of the fill.

CONSTRUCTING SPLINE CURVES

AutoCAD has two methods for drawing spline curves: (1) use the Spline option of the PEdit command on an existing polyline; and (2) use the Spline command to draw a NURBS-based spline curve.

Here we first look at polyline splines; later we'll look at NURBS (short for non-uniform rational Bezier spline).

Spline Anatomy

To properly construct spline curves, we must first understand the manner in which they work. The spline uses a frame of the spline curve.

This frame may be open or closed. In the case of the open frame, the spline is "connected" to the first and last vertex points (beginning and endpoints of the polyline). Each vertex point between these exerts a "pull" on the curve.

Notice that more central points about an area will exert more pull on the curve in that area.

The spline curve frame (if constructed from a polyline) may be displayed or not, and consequently, plotted or not. The Splframe variable is used to control the visibility of the frame. If Splframe is set to 0, frames are not displayed. If set to 1, the frames will be displayed after a subsequent regeneration.

The Decurve option allows you to restore the polyline to its original configuration. Decurve may be applied to curves constructed with either the Curve or Spline options.

Polyarc segments are straightened for spline frame purposes. If the polyline contains differing line widths, a resulting spline curve will contain a smooth taper that begins with the width of the first endpoint and ends with the width of the last endpoint. This differs from a curve constructed with the Fit Curve option, which maintains width information for each segment.

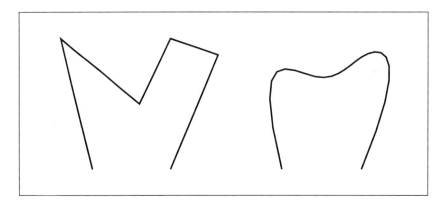

FIGURE 24-9 Spline-Curved Polyline

Spline curves construct curves in a different manner than the Fit Curve method. Fit Curves constructs curves using a pair of arcs at each central point. More central points are required to obtain each central point. More central points are required to obtain a reasonably accurate curve. The nature of the spline curve allows it to yield a more accurate curve with fewer central points. Figure 24-9 shows the relationship of applying both the Fit Curve and Spline method of curve construction to the polyline shown on the left.

Spline Segments

Each spline contains a specified number of line segments used to construct the spline. These are the number of lines between each central point. The variable Splinesegs (spline segments) is used to control this number. The default is eight segments.

    Command: **Setvar**
    New value for SPLINESEGS <8>:

Setting a larger Splinesegs value will yield a smoother curve. This should be balanced against the increased drawing size and longer regeneration times created by the larger number of segments. If a negative value is used for Splinesegs, the resulting curve will be constructed as a Fit Curve, with the number value determining the number of segments between central points.

Effects of Edit Commands on Splines

Edit commands cause spline curves to react in different ways. The following describes the effects of each.

> **Move/Copy/Erase/Rotate/Scale/Mirror:** Changes both the curve and the curve frame.
>
> **Explode/Break/Trim:** Deletes the frame and creates a permanent change to the curve.
>
> Decurve cannot be applied after using.
>
> **Offset:** Copies only the curve (without the frame) for the new offset object.
>
> **Stretch:** The frame itself is stretched and the curve is refitted to the new frame. The frame can be stretched whether it is visible or not.
>
> **Divide/Measure/Area (Entity)/Hatch/Chamfer/Fillet:** Applies only to the curve and not the frame.
>
> **Pedit (Join):** Decurves the polyline. After the Join process is completed, the Spline curve fit may be reconstructed.
>
> **Pedit (Vertex edit):** Moves the markers (denoted by "x") to the vertices on the frame (whether visible or invisible). When the spline curve is edited with Insert, Move, Straighten, or Width, the curve is automatically refit. When the Break option is applied, the polyline is decurved.
>
> **Object Snap:** Object snap recognizes only the curve.

THE SPLINE COMMAND

Unlike the Spline option of the PEdit command, the Spline command creates a true spline object, constructed as a cubic or quadratic NURBS (short for non-uniform rational Bezier spline). Like PEdit Spline, the result is a smooth curve that fits between three or more control points; see Figure 24-10. (A two-point spline results in a straight line.)

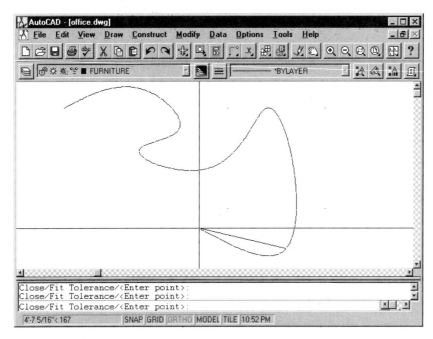

FIGURE 24-10 Drawing a NURBS (Spline) Curve with the Spline Command

Start the Spline command, as follows:

Command: **Spline**
Object/<Enter first point>: *(Pick)*
Enter point: *(Pick)*
Close/Fit Tolerance/<Enter point>: *(Pick)*
Close/Fit Tolerance/<Enter point>: *(Pick)*
Close/Fit Tolerance/<Enter point>: *(Press ENTER)*
Enter start tangent: *(Pick)*
Enter end tangent: *(Pick)*

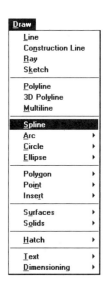

As you pick points on the screen, AutoCAD automatically draws the NURBS between the points. When done, the Spline command asks you for the starting and ending tangencies, which determine the "tangent angle" for the start and end of the spline. You can use the TANgent and PERpendicular object snaps to force the spline tangent and perpendicular to an existing object.

Object Option. The Object option is an addition to the PEdit Spline command: it converts a 2D or 3D polyline spline into a true NURBS curve.

Fit Tolerance Option. The Fit tolerance option prompts you, as follows:

Enter Fit tolerance:

The value of the fit tolerance determines how closely the curve fits to its control points. A tolerance of zero forces the spline to pass through the points.

Close Option. The Close option closes the spline: the end is joined with the start point.

EDITING SPLINES WITH SPLINEDIT

To edit a NURBS-based spline, you may use the SplinEdit command. DdModify and Grip commands also can be used to edit a spline.

Command: **Splinedit**
Select spline: *(Pick)*
Fit Data/Close/Move Vertex/Refine/rEverse/Undo/eXit <X>:

The SplinEdit command has six options, as follows:

Fit Data Option. The Fit data option lets you edit the control points that define the position of the spline.

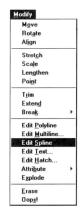

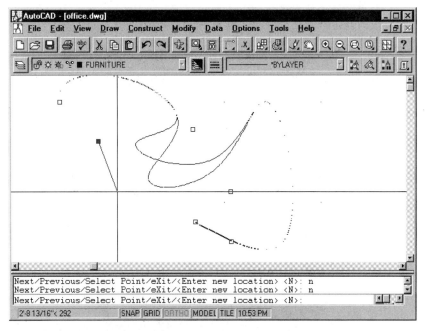

FIGURE 24-11 Editing a Spline with the SplinEdit Command

Fit Data/Close/Move Vertex/Refine/rEverse/Undo/eXit <X>:
Add/Close/Delete/Move/Purge/Tangents/toLerance/eXit <X>:

> **Add:** adds a fit point to the spline.
>
> **Close:** closes an open spline; if closed, the Open option opens up the spline.
>
> **Delete:** removes a fit point.
>
> **Move:** moves a fit point to another location.
>
> **Purge:** purges fit point data from the drawing.
>
> **Tangents:** changes the tangency of the start and end of the spline.
>
> **toLerance:** changes the tolerance value of the spline.

After each change, AutoCAD reflows the spline to fit the new parameters.

Close Option. The Close option closes an open spline. If the selected spline is already open, then the Open option appears instead.

Move Vertex Option. The Move vertex option moves the control vertices, one at a time.

Refine Option. The Refine option changes the primary spline parameters, as follows:

Fit Data/Close/Move Vertex/Refine/rEverse/Undo/eXit <X>:
Add control point/Elevate order/Weight/eXit <X>:

> **Add control point:** adds a control point nearest to the cursor position between two existing control points.
>
> **Elevate order:** uniformly increases the number of control points along the spline; maximum value is 26.
>
> **Weight:** change the distance between the spline and a control point.

Reverse Option. The rEverse option reverses the direction of the spline. The start point is now the ending point.

Undo Option. The Undo option undoes the most recent spline edit.

POLYLINE VERTEX EDITING

Vertex editing is used to modify the vertices of a polyline. The PEdit command edits polylines. Type PEdit and AutoCAD prompts:

Select polyline: *(Pick polyline.)*
Close/Join/Width/Edit vertex/Fit/Spline/Decurve/Ltype gen/Undo/eXit <X>: **E**

If you wish to edit the vertices in a polyline, choose the "E" (Edit vertex) option and you are presented with another option line.

Next/Previous/Break/Insert/Move/Regen/Straighten/Tangent/Width/eXit <N>:

ADVANCED OPERATIONS **24-19**

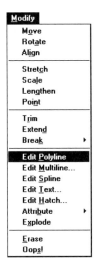

The first vertex is marked by an "X". If you have specified a tangent direction for that particular vertex, an arrow is displayed designating the direction.

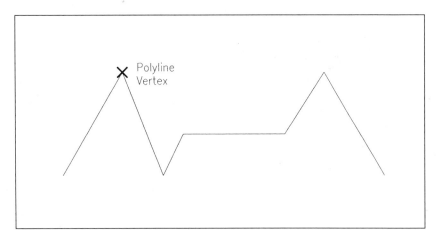

FIGURE 24-12 Vertex Editing

The following list explains the options on the vertex editing option line and their functions:

N (Next) & P (Previous): These options are used to move the identification marker "X". If you enter **N**, you may then step through the vertices of the polyline by pressing [ENTER] each time a move is desired. Choosing the P (Previous) option allows you to "back up" in the same manner. When you reach the last vertex in the polyline, you must back up to reach previous vertices. You may not wrap around the polyline, even if it is closed.

B (Break): The Break option performs the same function as the normal Break command, except that the break may only occur at vertices. To perform a break, move the marker to the vertices where you wish the break to begin and enter **B**. The following prompt then appears:

Next/Previous/Go/eXit <N>:

Use the Next and/or Previous options to move to the end of the desired break and enter **Go**. The segments between the points will be deleted. The example below shows the effects of the Break option:

If you wish to cancel the break while it is in progress, enter **X** (exit).

It is not possible to delete the entire polyline by entering break points at the first and last vertices.

If you break a closed polyline, it becomes "open" and AutoCAD removes the closing segment.

I (Insert): A new vertex may be added to the polyline by using the Insert option. The following prompt is displayed:

Enter location of new vertex:

The new vertex will be added at the location *after* the vertex that is currently marked by the "X". Figure 24-13 shows a vertex insert.

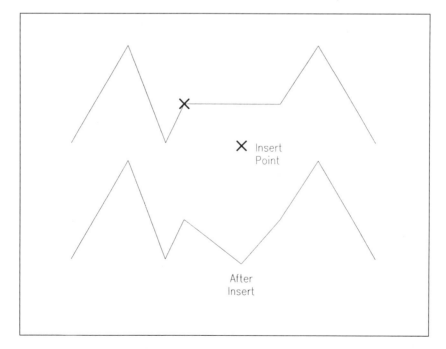

FIGURE 24-13 Inserting a Vertex

M (Move): Use the Move option to relocate a vertex. The following prompt appears:

Enter new location:

Enter the point that represents the new location for the current vertex.

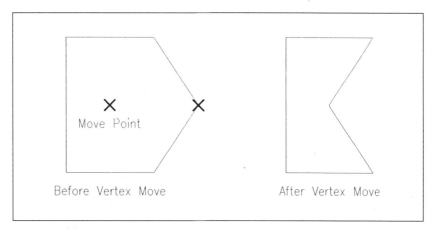

FIGURE 24-14 Move Option

R (Regen): Regenerates the polyline.

S (Straighten): The Straighten option is used in the same manner as the Break option described earlier, except the segments between the selected points are deleted and then replaced with a straight line segment. If you wish to straighten an arc segment, select the vertex immediately preceding the arc and enter both points on that vertex.

Figure 24-15 shows the effect of the Straighten option:

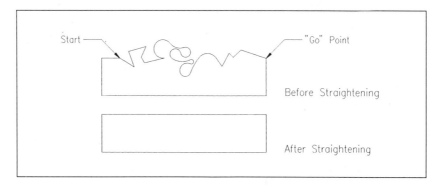

FIGURE 24-15 Straightening a Polyline

The eXit option cancels the operation if it is currently in progress.

T (Tangent): Used to assign a tangent direction to the current vertex for use at a later time in curve fitting. You are prompted:

Direction of tangent:

You may specify the direction by either entering a numerically described angle or showing AutoCAD the angle by entering a point in relation to the currently marked vertex.

W (Width): The Width option in the vertex editing portion of Pedit differs from the normal Width option. Whereas the normal Width command determines the width for the entire polyline (and all its segments), the Width option in the vertex editing option line changes the width of just the segment following the vertex currently marked by the "X". AutoCAD prompts:

Enter starting width <current>:
Enter ending width <start>:

The default starting width is equal to the current starting width for the segment being edited. As usual, the default ending width is shown equal to the starting width.

You must use the Regen option (see above) to redraw the screen and display the new segment width.

X (eXit): Exits vertex editing and returns to the Pedit option line.

EXCHANGE FILE FORMATS

AutoCAD drawing files may be converted to a form that can be read and used by external programs. For example, a database program can read a DXF (drawing interchange format file) and extract blocks for compilation. The file format is available to programmers for development purposes and is generated in ASCII format.

DXF files may also be produced by other programs for AutoCAD to read. Let's look at AutoCAD's file exchange formats.

Drawing Interchange File Format

Drawing interchange files (DXF) are a popular way to exchange graphics files between software programs. AutoCAD uses the Dxfin and Dxfout commands to use and produce DXF files.

Producing a Dxfout file. The Dxfout command produces a drawing interchange format file from an AutoCAD drawing for use by external programs. To produce a DXF file, issue the Dxfout command while in the drawing from which you wish to produce the file.

Command: **Dxfout**

AutoCAD displays a Create DXF File dialog box. Enter the name of the file you wish to produce and click the "OK" button. AutoCAD will prompt for the accuracy.

Enter decimal places of accuracy (0 to 16)/Objects/Binary <6>:

The file is written with a .DXF file extension (do not specify the extension, it is automatically appended).

The second prompt line permits the option of specifying Objects or Binary.

Selecting Objects will cause AutoCAD to request the entities that will be extracted. Block definitions will not be included.

Selecting Binary causes the file to be in binary format, as opposed to ASCII format.

You are finally prompted for the precision of the output file (0 to 16 decimal places).

The output file will be a separate file from the drawing; it will not replace it.

Using an Existing DXF File. A drawing interchange file may be converted to an AutoCAD drawing file by using the Dxfin command.

To convert a DXF file, first enter the drawing editor by starting a new drawing. The drawing may be of any name.

Next, issue the Dxfin command.

Command: **Dxfin**

AutoCAD will display a Select DXF File dialog box. Select the name of the DXF file and click on the "OK" button. It is not necessary to include the .DXF file extension.

If you wish to convert the entire drawing, only use the Dxfin command in a new drawing, before any entities are added. If a DXF file is loaded into a drawing that is not new, only the entities will be converted and the following message will be displayed:

Not a new drawing — only ENTITIES section will be input.

Drawing Interchange Binary Files

Some programs can use an abbreviated form of a binary format file. This special type of format is called a binary drawing interchange file (DXB).

AutoCAD can load a binary drawing interchange file in the same manner as a DXF file by using the Dxbin command.

Command: **Dxbin**

AutoCAD displays a Select DXB File dialog box. Select the name of the DXB file and click on the "OK" button.

DRAWING FILE DIAGNOSTICS

The Audit command is used to examine an existing drawing for damage. This diagnostics tool can also be used to correct damage to a file. The command sequence for the Audit command is as follows.

Command: **Audit**
Fix any errors detected? <N>:

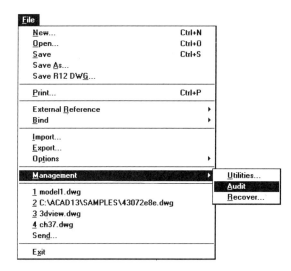

If you answer "No" to the "Fix any errors detected?" question, AutoCAD will display a report, but not fix the errors. Answering "Yes" will both display a report and fix the errors.

If a drawing contains no errors, AutoCAD will display a report showing the audit activity and the conclusion that no errors were found.

In addition to the screen report, the Audit command prints to file an audit report of the drawing file. This file has the same name as the drawing file with a file extension of ".ADT". You can read (with the Notepad text editor or a word processor) this file, or print it on a printer.

DRAWING FILE RECOVERY

In a perfect world, everything works as planned. In our real world, however, we sometimes encounter difficulties. One day you will see a message on the screen that says:

INTERNAL ERROR *(followed by a host of code numbers)*

or

FATAL ERROR.

This means that AutoCAD has encountered a problem and cannot continue. You will usually be given a choice of whether or not you wish to save the changes you have made since the last time you saved your work. The following message is displayed:

> AutoCAD cannot continue, but any changes to your drawing made up to the start of the last command can be saved.
>
> Do you want to save your changes? <Y>:

If you select "Yes", AutoCAD will attempt to write the changes to disk. If it is successful, the following message will be displayed:

> DRAWING FILE SUCCESSFULLY SAVED

If the save is unsuccessful, one of the following messages is displayed:

> INTERNAL ERROR

or

> FATAL ERROR

If you see this, you can wave goodbye to all the changes made since you last saved your work. (Of course every good CAD operator saves their work regularly.)

You may manually recover damaged drawing files by using the Recover command. When you use the Recover command, AutoCAD displays the Open Drawing File dialog box. Pick the damaged drawing file and click on the "OK" button. AutoCAD will attempt to recover the damaged file. If successful, the drawing will be displayed.

If a drawing file is detected as damaged when you open it with the Open command, AutoCAD will perform an automatic audit of the drawing. If the audit is successful, the drawing is loaded for use. If it is not, the drawing will usually be unrecoverable.

If you exit the drawing without saving, the "repair" performed by AutoCAD will be discarded. If the recovery is successful and you save the drawing, you can load it normally the next time.

EXERCISES

1. Let's draw some basic polylines!

    We'll start by drawing a box. Start a new drawing called "POLY". Set limits of 0,0 and 12,9.

    Command: **Pline**
    From point: **1,2**
    Current line width 0.0000
    Arc/Close/Halfwidth/Length/Undo/Width/ <Endpoint of line>: **W**
    Starting width: <0.0000>: **.2**
    Ending width: <0.2000>: [ENTER]

This sets a polyline with a width of .2. Now let's draw the box:

Arc/Close/Halfwidth/Length/Undo/Width/ <Endpoint of line>: **1,5**
Arc/Close/Halfwidth/Length/Undo/Width/ <Endpoint of line>: **3,5**
Arc/Close/Halfwidth/Length/Undo/Width/ <Endpoint of line>: **3,2**
Arc/Close/Halfwidth/Length/Undo/Width/ <Endpoint of Line>: **CLOSE**

This exits from the Polyline mode back into the command mode.

2. Let's draw and edit some multilines!

    Start a new drawing called "MLINES". Set limits to 24,10 and Zoom All.

    Start the Mline command and draw the five-sided shape, as follows:

    Command: **Mline**
    Justification = Top, Scale = 1.00, Style = STANDARD
    Justification/Scale/STyle/<From point>: **2,2**
    <To point>: **2,4**
    Undo/<To point>: **8,8**
    Close/Undo/<To point>: **14,4**
    Close/Undo/<To point>: **14,2**
    Close/Undo/<To point>: **CLOSE**

    Notice how the Close option automatically joins the start and end segments of the multiline.

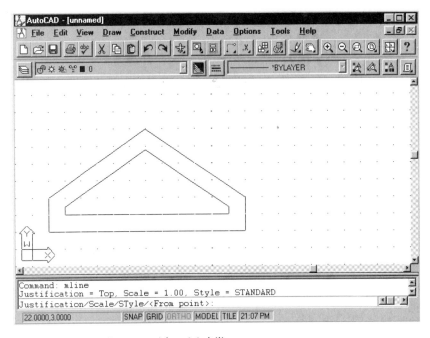

FIGURE 24-16 Drawn with a Multiline

You can use grips to edit a multiline, just like any other AutoCAD object. First, though, we'll use the MlEdit command to insert a vertex along the bottom of the multiline:

Command: **Mledit**

The Multiline Edit Tools dialog box appears. Select the Add Vertex icon (second row, third from the left). AutoCAD prompts you:

Select mline: *(Pick one of the multiline's bottom segments, in the middle.)*
Select mline(or Undo): *(Press ENTER.)*

AutoCAD adds a vertex (a joint) to the multiline, although you can't see it yet. To see the vertex, click on the multiline. AutoCAD displays the blue grip boxes, including one where you added the vertex.

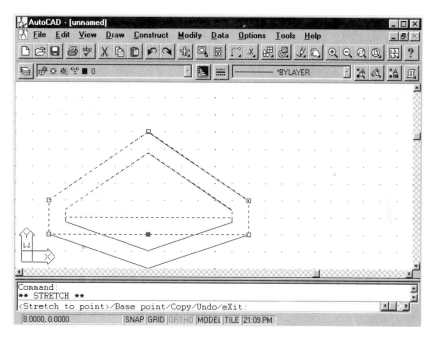

FIGURE 24-17 Stretching the New Vertex in the Multiline

Click on the blue grip box where you added the vertex. It changes to red.

Stretch the multiline to a new shape.

Press the ESC key twice to dismiss the grips.

Now let's see how AutoCAD handles intersecting multilines. With the MLine command, draw a second multiline so that it crosses the first multiline.

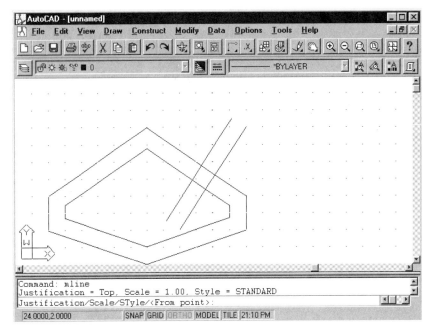

FIGURE 24-18 Drawing a Second Multiline

Start the MlEdit command again. This time select the Open Cross icon (second line, first icon). AutoCAD prompts you:

Command: **Mledit**
Select first mline: *(Pick the first multiline near the intersection.)*
Select second mline: *(Pick the second multiline.)*
Select first mline(or Undo): *(Press ENTER.)*

AutoCAD cleans up the intersection.

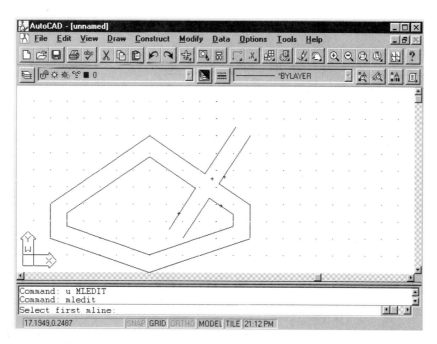

FIGURE 24-19 Cleaned-Up Intersection

CHAPTER REVIEW

1. What does the Pline option "Close" perform?

2. What option is used to create a tapered polyline?

3. What mode controls the display of solid areas in polylines?

4. What command is used explicitly to edit polylines?

5. What is a polyline vertex?

6. How do you edit a polyline vertex?

7. What types of vertex editing can you perform?

8. Can a spline curve be used on an entity other than a polyline?

9. What variable setting controls the number of segments used in a spline curve?

10. How do you create a DXF file?

11. What is the Audit command used for?

12. How do an ACIS spline and a polyline spline differ?

13. How many lines can MLine draw at one time?

14. Which command is used to edit:
    - ACIS splines? _____
    - Multiline vertices? _____
    - Polylines? _____

CHAPTER 25

VIEWPORTS AND WORKING SPACE

One of the most interesting aspects of AutoCAD is the viewports and working space capability. Learning to use these can be very helpful in many types of work. After completing this chapter, you will be able to:

- Construct and use multiple viewports.
- Save and retrieve viewport configurations.
- Understand the aspects of the two types of working spaces in AutoCAD.
- Manipulate viewports and working spaces proficiently.

USING VIEWPORTS IN AUTOCAD

The AutoCAD drawing screen may be divided into several "windows", called *viewports*. Each viewport may display a different view or zoom of the current drawing.

The following illustration shows a four-viewport screen, with different views and zooms of the same drawing.

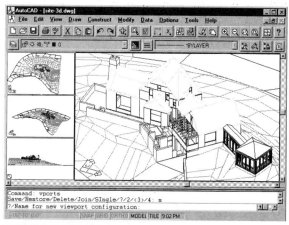

FIGURE 25-1 Viewports

25-1

The Current Viewport

Each viewport may be drawn in individually. However, only one viewport may be active at one time. When drawing in the active viewport, the crosshair will be displayed in that viewport, and all command activities are performed in the normal manner.

If the crosshair is moved into another view window, it becomes an arrow. To change another viewport to the active viewport, move the arrow into it and click the left button on the input device. The crosshairs will then be displayed and normal drawing activities may be performed.

You may alternately change the current viewport by pressing [CTRL] [R] on the keyboard (if you have AutoCAD's preferences set for "Classic Keystrokes", then press [CTRL] [V] to switch viewports). As the active viewport is changed, the border around the viewport is highlighted.

Drawing Between Viewports

It is sometimes helpful to draw between viewports. For example, you may want to zoom in to a particular area of a drawing and connect a line to another area of the drawing that is not displayed in the current viewport. You may start the line in the current viewport, then move the cursor into the other viewport and click the input device. You have now changed the current viewport. The crosshair is displayed in the new current viewport and you may continue the Line command by connecting to the desired point that is displayed in the new current viewport.

> **NOTE:** Drawing between viewports is especially helpful when drawing in 3D. You may display a 3D view in one viewport, and the plan view in another.

You cannot change viewports while some commands are active. These commands are as follows.

Dview	Vpoint
Grid	Vport
Pan	Zoom
Snap	

Setting Viewport Windows

You may create up to 48 viewport displays, depending on the type of display system. However, AutoCAD only displays the drawing in the 16 most recently accessed viewports.

To set the number and design of viewports displayed on the screen, use the Viewport command.

Command: **Viewports** *(You may alternately use the command VPORTS.)*
Save/Restore/Delete/Join/SIngle/?/2/<3>/4:

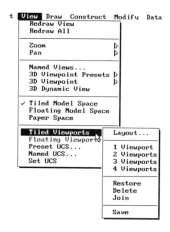

Let's look at each option.

Save: A current viewport configuration may be named and saved. It may then be recalled at any time in the future (you must, of course, be in the same drawing). The following information is stored with the Save option.

Number of viewports
Viewport positions
Grid and snap modes and spacings for each
Viewres mode
Ucsicon settings
Views set by either the Dview or Vpoint commands
Dview perspective mode
Dview clipping planes

The saved viewport configuration name may be up to 31 characters in length. You may use any letter and the characters dollar ($), hyphen (-), and underscore (_).

Restore: After one or more viewport configurations are saved, a desired configuration may be recalled by using the Restore option. The prompt will ask for the name of the configuration to restore. You may enter a question mark (?) in response to the prompt to obtain a listing of the saved viewport configurations.

Delete: Deletes a saved viewport configuration.

Join: Permits merging one viewport with another. One of the viewports to be merged is identified as the *dominant* viewport. The dominant viewport's display will inherit the new viewport created by the merging of the two.

When the Join option is selected, the following prompt is displayed:

Select dominant viewport <current>:
Select viewport to merge:

Select each viewport by moving the cursor into it and clicking. You may select the current viewport as the dominant viewport by pressing [ENTER] in response to the prompt for the dominant viewport.

If you select viewports that will not merge into one rectangular viewport, an error message will be displayed. AutoCAD then reissues the prompt.

SI: Returns to a single viewport display. The viewport that is current will be displayed in the single viewport.

?: Each viewport is identified by a number. Responding to the prompt with a question mark (?) displays a listing of the viewport numbers, with the following information.

Coordinates of each display window of the current configuration

Names of each saved configuration

Coordinates of each display window in the saved configurations (viewport identification numbers are not displayed for saved coordinates)

The number and coordinates of the currently active window are displayed first. The coordinates are not true absolute coordinates. The lower left of the screen is designated by the coordinates of 0,0. However, the upper right coordinates are always identified as a value up to 1,1.

2, 3, 4: To set the number of viewports, enter the desired number (2, 3, or 4) in response to the prompt. You are then prompted for the desired placement of the windows. The subsequent prompts depend on the number of viewports you request. The following illustration show some possible placements.

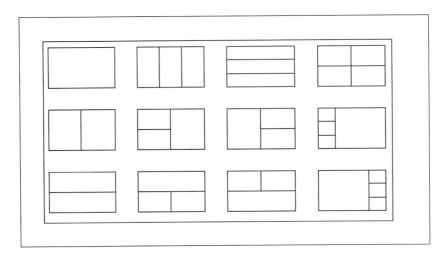

FIGURE 25-2 Sample Viewport Configurations

Let's look at how the selected number of viewports and placements are controlled.

2: Splits the screen in half, displaying two windows. You are prompted to designate whether the screen is divided horizontally or vertically.

3: Creates three windows: two small windows and one large, or three equal windows. You are prompted:

Horizontal/Vertical/Above/Below/Left/<Right>:

The Horizontal and Vertical options display three equal windows, situated either horizontally or vertically. The other options are used to create one large and two small windows. The selected option determines the location of the large window. For example, selecting Above places one large window above two small ones.

4: Creates four windows of equal size.

When multiple viewports are selected, each window initially displays the current view of the previous single-screen display. If the previous display contained multiple viewports, each of the windows in the new configuration will initially display the view of the previously current window.

Redraws and Regenerations in Viewports

The Redraw and Regen (regeneration) commands affect only the current viewport. If you wish to perform either a redraw or regeneration on all viewports simultaneously, use Redrawall or Regenall, respectively.

WORKING SPACES

In AutoCAD, you can work in either model space or paper space. Model space is the space you are used to working in now. Most of your work will still be performed in model space.

Paper space is used to arrange, detail, and plot views of your work. You arrange views in paper space by moving and sizing viewports. In paper space, viewports are handled as entities and can be edited with AutoCAD's standard edit commands. This allows you to place and plot different views of your work on the same drawing sheet (you can only plot the currently active viewport from model space).

In order to use paper space, you must first set the system variable Tilemode to zero (or "off"). You can do this by either entering the Setvar command, then entering Tilemode as the system variable to change or issue Tilemode as a command at the command prompt.

MODEL SPACE AND PAPER SPACE

You switch between paper space and model space with the Pspace and Mspace commands, respectively. Let's look at each command and how it works.

Switching to Paper Space

Use the Pspace command to switch to paper space. Before you can do this, the system variable Tilemode must be set to zero. The Pspace command has no options.

Command: **Tilemode**
New value for Tilemode <1>: **0**
Command: **Pspace**

Paper Space Icon

When you are in paper space, AutoCAD displays an icon at the lower left of the screen. A "P" is also displayed on the status line at the top of the screen.

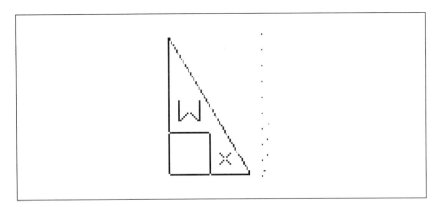

FIGURE 25-3 Paper Space Icon

Switching to Model Space

You switch to model space by issuing the Mspace command. Model space is the space you were used to working in before now. There are no options for Mspace.

Command: **Mspace**

MVIEW COMMAND

The Mview command is used to create and manage viewports in paper space. Mview only works in paper space. If you select Mview while in model space, AutoCAD automatically switches to paper space, while issuing a message to inform you of the change.

Command: **Mview**
Switching to Paper Space
ON/OFF/Hideplot/Fit/2/3/4/Restore <First Point>:

When you switch to paper space, you will not initially see any of your drawing. Before you do, you must first create a viewport with the Mview command. Your drawing will be displayed in this viewport. The drawing will be displayed in the last zoom or viewpoint (if in 3D).

In paper space, you can edit the viewports you created with the Mview command with AutoCAD's edit facilities. Let's look at the Mview command's options, then look at an example of using paper space.

Mview Options

When you issue the Mview command, you are presented with a selection of options.

Command: **Mview**
Switching to Paper Space
ON/OFF/Hideplot/Fit/2/3/4/Restore <First Point>:

Let's look at each of the options.

<First point>: This is the default. If you simply enter a point on the screen, you will set the first corner of a box that will become a new viewport. AutoCAD will prompt you for the other corner. Move the cursor away from the first point and you will see a box forming. Enter the second point and you will see the viewport form with your drawing contained within the viewport.

ON/OFF: Used to turn off or on the contents of a viewport. This is helpful if you do not want to regenerate all the viewports as you edit.

If you turn off all the viewports, you will not be able to work in model space.

If you create and attempt to turn on more viewports than the maximum number, AutoCAD will turn on the maximum number of viewports. The rest will be marked ON and will be plotted, but will not display.

The maximum number of viewports is determined by the operating system and your display system. The maximum number (up to the possible maximum) is set with the Maxactvp system variable.

Hideplot: This option allows you to specify which viewports will be plotted with the hidden lines removed. You can either choose to have the hidden lines removed in all the viewports or just in the ones you choose.

This is especially useful if you are plotting a sheet that contains a 3D view of an object in one of the viewports. When you select Hideplot, you can turn the Hideplot function either on or off. If you turn it off, you can choose individual viewport(s) to be plotted with hidden lines.

```
Command: Mview
Switching to Paper Space
ON/OFF/Hideplot/Fit/2/3/4/<First Point>: H
ON/OFF: OFF
Select objects:
```

You can now select the viewports to be plotted with hidden lines.

Note that the selection of a viewport does not create a hidden line view. It will only plot with hidden lines. If you wish to display a hidden line view, use the Hide command.

Fit: Selecting Fit will create a new viewport that will fill the current graphics screen. Note that this will not fill the entire limits of the drawing if you are zoomed in to a part of the drawing. It will fill the currently displayed area.

2, 3, 4: Creates two, three, or four viewports. The use of this option is similar to setting a specified number of viewports with the Vports command, except the viewports are created to fill the currently displayed screen area or an area you specify by building a box by placing two corner points. If you select 2, 3, or 4, AutoCAD responds with:

```
Fit/<first point>:
```

Fit will fill the currently displayed screen area with the specified number of viewports, while "<first point>" (the default) allows you to specify the area for the viewports by selecting two corner points that define the area.

Since the choices can conceivably build configurations of different arrangements (for example, side by side or over and under for two viewports), AutoCAD prompts for the arrangement.

The following is an overview of the prompts and responses used with each choice.

2: A prompt asks if you want a horizontal or vertical division. Vertical is the default.

3: The following prompt is displayed:

Horizontal/Vertical/Above/Below/Left/<Right>:

4: Creates four equally spaced viewports automatically.

Restore: Permits you to create a new set of viewports based on a previously saved configuration. AutoCAD prompts:

?/Name of window configuration to insert <*default*>:

Entering a question mark (?) will display the names of saved viewport configurations. Enter the name of the saved configuration and press [ENTER] and AutoCAD prompts:

Fit/<first point>:

As before, Fit fills the entire currently displayed area with the viewports, while "<first point>" permits you to define an area for the viewports by creating a box with two corner points.

CREATING RELATIVE SCALES IN PAPER SPACE

The obvious primary use of AutoCAD's paper space is to create drawings with various views and details in a single, plottable drawing. Many times this requires that part of the drawing be plotted at a different scale. Let's look at an example.

Suppose that we have a floor plan of a house. The scale of the floor plan is ¼" = 1'0". We want to place an enlarged plan of a bath, with dimensions, on the same drawing sheet. We could switch to paper space and create a new viewport. We can then switch back to model space, zoom into the bath area, and dimension it. The problem is that we wish the bath area to be plotted at a scale of ½" = 1'0".

We can create this scale by using a function of the Zoom command. If we are in model space, we can make a viewport active, then zoom at a factor of the paper scale. Let's look at an example in this tutorial.

TUTORIAL

Start the drawing from the work disk named "PSPLAN". This is a drawing of a floor plan as shown in the following illustration.

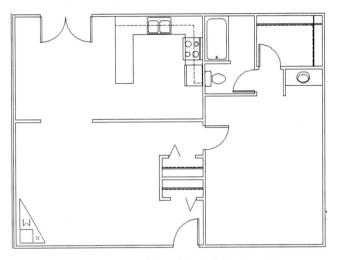

FIGURE 25-4 PSPlan Work Disk Drawing

The drawing is drawn in architectural units with limits of 0,0 and 96',72'. This is set up to be plotted at ¼" = 1'0" on a C-size (24" × 18") sheet of paper.

We wish to enlarge an area containing the bath and show this area at a scale of ½" = 1'0". Let's look at the steps for doing this.

First, let's set the Tilemode to 0.

   Command: **Tilemode**
   New value for TILEMODE <1>: **0**

AutoCAD will now switch to paper space automatically, displaying a message telling you it is doing so.

The limits for the model space have been set, but they have not been set for paper space. Let's set the limits.

   Command: **Limits**
   Reset paper space limits:
   ON/OFF/<Lower left corner> <default>: **0,0**
   Upper right corner <default>: **96',72'**

Now perform a Zoom All to display the entire drawing area.

Let's now build the first viewport, using the Mview command.

Command: **Mview**
ON/OFF/Hideplot/Fit/2/3/4/Restore/<First Point>: **F**

Your drawing should now be present on the screen. Next, use the Mview command again to place a new viewport in the upper right area of the drawing. Use the following command sequence and Figure 25-5 to place the viewport.

Command: **Mview**
ON/OFF/Hideplot/Fit/2/3/4/Restore/<First Point>: *(Enter point "1".)*
Other corner: *(Enter point "2".)*

Notice how the entire floorplan is contained in the viewport. Let's switch to model space.

Command: **Mspace**

Move the cursor into the new viewport and click to make it the active viewport. Now zoom window into the area of the bath.

Let's now use AutoCAD's Zoom command to zoom and scale the contents of the new viewport in relation to the paper space units.

Command: **Zoom**
All/Center/Dynamic/Extents/Left/Previous/Vmax/Window/<Scale (X/XP)>: **2XP**

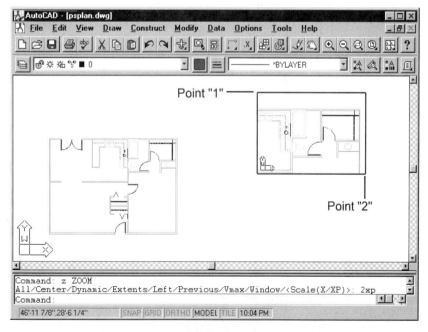

FIGURE 25-5 Completed Paper Space Drawing

25-12 USING AUTOCAD

Notice how the scale of the viewport changes. The area of the plan you wanted to show may not be correctly centered. While in model space, use the Pan command to reposition the drawing. After you have finished, change to paper space.

Command: **Pspace**

Now that you are in paper space, the viewports themselves are entities. Note that you cannot edit the actual floor plan, but you can edit the viewports. The viewport may be too large or small. Use the Stretch command to stretch the viewport (don't worry about getting the crossing window over the floor plan). You can stretch the viewport to the correct size to show the area of the bath you want.

After the bath area is properly positioned, use the Text command to create titles for each drawing. If this is performed in paper space, the text can be placed anywhere on the screen, even outside of the viewports.

You may want to dimension the floor plan. If you do, switch back to model space so you can edit the floor plan drawing.

EXERCISE

Start a new drawing named "PORTS". Let's use the Vports command to set up four viewports.

Command: **Vports**
Save/Restore/Delete/Join/SIngle/?/2/<3>/4: **4**

You should now see four viewports on the screen as shown in the following illustration.

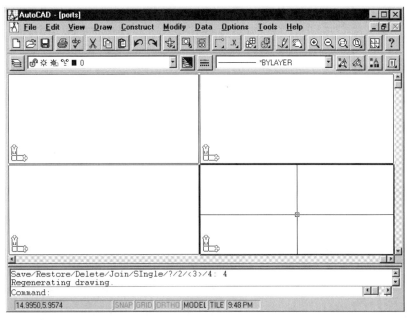

FIGURE 25-6 Setting Four Viewports

Move the crosshairs between the viewports. Notice how the crosshairs only show up in one of the viewports. When you move into the other viewports, the crosshairs turn into an arrow. Move into one of the viewports where an arrow is displayed and click the button on your input device. You should now see crosshairs in that viewport. This is how you change active viewports.

If you are not there now, move into the upper left viewport and make it active. Draw several circles in the drawing area. Notice how the circles appear in all the viewports.

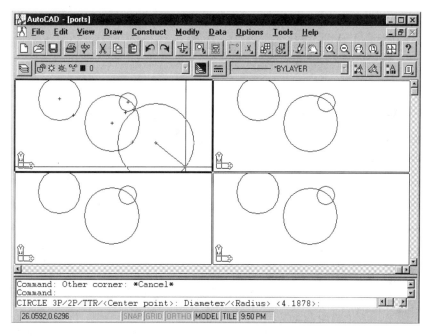

FIGURE 25-7 Drawing Circles

Move to the upper right viewport and make it active. Use the Zoom command to zoom in on one of the circles. Next, move to the lower left viewport and zoom in on another circle. Now select the Line command and with object snap, select the center of the circle. Before placing the second endpoint of the line, move to the upper right viewport and click to make it active. With object snap Center, place the line's second endpoint at the center of the circle on the screen. Notice how you can monitor the activity of the entire screen on the viewport at the upper left, since it is in a "zoom all" display.

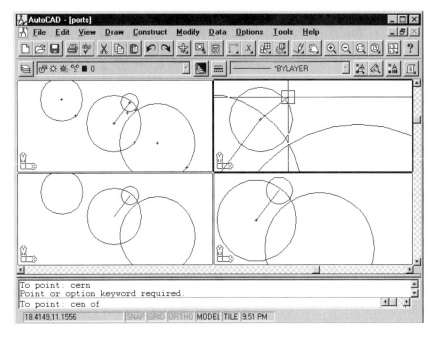

FIGURE 25-8 Drawing a Line from One Viewport to Another

CHAPTER REVIEW

1. How many viewports can be active at one time?

2. Can you switch between viewports while in a draw command?

3. How would you save a viewport configuration that you wanted to use again?

4. When you join two viewports, what is meant by the term *dominant* viewport?

5. How would you regenerate all the viewports in one operation?

6. What must the Tilemode variable be set to before you can enter paper space?

7. When can the Mview command be used?

8. How would you place a new viewport that would fill the entire screen in paper space?

9. Do model space and paper space have different limits?

10. How do you switch from paper space to model space?

CHAPTER 26

ATTRIBUTES

Many types of CAD drawings can use attributes. Once you are familiar with the concept and learn the use of some basic commands, attributes can be very helpful and beneficial. After completing this chapter, you will be able to:

- Demonstrate what attributes are and how they are used.
- Understand the anatomy of attributes.
- Manipulate attributes proficiently using AutoCAD commands.
- Produce attribute output files for printing and use with other computer programs.

ATTRIBUTES

AutoCAD allows you to add *attributes* to a block. An attribute may be considered a "label" that is attached to the block. This label contains any information you desire.

The information from each block may be taken from AutoCAD's database file and used in other places such as database programs.

Attributes are placed in the drawing with the block they are attached to. When the block is inserted, the values for the attributes are requested by AutoCAD. You determine what information is requested, the actual prompts, and the default values for the information requested.

Suppression of Attribute Prompts

The system variable ATTREQ is used to suppress attribute requests. If ATTREQ is set to 0, no attribute values are requested, and all attributes are set to their default values. A setting of 1 causes AutoCAD to prompt for attribute values.

TUTORIAL

Let's look at an example of using attributes. You are now the manager of an engineering department that utilizes CAD (see how CAD has already helped your career?). The department contains several CAD workstations, each with an employee, a computer, and a telephone. You use attributes to keep information on each of these items contained in the floor plan of the department.

Start a new drawing called OFFICE. Set limits of 0,0 and 24,18. Draw walls with the Pline command (with width = 0.1), as shown below:

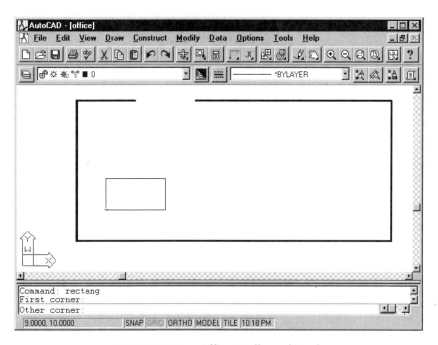

FIGURE 26-1 Office Walls and Desk

First, you must draw the desk. (Do this now with the Rectang command.) It might help to zoom a bit. Now to set up the attributes.

DEFINING ATTRIBUTES

The DDATTDEF command is used to set up a template for the attribute (or label). Each label is made up of different bits of information. Your attribute will include the name of employee, the type of computer used, and the telephone extension number of that station. These items are called *tags*.

Attribute Tag

A tag is the name of a part of an attribute. The following tags will be used in your attribute:

EMPLOYEE_NAME:
COMPUTER:
EXT._NO.:

Notice that we used underlines instead of spaces because blank spaces are not allowed in tag names. You may also use a backslash (\) as a leading character in lieu of spaces.

Let's jump in and execute the DdAttdef command:

Command: **DdAttdef**

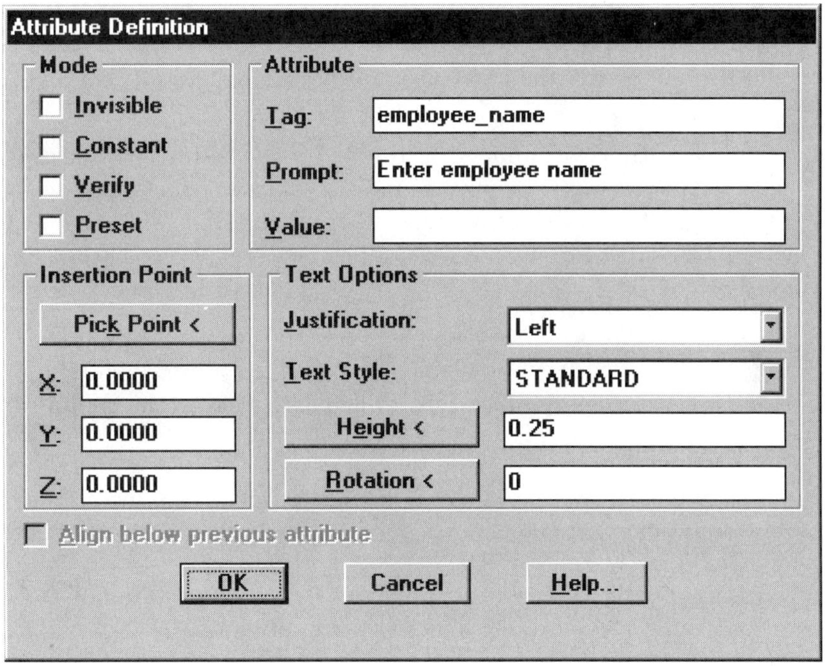

FIGURE 26-2 Attribute Definition Dialog Box

Let's look at the options for the attribute modes:

Invisible: The "I" option determines whether the labeling is visible when the block containing the attribute is inserted. If you later want to visibly display the attribute, you may use the Attdisp command.

Constant: The "C" option gives every attribute the same value. This might be useful if every computer on every desk is the same. Beware! If you designate an attribute to contain a constant value, you cannot change it later.

Verify: If you use the "V" option, you will be asked to verify that every value is correct.

Preset: Allows presetting values that are variable, but not prompted when the block is inserted. The values are automatically set to their preset values. The preset option is not active if the attributes are entered via the attribute dialog box.

Each preset option is either activated (check mark) or not activated (no check mark). To change the current setting for each, click on it.

Let's leave the modes as they are for now.

Let's look at the Attribute prompts:

Tag:

The attribute tag is the name of the attribute. We first enter the tag of **Employee_name**.

You are then prompted for the:

Prompt:

The prompt is the text that will appear on the text line when the block containing the attribute is inserted. If you want the prompt to be the same as the tag name, enter a null response (press [ENTER]). If the Constant mode was specified for the attribute, a default prompt is not requested. For the default prompt, we will use **Enter employee name**.

You are then prompted for the default attribute value:

Value:

The Attribute default value is the name that will be displayed on the default prompt line.

The response to this prompt will determine the constant value.

The remainder of the dialog box consists of a series of prompts that are similar to the text prompts, except that the text string is not requested. The attribute information is used as the text string. The location and text size you specify will become the location of the information in the inserted block. Enter a text height of **0.25** and a rotation of **0**. Click on the Pick Point button. AutoCAD clears the dialog box and prompts:

Start point: *(Pick.)*

Pick a point inside the desk. AutoCAD returns the dialog box. Click on the OK button.

Continue by entering the DdAttdef command again:

Command: **DdAttdef**

This time we want to enter the tag for the computer:

Tag: **Computer**
Prompt: **Enter computer name**
Value: **IBM-AT**

Notice that we used IBM-AT as the default attribute value. Every time the block is inserted, AutoCAD will show this as the default. This time, click on the "Align below previous attribute" option and click "OK".

Now, for the telephone extension number. Enter the DdAttdef command again:

Command: **DdAttdef**

Enter the following:

Tag: **Ext_no.**
Prompt: **Enter telephone extension number**

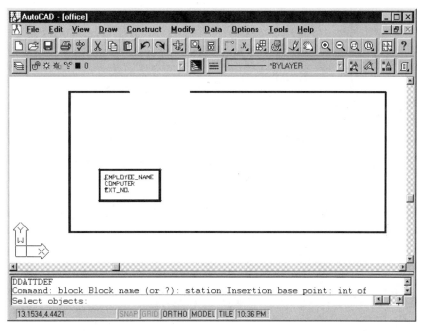

FIGURE 26-3 Defining the Block Attribute

Again, we did not specify a default value. Now, to block the desk:

Command: **Block**
Block name (or ?): **STATION**
Insertion base point: **INT** of *(Pick point "A".)*
Select objects: **W**
First corner: *(Pick corner.)*
Second corner: *(Pick corner.)*
Select objects: [ENTER]

The block disappears. You are now ready to insert your block with its attributes!

Insert the block named **STATION** and answer the prompts with different names, computer types, and extension numbers.

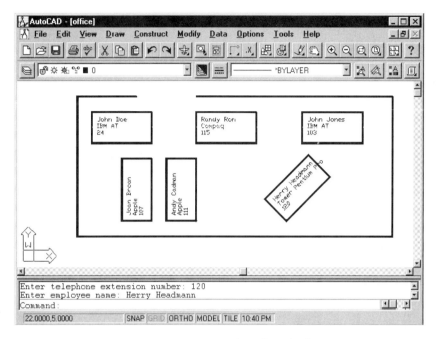

FIGURE 26-4 Inserting the Blocks

CONTROLLING THE DISPLAY OF ATTRIBUTES

You don't always want the attributes to show up on the display screen. You can use the Attdisp (attribute display) to determine the visibility of the attributes.

Command: **Attdisp**
Normal/ON/OFF/<current value>

The options are listed below:

> **Normal:** Attributes are visible, unless you specified them to be invisible on the attributes modes line in the original Attdef command when they were formed. This option is useful if you want some attributes displayed and not others.
>
> **On:** All attributes are visible.
>
> **Off:** All attributes are invisible.

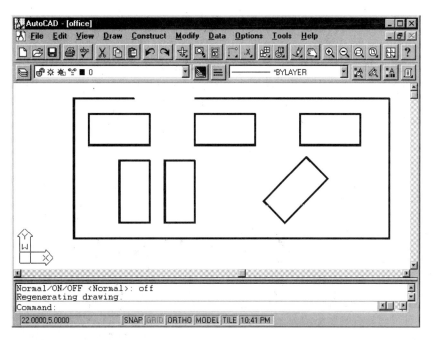

FIGURE 26-5 Attribute Display Turned Off

After changing the Attdisp, the display is regenerated to show the new state (unless the REGENAUTO is off).

EDITING ATTRIBUTES

Now that everything is all set up properly, John Smith leaves and Andy Cadman is hired to take his place. You have to make a change in your attribute base. The Attedit command is used to make changes in attributes.

Command: **Attedit**
Edit attributes one at a time? <Y>

The response determines the string of options which will follow:

Yes: Selects *individual* (one by one) editing. The attributes that are currently visible on the screen may be edited. The attributes to be edited may be further restricted by object selection or block names, tags, and values of the attributes to be edited.

No: Used for *global editing* of attributes (all). You may also restrict the editing to block names, tags, values, and on-screen visibility.

You will next be asked to select the method of editing. You may choose the attributes to be edited by global (*), or using the ? symbol to replace common characters. AutoCAD prompts for the parts of the attributes to be edited:

Block name specification <*>:
Attribute tag specification <*>:
Attribute value specification <*>:

Your reply to each prompt will determine the parts of the attribute that may be edited.

You may choose to edit individual attributes or all attributes.

Individual Editing

You may choose the individual attributes to be edited by using the object selection process. The prompt issued after you have selected the block, tag, and values that are possible to edit is:

Select Attributes:

Use the standard object selection process to choose the attributes to be edited. (The attribute set selection will limit the attributes to be selected. If you entered an ENTER to each prompt at that time, all attributes will be edited.)

After you have selected each attribute to be edited with the object selection process, an "X" will mark the first item that may be edited. The "X" marks the current spot to be edited until you enter **Next** (or ENTER) and a new spot is marked. You are then prompted:

Value/Position/Height/Angle/Style/Layer/Color/Next <N>:

The options are:

- **Value** — attribute value
- **Position** — text position
- **Height** — text height
- **Angle** — text angle
- **Style** — text style
- **Layer** — layer
- **Color** — color
- **Next** — next

The first letter may be used to select the appropriate option or press ESC to cancel.

If you enter **Value**, AutoCAD will prompt with:

Change or Replace? <R>

Change is used to change a few characters, as for a misspelling. If you choose C, the following prompt appears:

String to change:
New string:

You should respond to the first string with the string of characters to be changed and to the second string with the string you want it replaced with.

Replace is used to change the attribute value. You are prompted:

New Attribute value:

Responding with Position, Height, Angle, Style, Layer, or Color, will result in prompts that request the new text parameters and layer location.

Editing Attributes with a Dialog Box

Attribute values may be edited by use of a dialog box. The Ddatte command is used to access the dialog box.

Command: **Ddatte**
Select block:

Select the block with attributes to be edited and the dialog box is displayed.

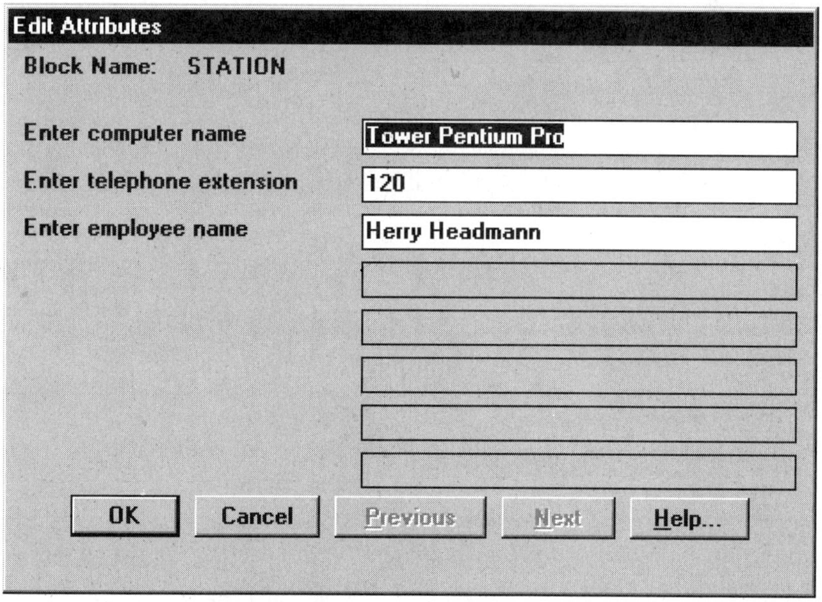

FIGURE 26-6 Edit Attributes Dialog Box

Global Editing

Global editing is used to edit all the attributes at one time. As usual, the limits you set for editing will be used.

You may choose global editing by responding to the initial prompt:

> Command: **Attedit**
> Edit Attributes one at a time? <Y>: **N**

AutoCAD then prompts:

> Global edit of attribute values.
> Edit only Attributes visible on screen? <Y>

An **N** response to this prompt will result in the comment:

> Drawing must be regenerated afterwards.

All this means is that your changes will not be immediately shown on the screen. You will have to Regen to see them.

You must then restrict the set of attributes to be edited to the specific tags, values, or blocks.

Visible Attributes

If you responded with a **Y** to the prompt, you will only edit visible attributes. You are prompted:

> Select Attributes:

Use the standard object selection process to choose the group of attributes to edit. An **X** is displayed at the starting point of the selected attributes. You are then prompted:

> String to change:
> New string:

Respond to the prompts with the string you wish to change and the changes you wish to make.

ATTRIBUTE EXTRACTIONS

Attributes are a great feature for keeping records of your inserted blocks. You could maintain a database on furniture; model and cost figures on parts used in a design; or the number, type and cost of windows in a plan.

It would be useful if you could print out all these items in a report. The Attext command can!

The Attext (ATTribute EXTract) command is used to extract database information from the drawing in a specified form.

    Command: **Attext**
    CDF, SDF or DXF Attribute extract(or Entities)?<C>:
    Template file <*default*>:
    Extract filename <name>:

The attributes may be extracted in three formats:

    CDF: The CDF (Comma Delimited Format) produces a file that contains delimiters (commas) that separate the data fields. The character fields are enclosed in quotes. This format may be read directly by some database programs.

    SDF: The SDF (Space Delimited Format) format is similar to the CDF form, except it does not use commas and requires fixed field lengths. The SDF format is the standard for input to database systems on mini-computers.

    DXF: Similar to the AutoCAD drawing interchange file. This format contains the block reference, attribute, and end-of-sequence entities only.

Entities. Allows you to select the specific blocks whose attributes you wish to extract.

We use the SDF format to prepare an attribute extraction of information from our drawing.

Using a Dialog Box to Extract Attributes

The Ddattext command displays a dialog box for use in extracting attribute values.

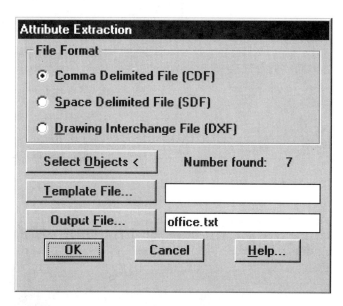

FIGURE 26-7 Attribute Extraction Dialog Box

Creating Template Files

To extract our information, we must first prepare a template file. A template file is a guide by which the information is extracted from the drawing.

You may use a text editor or the appropriate commands from a database program to prepare the template file. Each line of the template represents one field to be listed in the extract file. You may also specify the width of the field (in characters), and the number of decimal places to be displayed in numerical fields. Each field will be listed in the order shown in the template file. The following table shows the possibilities of field choices:

Field	Format	Description
BL:LEVEL	Nwww000	Block nesting level
BL:NAME	Cwww000	Block name
BL:X	Nwwwddd	X-coordinate of block
BL:Y	Nwwwddd	Y-coordinate of block
BL:NUMBER	Nwww000	Block counter
BL:HANDLE	Cwww000	Block's handle
BL:LAYER	Cwww000	Block insertion layer name
BL:ORIENT	Nwwwddd	Block rotation angle
BL:XSCALE	Nwwwddd	X scale factor of block
BL:YSCALE	Nwwwddd	Y scale factor of block
BL:ZSCALE	Nwwwddd	Z scale factor of block
BL:XEXTRUDE	Nwwwddd	X component of extrusion
BL:YEXTRUDE	Nwwwddd	Y component of extrusion
BL:ZEXTRUDE	Nwwwddd	Z component of extrusion
tag	Cwww000	Attribute tag (character)
tag	Nwww000	Attribute tag (numeric)

TABLE 26-1

Each field may be a character field (C) or a numerical field (N). The first character designates the type. The next three numbers represent the field width. The last three represent the number of decimal places in a numeric field. Thus, a character field representing a tag with a field spacing of 18 characters would be represented as "C018000".

EXERCISES

1. Consider the following workstations and their corresponding attribute values:

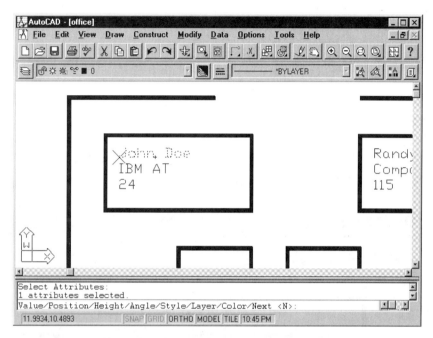

FIGURE 26-8 Changing the Employee Name

Let's assume that we want to change John Doe's name to Jane Doe. Execute the Attedit command:

Command: **Attedit**
Edit Attributes one by one? <Y>: [ENTER]
Block name specification <*>: [ENTER]
Attribute tag specification <*>: [ENTER]
Attribute value specification <*>: [ENTER]
Select Attributes: *(Select.)*

Value/Position/Height/Angle/Style/Layer/Color/Next <N>: **V**
Change or Replace <R>: **C**
String to change: **JOHN DOE**
New string: **JANE DOE**

The name change will automatically be reflected on the screen. Press [ENTER] to end command.

2. The two workstations below show IBM-AT as the type of computer. We want to change it to IBM-PS/2. Let's use global editing to do this.

Command: **Attedit**
Edit Attributes one by one? <Y>: **N**
Global edit of attribute values.
Edit only attributes visible on ? <Y>: [ENTER]
Block name specification <*>: **STATION**
Attribute tag specification <*>: **COMPUTER**
Attribute value specification <*>: [ENTER]
Select Attributes or Window or Last: **W**
First point: *(Select.)*
Second point: *(Select.)*
String to change: **IBM AT**
New string: **IBM PS/2**

Each station should now contain the attribute value of IBM-PS/2 in place of IBM-AT.

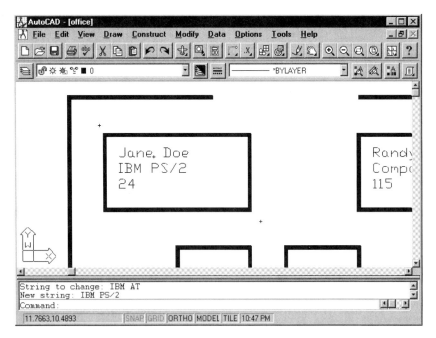

FIGURE 26-9 New Workstation Name

3. Let's use our OFFICE drawing to extract the attributes from. We want to obtain the block name, X-scale factor, Y-scale factor, employee name, computer type, and telephone extension number.

First, we must prepare a template file. Use a text editor to prepare a template like the one shown in Figure 26-10. Call the file TEMPLATE.TXT. The template file must have a file extension of "TXT".

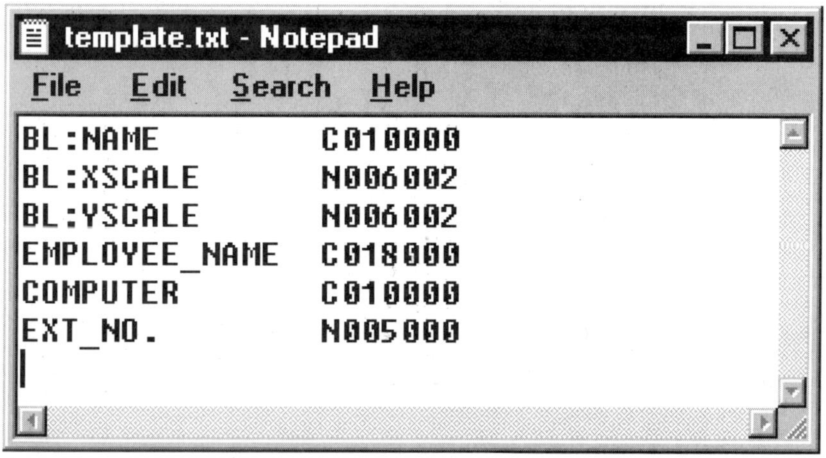

```
template.txt - Notepad
File  Edit  Search  Help
BL:NAME           C010000
BL:XSCALE         N006002
BL:YSCALE         N006002
EMPLOYEE_NAME     C018000
COMPUTER          C010000
EXT_NO.           N005000
```

FIGURE 26-10 Template File

Notice that the tag listings in the template file must exactly match the tag names in the attribute. Do not use tabs. The file must have a blank line at the end.

Now, proceed to AutoCAD and obtain your drawing called OFFICE. Execute the Attext command:

Command: **Attext**
CDF, SDF or DXF Attribute extract <C>?: **SDF**
Template file <*default*>: **TEMPLATE**
Extract filename <*default*>: **OFFICE**

You have now created an extract file with the name Office.Txt. You may use the print utility of your word processor to obtain a copy of the listing. Your listing should look roughly like the following:

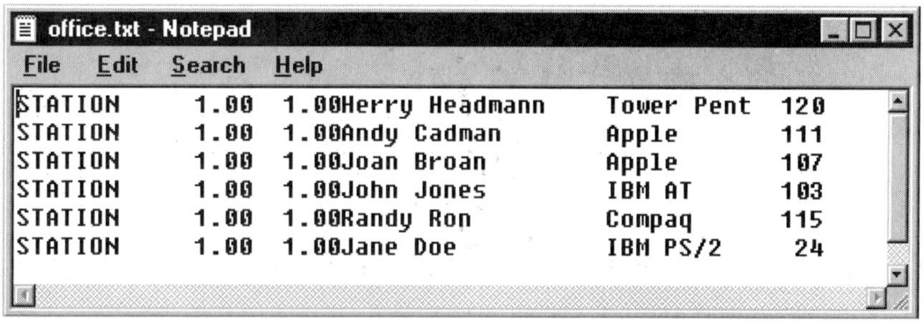

```
office.txt - Notepad
File  Edit  Search  Help
STATION    1.00   1.00Herry Headmann    Tower Pent   120
STATION    1.00   1.00Andy Cadman       Apple        111
STATION    1.00   1.00Joan Broan        Apple        107
STATION    1.00   1.00John Jones        IBM AT       103
STATION    1.00   1.00Randy Ron         Compaq       115
STATION    1.00   1.00Jane Doe          IBM PS/2      24
```

FIGURE 26-11 Attributes Extracted

CHAPTER REVIEW

1. What is an attribute?

2. What is an attribute tag?

3. What command do you use to create attributes?

4. What is the attribute prompt?

5. How would you suppress attribute prompts?

6. How do you control whether the Attribute dialog box is displayed?

7. What determines whether attributes are displayed in the drawing?

8. How are attributes edited?

9. What parts of an individual attribute can be changed with the attribute edit capabilities?

10. What command will prompt the Attribute Edit dialog box?

11. How could you efficiently edit successive blocks with attributes?

12. How would you globally edit all the attributes?

13. How would you obtain a file of all the attribute values in your drawing?

14. What is a template file?

15. What are the three types of attribute extract file formats?

CHAPTER

27

EXTERNAL REFERENCE DRAWINGS AND OLE OBJECTS

External reference drawings are extremely useful for certain types of CAD work. Those who assemble a large part of their work from library symbols and other drawings should consider the use of external reference drawings. After completing this chapter, you will be able to:

- Develop the concept and use of external reference drawings.
- Perform the AutoCAD commands used with the external reference functions.
- Manage externally referenced files.
- Use object linking and embedding with drawings.

OVERVIEW

One of the strengths of CAD is its ability to draw small components, then assemble them into a larger, more complex drawing. In Chapter 20 you learned how to do this with the Block and Insert commands. The Insert command allows you to merge a drawing into another drawing, controlling the placement and scale of the insertion.

When you insert a drawing, AutoCAD creates a block reference of that drawing in the destination drawing file. If you insert that drawing again, AutoCAD uses the block reference to obtain a copy of the previously inserted drawing. It is not an easy task to change the part of the drawing that was inserted. You can delete all copies of the block, then use the Purge command to delete the block reference and reinsert the drawing with the changes, or you can use the Explode command to break the inserted block into its simple entities, then edit the entities. This means that any changes made to a component drawing must also be changed in the drawing into which it was previously inserted.

AutoCAD's external reference capabilities can be useful when you have a situation such as this. The external reference capabilities allow you to insert a drawing into another in a similar way as you would with the Insert command. If the drawing is inserted as an external reference drawing, however, AutoCAD does not load a block reference. Each time the master drawing is loaded, the component drawing is scanned, then loaded at that time. Thus, any changes made to the component drawing are updated in the master drawing automatically.

Let's look at an example. Let's suppose that you have a design firm that designs machinery. Your machinery drawings are made up of many standard and nonstandard parts. You normally insert the parts (created and stored as component drawings) into a master drawing to create the finished machine drawings. One day you decide that one of the component parts would be better if a change were made. The problem is that you have 25 master drawings that contain that part. If you use the Insert command to place the component drawings into the master drawings, you have 25 master drawings to correct. If you use external references, each master drawing is automatically updated when it is next started or plotted.

External references can be useful for any type of application that uses component drawings, or even for multistation offices that have several people drawing parts of the work. There are several particular features of external references that we need to be aware of. Let's review these.

- The Xref command is used to insert a drawing as an external reference. The process is very similar to the Insert command.
- An external reference drawing is not stored as a block reference in the master drawing. Because of this, the original component drawing must be available for AutoCAD to scan and load into the master drawing. The drive and path to the component must be either maintained or, if it is moved, redefined.
- Since no block reference is loaded, the master drawing can have many inserted drawings without the disadvantage of excessive single drawing file size. Note, however, that the component drawing must also reside on disk or network for the master drawing to load.
- You can choose to "bind" the externally referenced drawing into the master drawing. This has the effect of turning the component drawing into a block, with all the standard aspects of a block.
- AutoCAD codes the external referenced drawing's layers for identification. Let's look at an example. Let's assume that we have a component drawing named "WIDGET". The Widget drawing has a layer named "DETAILS". If we insert the Widget drawing into a master drawing, a layer will be created with the name "WIDGET|DETAILS". This is a combination of the component drawing name and its layer name, separated by the vertical bar (|). This allows layers of the same name to coexist with other component drawings.

- External reference drawings may themselves contain other referenced drawings. For example, the Widget drawing may contain an external reference to another drawing named "COG". Thus, if you place the Widget drawing into a master drawing, you will also include the Cog drawing as an external reference. Both drawings will have to be "available" to the master drawing whenever it is started or plotted.
- Changes made in externally referenced drawings will be reflected in the master drawing either the next time it is started or at the time it is updated if the master drawing is current. This will continue until you bind the externally referenced drawing.

Let's look at the commands used with external references.

XREF COMMAND

The Xref command is used to insert externally referenced drawings into a master drawing. It is also used to bind a referenced drawing, remove a referenced drawing, reset the path to a referenced drawing, and update a referenced drawing.

When you select the Xref command, the following prompt is displayed:

Command: **Xref**
?/Bind/Detach/Path/Reload/Overlay/<Attach>:

Let's look at each of the Xref command's options and see how they work.

Attach (Adding an External Reference)

Attach is the default when you use the Xref command. Attach is used to insert a drawing. If you select Attach, AutoCAD prompts:

Xref to Attach <*default*>:

The default is the last drawing attached. If you would like to use a dialog box from which to choose the drawing name, enter a tilde (~) at the prompt. If AutoCAD detects a block name that matches, it issues an error message and terminates, since you cannot have a block reference and an external reference by the same name.

27-4 USING AUTOCAD

If AutoCAD detects that an external reference by that name is already present, it alerts you and proceeds. For example, the following command sequence shows an external drawing named "FLANGE" being reloaded as an external reference drawing.

Command: **Xref**
?/Bind/Detach/Path/Reload/<Attach>: **ATTACH**

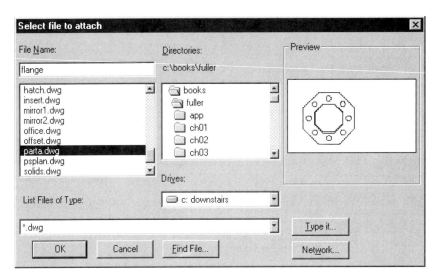

FIGURE 27-1   Attach an External Reference

Xref to Attach <FLANGE>: **FLANGE** (or [ENTER])
Xref FLANGE has already been loaded.
Use Xref Reload to update its definition.
Insertion point:

The remaining prompts are identical to the Insert command.

? (List External Reference Information)

Entering a question mark (?) in response to the prompt allows you to display a list of existing external references. AutoCAD prompts:

Xref(s) to list <*>:

The listing will include the name of the external reference, the pathname for the reference, and the total number of external references. The following is a sample listing.

Xref Name	Path
SPROCKET	/ACAD/DWGS/SPROCKET
COG21	/ACAD/PARTS2/COG21

Total Xref(s): 2

Bind (Bind an Xref to the Drawing)

Using Bind will bind the external reference drawing to the master drawing, causing it to become a regular block. AutoCAD prompts:

   Xref(s) to bind:

List the name(s) of the external references to bind.

> **NOTE:** You can bind all the external references by responding to the prompt with the wild-card asterisk (*).

If there are nested Xrefs (Xrefs to Xrefs), they will also be bound.

Detach (Remove an Xref from the Drawing)

The Detach option allows you to remove an external reference from the master drawing. This is equivalent to erasing all occurrences of a block, then purging its reference. When you select Detach, AutoCAD prompts:

   Xref(s) to detach:

Enter the name(s) of the Xrefs to detach. You can also respond with an asterisk (*) to detach all Xrefs.

Path (Change Path to an Xref)

The Path option is used to either view the path of an existing Xref or specify a new path for an Xref. As we learned earlier, an attached Xref must remain on the disk drive and directory where it was located when it was attached. The Path option lets you review or change it.

When you select the Path option, AutoCAD displays the prompt:

Edit path for which Xref(s):

You can enter a single Xref name or a list of names. If you enter an asterisk (*), AutoCAD lists all the Xrefs. If you respond with an [ENTER], the command is canceled and you will be returned to the Command: prompt.

Let's look at how you would change the path for an Xref.

Command: **Xref**
?/Bind/Detach/Path/Reload/<Attach>: **P**
Edit path for which Xref(s): **TEE?**
Xref name: **TEE7**
Old path: **C:\DWGS\PARTS\TEE7**
New path: **D:\NEWDWGS\PARTS\TEE7**

After you are finished, AutoCAD performs an automatic reload of Xrefs and updates the drawing.

Reload (Update External References)

When you first start a drawing that contains external references, each Xref is automatically reloaded. The Reload option is used to update one or more Xrefs without exiting and reentering the drawing. When you select the Reload option, AutoCAD prompts:

Xref(s) to reload:

You can enter a single Xref or a list. Entering an asterisk (*) reloads all the attached Xrefs. When you reload an Xref, AutoCAD scans the Xref(s), looking for updates. This can take a period of time. AutoCAD displays the message "Scanning..." while the scanning process is in progress.

Overlay (Unrepeated Xrefs)

When you overlay a drawing (rather than attach it) you avoid the possibility of recursive Xrefs. An overlaid Xref is not displayed when your drawing itself is Xref'ed.

When you select the Overlay option, AutoCAD prompts:

Select file to overlay:

and displays the Select File to Overlay dialog box.

XREF LOG

Your xref operations are recorded in a log that is kept by AutoCAD. The log file is written to disk as an ASCII file. You can view this log by using the Notepad text editor, or with a word processor. (You can also use the Utility/External Commands/Type facility in the screen menu.) The log file is named the same as the drawing file, except it has a file extension of .XLG. Thus, if your drawing file is named "MASTER1", the log will be named "MASTER1.XLG".

XBIND COMMAND

You have learned how to use Xref/Bind to bind an Xref drawing to a master drawing. There are times, however, when you may only want to bind a part of a drawing. For example, you may want to bind a linetype of layer to the drawing, without binding the rest of the externally referenced drawing. You can do this with the Xbind command.

When you select the Xbind command, AutoCAD prompts:

Command: **Xbind**
Block/Dimstyle/LAyer/LType/Style:

Enter the capitalized letter of the part of the Xref you wish to bind. When prompted for the name of the symbol, style, etc., you can enter a single name, or several names separated by commas.

The items are renamed when they are bound. You learned earlier how AutoCAD lists xref items, such as layers, with a vertical bar (|). AutoCAD removes the vertical bar and replaces it with two dollar signs ($) and a number that is usually 0. For example, a layer named "PARTA|HEXAGONS" will be named "PART0HEXAGONS." If there is already a layer by that name, AutoCAD will try "PARTS1HEXAGONS", and so forth. The number of characters must be 31 or less, or AutoCAD will terminate the command and undo the effects of the Xbind command.

EXERCISE

Let's use an external reference. The work disk contains a drawing named "PARTA". Figure 27-2 shows the drawing.

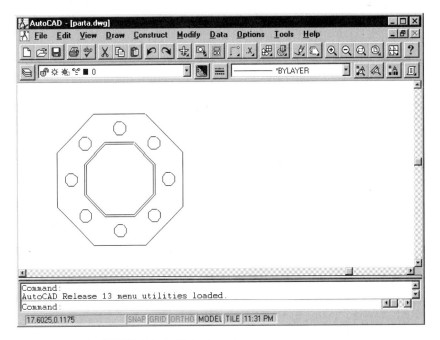

FIGURE 27-2 PARTA Work Disk Drawing

Let's start a new drawing first. Begin a new drawing named "MASTER1". Set the units to decimal, and the limits to 24,18. Let's now attach the "PARTA" drawing as an external reference. Use the Xref command as follows.

Command: **Xref**
?/Bind/Path/Reload/<Attach>: **A**
Xref to Attach<*default*>: **PARTA** *(Be sure to include any drive specifier and directory.)*
Insertion point: *(Place anywhere in drawing.)*
X scale factor <1>/corner/XYZ: *(ENTER)*
Y scale factor (default=X: *(ENTER)*
Rotation angle <0>: *(ENTER)*

The externally referenced drawing is now attached to the "MASTER1" drawing.

Now use the Save command to save the drawing. Open the drawing named "PARTA". This is the externally referenced drawing. Use the Erase command to erase the hexagons so the drawing looks like Figure 27-3.

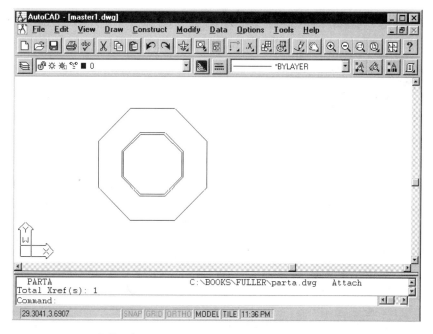

FIGURE 27-3 Edited PARTA Drawing

Use the End command to save and exit the drawing. Now start the "MASTER1" drawing again. Notice the messages about scanning the xref PARTA. When the drawing appears on the screen, you will notice that the edits you performed in the "PARTA" drawing were scanned and incorporated into the "MASTER1" drawing.

Continue and use the Xref/Bind command to bind the xref drawing "PARTA". End the drawing and start the "PARTA" drawing. Edit the drawing again in any way you want. End the drawing and start the "MASTER1" drawing again. Were the last edits incorporated?

OBJECT LINKING AND EMBEDDING

The Xref and XBind commands only work with other AutoCAD drawings. The Windows version of AutoCAD also lets you attach any kind of file using OLE, short for object linking and embedding. The file can be a text file, a spreadsheet, an image from any source (raster or vector), an animation, or even a sound file. Figure 27-4 shows an AutoCAD drawing with four OLE objects: (clockwise from upper left) a multimedia object (double-click to run it), a floor plan from another CAD package, a WordArt object, and a raster image from a paint program.

27-10 USING AUTOCAD

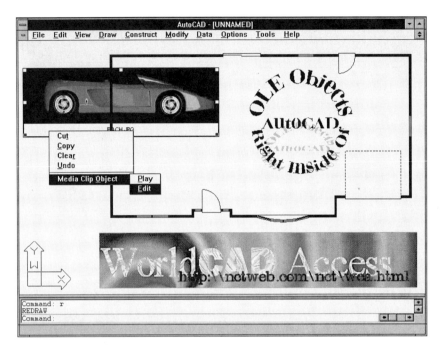

FIGURE 27-4 OLE Objects Placed in AutoCAD Drawing

Be careful, though: OLE objects in an AutoCAD drawing are only visible when loaded into Release 13 for Windows. The objects disappear when loaded into any other version of AutoCAD, such as the DOS version.

OLE allows you to place any AutoCAD drawing in just about any other Windows application. The drawing can be placed in a Word document (see Figure 27-5), an Excel spreadsheet, a Cardfile card, a paint program, or another CAD package. This feature is available in the Windows versions of AutoCAD Release 12, 13, and LT.

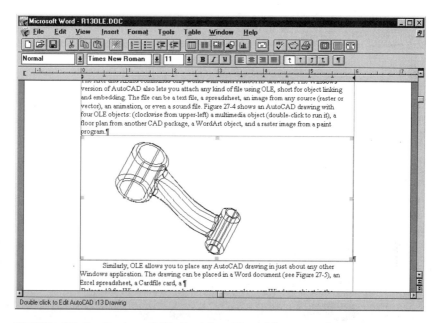

FIGURE 27-5 An AutoCAD Drawing Placed in a Word Document

The commands for placing an OLE object into the drawing are found in the Edit menu:

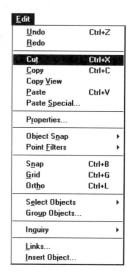

Paste [Ctrl+V]: Paste the object in the Windows Clipboard into the drawing, without your control.

Paste Special: Displays the Paste Special dialog box and gives you control over the paste process.

Insert Object: Select a Windows application to open, then place its document into the AutoCAD drawing.

Links: Change the settings of linked OLE objects.

Similarly, the commands for exporting the AutoCAD drawing as an OLE object are in the Edit menu:

Copy [Ctrl+C]: Copy a portion of the drawing to the Windows Clipboard.

Copy View: Copies all objects visible in the current viewport to the Windows Clipboard.

Cut [Ctrl+X]: Cut objects out of the drawing and place them in the Clipboard.

TUTORIAL

Placing an OLE Object in AutoCAD

Let's place an object into a drawing. Here's how:

1. Start AutoCAD with a new drawing.
2. From the menu bar, select "Edit | Insert Object".
3. The Insert New Object dialog box appears. You can select objects in one of two ways:
    - **Create New:** Start the application, then create the object or load a file.
    - **Create from File:** Select the file name, which launches an associated program.

   With the Create New option, the dialog box lists the names of all applications that Windows knows can provide objects. Scroll through the list to find a suitable application, such as "Equation" for placing formula text in the drawing and "ClipArt Gallery" for selecting a piece of clip art, such as your firm's logo. (The applications listed by the dialog box depend on the applications loaded into your computer.)

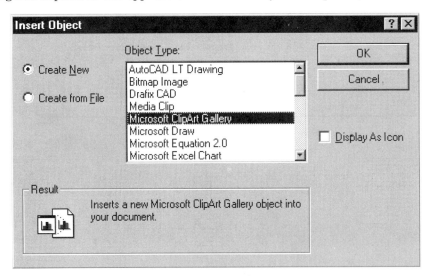

FIGURE 27-6 Insert Object Dialog Box

   (With the Create from File option, the dialog box lets you select a file name. When you do, Windows launches the software application that it thinks is best suited for the file.)

4. After selecting an application, click OK.
5. The object appears in the AutoCAD drawing. AutoCAD always places OLE objects in the upper left corner.

   (For some older applications, Windows launches the application, letting you open and edit the object. When you are ready to leave the application, select "File | Update" from the menu bar; some applications have a "File | Exit and Return to AutoCAD" option.)

Working with OLE Objects in AutoCAD. Once the OLE object has been placed in the drawing, AutoCAD provides a few editing commands.

Resize the Object. Click on the OLE object. Its border is surrounded by a dashed rectangle (OLE objects can only be rectangular). Notice the eight small black (or white) squares that surround the object. These grips let you resize the object. Move the cursor over one of the squares, hold down the left mouse button, and drag the rectangle to a different size.

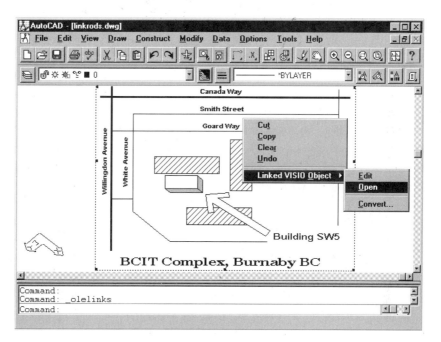

FIGURE 27-7 OLE Object in Drawing

Move the Object. To move the object from the upper left corner to another location in the drawing, click on the object, then drag it into place.

Additional Editing Commands. Move the cursor over the object, then press the rightmost mouse button. AutoCAD displays a cursor menu with five primary options:

1. Cut the object out of the drawing to the Windows Clipboard.
2. Copy the object to the Clipboard.
3. Clear the object (erase it).
4. Undo the previous cursor menu operation.
5. The fifth option on the cursor menu is specific to the object. In Figure 27-7, the object is a drawing created by the Visio Technical software. There are three options: edit or open the drawing (start Visio and load the file) or convert the drawing to AutoCAD format. Because the Visio drawing is "linked", any changes you make to the drawing while in Visio are automatically updated back in AutoCAD.

If the object were a multimedia object, it could be either played (view the animation) or edited (launch the source application, a morphing program in this case).

Changing OLE Links. To change the nature of the link between the OLE object in AutoCAD and the originating software, select the Links option from the Edit menu. AutoCAD displays the Links dialog box and the names of all linked OLE objects.

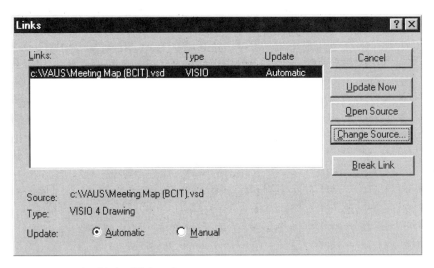

FIGURE 27-8 Links Dialog Box

Just as the Xref command lets you change the links to externally referenced drawings, the Links dialog box lets you change the links to OLE objects. The dialog box has these options:

> **Update:** Automatic updates occur whenever the source document is changed; manual updates require you to click on the Update Now button.
>
> **Open Source:** Starts the application that provided the OLE object.
>
> **Change Source:** Lets you change where the object comes from, such as a new file or application or subdirectory.
>
> **Break Link:** Removes the information that links the object back to its source.

Converting OLE Objects to AutoCAD Format. You must break the link if you want to covert the object into AutoCAD format. This lets you see the object in versions of AutoCAD other than Release 13 for Windows.

To do this, first break the link via the Links dialog box. Then, right-click on the object and select "Convert Picture Object." AutoCAD displays the Convert dialog box.

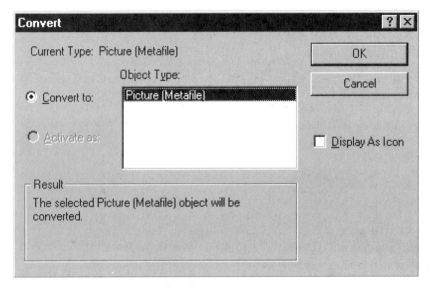

FIGURE 27-9 Convert Dialog Box

This dialog box typically gives you one or more choices of how to handle the conversion.

TUTORIAL

Placing an AutoCAD Drawing as an OLE Object

OLE can be used in reverse; an AutoCAD drawing can be placed into the document of another Windows application. For example, you can use Cardfile to create a simple drawing manager by placing each AutoCAD drawing on a card, along with the drawing's file name. Here's how:

1. Start Cardfile.

2. From the menu bar, select "Edit | Index" to give the card a name on the index line, such as the drawing's file name: "Office.Dwg."

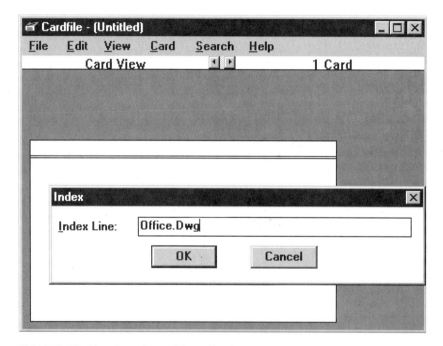

FIGURE 27-10 Creating a New Card

3. Switch Cardfile from text to graphics mode with **Edit|Picture**. The cardfile won't look any different at this point.

4. Place the AutoCAD drawing on the card by selecting **Edit|Insert Object**.

5. When the Insert New Object dialog box appears, select "AutoCAD Drawing" and click OK.

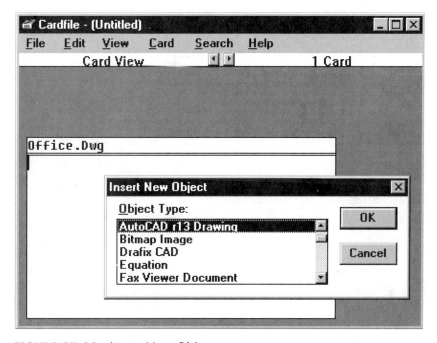

FIGURE 27-11   Insert New Object

6. Windows starts AutoCAD. Open the drawing Office.Dwg provided on the diskette.

7. After AutoCAD finishes loading the drawing, select "Update (Untitled) in Cardfile" in the File menu. The Update command only appears when AutoCAD is being operated by "remote control" from another Windows application via OLE. And via OLE, Windows places an image of the drawing in Cardfile.

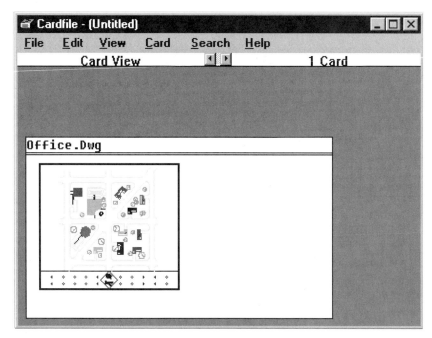

FIGURE 27-12 AutoCAD Drawing in Cardfile

8. Now that Cardfile is set up, you launch AutoCAD (together with the drawing) by simply double-clicking on the drawing's picture in Cardfile. The OLE link feature saves you from the manual updating process required when using traditional cut and paste.

CAUTIONS WITH OLE

While Release 13 has made great strides in compatibility with Microsoft's OLE, there are a couple of drawbacks to using it.

OLE is resource hungry. That means it needs lots of RAM and CPU power to work properly. Even 16MB of RAM can't prevent AutoCAD from crashing occasionally during OLE operations.

OLE is specific to Windows. That means OLE objects you place in an AutoCAD Release 13 drawing under Windows cannot be viewed when you bring the drawing up in the DOS or Unix versions of AutoCAD, nor can OLE objects be seen in earlier releases of AutoCAD.

Still, the potential of OLE makes it worthwhile to spend some time experimenting with AutoCAD and other Windows applications. For example, AutoCAD LT and AutoSketch for Windows have some OLE capabilities, as do Write and many other Windows applications.

CHAPTER REVIEW

1. Explain the difference between an inserted drawing and one placed as an external reference.

2. Is an external reference stored as a block?

3. Can an external reference contain another external reference?

4. What are some advantages of using external references?

5. What command is used to place an external reference drawing?

6. How would you obtain a listing of the external references placed in a drawing?

7. How would you convert an external reference to a block?

8. Under what conditions would you want to convert the external references to blocks?

9. How would you remove an external reference from a drawing?

10. Why must an externally referenced drawing remain in its original drive and directory?

11. How would you change the drive and directory of an externally referenced drawing?

12. What option is used to update externally referenced drawings in the master drawing?

13. How would you write a log file that describes the external reference activity to disk?

14. Can you bind only part of an externally referenced drawing to the master drawing? Explain.

15. What does OLE stand for?

16. What purposes can OLE be used for?

17. Can OLE objects be viewed in any version of AutoCAD?

18. What are some drawbacks to placing OLE objects in an AutoCAD drawing?

CHAPTER

28

CUSTOMIZING AUTOCAD

At some point, all serious AutoCAD users will want to learn to customize their program. Increased drawing ease and performance will result from a properly customized system. After completing this chapter, you will be able to:

- Control the look and performance of the AutoCAD program with the many variable settings.
- Control many of the display and adjustment settings such as the display of drawing blips, and the target box sizes.
- Control display characteristics such as the view resolution and regeneration rules.
- Redefine existing commands.

SETTING SYSTEM VARIABLES

The Setvar (Set System Variables) command allows you to change AutoCAD's system variables.

The system variables control default settings for the drawing editor. Values for such items as default aperture size, global linetype scale factor, and other items are stored here. Some variables are changed by commands, some are "read only", and some may be set by using the Setvar command.

To change a setting, issue the Setvar command.

Command: **Setvar**
Variable name or ? <default>: *(Enter variable name.)*

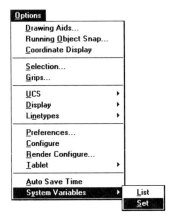

NOTE: Many variables can be executed directly from the command line without using the Setvar command first. This can be performed with variables that do not match AutoCAD command names.

After entering the variable name, you may enter a new value. If you respond to the prompt with a "?", AutoCAD prompts:

Variable(s) to list <*>:

If you respond with an ENTER, all the variables are displayed. Some values are defined as "read only". Entering one of these values will result in a response in the form of:

variable-name = current (read only)

An example of this would be:

Command: **Setvar**
Variable name or ?: **AREA**
AREA = 0.0000 *(Read only.)*

If you enter a changeable variable, AutoCAD will prompt:

New value for (variable name) <*default*>:

Using Setvar While in a Command

There are times when it is desirable to change a system variable while in a current command. AutoCAD allows you to do this by issuing the special code of an apostrophe (').

An example would be:

Command: **Line**
From point: **'SETVAR**
>>Variable name or ?: **TRACEWID**
>>New value for TRACEWID <default>: *(Enter new value.)*

You may maintain the default value by entering a null response to the prompt.

After you have entered the new value, AutoCAD returns you to the current command:

Resuming LINE command.
From point:

You may now resume the current command.

DISPLAYING BLIP MARKS

The marker that is left on the screen by your drawing activity is called a *blip*. Blips do not plot and may be thought of as "push pin" reference points to aid you. You turn the drawing of blips on and off by using the Blipmode command. To use the command, enter the following:

Command: **Blipmode**
On/Off <current>:

If the blipmode is on, you may remove blips on the screen by issuing a Zoom, Pan, Redraw, or Regen command.

The initial setting of the Blipmode command is determined by the prototype drawing. You may change the setting at any time and as often as you desire. AutoCAD remembers the setting when you end the drawing and retains it as the initial setting when you reenter the drawing at a later time.

SETTING APERTURE SIZE

The Aperture command is used to adjust the size of the target box used in object snap modes.

The box size is described by the number of screen pixels specified. To set this number, enter:

Command: **Aperture**
Object snap target height (1-50 pixels) <default>:

The last aperture size will be remembered when you reenter the drawing.

The aperture is only used with Object Snap and is not the same as the pick box used with object selection. The pick box can be changed by using the PICK BOX system variable.

CONTROLLING DRAWING REGENERATIONS

Some functions performed with AutoCAD cause an automatic regeneration. This regeneration reorganizes the data on the screen to ensure that all information is current. It is possible that you may have need to perform several operations at one time that will force a regeneration each time. Since each regeneration takes some time to perform, AutoCAD provides a command to warn you before regenerations are performed by some operations.

Command: **Regenauto**
On/Off <current>:

The initial status of the Regenauto mode is determined by the prototype drawing. You may change it at any time and as many times as necessary. The last setting is remembered by AutoCAD and is restored when you reenter the drawing at a later time.

If Regenauto is Off and a regeneration is required, AutoCAD will prompt:

About to regen — proceed? <Y>

SETTING THE VIEW RESOLUTION

The Viewres command controls the fast zoom mode and resolution for circle and arc regenerations.

AutoCAD regenerates some zooms, pans, and view restores. This regeneration can sometimes, depending on the complexity of the drawing, take a great amount of time. The Fast Zoom mode allows AutoCAD to simply redisplay the screen wherever possible. This redisplay is performed at the faster redraw speed. (Some zooms that are more extreme still require a Regen.)

AutoCAD also calculates the number of segments that is required to make circles and arcs "smooth" at the current zoom. (Circles and arcs are made up of many short line segments for display purposes.) The Viewres command allows you to control the number of segments used. Using fewer segments speeds regeneration time, but trades off screen resolution. Although the displayed circles and arcs are not as smooth, plotter and printer plots are not affected.

To use the Viewres command, enter:

Command: **Viewres**
Do you want fast zooms? <Y>: *(Choose Y or N.)*
Enter circle zoom percent (1-20000) <default>:

Entering **N** at the first prompt will cause all zooms, pans, and view restores to regenerate.

The default value for the circle zoom percent is 100. A value less than 100 will diminish the resolution of circles and arcs, but will result in faster regeneration times.

A value greater than 100 will result in a larger number of vectors than usual to be displayed for circles and arcs. This is not important unless you zoom in a great amount.

For example, if you will be zooming in at a factor of 10, setting the circle zoom percent to 1,000 will result in smooth circles and arcs at that zoom. If you want to maintain a smooth display and still achieve the optimum regeneration speed, set the percent equal to the maximum zoom ratio you intend to use.

Regardless of the setting, AutoCAD will never display a circle with fewer than eight sides. On the other hand, AutoCAD will not display any more circle or arc segments than it calculates to be necessary for the current zoom. If the circle uses less than two screen pixels at the maximum zoom magnification that does not require a regeneration, the circle will be displayed as a single pixel.

If you wish to show the drawing to others at the maximum resolution, use a smaller percentage while drawing, then change the percentage and perform a regen. This allows you to perform the drawing at fast regeneration times, then redisplay the drawing at its optimum resolution for presentation.

REDEFINING COMMANDS

An existing AutoCAD command may be redefined to suit customized purposes. The UNDEFINE command is used to facilitate this. Undefine is used extensively by AutoCAD third-party programmers to create new commands for their applications. You may want to use Undefine to customize some existing commands, making the use of AutoCAD more applicable to your particular tasks. See an example of using the Undefine command in the Exercise section below.

EXERCISE

This chapter has been an introduction to customizing AutoCAD. Typically, customizing requires the use of AutoLISP, the programming language of AutoCAD (further discussed in chapters 40 and 41).

Let's say you want to add a line to remind one of your CAD operators to place blocks in the proper subdirectory. You could use AutoLISP and the Undefine command to accomplish this. Let's see how this would work. Start a new drawing of any name. From the "Command" prompt, enter the following as indicated in boldface type:

Command: **(defun c:BLOCK ()**
1>**(princ** *"Remember to specify the correct subdirectory* **\n"**)
1> **(command ".BLOCK"))**
C:**BLOCK**
Command:

The Block command has been redefined to display the reminder message. The original Block command is not actually destroyed. You may still use it by simply preceding it with a period, such as .BLOCK. NOTE: After the above message was prompted, the original Block command was initiated .BLOCK.

Now undefine the Block command.

>   Command: **Undefine**
>   Command name: **Block**

Now issue the Block command to see the message.

>   Command: **Block**
>   Remember to specify the correct subdirectory
>   .BLOCK Block name (or?): nil
>   Block name (or?):

Let's now redefine the Block command. This will return it to its original form.

>   Command: **Redefine**
>   Command name: **Block**
>   Command:

The Block command is now redefined to its original form and the message will not be displayed if it is subsequently issued.

CHAPTER REVIEW

1. What command can be used to set AutoCAD's system variables?

2. Under what condition can variables be entered directly from the command line?

3. How could you change a variable while actively in a command?

4. What is a blip?

5. How do you turn the blips on or off?

6. In what increment is the aperture setting measured?

7. Why would you sometimes want to set a low Viewres value?

8. What difference in the displayed drawing would you notice if you set the Viewres low?

9. Describe a good example of redefining a command.

CHAPTER 29

CUSTOMIZING MENUS AND ICONS

Using custom menu systems can multiply the productivity of any CAD system. You can buy expensive systems or you can become proficient at designing your own. After completing this chapter, you will be able to:

- Utilize AutoCAD menu systems.
- Understand the anatomy of a menu system.
- Write the files necessary for the different types of AutoCAD menus.
- Construct special types of AutoCAD menus such as icon boxes and custom pull-down menus.
- Change icon buttons on the toolbar.

CUSTOM MENUS

Custom menus are an exciting part of AutoCAD. You may prepare a menu that is particular to your type of work, or one that just suits your style (or both!).

A menu is nothing more than a text file. AutoCAD reads the item from the menu and executes it as though it were entered from the keyboard. Before we jump all the way in, let's learn two basic rules and look at a simple menu.

SIMPLE MENUS

Menu items are arranged one to a line. Each line contains one or more items that will be "typed" if chosen.

Menus are constructed with a word processor in non-document mode or a line editor. The file must have a .MNU extension in order to be loaded by AutoCAD. The following is a simple menu that could be written for AutoCAD:

```
LINE
ARC
CIRCLE
TRACE
REDRAW
ZOOM
COPY
MOVE
```

If you choose any of these from the screen, it will be executed as though you typed it from the keyboard.

SCREEN DISPLAY

Sometimes it is desirable for the menu item to have a special listing on the screen menu. Consider the case of the Quit command. You may want the screen to display DISCARD instead of QUIT.

You may include text that AutoCAD will not execute by enclosing it in brackets ([ ]). The first eight characters within the brackets will be displayed on the menu screen and the items immediately following the closing bracket will be executed. Consider the following menu items:

```
[FAT LINE]TRACE
[TARGET]APERTURE
[RULER]AXIS
[BACKWARD]MIRROR
```

In the second line, TARGET will appear on the screen and APERTURE will be executed. Be sure that you do not put a space between the second bracket and the command, otherwise AutoCAD will interpret the space as if you had pressed ENTER.

MULTIPLE MENUS

AutoCAD allows you to store, concurrently, several device menus in one menu file. Each device menu is compartmented in its own section and marked by a beginning label. The following table shows the section labels and the associated devices:

***SCREEN	Menu area on screen
***BUTTONS	Pointing device buttons
***TABLET1	Tablet menu area one
***TABLET2	Tablet menu area two
***TABLET3	Tablet menu area three
***TABLET4	Tablet menu area four
***AUX1	Auxiliary device
***ICON	Icon menu area
***POPn	Pull-down menu (where *n* is a number between 0 and 16).

Notice that each label starts with three asterisks. The asterisks, along with the label name, tell AutoCAD that this is the start of the menu items for that particular menu area or device.

The items that follow will be contained in the associated section until another label is listed or the end of the file occurs. The following short menu shows two sections: a screen menu and an auxiliary menu for a device such as a function box:

```
***SCREEN
LINE
ARC
CIRCLE
[BACKWARD]MIRROR
***AUX1
LINE
TRACE
ZOOM
```

SUBMENUS

You have already noticed that the default AutoCAD menu is actually made up of many menus that are displayed almost magically on the screen as you make your choices. Consider what occurs when you choose Line from the screen menu. You must first choose Draw from the Main Menu. You are then presented with a special Draw menu. You then choose Line from this menu and are presented with a special Line menu. These menus are called submenus.

In order to construct a submenu, you must use another section label. The structure for a submenu is:

    **name

Every menu item that is listed after the submenu section label through the next section label belongs to that section. Each submenu label must be named differently, even though it occurs under a different menu or device label.

When a submenu is displayed, it will replace every item of the currently displayed menu down to the end of the submenu file. It is therefore possible that a submenu may only replace a part of the previous menu. At times, this may be desirable. If you do not want this to occur, fill up the number of screen items on your display with blank lines.

You may specify which item number (down from the top) the submenu replaces by designating a number describing the position after the submenu name. For example:

    **SUB-A 5

will display the submenu labeled **SUB-A starting at the fifth position down on the currently displayed menu, replacing the remaining items to the bottom of the submenu file. Using a negative number will cause the same result, except from the bottom of the displayed menu.

LINKING MENUS

You have learned about section labels that separate menus in the same file and submenus that are separate screen lists. Now we need to learn how to navigate.

All those submenus are nice, but you need a way to move between them. If you select the Circle command from the screen menu, you will then want a submenu that contains all the commands for working with circles.

AutoCAD provides a way to "jump" to a named submenu. The following format is used:

    $section=submenu

The section refers to the section label. The sections may be referenced by using the following letters:

S	SCREEN menu
B	BUTTONS menu
T1	TABLET area one
T2	TABLET area two
T3	TABLET area three
T4	TABLET area four
A1	AUXILIARY device
Pn	POPDOWN menu n
P0	CURSOR menu

If you wanted to access a submenu named **subline in the Screen section, the correct entry would be:

$S=SUBLINE

Notice that the double asterisk is not used when referring to the submenu.

AutoCAD also provides a way to return to the last menu. The format for this is:

$S=

This returns you to the last screen menu item. The number of *last* (nested) menus allowed is eight.

At first, submenus can be confusing. You may use a word processor to print out AutoCAD's menu for an example. The name of the file is ACAD.MNU. If you do not have a word processor, use the print echo from DOS (CTRL P) and enter TYPE ACAD.MNU to print the menu. (Be sure to turn off the echo by entering CTRL P again after finishing.)

MULTIPLE COMMANDS IN MENUS

Commands may be linked together to perform several functions at once. This is called a macro. To do this, you must have a good understanding of AutoCAD's command sequences. The following special input items are used for this purpose:

Space: A space (or blank) is read as an ENTER.

End of line: AutoCAD automatically inserts a blank at the end of each menu line. A blank is used interchangeably with the ENTER in most commands.

Semicolon (;): The semicolon is used if an ENTER is desired instead of a blank.

Backslash (\): A backslash is used where user input is desired. The command will pause and await your input.

Plus mark (+): A plus mark is used to continue a long command string to the next line. If a plus mark is not present at the end of the line, AutoCAD will insert a blank.

Let's look at an example. Suppose you wanted to set up a new drawing by setting limits of 0,0 and 36,24, perform a zoom-all, and turn the grid on. The following menu sequence would perform this:

[SET LIM]LIMITS 0,0 36,24 ZOOM A GRID ON

AutoCAD would read this as:

LIMITS<enter>0,0<enter>36,24<enter>ZOOM<enter>A<enter>GRID<enter>ON<enter>

Special functions may also be performed. Let's suppose that you wanted to insert a window in a solid wall made of a Trace line. The window is a drawing called "WIN-1" and is stored in the C: drive. The window is 36 units in length, and the base point is at the right end of the window and four units down.

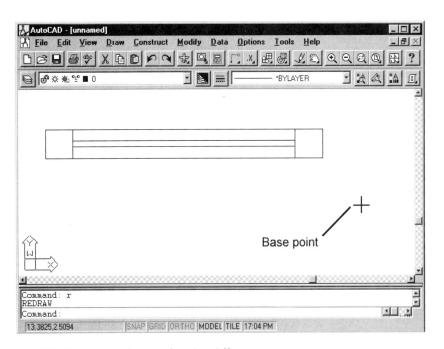

FIGURE 29-1 Window with 4 in. Offset

The following menu string could be used to insert the window in the wall:

[36" WIN]BREAK \@36,0 INSERT C:WIN-1 @ 1 1 0

AutoCAD would break the wall, leaving a 36-unit opening and insert the window in it.

LOADING MENUS

To load a menu, use the Menu command:

Command: **Menu**

A Select Menu File dialog box is displayed.

A menu has a file extension of .MNU until the first time it is loaded. When loaded, AutoCAD compiles a menu for faster operation. A new copy of the menu is made with a .MNX file extension. This is the menu used by AutoCAD. If you edit the menu, you will need to edit the one with the .MNU extension. To use the reedited menu, you do not need to delete the menu of the same name with the .MNX extension. AutoCAD senses the change and automatically recompiles the menu.

TABLET MENUS

A digitizing pad may be used with a tablet menu. You may specify up to four tablet menus and a drawing area. Figure 29-2 shows a typical tablet menu set up.

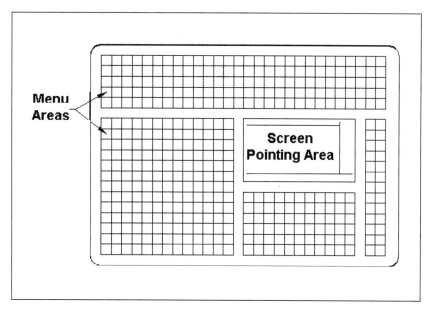

FIGURE 29-2 Tablet Menu Template

The four menu areas are designated TABLET 1 through 4. The menus are made up of smaller boxes of equal dimension in rows and columns. The menu boxes are labelled from left to right and top to bottom. The label corresponds to the item in the menu under the appropriate section label. That is, the first command to appear will be contained in box A1, and so forth. You may arrange the menus in any fashion you wish. The CFG option in the TABLET command is used to configure the menus and drawing.

> Command: **Tablet**
> Option (ON/OFF/CAL/CFG): **CFG**
> Enter number of tablet menus desired (0-4) <*default*>:

If you have already used the tablet menu, AutoCAD will prompt:

> Do you want to realign tablet menu areas? <N>

AutoCAD will proceed and prompt:

> Digitize upper left corner of menu area x:
> Digitize lower left corner of menu area x:
> Digitize lower right corner of menu area x:

After entering the descriptive points, AutoCAD prompts:

> Enter the number of columns for menu area x:
> Enter the number of rows for menu area x:

You may now define the drawing area. You are prompted:

> Do you want to respecify the screen pointing area? <N>

If you reply **Y**, AutoCAD prompts:

> Digitize lower left corner of screen pointing area:
> Digitize upper right corner of screen pointing area:

PULL-DOWN MENUS

Pull-down menus may be written that are accessed from the top screen menu bar. These are obtained by moving the cursor to the status line area at the top of the screen. When you do this, the status lines are replaced by a listing of menus across the top of the screen. Placing the cursor over one of these and clicking pulls down the corresponding menu.

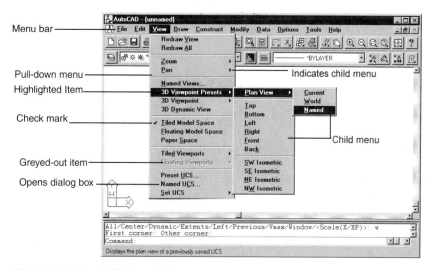

FIGURE 29-3 Pull-Down Menu

Pull-down menus are designated as "POP1" through "POP16". They are written in the same fashion as screen and tablet menus, with some exceptions. Each menu is handled as a separate menu; using, for example, ***POP1 as the menu area for the first pull-down menu. The equivalent to the $S= command in the pull-down menu is the $Pn= command, where *n* is the name of the pull-down menu to access. In addition, the special command $Pn=* line is used to automatically pull down the menu desired.

Let's look at an example of a pull-down menu (designated as POP3) that is to be automatically pulled down when accessed from a screen menu. You could use the following line in the screen menu area.

[PARTS]$p3=parts $p3=*

This would pull down the submenu named "PARTS" from the POP3 menu (listed in the written menu under ***POP3). The menu would then be forced down by the $p3=* command.

When writing the POPn menu sections, the first line under each area is used as the header bar title (listing in the top screen area). These listings will be displayed across the top of the screen. The menu under each will be as wide as the longest item in the corresponding menu. Keep in mind that lower resolution screens are capable of displaying only 80 columns in width. If the total width of the menus is longer, they will be truncated.

You may provide separation between items in the menu by placing a separator line between them. This is done by placing a line as follows.

[--]

The two hyphens expand to the width of the menu and provide the separator line.

A menu item label may be displayed as "not available" by beginning it with a tilde (~). An unavailable item typically denotes an item that is not currently active. This could be used if a menu is in progress and will be completed at a later time. It could also be used for minor selections. If the item is followed by a valid string, it will execute normally.

ICON MENUS

Icon menu listings are placed under a ***ICON heading in the menu text file.

Icon menus use slides to display the graphic part of the icon box. The slides are normal AutoCAD slides made with the Mslide command. Some displays are not capable of displaying icon boxes on the screen. On these systems, selecting a box will have no effect.

The first line of the icon menu will be the title of the icon box. The title is displayed at the top of the box.

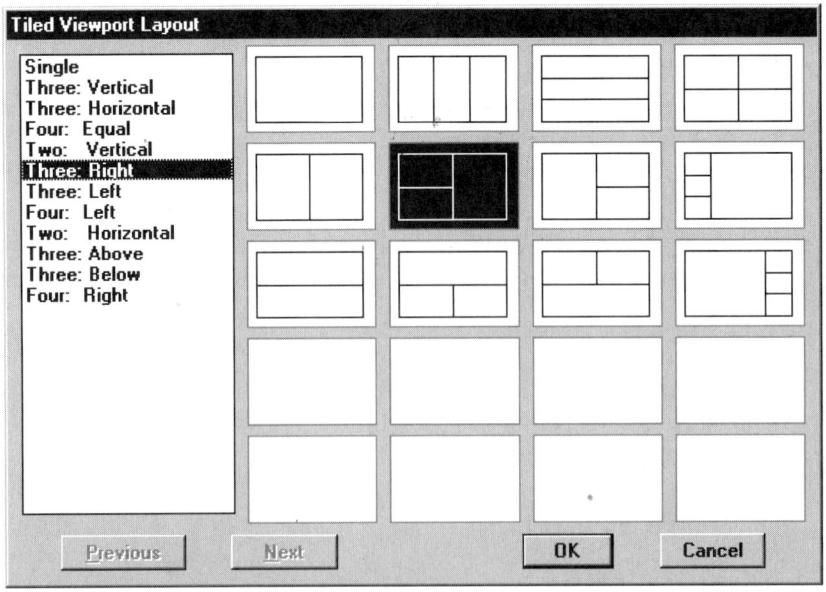

FIGURE 29-4 Icon Menu

Let's look at an example of a menu text listing for an icon menu.

```
***ICON
**FURNITURE
[Living Room]
[sofa]^Cinsert sofa
[lchair]^Cinsert lchair
[ctable]^Cinsert ctable
[lamp]^Cinsert lamp
[bookcase]^Cinsert bcase
[ Cancel]^C
[ Bedroom]$i=bdrm $i=*

**BDRM
[Bedroom]
[Bed]^Cinsert bed
```

Notice how the format is similar to the standard screen menus and icon menus. Submenus are used to create different icon boxes under the main icon heading of ***ICON. The slide names of the items to be displayed in the icon box are enclosed in brackets ([ ]). The slide names should be the same as the name listed if it were viewed with the Vslide command. An executable string may be prefaced by ^C (executes as [ESC]) to cancel any command that is in progress when the item is selected.

If the first character of an item is blank, the balance of the string in the brackets will be displayed as text characters instead of a slide. Thus, the lines in the preceding sample menu listed as Cancel and Bedroom will be displayed in the icon box as text items. You may want to use Cancel, Exit, Next, Previous, or the name of another icon box in the design of an icon box and display only text for that purpose.

The special string $i=* causes the icon box to be displayed on the screen. A prompt to display an icon box may be placed in any menu section (***SCREEN, ***POPn, etc.), but may not be executed from the keyboard. For example, the following line could be placed in a pull-down menu to execute the icon box described by the previous menu.

[Living Room]$i=furniture $i=*

The number of items displayed in the icon box is dependent on the number of items in the menu section for that box. You may have 4, 9, or 16 items displayed. A selector box will be displayed next to each item in the icon box. When an icon box is displayed, an arrow appears. The arrow may be moved by the input device. Moving the arrow over a selector box causes a frame to be displayed around the item. Clicking the input device causes the item to be selected and the corresponding text string in the text menu is executed.

Selecting Slides from Libraries

Slides for icon menus may be selected from slide libraries. This is an efficient way to manage groups of slides for icon menus.

In order to select a slide from a slide library, the name of the library must be listed in the menu before the slide name. Let's assume the slides for the icons we used in the earlier menu were stored in a library named "FURNLIB" (furniture library). The same menu listing would look like the following.

```
***ICON
**FURNITURE
[Living Room]
[furnlib(sofa)]^Cinsert sofa
[furnlib(lchair)]^Cinsert lchair
[furnlib(ctable)]^Cinsert ctable
[furnlib(lamp)]^Cinsert lamp
[furnlib(bookcase)]^Cinsert bcase
[Cancel]^C
[Bedroom]$i=bdrm $i=*

**BDRM
[Bedroom]
[furnlib(Bed)]^Cinsert bed
```

Designing Icon Boxes

There are several guidelines that ensure good icon box design.

First, always be sure to design a way out. If an icon box is displayed by mistake, or you change your mind about making a selection, you need to be able to cancel the box.

If there are more selections under a heading than an icon box has room for (the maximum icons in one box is 16), you may place a "Next" button to display a subsequent box with more choices. In this case, you may also want to put a "Previous" choice.

When preparing slides, keep the slides simple. The icon box will take less time to generate onto the screen. You may want to construct outline frames of complex slides just for this purpose. You can also turn off the fill when preparing the slides to cut down on display time.

Make the slides as large as possible. The screen area for the slide is very small. Keeping the slide big and simple makes selection easier.

CUSTOMIZING THE TOOLBAR

The Windows version of AutoCAD Release 13 has toolboxes, which are collections of buttons with icons for buttons. Clicking on an icon executes a command; pausing the cursor over the icon displays a tool tag explaining the icon's purpose. You can create your own customized icons. To do this, AutoCAD has two commands: Toolbar and TbConfig.

The Toolbar command displays (and hides) one or more or all toolbars. Be careful, though: the ALL Show option completely fills your screen! The Toolbar ALL Hide command gets rid of them again. To open just a couple of toolbars, select "Tools I Toolbars" from the menu bar.

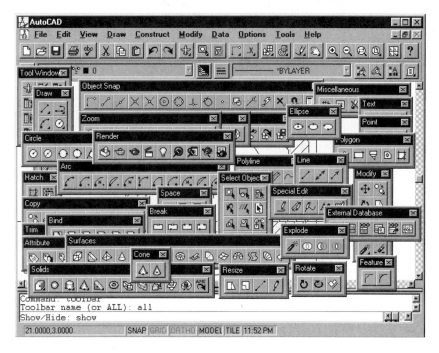

FIGURE 29-5 Opening All of Release 13's Toolbars

THE TBCONFIG COMMAND

The TbConfig command handles everything about customizing toolbars:

- macro behind the icon
- tooltip
- location of the icon in the toolbar
- look of the icon

Unfortunately, the customization process is not at all intuitive and involves several dialog boxes. Here's how to do it, step by step:

1. Type the TbConfig command. AutoCAD displays the Toolbars dialog box, which lists the names of all 50 currently defined toolbars. At the bottom of the dialog box, you can change whether AutoCAD displays small or large icons, and whether tooltips are displayed.

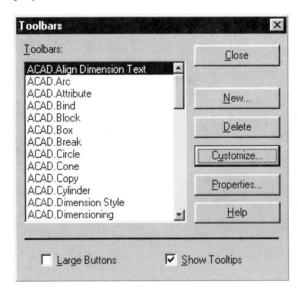

FIGURE 29-6 Toolbars Dialog Box

2. Click on the New button to create a new toolbar. The New Toolbar dialog box appears.

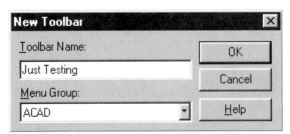

FIGURE 29-7 New Toolbar Dialog Box

3. Type a name for the toolbar, such as "Just Testing" and click the OK button. A tiny, empty toolbar appears on the screen with the name Just Testing on the title bar; you'll probably see the first three letters, "Jus".

4. You've created a new toolbar but it is empty. Now you fill it with icons. Click on the Toolbar dialog box's Customize button. AutoCAD displays the Customize Toolbars dialog box.

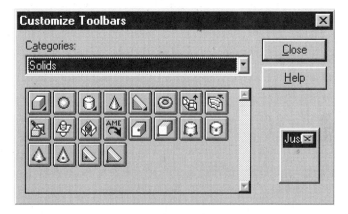

FIGURE 29-8 Customize Toolbars Dialog Box

5. The Categories drop list box has the names of AutoCAD's groups of icons, such as Object Properties, Standard, and Solids. You may have to hunt around for an icon by selecting one category after another. AutoCAD displays the group of icons associated with each category.

6. Drag an icon from the Customize Toolbars dialog box to the Just Testing toolbar. ("Drag" means to hold down the left mouse button over the icon, then move the icon to its destination, and let go of the button.) As you drag icons, your custom toolbar expands to accommodate the icons.

FIGURE 29-9 Just Testing Toolbar

If the icon has a small triangle, that means it is a "flyout." A flyout displays one or more additional icons.

7. You now have a new toolbar with several icons in it. If you want, you can now customize the look and meaning of each icon. To do so, click on the icon using the right mouse button. AutoCAD displays the Button Properties dialog box.

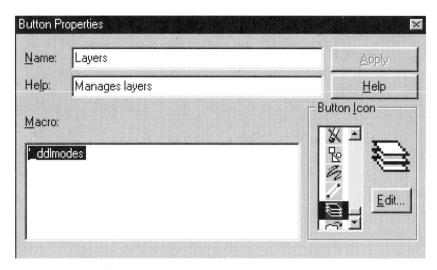

FIGURE 29-10 Button Properties Dialog Box

8. To change the action AutoCAD performs when you click on the icon, change the text of the Macro area.

CUSTOMIZING MENUS AND ICONS **29-17**

9. To change the look of the icon, click on the Edit button. AutoCAD displays the Button Editor. The button editor lets you edit the icon or create a new one. The tools along the top let you (from left to right) color individual pixels, draw a straight line, draw a circle, and change a pixel to make it unavailable (erase). You can insert any BMP (bitmap) file as the icon. When done, click on the Save and Close buttons.

FIGURE 29-11 Button Editor Dialog Box

10. Back in the Button Properties dialog box, click on the Apply button. AutoCAD saves the changes. Test your customization by clicking on the icon in the Just Testing dialog box to make sure it works as you expect.

When done, close all remaining dialog boxes related to toolbox customization.

CHAPTER REVIEW

1. How many characters can be displayed on a single line on an AutoCAD screen menu?

2. How many different drawing devices can be served by an AutoCAD menu?

3. How do you designate a submenu within the menu file?

4. What is a macro?

5. What command is used to load a new menu?

6. What is a tablet menu?

7. What are the maximum menu areas on a tablet menu?

8. How are the graphics in an icon menu produced?

9. What is meant if a pull-down menu selection is unavailable?

10. What is the range of drawings that can be displayed in an icon box?

11. What is a toolbar for?

12. How is a Tooltip helpful?

13. What command is used to open all toolbars?

14. What is a flyout?

CHAPTER 30

INTRODUCTION TO AUTOCAD 3D

One of the most exciting parts of AutoCAD is 3D. Before you can start constructing 3D drawings, you must first become familiar with the many concepts of three-dimensional drawing construction. After completing this chapter, you will be able to:

- Utilize the three-dimensional coordinate system.
- Identify the differences between 3D and perspective.
- Apply the concepts used with clipping planes.

INTRODUCTION TO AUTOCAD 3D

AutoCAD contains the capabilities to produce true 3D and perspective models. These capabilities may be used interactively with 2D drawings.

The following chapters explain the use of AutoCAD 3D. Chapter 30 introduces you to 3D drawing. Chapter 31 covers the methods of constructing views. This is important both while constructing the drawing and for modeling a final product. Chapter 32 covers the User Coordinate System. This is an important principle in creating 3D drawings. Finally, Chapter 33 explores the many methods of drawing in 3D.

These chapters have been organized in a learning order. The best results will be achieved by reviewing the information in the order presented. For example, you will need to become skillful in creating 3D views before learning to draw in 3D, since much of the drawing construction process is performed while in 3D views.

HOW TO APPROACH 3D

It is first assumed that you have a good working knowledge of the 2D drawing functions of AutoCAD. Using AutoCAD 3D requires expertise in many of the commands you have learned in the previous chapters. If you are not familiar with some of the commands or concepts in certain areas, review these in the previous chapters.

Although AutoCAD 3D uses many of the commands you are familiar with, the approach to 3D drawing requires knowledge of some unique concepts. The primary difference is the *X,Y,Z* coordinate system. Other concepts such as camera positions, target points, clipping planes, and the User Coordinate System are used to create and view your 3D drawings.

Take time not only to read and study these concepts, but to understand them. Practice each concept with simple shapes before proceeding to more advanced problems.

Three-dimensional drawing takes time and practice to learn, but with time the results will be well worth the effort.

3D THEORY

Let's look at some of the concepts associated with three-dimensional drawing.

Coordinate System

In order to comprehend the construction of 3D drawings, the concept of the *X,Y,Z* coordinate system must be understood. This is the same *X,Y* (Cartesian) coordinate system you work with when constructing a 2D drawing, with the Z-axis representing the "height" of the entity added.

Before we proceed to learn about the Z-axis, it is necessary to understand that the 2D drawings you have constructed have used the *X,Y,Z* coordinate system all along. You have only drawn in the *X,Y* plane of this system. We will now expand to also draw in the Z-axis planes.

It is easy to understand the *X,Y,Z* coordinate system if you think of the "plan" view of your drawing as lying in the *X,Y* plane. Figure 30-1 shows the different axes.

INTRODUCTION TO AUTOCAD 3D **30-3**

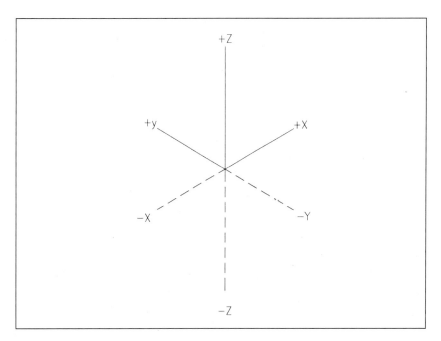

FIGURE 30-1 X,Y,Z Axes

You may, at this point, wish to review the section on coordinates in Chapter 9.

When working with a 3D drawing, you must also draw in a manner that can be thought of as "up from the page." Let's take an example of a box with dimensions of 4 × 3 × 2 (in current units). Figure 30-2 shows the box sitting in the positive quadrants of an *X,Y,Z* coordinate system, with four units along the X-axis, three units along the Y-axis and two units along the Z-axis.

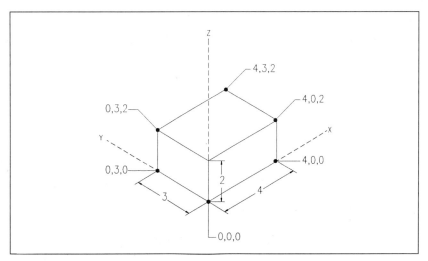

FIGURE 30-2 Box on 3D Axis

As in 2D drawing, each intersection has coordinates. In the case of a 3D drawing, the coordinates are represented in an X,Y,Z format. Notice the coordinates listed in the previous illustration.

3D Versus Perspective

AutoCAD is capable of displaying your drawings in either 3D or 3D perspective views. A 3D view may be perspective or non-perspective. A perspective view shows the drawing in 3D form with faces diminishing to a vanishing point and all forms shortened in distance. This is similar to the way the human eye perceives forms in the environment.

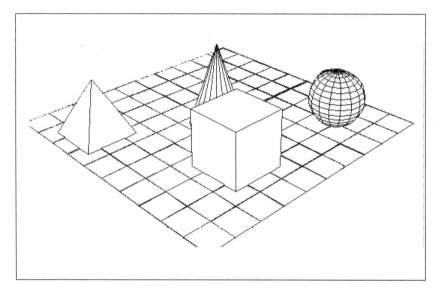

FIGURE 30-3   3D Perspective Views

A non-perspective view displays the drawing with "true lengths" for all lines in the drawing, similar to an isometric drawing. Figure 30-3 shows 3D perspective views.

Clipping Planes

Some applications require that a "cutaway" type of view be displayed for the purpose of showing parts of the drawing that may otherwise be concealed. This can be accomplished by a clipping plane.

INTRODUCTION TO AUTOCAD 3D   **30-5**

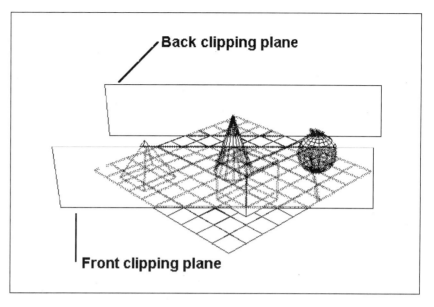

FIGURE 30-4 Clipping Planes

You can think of a clipping plane as an imaginary plane or surface that can "cut" through the drawing and eliminate every part of the drawing either in front or behind the plane. Figure 30-5 illustrates the effect of the clipping plane shown in Figure 30-4.

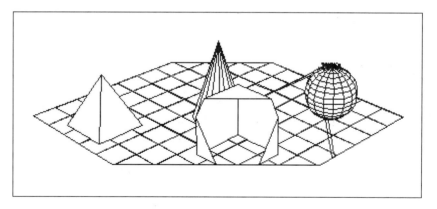

FIGURE 30-5 Results of Clipping Planes

EXERCISE

Refer to Figure 30-6 and write the coordinates for each corner in the space provided.

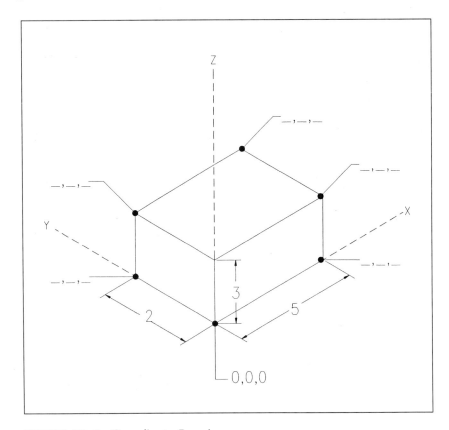

FIGURE 30-6 Coordinate Exercise

CHAPTER REVIEW

1. What are the axes in the Cartesian coordinate system?

2. From which direction from a page would the Z-axis normally project?

3. What is the difference between normal 3D and perspective?

4. What is a *clipping plane*?

5. How many clipping planes are in AutoCAD? Name each.

6. What form of 3D most closely approximates the view of the human eye?

7. In AutoCAD, how would you show a coordinate that has values of X = 3, Y = 5, Z = 9?

8. Which axes are two-dimensional drawings constructed in?

9. Which axis would normally represent the *height* of an object?

10. What is meant by *true lengths* when referring to a 3D drawing?

CHAPTER 31

VIEWING 3D DRAWINGS

To construct a 3D drawing, you must be capable of manipulating the drawing in the three planes so you can work with the different parts. After the drawing is completed, you can use the viewing commands to observe the drawing from different viewpoints. After completing this chapter, you will be able to:

- Utilize the commands to view AutoCAD 3D drawings.
- Place the imaginary camera and target to obtain the view you desire.
- Manipulate the Dview command options to control the look of the 3D view.
- Produce hidden line and shaded views of the 3D objects in the drawing.

VIEWING 3D DRAWINGS

Learning the methods to model your drawing in 3D is not only important for presentation purposes but is important during the drawing process. In this chapter, we will review the methods of displaying a drawing in 3D.

METHODS OF VIEWING 3D DRAWINGS

There are two methods of creating views of your drawings: using the Vpoint or Dview commands. The Vpoint command may be referred to as a "static" method, while the Dview command is a "dynamic" method of viewing. The Dview's dynamic method allows you to see the drawing as it is being rotated into view. The Vpoint's static method is a simpler shortcut method for creating quick view points. Let's look at each method of creating views.

SETTING THE 3D VIEWPOINT

The Vpoint command is used to select a viewpoint for the current viewport by setting a direction and elevation of view.

When you select the Vpoint command, the following prompt appears:

Command: **Vpoint**
Rotate/<View point> <current>:

The coordinates listed as <current> show the current view coordinates.

You can construct the viewpoint in three ways. Let's look at each way.

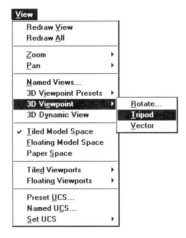

Creating a View by Coordinates

You may enter an X,Y,Z coordinate in response to the prompt to stipulate a point from which to view the drawing. The coordinate will be the point to look from, with the coordinate of 0,0,0 always the point to look at.

A viewpoint of 0,0,1 looks at the drawing in plan view (directly down along the Z-axis). A negative coordinate value places the viewpoint at the negative end of the axis. Thus, a viewpoint of 0,0,-1 would look at the drawing directly from below.

Creating a View by Axes

If you respond to the prompt by pressing [ENTER] on the keyboard, the drawing screen will temporarily display a special axes diagram. The visual diagram consists of a tripod, representing the X, Y, and Z axes and a flattened globe. Figure 31-1 shows the tripod and globe.

VIEWING 3D DRAWINGS 31-3

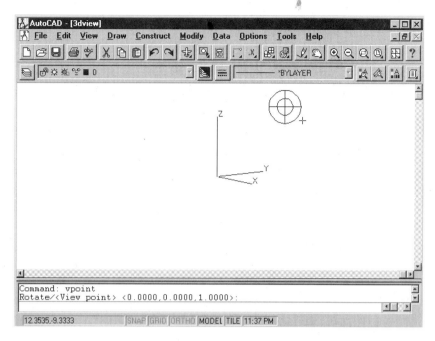

FIGURE 31-1 Vpoint Axes

To set the desired view, move the mouse. This causes the tripod to rotate, representing the rotation of each axis. It helps to think of your drawing as lying in the X,Y plane of the tripod when visualizing the desired view.

The compass-appearing icon to the right is a 2D representation of a globe. The center point of the crosshairs is the north pole, the middle circle is the equator, and the outer circle is the south pole. The four quadrants of the globe may be thought of as the direction of the view. For example, the lower right quadrant would produce a view that is represented from the lower left of the plan view. Moving above the equator produces an "above-ground" view; below the equator, a "below-ground" view.

When you wish to return to the plan view, type **Plan**.

Don't be afraid to experiment with different views. Try to associate the globe and tripod with the results of the 3D view. With a little practice, you will be able to obtain whatever results you desire.

SETTING VIEW BY DIALOG BOX

You can use a dialog box to set standard views of your 3D drawing. Select "View/3D Viewpoint/Rotate" from the pull-down menus. Click on the degree settings shown in the dialog box illustration to change the degrees. You may alternately set the degrees in the text box.

31-4 USING AUTOCAD

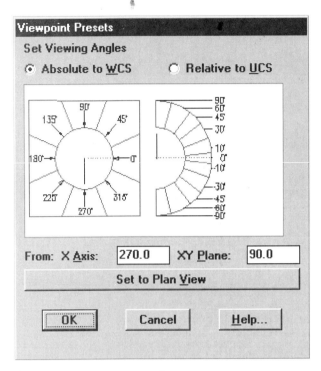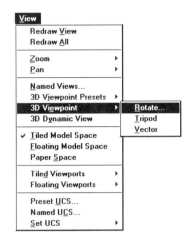

FIGURE 31-2 Vpoint Dialog Box

DYNAMIC VIEWING

The Dview command is a very powerful tool for modeling the drawing dynamically. This means you can rotate the drawing and see the results as you go.

The Dview command is also used to set clipping planes, set the distance from which the drawing is to be viewed, and other functions that allow you to precisely control the appearance of the 3D view.

Dview Options

When you issue the Dview command, the following prompt is displayed:

Command: **Dview**
Select objects: *(Select.)*
CAmera/TArget/Distance/POints/PAn/Zoom/TWist/CLip/Hide/Off/Undo/<eXit>:

VIEWING 3D DRAWINGS **31-5**

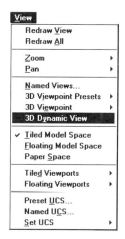

You must have selected objects and be in the Dview command in order to use any of the options listed.

If you do not select objects and just press [ENTER], AutoCAD displays a house (derived from a block named "DVIEWBLOCK") and displays it for a viewing model.

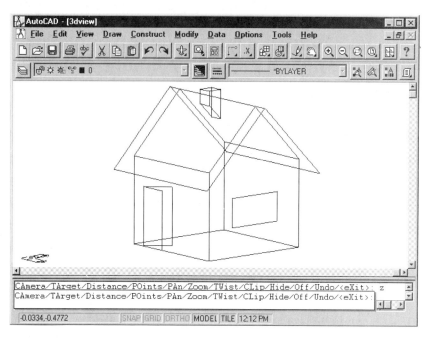

FIGURE 31-3   Dviewblock Drawing

The following is a brief description of the options.

CAmera: Dynamically sets a camera position. This is the position from which you will view the drawing.

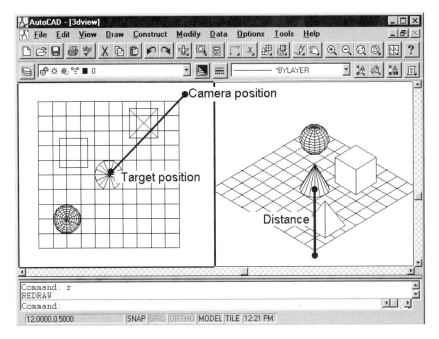

FIGURE 31-4 Camera Positioning

TArget: Dynamically sets a target position. This is the point in the drawing the camera will look at.

Distance: Moves the camera closer or further from the target along a path. This path is defined by a line between the camera and target point known as the "sight line."

POints: Allows you to position the camera and target points using X,Y,Z coordinates.

PAn: Pans the camera from its present position.

Zoom: If in perspective mode, Zoom is used to change the focal length of the camera. This is similar to changes in a camera image created by different lenses. For example, a wide angle or telephoto-type lens could be used to obtain different results.

If perspective mode is off, the Zoom option is used similarly to the standard Zoom command.

TWist: Tilts the 3D view around the line of sight.

CLip: Sets clipping planes in the view. This has the effect of a cutaway type of view.

Hide: Creates a hidden line view.

Off: Turns off the perspective mode.

Undo: Reverses the previous command; same as the standard U command.

eXit: Exits the Dview command.

Let's learn how to use Dview to dynamically view your drawings.

CAMERA AND TARGET POSITIONING

When you view objects in 3D, there is a point you look from and a point you are looking to. AutoCAD uses a camera and target imagery to assist in placing these points. Figure 31-4 shows a camera location and views of the resulting 3D viewpoints are in Figures 31-5 and 31-6.

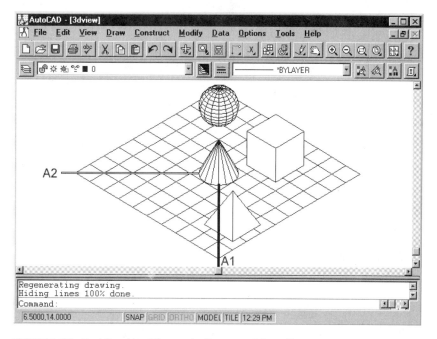

FIGURE 31-5 Viewing Through Camera A1 to Target

31-8 USING AUTOCAD

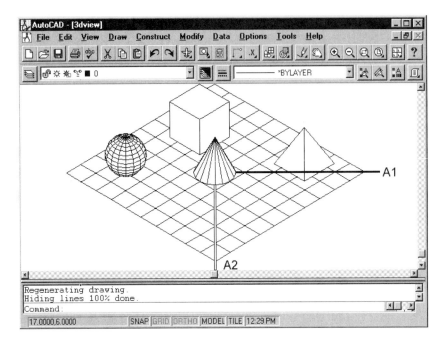

FIGURE 31-6 Viewing Through Camera A2 to Target

Positioning the Camera and Target

The Dview command is used to place the camera and target. Let's look at how to do this. Type the Dview command.

Command: **Dview**
Select objects: *(Select.)*
CAmera/TArget/Distance/POints/PAn/Zoom/TWist/CLip/Hide/Off/Undo/<eXit>:

You must first select the entities that will be dynamically moved. The failure to select entities will not exclude them from the subsequent change of view, but will exclude them from being displayed dynamically. Selecting too many entities to be dynamically viewed will result in slow operation. It is prudent to select representative "frame" entities in the amount suitable for visualizing the overall view. When you accept the final view, the entire drawing will be generated in that view.

If you respond to the prompt with a null response (ENTER), a house will be displayed for viewing purposes only. You may also select Dviewblk from the screen menu in response to the "Select objects" prompt to display the house.

After the desired entities for dynamic viewing are selected, choose the CAmera option. Only the selected entities are displayed. The command line displays:

> Toggle angle in/Enter angle from XY plane <>*default*>:

You may set the angle in two ways. This prompt is asking for the angle "up or down". You can set the angle, in degrees, by entering the numerical degrees from the keyboard. You may also set the angle "dynamically" by moving the crosshairs around the screen, rotating the selected entities of the drawing.

The selected entities change as you move the crosshairs, showing the new camera position. Although the drawing appears to be actually rotating, the effect is caused by the "camera" moving around the entities in the drawing.

When you are satisfied with the vertical view rotation, press the enter button on your input device.

If you enter the number of degrees from the keyboard, AutoCAD continues and prompts you for the number of degrees of rotation "around" drawing made up from the selected entities.

> Toggle angle from/Enter angle in XY plane <>*default*>:

You can toggle between the two prompts by entering a **T** (for *toggle*) from the keyboard.

If you choose to dynamically set the camera angle by moving the crosshairs, AutoCAD will not display the second prompt, since both display angles are set at one time.

After you have set the new camera angle, AutoCAD redisplays the Dview prompt.

> CAmera/TArget/Distance/POints/PAn/Zoom/TWist/CLip/Hide/Off/Undo/ <eXit>:

You may now select another option and continue to model the drawing. If you wish to reposition the camera or target, select the desired option. It is not necessary to reselect objects unless you exit the Dview command and the Dview prompt is no longer displayed.

You can set the camera target by selecting the TArget option.

> CAmera/TArget/Distance/POints/PAn/Zoom/TWist/CLip/Hide/Off/Undo/ <eXit>: **TA**

Remember, the target is the point at which the camera is pointed. The TArget option and the CAmera option perform in the same manner.

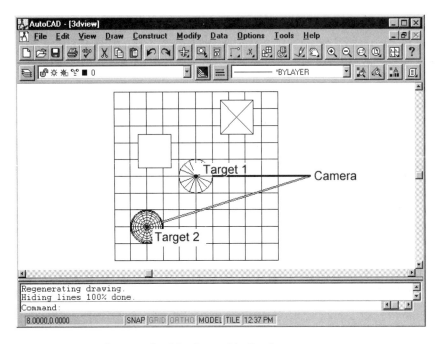

FIGURE 31-7 Camera Positioning with Options

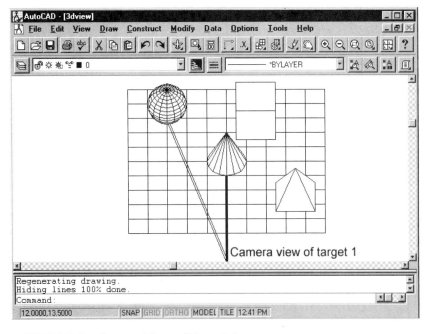

FIGURE 31-8 Camera View of Target 1

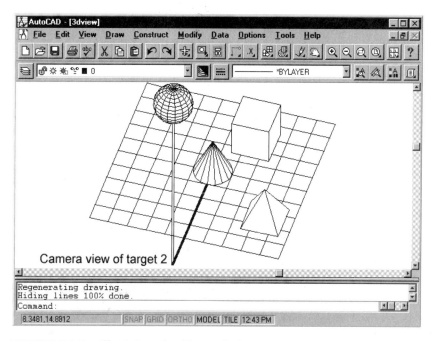

FIGURE 31-9 Changing the Target Point

Setting the Distance to View From

When the camera and target are placed, the line extending between them is called the "line of sight." The Distance option moves the camera along the line of sight, changing the distance between it and the target point. To select the Distance option, enter **D** in response to the prompt.

CAmera/TArget/Distance/POints/PAn/Zoom/TWist/CLip/Hide/Off/Undo/ <eXit>: **D**

Selecting Distance also activates the perspective mode, causing the view to appear in perspective form. A perspective icon appears in the lower left corner of the screen to denote this.

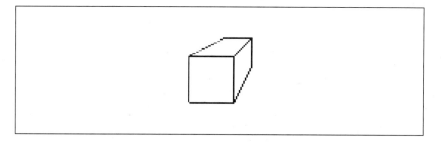

FIGURE 31-10 Perspective Indicator Icon

A single slider bar is used to dynamically set the camera-to-target distance. The slider bar contains graduations from 0× to 16×. This represents the distance factor, relative to the current distance between the camera and target. The current distance is, of course, represented by the value of 1. Moving the diamond marker to 4×, for example, would move the camera four times as far from the target point. This would have the visual effect of moving away from the objects in the view. Moving the slider bar to a value less than zero will shorten the distance, as though you approached the objects in the view.

When you are satisfied with the view distance, press [ENTER] on the keyboard.

Certain commands cannot be used in perspective mode. If you use one of these commands with Regenauto turned off, AutoCAD will display the message:

> About to regen with perspective off — proceed?<Y>

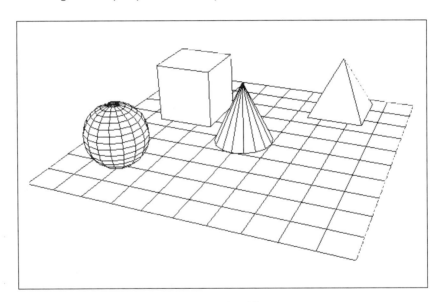

FIGURE 31-11 Drawing in Perspective View

Entering a **Y** or pressing [ENTER] will turn perspective mode off and continue the command. Entering an **N** will cancel the command and leave perspective mode on.

You may manually turn perspective mode off (and return to non-perspective mode) by selecting the Off option from the Dview prompt.

PANNING THE VIEW

Selecting the PAn option allows you to pan the camera from its present location. You can think of this as a "look around" option.

The operation is very simple. Just enter two points on the screen, in the same manner as a 2D pan, and the scene will change accordingly. If you are not in perspective mode, you may alternately enter keyboard coordinates.

ZOOMING THE VIEW

The Zoom option, like the PAn option, operates differently, depending on whether the perspective mode is on or off.

If perspective mode is off, a slider bar is displayed at the top of the screen, allowing you to move the marker and view the resulting zoom. The slider bar is marked in graduations from 0 to 16×, with 1× representing the current zoom level.

If perspective mode is on, you may either use the slider bar, or stipulate the camera lens focal length. Changing the focal length is an interesting feature. The effect on the view is the same as if you changed the actual camera lens. For example, a 28mm setting displays a wide angle shot and a 100mm setting displays a telephoto shot towards the target point. The default is 50mm, which closely resembles the view from the human eye.

Figure 31-12 shows the effects of changing the focal length of the lens (see continuation on following page).

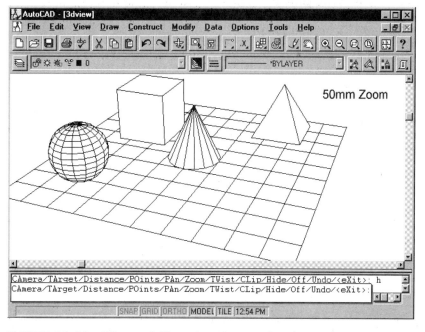

FIGURE 31-12 Effects of Changing Camera Focal Length

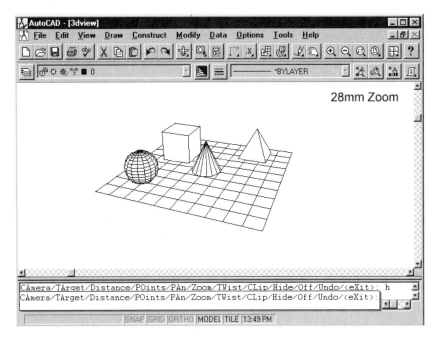

FIGURE 31-12 (Continued)

Twisting the View

The TWist option is used to "twist" the view around the current line of sight. You can dynamically twist the view by entering a point on the screen, or by entering a twist angle.

When you select TWist, a rubber-band line is drawn from the center of the screen to the crosshairs. Moving the cursor allows you to dynamically twist the view.

You may alternately enter the twist angle from the keyboard. The angle is measured counterclockwise, with 0° to the right at 3 o'clock.

Setting Clipping Planes

Clipping planes are used to cut away a part of the drawing. For example, you may want to cut away part of a building to see the interior.

You can think of clipping planes as transparent walls that slice through part of the drawing and remove everything that is either in front of or behind them.

There is a front and a back clipping plane. Either or both of these planes can be positioned in the drawing. Positioning is accomplished by either specifying a distance from the target point, or dynamically with slider bars.

The clipping planes may be positioned in either perspective or non-perspective mode. When perspective mode is activated by using the Distance option, the front clipping plane is automatically turned on and positioned at the location of the camera by default.

Figure 31-13 shows a perspective view. The clipping planes are shown as "walls." The dotted parts of the drawing are the areas in front of the front clipping plane and to the rear of the back clipping plane. The second illustration shows the effect of placing the clipping planes at these locations.

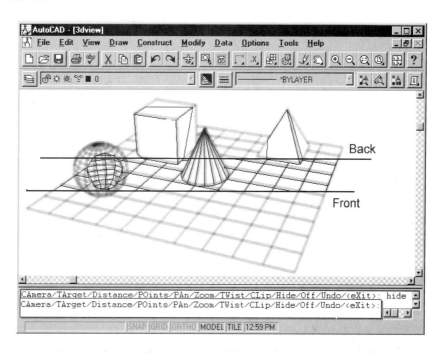

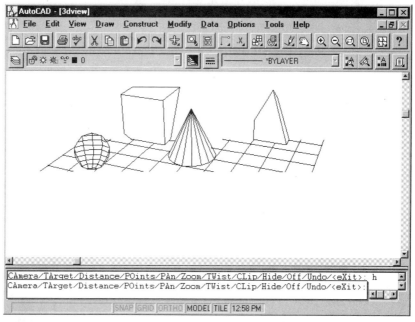

FIGURE 31-13 Clipping Planes

To set one or both clipping planes, select the CLip option from the Dview menu.

CAmera/TArget/Distance/POints/PAn/Zoom/TWist/CLip/Hide/Off/Undo/ <eXit>: **CL**

The following prompt appears:

Back/Front/<Off>:

Let's look at each of the Clip options.

Back: Sets the location of the back clipping plane.

Selecting the Back option displays the prompt:

ON/OFF/distance from target <current>:

If you enter a positive numerical value, the back plane is placed at that distance (in current units) in front of the target point. A negative value places the plane that distance behind the target point.

A slider bar is also displayed that alternately allows dynamic placement. You can see the effects of the clipping plane on the drawing as you move it with the slider bar.

The ON/OFF options allow you to turn the effects of the plane on or off. The ON option uses the current distance value.

FRont: Sets the location of the front clipping plane.

The front clipping plane is positioned in the same manner as the back clipping plane.

Selecting the FRont option displays the prompt:

ON/OFF/Eye/distance from target <current>:

The Eye option simply places the front plane at the current camera position.

PRODUCING HIDDEN LINE VIEWS

The Hide option is used to produce hidden line drawings of the current view. A similar Hide command is available when using the Vpoint command to produce a 3D view.

The 3D views are drawn and modeled in "wireframe." A wireframe is a drawing that you can see through the sides of. (Think of your object being made of wires.) The Hide command removes the parts of the drawing that you normally would not see in a 3D view.

Hidden lines must be regenerated at the time of the plot if you wish to obtain a hard copy drawing with hidden lines removed. This is performed in the plot routine. See Chapter 18 for detailed instructions on plotting.

VIEWING 3D DRAWINGS 31-17

PRODUCING SHADED IMAGES

The Shade command is used to produce shaded images, similar to AutoShade, of your drawing. To use Shade, enter the Shade command:

Command: **Shade**

There are no options in the Shade command; the shading process begins immediately.

On machines with limited memory, AutoCAD will produce the shaded image with several passes. (Make sure you have turned off the Monochrome Vectors option in Preferences/Color.)

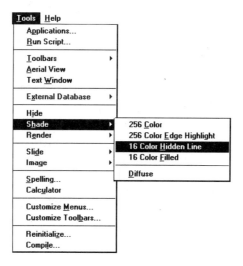

Shading Types

This method of shading produces an image with one light source from the eyepoint. The shading type is controlled by the Shadedge and Shadedif system variables.

If Shadedge is set to 0 or 1, the faces are shaded based on the angle the faces form with the viewing direction. The percentage of diffuse reflection and ambient light level is set with the Shadedif variable. Shadedif is set to 70 by default. This means that 70 percent of the light is diffused light from the light source and 30 percent is ambient (overall general level) light. These numbers will always add to 100 percent of the light. The Shadedif variable sets the diffused light, while the balance of 100 percent is ambient light. Higher diffused light levels produce more contrast in a shaded image.

Table 31-1 lists the four settings for the Shadedge variable and their effect.

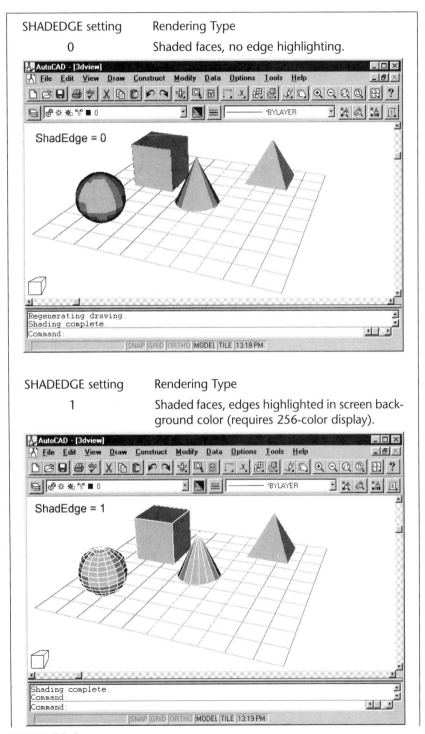

TABLE 31-1

SHADEDGE setting	Rendering Type
2	Simulates hidden line rendering. Centers of polygons are painted in background color. The color of visible edges is determined by the entity color.

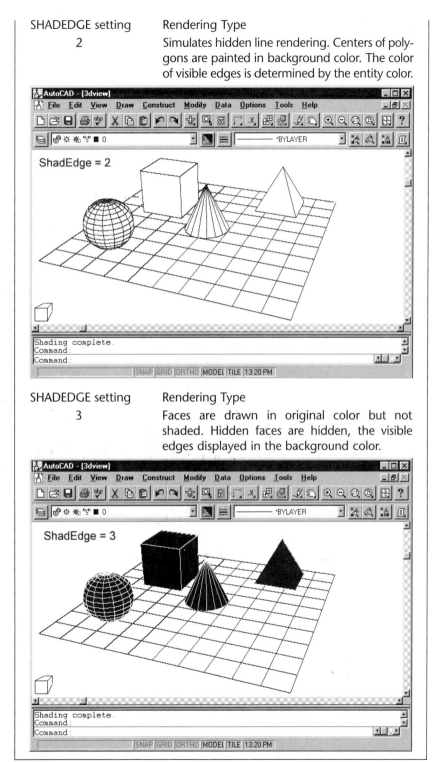

SHADEDGE setting	Rendering Type
3	Faces are drawn in original color but not shaded. Hidden faces are hidden, the visible edges displayed in the background color.

TABLE 31-1 (Concluded)

CHAPTER REVIEW

1. What commands are used to generate a third dimension of a drawing?

2. What is meant by *dynamic modeling*?

3. What does the Dview/TArget command do?

4. How do you turn on perspective mode?

5. What are the conditions under which you could set either the zoom or camera lens when using Dview/Zoom?

6. Why would you only select part of the drawing in the Dview selection set?

7. In the Dview command, how would you display a model house for viewing?

8. What Dview option would you use to specify X,Y,Z coordinates to set a view from?

9. How does setting a smaller millimeter camera lens affect the view?

10. What is meant by *wireframe*?

11. What does the Shade command do?

12. What is the sum of the ambient light and diffused light levels in AutoCAD?

CHAPTER 32

THE USER COORDINATE SYSTEM

To effectively construct a 3D drawing, you must become proficient with the User Coordinate System. After completing this chapter, you will be able to:

- Comprehend the concept and use of the User Coordinate System.
- Manipulate the UCS.
- Identify the effects of the UCS on AutoCAD commands.
- Save, restore, and manage UCS systems you create.

THE USER COORDINATE SYSTEM

When you are drawing in 2D plan, you are working in the X,Y plane. Drawing is simplified, since you are working in a single plane. Drawing in 3D, however, is more complicated. There may be many planes in which you wish to work. The User Coordinate System is designed to make this process simpler. We'll refer to the User Coordinate System as the "UCS."

In order to effectively draw in AutoCAD 3D, it is essential to understand the UCS. The primary purpose of the UCS is simplification of the 3D process. Mastering this system will allow you to construct 3D drawings efficiently.

Let's consider the example of a sloped barn roof that has a graphic painted on it. It would be simple to draw the graphic in plan view, but placing it on the slope of the roof is quite different. Being able to draw on the slope of the roof as if it were in plan would be quite efficient.

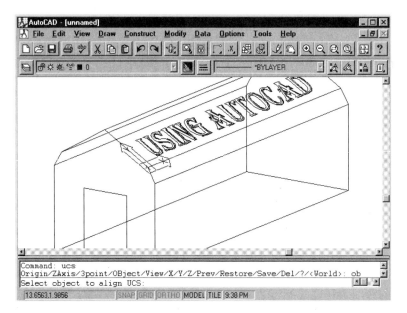

FIGURE 32-1 Drawing on Sloped Roof

The UCS allows you to do this. You may slope and rotate the UCS to change the plan view to match the slope of the roof. It may help to think of the UCS as the X,Y plane that you are used to seeing when you draw in 2D plan view. Then imagine the ability to move, turn, and/or rotate it to any position on the 3D object you are drawing. Now you can redisplay the drawing with the new plan view "flat on the paper" and draw on it as though you were in plan.

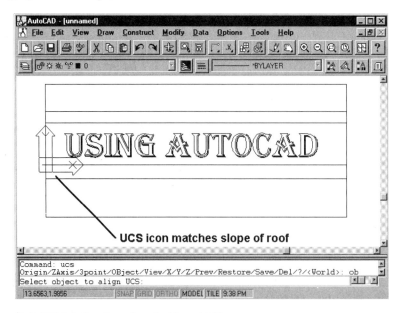

FIGURE 32-2 Drawing in New UCS

Actually, this is exactly what you are doing. You can always return to "true" plan. The true plan is called the World Coordinate System. The User Coordinate System is a temporary, user-defined drawing plane to make drawing on the sides, slopes, etc. of the 3D object simpler. Let's continue and see how to manipulate the UCS to make 3D drawing simple.

THE UCS ICON

The UCS icon is used to denote the orientation of the current UCS. The icon is displayed at the lower left of the drawing screen.

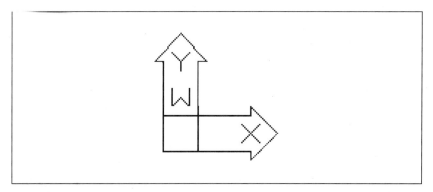

FIGURE 32-3 UCS Icon

The icon indicates the current UCS axes orientation, the origin of the UCS, and the viewing direction. If the icon contains a "W," the World Coordinate System is in effect. The box indicates plan view. A plus (+) indicates the drawing is being viewed from above. If the plus is absent, the drawing is being viewed from below. The UCSicon command is used to control certain display characteristics of the icon. We will learn about the command later in this chapter.

Head-On Indicator

It is possible to rotate the UCS, or the view relative to the current UCS, to a position that looks head-on into the edge of the drawing.

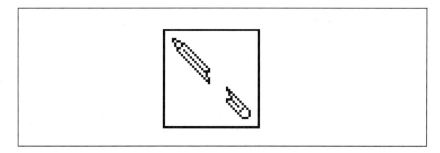

FIGURE 32-4 Head-on Indicator

Such a position is almost useless as a drawing view. When such a view is current, AutoCAD displays a "head-on" indicator.

CHANGING THE UCS

The UCS command is used to change the current UCS. When you issue the UCS command the following prompt is displayed.

Command: **UCS**
Origin/ZAxis/3point/OBject/View/X/Y/Z/Prev/Restore/Save/Del/?/<World>:

Let's look at each option under the UCS command. It may be helpful to use a work disk drawing to practice each option as you study it.

Origin: Used to change the origin of the UCS. The axes are left unchanged. (The location of the icon will not change unless the Origin option of the UCSicon command is on. The UCSicon command is covered later in this chapter.) When you select the Origin option under the UCS command, the following prompt is displayed:

Origin/ZAxis/3point/OBject/View/X/Y/Z/Prev/Restore/Save/Del/?/<World>: **O**
Origin point <0,0,0>:

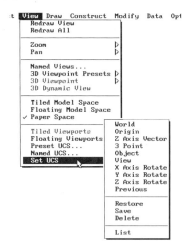

Enter an X,Y,Z coordinate point to describe a new origin point. If you enter only an X,Y point, the Z coordinate will be equal to the current elevation.

You may alternately enter a point on the screen to designate the new origin.

Object snap is especially useful in placing the new origin at a 3D point that can be described by a point on an entity.

NOTE: The use of object snap is very helpful when using many of the UCS options.

ZAxis: Used to define a new UCS defined by an origin point and a point along the Z-axis.

AutoCAD prompts:

Origin point <0,0,0>:
Point on positive portion of the Z axis <default>:

Responding to the second prompt will cause the Z-axis of the new origin to be parallel to the previous system.

3point: Aligns the UCS by placing three points that define the origin and rotation of the X,Y plane of the new UCS. This is particularly helpful when you align the UCS with existing entities using object snap. Selecting the 3point option displays the following prompts.

Origin/ZAxis/3point/OBject/View/X/Y/Z/Prev/Restore/Save/Del/?/<World>: **3**
Origin point <0,0,0>:
Point on positive portion of the X axis <default>:
Point on positive-Y portion of the UCS X-Y plane <default>:

The point given for each may be either a numerical coordinate entered from the keyboard, or a point entered on the screen.

The point entered for the origin point prompt designates the new origin point. The second point defines the direction of the X-axis. The third point will lie in the X,Y plane and define the direction of the positive Y-axis. Simply pressing [ENTER] in response to any of the prompts will designate a value equal to the existing origin or direction.

Figure 32-5 shows the points entered to align the UCS with our barn roof.

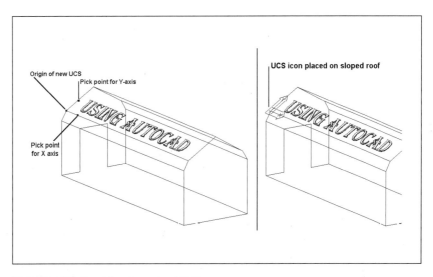

FIGURE 32-5 Aligning the UCS

OBject: Places the UCS relative to an existing entity. Selecting OBject results in the following prompt.

Origin/ZAxis/3point/OBject/View/X/Y/Z/Prev/Restore/Save/Del/?/<World>: **E**
Select object to align UCS:

You must use object pointing to select the object. No other object selection method is allowed.

You cannot select the following objects for UCS alignment:

    3D solid

    3D polyline

    3D mesh

    Viewport

    Mline

    Region

    Spline (created by Spline command)

    Ellipse (when PEllipse = 0)

    Ray

    Xline

    Leader

    Mtext

Using this option will position the X,Y plane of the UCS parallel to the entity selected. The direction of the axes, however, will depend on the object selected. Table 32-1 explains the effect of each.

Arc	The center of the arc becomes the new UCS origin with the X-axis passing through the point on the arc visually closest to the pick point.
Circle	The circle's center becomes the new UCS origin, with the X-axis passing through the point on the arc visually closest to the pick point.
Dimension	The new UCS origin is the middle point of the dimension text. The direction of the new Y-axis is parallel to the X-axis of the UCS in effect when the dimension was drawn.
Line	The endpoint visually nearest the pick point becomes the new UCS origin. The X-axis is chosen such that the line lies in the X,Z plane of the UCS (i.e., its second endpoint has a Y coordinate of zero in the new system).

TABLE 32-1

Point	The new UCS origin is the point's location. The X-axis is derived by an arbitrary but consistent algorithm.
Polyline (2D only)	The polyline's start point is the new UCS origin, with the X-axis extending from the start point to the next vertex.
Solid	The first point of the solid determines the new UCS origin. The new X-axis lies along the line between the first two points.
Trace	The "from" point of the trace becomes the UCS origin, with the X-axis lying along its center line.
3D Face	The UCS origin is taken from the first point, the X-axis from the first two points, and Y positive side from the first and fourth points. The Z-axis follows by application of the right-hand rule. If the first, second, and fourth points are colinear, no new UCS is generated.
Shape, Text, Block Reference, Attribute Definition	The new UCS origin is the insertion point of the entity, while the new X-axis is defined by the rotation of the entity around its extrusion direction. Thus, the entity you pick to establish a new UCS will have a rotation angle of 0° in the new UCS.

TABLE 32-1 (Continued)

View: It is often convenient to set the current UCS to be oriented to the current view. This view may have been set by using either the Vpoint or Dview commands. Selecting the View option resets the UCS to be perpendicular to the current viewing direction. That is, the current view will become the plan view under the new UCS created by the option. This results in a new UCS that is parallel to the computer screen.

Origin/ZAxis/3point/OBject/View/X/Y/Z/Prev/Restore/Save/Del/?/<World>: **V**

X/Y/Z: Used to rotate the UCS around any of the three axes. This is actually three options. You may enter either X, Y, or Z in response to the UCS command prompt. Entering any of the three will result in the following prompt:

Origin/ZAxis/3point/OBject/View/X/Y/Z/Prev/Restore/Save/Del/?/<World>: *(n)*
Rotation angle about n axis <0.0>:

The *n* is either X, Y, or Z. Entering an angle will cause the current UCS to rotate about the specified axis the designated number of degrees. You may also enter a point on the screen to show the rotation. AutoCAD provides a rubber-band line from the current origin point to facilitate this.

The direction of the angle is determined by the standard engineering "right-hand rule" method of determining positive and negative rotation angles. Figure 32-6 shows the method of determining angle directions.

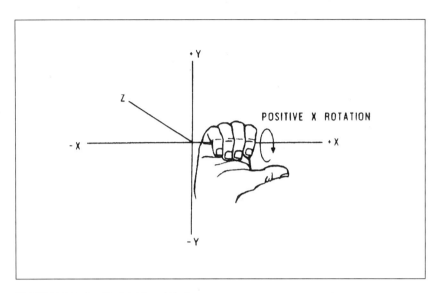

FIGURE 32-6 Right-Hand Rule

Imagine placing your right hand around the axis which the UCS is to rotate, with the thumb pointing in the positive direction. The direction your curled fingers point is the direction of positive angle of rotation of that axis. (Be especially careful of the angle settings under the Units command, since they affect this procedure.)

If Tilemode is off, the last 10 coordinate systems in both paper space and model space are saved. The "previous" coordinate system derived depends on the space in which you use the Previous options.

Previous: Returns the last UCS setting. Up to 10 settings are stored. When repeatedly selected, the Previous option will "step back" through each. The operation of this option is similar to the Zoom Previous function.

Origin/ZAxis/3point/OBject/View/X/Y/Z/Prev/Restore/Save/Del/?/<World>: **P**

Restore: Restores a UCS saved with the Save option (explained next). The following prompt is displayed:

Origin/ZAxis/3point/OBject/View/X/Y/Z/Prev/Restore/Save/Del/?/<World>: **R**
?/Name of UCS to restore:

Enter the name of the desired UCS to be restored. The named UCS then becomes current.

If you enter a question mark (?), AutoCAD prompts:

UCS name(s) to list <*>:

Press [ENTER] to display all the saved coordinate systems.

NOTE: The previously saved UCS becomes current, but the previous view does not. You may want to use the Plan command (explained later in this chapter) to restore the plan view of the now current UCS.

Entering a question mark (?) in response to the prompt will display a list of the saved coordinate systems.

Save: Permits naming and saving of the current UCS and return to it using the Restore option. The name may be up to 31 characters in length and may contain letters, numbers, dollar signs ($), hyphens (-), and underscores (_).

NOTE: Use names that describe the location of the UCS, such as "ROOF-TOP" or "FRONT-SIDE".

Origin/ZAxis/3point/OBject/View/X/Y/Z/Prev/Restore/Save/Del/?/<World>: **S**
?/Desired UCS name:

Del: Deletes a previously saved UCS name. The following prompt is issued:

Origin/ZAxis/3point/OBject/View/X/Y/Z/Prev/Restore/Save/Del/?/<World>: **D**
UCS name(s) to delete <none>:

The DOS-similar wild-card characters of "?" and "*" may be used to delete several UCS names at a time. Entering the name of a single UCS name will delete only that system. Several names may be entered by separating them with commas.

?: Lists the saved coordinate systems. This performs the same function as the ? option under Restore. A listing of each saved UCS and the coordinates of the origin and X,Y,Z axes is displayed.

Origin/ZAxis/3point/OBject/View/X/Y/Z/Prev/Restore/Save/Del/?/<World>: **?**
UCS name(s) to list <*>:

World: Resets the UCS to the World Coordinate System.

Origin/ZAxis/3point/OBject/View/X/Y/Z/Prev/Restore/Save/Del/?/<World>: **W**

PRESET UCS ORIENTATIONS

AutoCAD comes with several preset UCS orientations that create isometric views. The DDUCSP command (short for "dialog UCS preset") displays the UCS Orientation dialog box, as shown below:

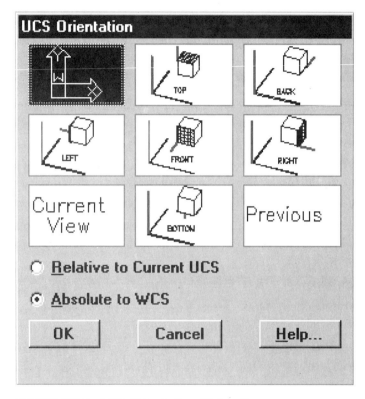

FIGURE 32-7 UCS Orientation Dialog Box

Click on an icon tile to change the UCS to one of the six isometric views. Two other tiles let you switch the UCS to the previous view and the current UCS.

UCSFOLLOW SYSTEM VARIABLE

The UCSFOLLOW system variable allows you to set the UCS plan view to follow a newly set User Coordinate System. Normally, when you set a new UCS, you will select the Plan command to reset the display to align in true plan view with the new coordinate system. Turning on (setting to "1") UCSFOLLOW will cause the plan view to be set automatically each time the UCS is changed.

THE USER COORDINATE SYSTEM 32-11

UCSICON COMMAND

The UCSicon command is used to control the display of the icon indicator. The icon is located, by default, at the lower left corner of the drawing screen. The UCS icon is different for model space and paper space. Figure 32-8 shows the UCS icon for each.

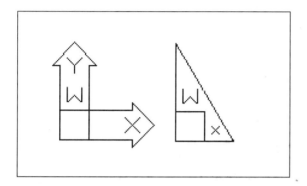
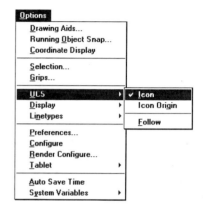

FIGURE 32-8 UCS Icon

Issuing the UCSicon command results in the following prompt.

Command: **UCSICON**
ON/OFF/All/Noorigin/ORigin <ON>:

The following is a listing of the options for the UCSicon command.

OFF: Turns off the display of the icon.

ON: Turns on the icon display if it is off.

All: Selecting the All option allows you to simultaneously apply the changes made to the UCS icon to all the viewports currently in effect. Normally, the UCS icon only affects the icon in the current viewport.

To use the All option, first select All. The prompt will repeat, allowing the choice of the option that will affect the icons in each viewport.

Noorigin: Causes the icon to be displayed at the lower left of the drawing screen, without respect to the actual current UCS origin (default setting).

ORigin: Causes the icon to be displayed at the actual UCS origin point of 0,0,0. If the origin point is not within the screen display, the icon is displayed at the lower left of the screen.

DDUCS COMMAND (UCS DIALOG BOX)

The DDUCS command displays a dialog box that assists management of the UCS system.

The dialog box contains a listing of the coordinate systems that have been defined (saved), as well as the World Coordinate System, the previous system, and the current system, if it has not been saved and named. Figure 32-9 shows a typical DDUCS dialog box, along with the UCS List sub-dialog box.

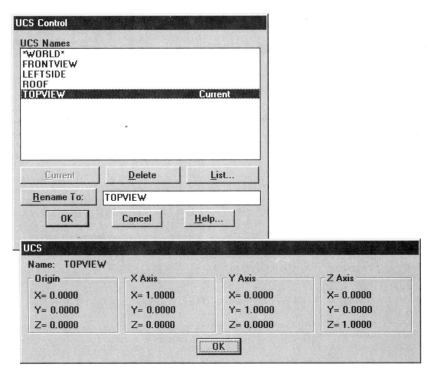

FIGURE 32-9 UCS Control Dialog Box (DDUCS)

The dialog box contains "scroll bars" that allow the listing to be scrolled if there are more listings than the box can display at one time. The UP and DOWN buttons scroll the list continuously, while the PAGE UP and PAGE DOWN buttons scroll the listing a page at a time. If there is less than one page of listings, the buttons have no effect. To operate the scroll buttons, place the arrow marker over the box and press the input key.

The first entry in the dialog box is always the World Coordinate System and is listed as *WORLD*. If you have changed the UCS, a listing named *PREVIOUS* is shown. A current listing that has not been named is shown as *NO NAME*. Additional listings show the name of each saved UCS.

Changing the Current UCS

The current UCS is noted with "Cur" in the listing. To select a new UCS, click on another name.

Listing UCS Information

You may obtain UCS information by clicking on the "List..." button in the dialog box. The List sub-dialog box displays the X,Y,Z vectors of the origin and each axis.

> **NOTE:** Deleting a UCS that is no longer in use makes UCS management easier.

Deleting a UCS

You may delete a UCS by selecting the UCS name, then the Delete button.

You cannot delete either the World or Previous UCS listings.

CHAPTER REVIEW

1. In what plane does the User Coordinate System exist?

2. What is the World Coordinate System?

3. What does the *head-on indicator* mean?

4. What command is used to manipulate the UCS?

5. What command is used to display the UCS icon?

6. What UCS option would you use to align the UCS with the current drawing rotation, placing the UCS parallel with the view?

7. What rule is used to determine the rotation of axes?

8. How would you store and recall a UCS?

9. What is the function of the UCSFOLLOW system variable?

10. What are the two possible positions for the UCS icon?

CHAPTER 33

DRAWING IN 3D

After you have become proficient with the use of 3D viewing and UCS commands, you must learn the methodology of constructing 3D drawings in AutoCAD. After completing this chapter, you will be able to:

- Comprehend the concepts of elevation and thickness.
- Draw in different UCS systems.
- Create solid faces for objects.
- Construct various types of 3D surface meshes.
- Use AutoCAD's functions to create basic 3D objects.

DRAWING IN 3D

Now that you have learned about the User Coordinate System and how to set up views of 3D drawings, it is time to learn the components of 3D drawing construction.

There are several manners in which to construct the components that make up a 3D drawing. Some are used to draw a unique shape, while some are basic shapes that can be placed in the drawing. Many drawings are composed of a combination of specially drawn objects and basic shapes.

The first step in constructing a new 3D drawing is studying the shapes to be drawn. Determine the best method of approach. It is often easier to construct each individual shape, than to insert it into a master drawing which is composed of many objects.

The ways to construct a 3D drawing that we will study are as follows:

- Elevation and thickness to create extruded entities.
- Standard drawing methods in different coordinate systems.
- 3D polygon meshes
- 3D objects

Let's look at how we can use each of these to create 3D drawings.

EXTRUDED ENTITIES

This method of placing 3D shapes forms the components by applying extrusions to drawing entities. An extrusion may be thought of as applying a thickness to an entity. Think of the entity as "growing" up or down from the flat drawing plane (X,Y drawing plane).

Thus, a line appears as a sheet of paper on edge. A circle appears as a tube, etc. Figure 33-1 shows several drawing entities on the left and the same entities in extruded form on the right.

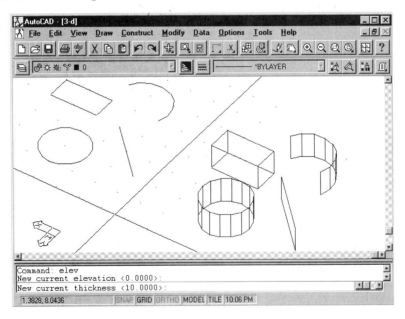

FIGURE 33-1 Extruded Objects

Elevation

The base, or bottom, of the extrusion can be set at different elevations. Consider an extruded tube (made from a circle) sitting on a table top. The top of the table could be considered as zero elevation. If you placed the bottom of the tube above the zero elevation, the tube would appear to be hovering above the table top. If the elevation is negative, the tube will appear to be shoved downward through the table top.

Of course, the floor on which the table sits could be placed at the zero elevation. If the table was 30 inches to the top, the table top elevation would be 30 inches. If you wanted the tube to sit on the table, the elevation of the tube would also be 30 inches.

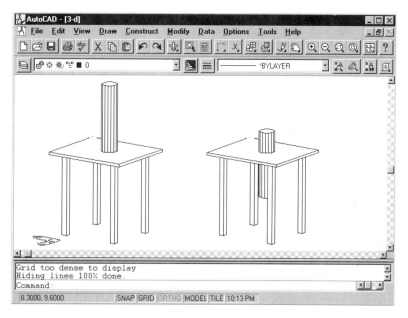

FIGURE 33-2 3D Elevation

Thickness

The thickness of an extruded entity is the distance from the base elevation of the entity. For example, a thickness of 6 inches would make our tube 6 inches high. The thickness is measured from the base elevation of the entity. Note that this is not necessarily zero, as in the previous example of the tube sitting on a table top with a 30 inch elevation.

Setting Elevation and Thickness

To draw an entity with a specified elevation and/or thickness, the Elevation command is used. This command presets both values, and every entity that is subsequently drawn has these values. The default value in the ACAD prototype drawing is zero for each, resulting in a flat entity in 3D.

When you issue the Elevation command, the following prompts are displayed:

Command: **Elev**
New current elevation <current>:
New current thickness <current>:

Setting new values only affects the entities drawn after the change. It is not retroactive, and changing either setting has no immediate visual effect on the drawing at the time the change is made.

TUTORIAL

Let's try constructing a simple drawing using the Elevation command to set the base elevation and the thickness of the extrusion.

Begin a new drawing called "3D." Set limits of 0,0 and 12,8. It is helpful to turn on snap and grid.

Now let's set the first elevation and thickness.

    Command: **Elev**
    New current elevation <current>: **0**
    New current thickness <current>: **4**

Draw a box (see Figure 33-3) using the Rectang command.

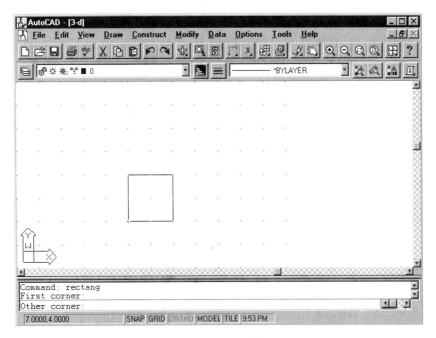

FIGURE 33-3 Box

Now change the thickness again:

    Command: **Elev**
    New current elevation <0>: (ENTER)
    New current thickness <4.00>: **10**

Notice how the currently set values show up as defaults. We pressed enter to accept the current elevation of zero but changed the thickness to 10.

Now use the Circle command to draw a circle like the one shown in Figure 33-4.

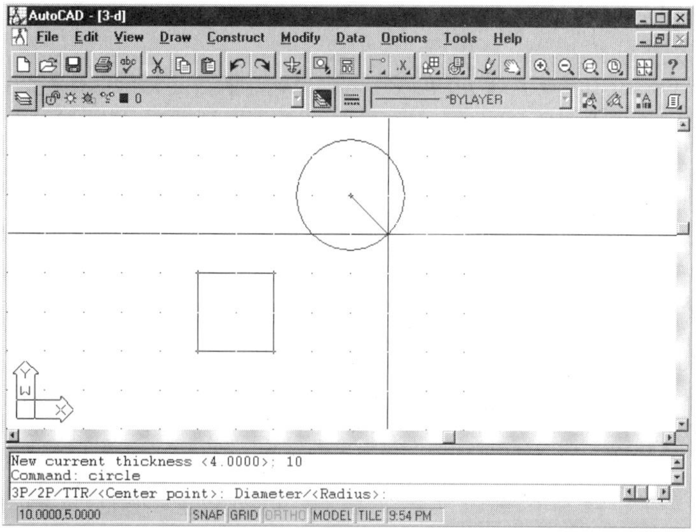

FIGURE 33-4 Circle and Box

Use the Vpoint -1,1,1 command to view the 3D shapes! Your drawing (depending on the viewpoint) should appear similar to the one in Figure 33-5.

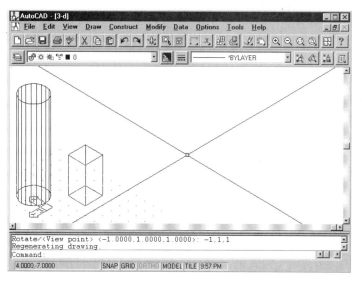

FIGURE 33-5 3D View

Try changing the elevation and thickness with the Change command.

Changing Existing Entities

In order to create several entities of different elevations and/or thicknesses, the Elevation command is used to change each value before drawing the new entities.

Many times, entities are drawn with an incorrect thickness. If this happens, it is not necessary to erase the entities and redraw them with a new thickness. The Change command is very convenient for changing this.

Simply issue the Change command.

> Command: **Change**
> Select objects: *(Select.)*
> Properties/<Change point>: **P**

At the last prompt, you may choose "Thickness" from the screen menu. If you are typing the commands from the keyboard, enter **P** to select the Properties option, and the following prompt appears.

> Change what property (Color/Elev/LAyer/LType/LtScale/Thickness)?:

Enter either **E** or **T** (for Elevation or Thickness) and you are prompted for the new thickness for the entity or entities you selected.

> **NOTE:** It is often easier to construct all or part of a drawing at a single thickness, than to use the Change command to reset one entity or groups of entities when you are through.

DRAWING IN COORDINATE SYSTEMS

If we wish to draw more complex objects, we must use the User Coordinate System to relocate the current system. This allows you to draw in a stipulated plane, as defined by the current UCS, as though you were in plan view.

This is especially useful, since most 3D drawings require detailing on the different faces (planes) of the objects in the drawing.

In addition to drawing in different planes, the thickness and elevation settings may be used. Note that these are relevant to the current UCS when the entity is drawn.

Let's look at an example of drawing in 3D by using different coordinate systems.

TUTORIAL

Let's draw a small shop building so we can become familiar with the techniques of drawing in AutoCAD 3D.

Starting the Drawing

Start a new drawing called "SHOP." Create the following setup parameters.

Units: Architectural
Limits: 0,0 / 80 ft.,50 ft. (remember to Zoom All)
Snap: 1 ft.
Grid: 10 ft. (initially)

In Release 13, the coordinates are set for relative display by default. Use [CTRL] [D] to toggle between the modes.

Drawing the Basic Shape

Select Line and draw the outline of the floor. The dimensions are 50 feet in length and 30 feet in width. Figure 33-6 shows the dimensioned plan. Do not dimension the plan.

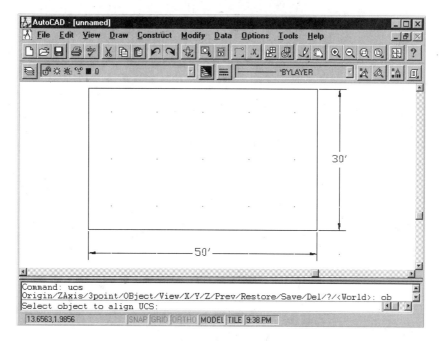

FIGURE 33-6 Floor Outline

Use the Vpoint -1,1,.3 command to create a view similar to the one shown in Figure 33-7.

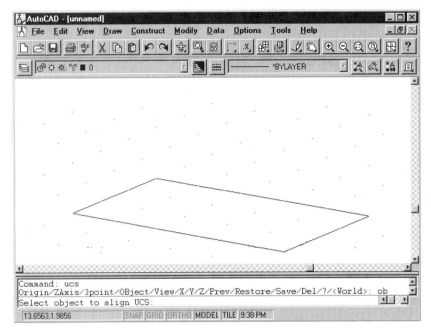

FIGURE 33-7 VPoint View

Let's save this view for recall later. Use the View Save command. Enter the name **view1** for this view.

We will now set the UCS icon so that it moves to each new origin as we set it. Select Options | UCS | Icon Origin from the menu bar.

Let's set the first UCS. Select View | Set UCS | Object from the menu bar. Select the nearest end wall line. Notice how the UCS icon moves to the corner of the building. Observe the orientation of the axes.

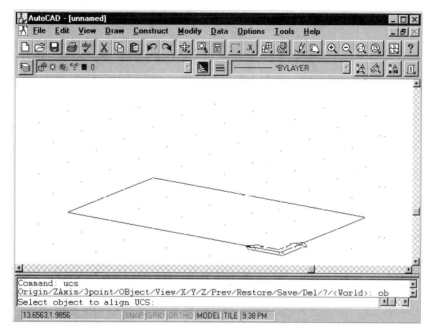

FIGURE 33-8 New UCS Origin

Let's now draw the first vertical corner line. Type **Line**, then choose the .XY filter from the cursor menu. Now select INTERSECtion snap. Snap to the intersection of the front corner. When the "need Z" prompt appears, enter a Z (height) value of **12 feet**. When AutoCAD prompts "To Point:" use INT to snap to the same corner.

Use the Copy command and Osnap to copy the vertical line to the other end wall corner.

Let's now rotate the UCS so we can draw directly on the end wall.

Type **UCS X**. The "X" option is used to rotate the UCS around the X-axis. By the right-hand rule, we want to rotate the UCS 90°. Enter **90** in response to the prompt. Notice how the UCS icon rotates.

We may want to save this UCS for recall later. Type **UCS Save**. Enter the name **END_WALL**.

Let's now change to the plan view for the current UCS. Type **Plan**. Press [ENTER] to default to the plan view for the current UCS.

We want to construct the roof angles as shown in Figure 33-9.

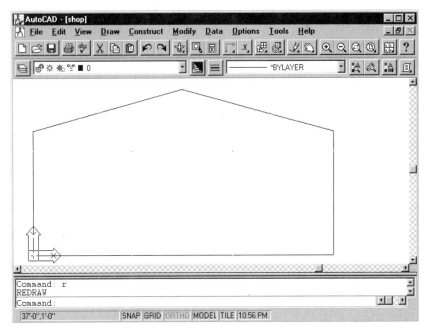

FIGURE 33-9 Endwall

You may want to zoom to a comfortable working size.

Type Line and use Endpoint object snap to connect to the top of the left vertical line. Enter the polar coordinate of **@16'<15** to define the endpoint. Connect a line to the other vertical line and enter the polar coordinate of **@16'<165**. Finish by filleting the ridge point to form a perfect intersection at the peak of the roof.

Restore the previously stored view by typing View Restore, then responding **VIEW1** to the prompt. Copy the end wall "panel" to the opposite end (using object snap aids in exact placement).

Enter VPOINT -2,-2,.75 to obtain the view like Figure 33-10. Now use the Line command and object snap to connect the ridge points and the edges of the roof as shown in Figure 33-10.

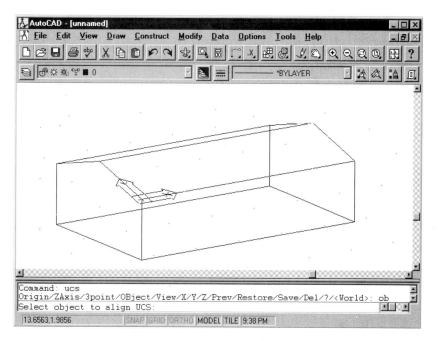

FIGURE 33-10 Main Building Lines

Drawing on the Roof

Let's now draw some lines on the slope of the roof to represent siding. Type UCS 3point. Use object snap to snap to the origin, X-axis line, and Y-axis line as shown in Figure 33-10. Notice how the UCS icon relocates to the edge of the roof and rotates to match the slope.

Type Plan and press ENTER to display the plan view of the new UCS. The plan view should look similar to the following illustration. You may want to zoom to a comfortable working size.

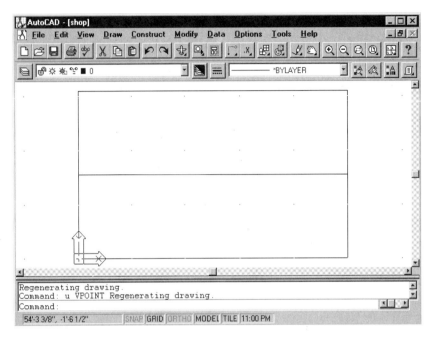

FIGURE 33-11   Roof in Plan View

Array the edge line of the roof (shown in the previous illustration). Type Array Rectangular, 1 row, 26 columns, and 24 inches between columns.

Type VPOINT 2,2,.75.

Type UCS 3point and reset the UCS to the opposite roof in accordance with the points shown in Figure 33-12.

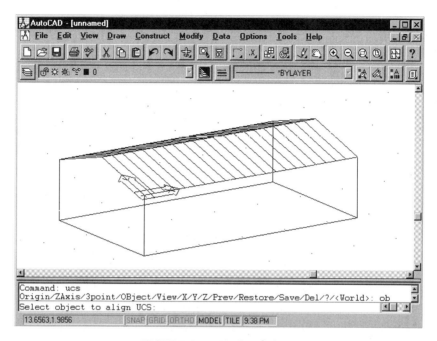

FIGURE 33-12 Roof Lines

Type Plan and press ENTER to default to the plan view of the new UCS. Array the roof line in the same manner as the opposite side of the roof.

Type View Restore View1 to restore our standard working view.

Drawing Doors on the Walls

Let's now proceed and draw some doors on one wall. Type UCS 3point. Select points as shown in the following illustration to set a new UCS. Type Plan, then press [ENTER] to set the plan view for the new UCS. The new plan view should look like Figure 33-13.

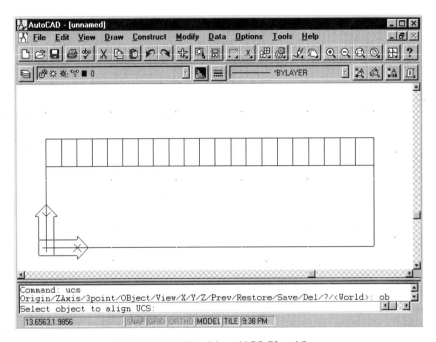

FIGURE 33-13 New UCS Plan View

Use the Line command to draw doors on the side wall of the shop. You can draw any type of doors and/or windows you wish. Figure 33-14 shows some doors as you may want to draw them.

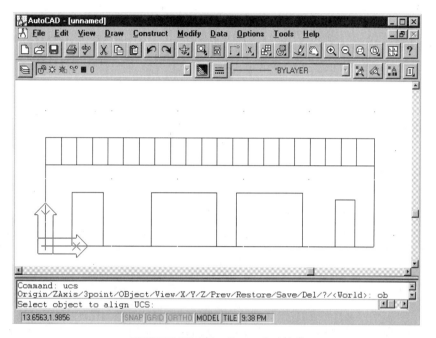

FIGURE 33-14 Doors in Wall

Let's now set the UCS system back to the World system before we model the drawing. Select UCS/World, then Plan/ENTER.

Viewing the Drawing in 3D

Use the Dview command to model the drawing. This is a good time to also practice using the Dview options such as Camera, Target, and Distance.

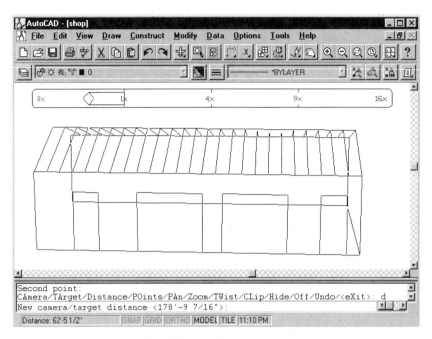

FIGURE 33-15 3D View

CREATING SOLID 3D FACES

It is often desirable to create solid faces on objects that will obscure objects behind them when the Hide command is used.

As you have already learned, many types of extruded entities naturally create solid faces. When you are drawing single entities (such as lines, circles, arcs, etc.), however, solid faces are not automatically formed.

3DFACE Command

The 3Dface command is used to create solid faces. The 3Dface command is similar to a solid face in that it creates a "face plane." It is dissimilar, however, because it can be defined with different Z coordinates for each corner of the face, creating the possibility of a nonplanar or warped plane. It is also void of solid fill.

Placing 3D Faces

The method of placing 3D faces is similar to that used with the Solid command. When stipulating the corner points, however, the points are entered in a consecutive clockwise or counterclockwise fashion. This differs from the method of entry for a solid, where such a sequence would create a "bow tie" effect.

The placement of 3D faces creates a visual "edge frame." This can be undesirable in some situations. Take, for example, an area of irregular shape. Placing a 3D face creates several planes that fill in the irregular area. Normally, each frame would contain an edge that is visible in the drawing. Figure 33-16 shows the same drawing with and without edge frames on the 3D faces.

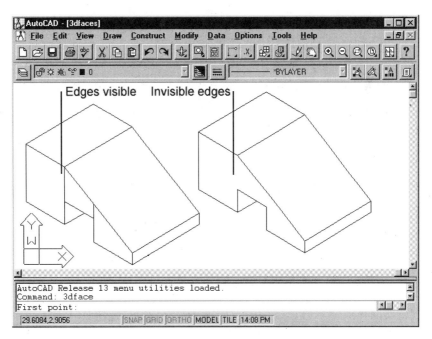

FIGURE 33-16 3D Face Edge Frames

TUTORIAL

Use the Line command to draw the polygon shown in Figure 33-17. Issue the 3Dface command and enter the following sequence.

Command: **3DFACE**
First point: *(Enter point 1.)*
Second point: *(Enter point 2.)*
Third point: **I** (ENTER) *(Enter point 3.)*
Fourth point: *(Enter point 4.)*
Third point: *(Enter point 5.)*
Fourth point: *(Enter point 6.)*
Third point: *(Press* ENTER *)*

After the final ENTER, the second 3D face plane is closed.

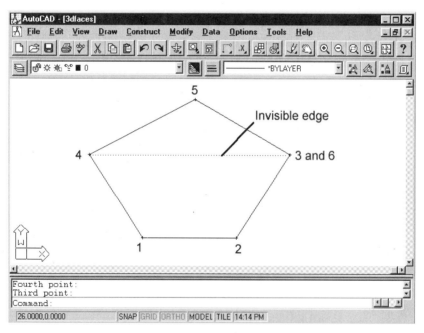

FIGURE 33-17 3D Face Sequence

Notice how the **I** (invisible) option is entered prior to the segment of the frame you wish to be invisible. Of course, all frame segments may be stipulated to be invisible.

The frame segments that are constructed with the **I** option may be changed to visible with the use of the Splframe system variable. A zero setting maintains invisibility, while a nonzero (such as 1) setting makes them visible. This is convenient when you wish to edit sections that were constructed as invisible. The change is not apparent until the next regeneration is performed.

The 3Dface command uses an option called "invisible" as a modifier to delete the appearance of the edge frame. Let's look at an example of both placing a 3D face and making part of the frame area invisible.

3D POLYGON MESHES

A 3D polygon mesh can be thought of as a "blanket" type of entity that can be curved and warped into shapes that cannot be described by other entities.

Mesh Density

There are several methods of constructing polygon meshes. All the methods of construction create a mesh that is defined by a density. The density is the number of vertices used to describe the surface of the mesh. Figure 33-18 shows an object described by a mesh in two different densities.

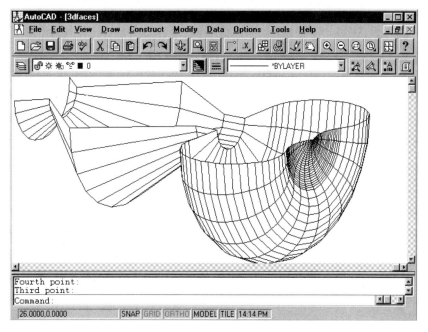

FIGURE 33-18 3D Polygon Meshes

The density is described by the system variables named Surftab1 and Surftab2. One-directional meshes are controlled by Surftab1, while two-directional meshes are controlled by both variables; one for each direction. The directions are defined in AutoCAD as "M" and "N."

The methods of mesh construction are as follows:

>3DMESH
>PFACE
>RULESURF
>TABSURF
>REVSURF
>EDGESURF

Let's look at each and study the construction method of each.

3DMESH COMMAND

A 3Dmesh is constructed by stipulating the number of vertices in each (M and N) direction, then specifying the X,Y,Z coordinate of each of the vertices.

It should be noted at this time that construction of 3D meshes is very tedious. In most situations, it is more efficient to construct a mesh of the same type with one of the other construction methods. The use of the 3Dmesh command is best utilized in LISP routines. With this in mind, let's proceed to study the method by which it works.

Constructing a 3D Mesh

To construct a 3D mesh, first issue the 3Dmesh command, then the number of vertices in the M and N directions, and finally the vertex coordinate of each. Let's look at an example.

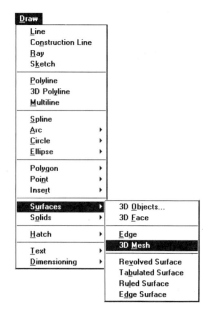

>Command: **3Dmesh**
>Mesh M size: **3**
>Mesh N size: **3**
>Vertex (0,0): **10,10,-1**
>Vertex (0,1): **10,20,1**
>Vertex (0,2): **10,30,3**
>Vertex (1,0): **20,10,1**
>Vertex (1,2): **20,20,0**
>Vertex (1,3): **20,30,-1**
>Vertex (2,0): **30,10,0**
>Vertex (2,1): **30,20,1**
>Vertex (2,2): **30,30,2**

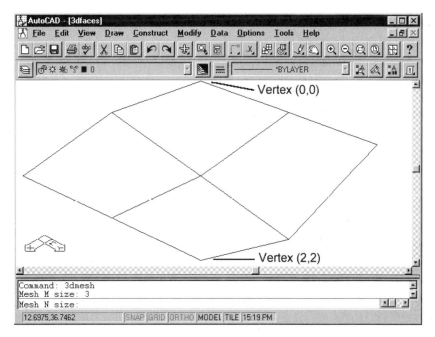

FIGURE 33-19 3D Polygon Mesh

The order of entry is one column at a time, then to the next column and up.

3D meshes are displayed as wireframes; when they are coplanar, they are considered opaque when the Hide command is used. 3D meshes cannot be extruded.

PFACE COMMAND (POLYFACE MESH)

A mesh constructed with the Pface command will draw a polygon mesh that is independent of a continuous surface. Pface meshes are defined by entering individual vertex values in X,Y,Z format.

Selecting the Pface command results in the prompt:

Command: **Pface**
Vertex 1:
Vertex 2:
Vertex 3:
Vertex 4:

and so forth, until you press [ENTER] on a blank line to close the mesh. Pressing [ENTER] again will terminate the construction of the mesh.

Polyface meshes can be used for simple mesh construction or be incorporated in a program to automatically enter many data points for a specialized application.

RULESURF COMMAND (RULED SURFACE)

A mesh created by the Rulesurf command creates a mesh that represents a ruled surface between two entities. The entities may be lines, points, arcs, circles, 2D and 3D polylines.

A similar effect could be achieved on a drawing board by dividing two entities into the same number of segments and drawing a line between the corresponding segments. Figure 33-20 shows ruled surfaces between entities.

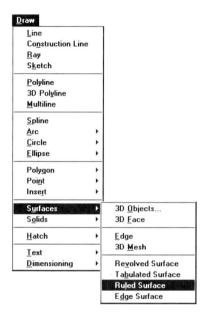

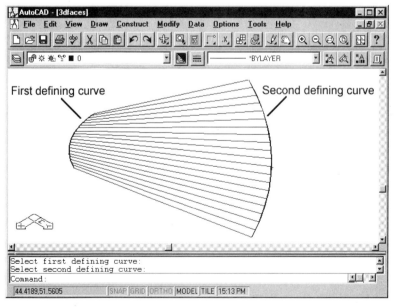

FIGURE 33-20 Ruled Surface

Constructing Ruled Surfaces

To construct a ruled surface, select the Rulesurf command, then select the two entities you wish the surface to be drawn between.

Command: **Rulesurf**
Select first defining curve: *(Select.)*
Select second defining curve: *(Select.)*

AutoCAD draws the ruled surface from the endpoint nearest to the selection point on the entity. Thus, selecting opposite ends of the entities will create a ruled surface that is crossed.

If an entity is a circle, the selection point has no special effect. The ruled surface will begin at the 0° quadrant point. If a point is selected, the mesh lines will be drawn from the single location of the point. Thus, a circle and a point would create a cone-type object as in Figure 33-21.

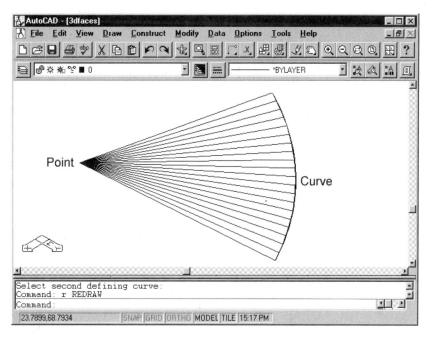

FIGURE 33-21 Cone Constructed with RuleSurf

TABSURF COMMAND (TABULATED SURFACE)

The TABSURF command constructs a mesh surface that is defined by a path and a direction vector.

The effect is similar to an extrusion, except for the mesh surface.

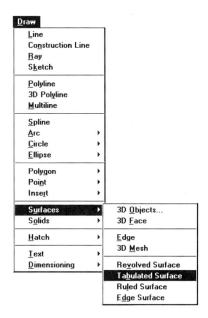

Constructing Tabulated Mesh Surfaces

To construct a tabulated surface, select a path curve (entity from which the surface will be calculated) and a direction vector (entity describing the direction and length of the mesh from the path curve).

The following command sequence is displayed.

    Command: **Tabsurf**
    Select path curve:
    Select direction vector:

The path curve may be a line, arc, circle, 2D or 3D polyline. The direction vector may be either a line or open polyline (2D or 3D).

Use object pointing to identify each. The closest end to the pick point on the direction vector defines the direction in which the mesh is projected. Picking the direction vector at one end causes the mesh to project in the direction of the opposite endpoint. Figure 33-22 shows the curve path, direction vector, and the pick point on the direction vector for the result shown.

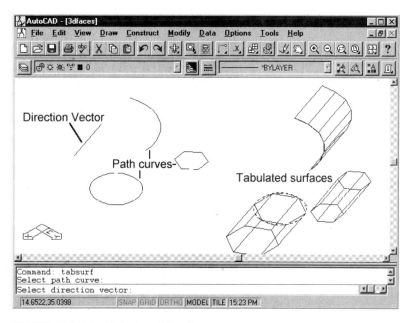

FIGURE 33-22 Tabulated Surfaces

The heavier line identifies the original path curve.

REVSURF COMMAND (REVOLVED SURFACE)

The REVSURF command is used to create meshes from entities revolved around an axis.

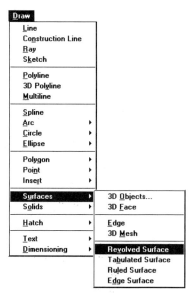

Creating Revolved Surfaces

To create a revolved surface, you must first have a path curve (entity to be revolved) and an axis of revolution (an entity that defines an axis around which the entity will revolve). The path curve can be a line, circle, arc, 2D or 3D polyline. The axis of revolution may be a line or open polyline (2D or 3D).

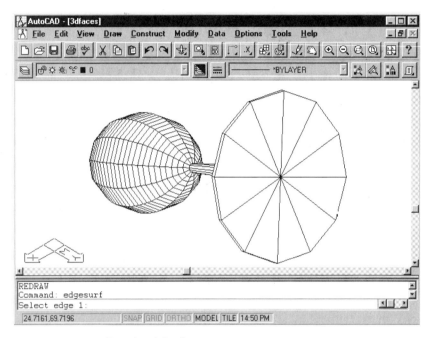

FIGURE 33-23 Revolved Surface

The displayed command prompts are:

Command: **Revsurf**
Select path curve:
Select axis of revolution:
Start angle <0>:
Included angle (+=ccw, -=cw) <Full circle>:

The start angle point of zero is defined by the location of the path curve. If you specify a start angle other than zero, the starting point of the revolution will offset the specified number of degrees from the path curve.

The included angle designates the degrees of rotation of the path curve around the axis of revolution and from the start point.

The direction to which the revolution emanates is dependent on the pick point when the axis of revolution is selected. The right-hand rule is used to determine the direction. The revolution is in a positive rotation direction around the axis of revolution; the positive direction of the axis defined as being from the end nearest the pick point to the following end. Thus, if your right hand is curved around the axis of revolution, with the thumb pointing toward the end of the axis furthest from the pick point, the direction of your fingers designates the positive rotation angle.

Figure 33-24 shows the effect of the location of the pick point on the axis of revolution in determining the direction of revolution. Each revolution has a starting angle of 0 degrees and an included angle of 90 degrees.

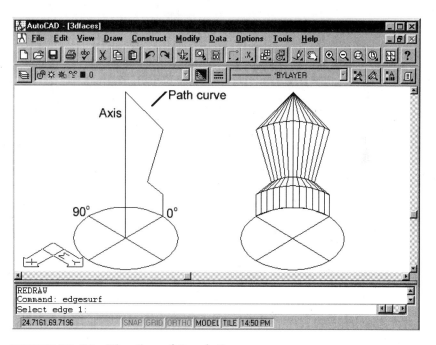

FIGURE 33-24 Direction of Revolution

EDGESURF COMMAND (EDGE-DEFINED SURFACE)

The EDGESURF command is used to construct a Coons surface patch from four edge surfaces. Figure 33-25 shows a Coons surface patch.

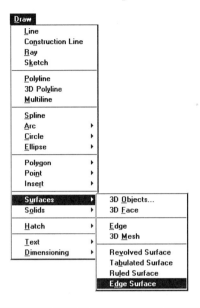

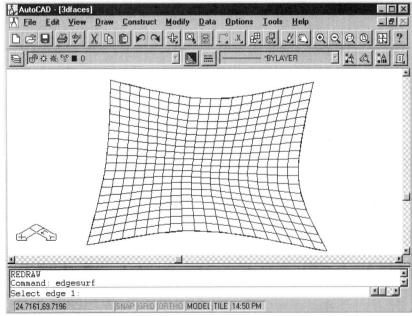

FIGURE 33-25 Coons Surface Patch

The patch follows the contour of each edge surface, forming a mesh that approximates a surface attached to all points of each edge.

Constructing Edge-Defined Surfaces

To construct a Coons patch, you must first have four edges. These edges may be lines, arcs, or open 2D or 3D polylines. They must connect at their endpoints. The command sequence is as follows:

Command: **Edgesurf**
Select edge 1:
Select edge 2:
Select edge 3:
Select edge 4:

If one edge is not connected to the adjacent edge, the message

Edge x does not touch another edge

is displayed, where x defines the number of the edge that does not touch.

3D OBJECTS

Several basic object shapes have been included for convenient use in drawings. These objects are actually constructed from the tools you have already learned to use. The sequence to construct them has been preprogrammed. Prompts question you for pertinent information for proper construction to your specifications. The following objects are included for your use.

BOX
CONE
DOME
DISH
SPHERE
TORUS
WEDGE

Let's look at each of these and the information you will need to provide.

Box

The AI_Box command constructs a box of given length, width, and height, or a cube of a given edge dimension. The command sequence is as follows:

Command: **AI_Box**
Corner of box:
Length:
Cube <Width>:
Height:
Rotation angle about Z axis:

Figure 33-26 shows the components of box construction.

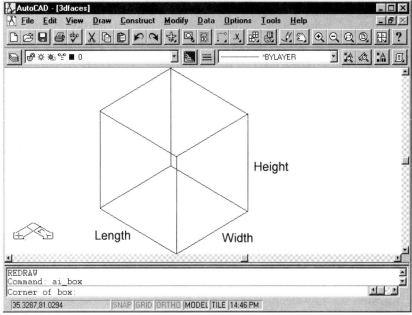

FIGURE 33-26 Components of a 3D Box

If you select the Cube option (on the fourth prompt line), a cube will be constructed with the edge length equal to the length you previously defined.

Cone

The AI_Cone command is used to construct cones. You define a cone by specifying the radius or diameter of the top and bottom and the height.

The command prompts are:

    Command: **AI_Cone**
    Base center point:
    Diameter/<radius> of base:
    Diameter <radius> of top <0>:
    Height:
    Number of segments <16>:

Figure 33-27 labels the parts of a cone.

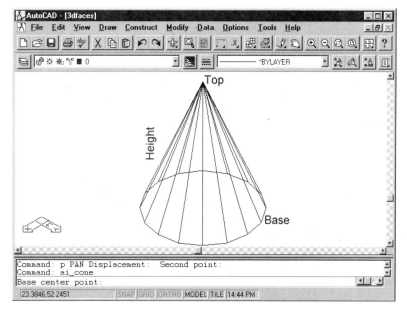

FIGURE 33-27 Components of a 3D Cone

Dome

Use the AI_Dome command to create domes for your drawing.

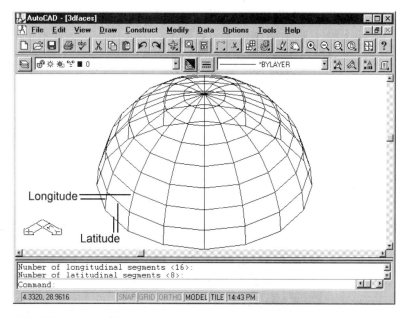

FIGURE 33-28 Dome

The command sequence is as follows.

Command: **Al_Dome**
Center of dome:
Diameter/<Radius>:
Number of longitudinal segments <16>:
Number of latitudinal segments <8>:

Dish

A dish is simply an inverted dome. The construction of a dish is similar to that of a dome.

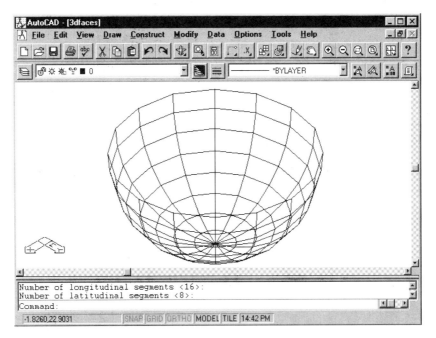

FIGURE 33-29 Dish

The following command sequence is displayed.

Command: **Al_Dish**
Center of dish:
Diameter/<Radius>:
Number of longitudinal segments <16>:
Number of latitudinal segments <8>:

Sphere

Spheres are created with the AI_Sphere command.

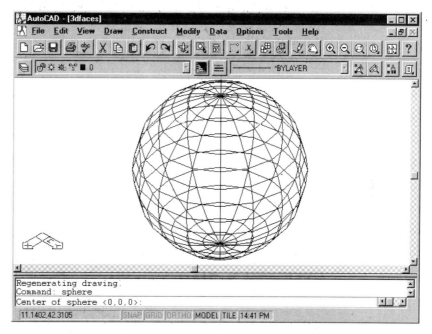

FIGURE 33-30 Sphere

The command sequence is as follows:

Command: **AI_Sphere**
Center of sphere:
Diameter/<radius>:
Number of longitudinal segments <16>:
Number of latitudinal segments <16>:

Torus

A torus is a closed tube, rotated around an axis.

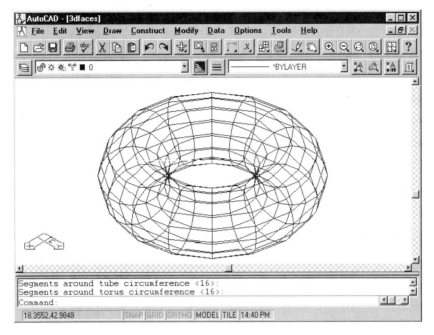

FIGURE 33-31 Torus

The command sequence for creating a torus is:

Command: **Al_Torus**
Center of torus:
Diameter/<radius> of torus :
Diameter/<radius> of tube:
Segments around tube circumference <16>:
Segments around torus circumference <16>:

If you enter a tube radius or diameter that is greater than the torus radius or diameter, the following message is displayed:

Tube radius cannot exceed torus radius

You will then be given the opportunity to reenter the data.

Wedge

A wedge could be thought of as a block, cut diagonally along its width.

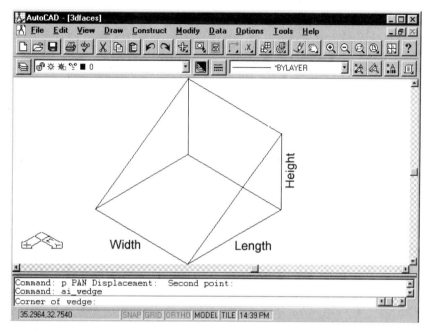

FIGURE 33-32 Wedge

The command sequence when constructing a wedge is:

> Command: **AI_Wedge**
> Corner of wedge:
> Length:
> Width:
> Height:
> Rotation angle about Z axis:

SUMMARY

Mastering the principles of 3D drawing takes time and effort. Try to simulate the examples in the text and experiment on projects of your own interest to build expertise in each area of 3D drawing.

Understand the principles of 3D, as discussed in Chapter 30.

Master the principles of viewing 3D drawings covered in Chapter 31. This is not only important for viewing, but to display the drawing for effective 3D drawing construction while the drawing is in progress.

The User Coordinate System explained in Chapter 32 is extremely important, since it is one of the primary methods of construction used in 3D drawings. You can not obtain the ultimate results from AutoCAD 3D without a complete understanding of this system.

The drawing methods described in this chapter should be considered as a toolbox to be used for 3D construction. Study each drawing to determine the best method of creating the objects in your drawing.

And most of all, have fun!

CHAPTER REVIEW

1. What is an *extrusion*?

2. Which axis direction is an extrusion normally projected into?

3. What is meant by the elevation and thickness of an entity?

4. How could you change the thickness of an existing entity?

5. How do you return to the plan view after rotating your drawing?

6. What does the 3Dface command perform?

7. Compare the placement of a 3D face with a solid's face that would be placed with the Solid command.

8. What procedure is used if you want to place a 3D face without showing the edges?

9. Describe a use for the Rulesurf method of constructing a surface.

10. The density of a two-direction 3D surface is described by the values of M and N. Which system variable controls the values of each?

CHAPTER

34

INTRODUCTION TO SOLID MODELING

Solid modeling can be easy, fun, and useful. This chapter introduces you to the fundamentals of solid modeling. After completing this chapter, you will be able to:

- Comprehend the purpose and use of the AutoCAD ACIS modeler.
- Differentiate the differences between solid models and 3D models.
- Display a solid model in different ways.
- Demonstrate the methods of constructing solid objects.

OVERVIEW

AutoCAD's ACIS modeler is another way to create solid three-dimensional (3D) objects. The solid objects created are similar in appearance to wireframe models created in AutoCAD 3D. Solid models, however, contain more information than a wireframe model. Let's look at some of the differences.

Wireframe models created in AutoCAD 3D are composed of solid faces. Imagine a cube that is constructed from thin pieces of board. The cube has six faces, but no solid interior. A solid model, however, is not constructed of faces; it is a solid object. The same cube constructed with a solid modeling program would have a solid interior.

Creating objects as solids allows you to scientifically analyze the object. Object properties such as center of gravity, mass, surface area, and moments of inertia can be calculated. Since different materials (such as steel, aluminum, etc.) differ in their properties, AutoCAD includes several material definitions. Each definition can be assigned as the material for the solid model. In addition, you can create a new definition by entering the material properties of any material.

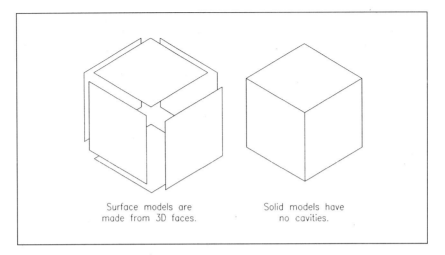

FIGURE 34-1 3D Versus Solid Objects

Who Uses Solid Modeling

Solid modeling is useful to many disciplines. The mechanical engineer who needs to calculate properties of objects will find the ability to scientifically analyze an object of different materials invaluable.

Architects can use solid modeling to visually represent complex intersections of building roofs. After the model is constructed, the drawing can be converted to a "standard" 3D form and embellished for presentation purposes.

Those who model the objects they create will find that solid modeling offers the very useful possibilities of intersection construction, scientific analysis, and the simplicity of "building block" construction.

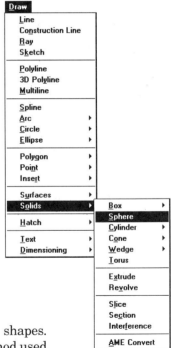

Drawing with Solids

Solid models are created from solid three dimensional shapes. This "building block" approach is different from the method used with the AutoCAD 3D program. Objects created in 3D, however, can be converted to solids. This allows the flexibility to create an object in the most efficient manner. Before you begin your work with solid modeling, you should be proficient in the AutoCAD 3D program.

Solid Primitives. AutoCAD provides commands that create basic building block shapes referred to as solid "primitives." These simple 3D solid shapes are the box, wedge, cone, cylinder, sphere, and torus. These shapes can be "added" or "subtracted" from each other to create more complex shapes. In addition, you can edit with commands that revolve, extrude, chamfer, or fillet the shapes.

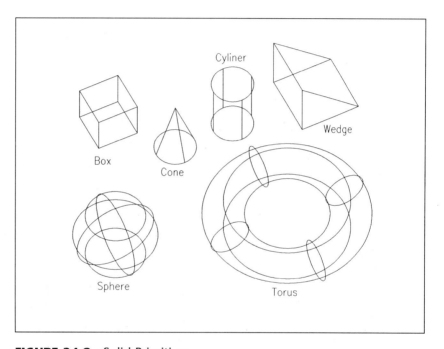

FIGURE 34-2 Solid Primitives

Composite Solids. When you create a solid object from several primitives, it is referred to as a composite solid. The primitives that combine to make the object can be consolidated to create a single object, enclosing the entire volume as one solid.

You may also add or subtract solids from each other. For example, a drill hole can be created from a cylinder that has been placed in a solid plate. Subtracting the cylinder from the plate creates the "drill hole."

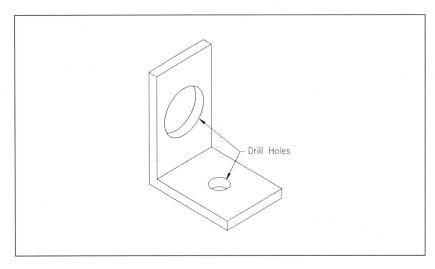

FIGURE 34-3 Solid with Drill Holes "Subtracted"

Solid Modeling Commands. AutoCAD's solid modeling commands are indistinguishable from other commands. For example, to draw a box, you use the Box command.

CHAPTER REVIEW

1. What is the primary difference between an object created with 3D and one created as a solid model?

2. What is meant by a *building block* approach?

3. What is a *solid primitive*?

4. What is a *composite solid*?

CHAPTER 35

CONSTRUCTING SOLID PRIMITIVES

Solid primitives are used to create solid models from building block shapes. Many solid models can be created from these basic shapes. In order to become proficient in the solid modeling program, you should master the construction of each of these primitive shapes. After completing this chapter, you will be able to:

- Comprehend the Solids menu organization.
- Utilize each of the solid primitive construction commands.

SOLID PRIMITIVES

Solid primitives are the "building blocks" of solid modeling. The solid primitives available are as follows:

| Box & Cube | Cylinder | Torus |
| Cone | Sphere | Wedge |

35-2 USING AUTOCAD

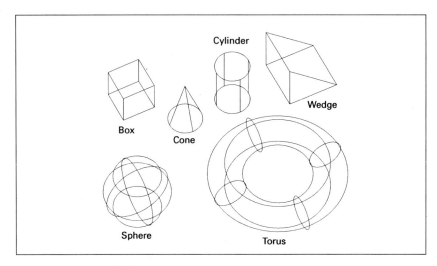

FIGURE 35-1 Solid Primitives

Solid primitives are created by providing the dimensions of the object. Let's draw each of the primitives. Start a new drawing of any name and follow the short tutorial for each.

Drawing Solid Primitives

If you are using a digitizer tablet with the AutoCAD tablet overlay, the ACIS Solids commands can be accessed from the tablet. Figure 35-2 shows the tablet area containing the ACIS Solids commands.

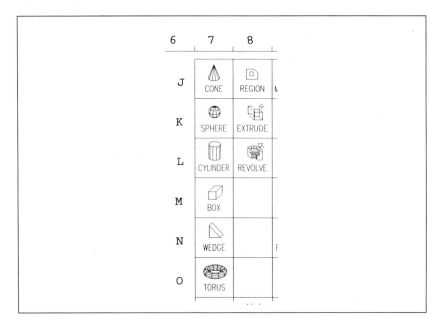

FIGURE 35-2 ACIS Commands Found on the Tablet Overlay

CONSTRUCTING SOLID PRIMITIVES **35-3**

The Solids icon toolbar can be accessed through the Tools/Toolbars/Solids pull-down menu.

Note also that you can use the "Draw/Solids" pull-down menu. We will use these menus to select the commands to construct our solid primitives.

Let's use the solid primitive commands to construct some solid objects. Let's start with the Box command.

Drawing a Solid Box

The Box command is used to draw a solid 3D box.

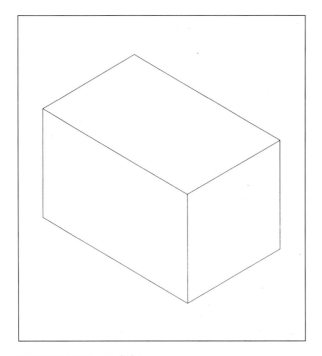
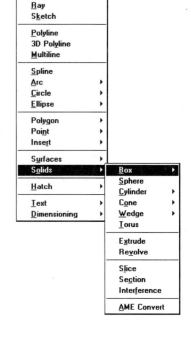

FIGURE 35-3 Solid Box

There are three methods of creating a box with this command. Let's look at each separately.

Creating a Solid Box Specifying Opposite Corners and Height. This is the default method of creating a box. Start by selecting the Solbox command. Enter the first point, then the opposite corner. Refer to the following command sequence and Figure 35-4.

Command: **Box**
Center/<Corner of box>: *(Enter point "1".)*
Cube/Length/<Other corner>: *(Enter point "2".)*

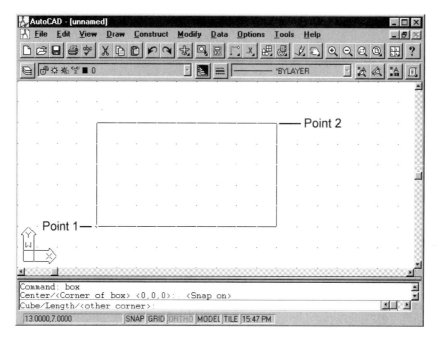

FIGURE 35-4 Drawing a Solid Box

The two points you have just entered specify the length and width of the box. You must now specify the height. You could also enter a relative coordinate to locate the opposite corner of the box after entering the first corner. The command line will now prompt you for that height. In this example, let's make our box 3 units high.

Height: **3**

You may want to stop at this point and use AutoCAD's Vpoint command to display a 3D representation of the box. Return to the Plan view and use a Zoom All after you are finished.

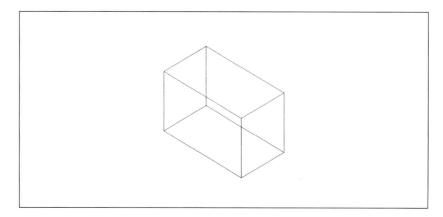

FIGURE 35-5 3D View of Box

Note that the length and width of the box are constructed relative to the current UCS and the height is perpendicular to the current UCS.

Let's continue and look at the other two ways to construct a solid box.

Length Option of Constructing a Solid Box. The Length option allows you to construct a box by designating the actual length, width, and height of a box in numeric values. Let's construct a box that is 4 units long, 3 units wide, and 2 units high. The following command sequence will guide you through the process.

> Command: **Box**
> Corner of box: *(Enter a point on the screen.)*
> Cube/Length/<Other corner>: **LENGTH**
> Length: **4**
> Width: **3**
> Height: **2**

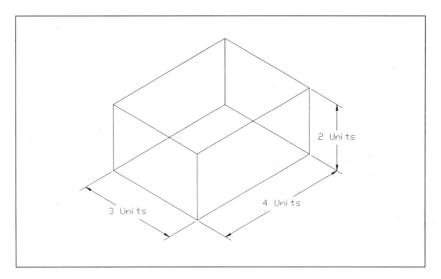

FIGURE 35-6 Constructing a Box with the Length Option

Creating a Solid Cube. The Cube option is used to create a solid cube. Let's draw a cube.

> Command: **Box**
> Corner of box: *(Enter a point showing the corner of the box.)*
> Cube/Length/<Other corner>: **CUBE**
> Length: **3**

AutoCAD will now construct the cube, using the value entered for the length (in this case, the value of 3 you entered) to construct a cube with all sides equal to the entered value.

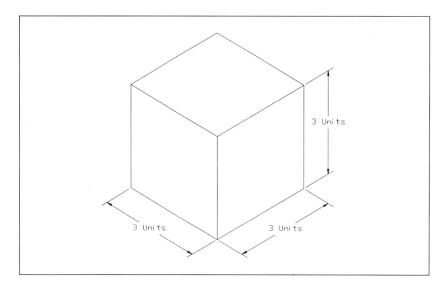

FIGURE 35-7 Solbox Cube

Creating a Solid Cone

The Cone command is used to create a solid cone.

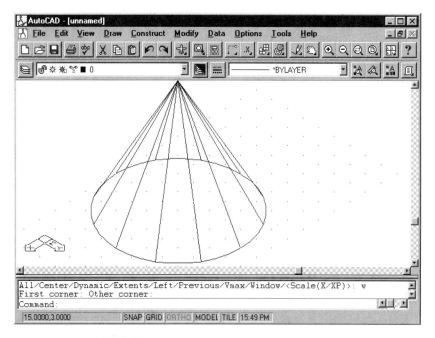

FIGURE 35-8 Solid Cone

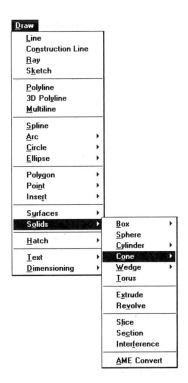

A cone can be constructed with either a circular or elliptical base. The cone's base will lie in the *X,Y* plane of the current UCS. The height describes the distance between the base and the point, and is perpendicular to the current UCS.

Let's construct a cone with a circular base. Select the Cone command and use the following command sequence.

> Command: **Cone**
> Elliptical/<Center point>: *(Enter a point on the screen.)*
> Diameter/<Radius>: **2**

Note that you can construct the base by specifying either the diameter or the radius of the base. In our case, we entered a value of 2 units for the radius, since the radius is the default (in <brackets>). If you wish to enter the diameter, enter **D** at this prompt and you will be prompted for the diameter. Let's continue with the command sequence:

> Apex/<Height>: **3**

Figure 35-9 shows the completed cone in 3D.

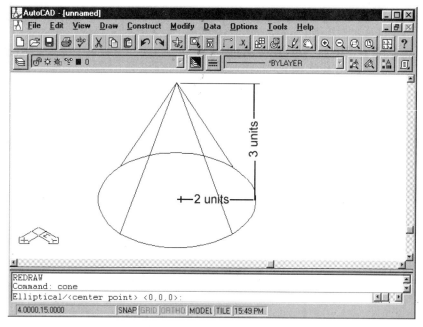

FIGURE 35-9 Completed Cone in 3D

Constructing a Cone with an Elliptical Base. The Elliptical option allows you to construct a cone with an elliptical base. Let's look at an example. Refer to Figure 35-10.

Command: **Cone**
Elliptical/<Center point>: **ELLIPTICAL**
Center<Axis endpoint>: *(Enter a point for the first axis endpoint — shown as "Point 1".)*
Axis endpoint 2: *(Enter "Point 2".)*
Other axis distance: *(Enter "Point 3".)*
Apex/<Height>: **3**

CONSTRUCTING SOLID PRIMITIVES **35-9**

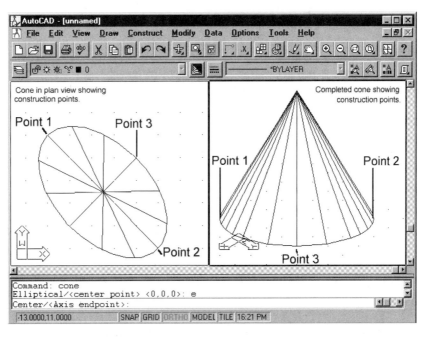

FIGURE 35-10 Drawing a Cone with an Elliptical Base

The base ellipse is constructed in the same manner as a standard ellipse when using the AutoCAD Ellipse command. You can also construct an ellipse using the Center option (see the third line of the previous command sequence). The following command sequence shows the prompts for this type of solid cone construction:

Command: **Cone**
Elliptical/<Center point>: **ELLIPTICAL**
<Axis endpoint 1>Center: **CENTER**
Center of ellipse:
Axis endpoint:
Other axis distance:
Apex/<Height>:

Notice, again, the similarity to the AutoCAD Ellipse command.

Creating a Solid Cylinder

The Cylinder command is used to construct solid cylinders.

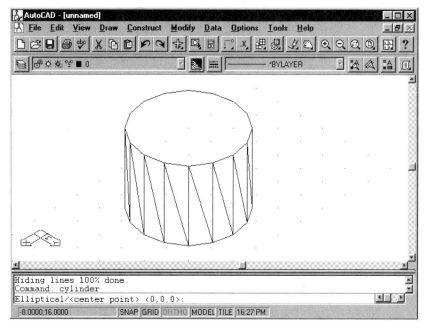

FIGURE 35-11 Solid Cylinder

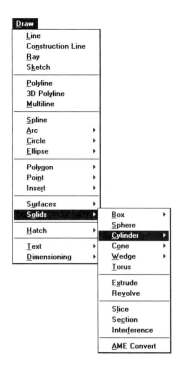

Solid cylinders are constructed in exactly the same manner as solid cones. The difference, of course, is that the cylinders are not tapered to a point.

Cylinders can be constructed with either circular or elliptical bases. Elliptical base cylinders are also constructed in the same manner as cones with elliptical bases.

Let's construct a solid cylinder with a circular base. Refer to Figure 35-12 for the points to enter.

Command: **Cylinder**
Elliptical/<Center point>: *(Enter "Point 1".)*
Diameter/ <Radius>: *(Enter "Point 2".)*
Center of other end<Height>: **3**

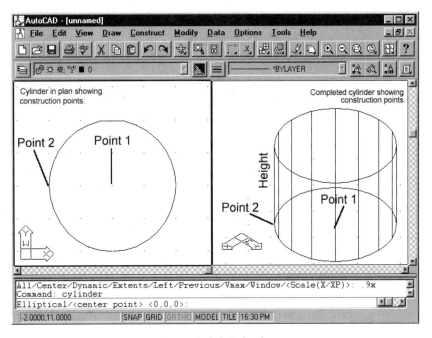

FIGURE 35-12 Constructing a Solid Cylinder

Creating a Solid Sphere

The Sphere command is used to create a solid 3D sphere (ball).

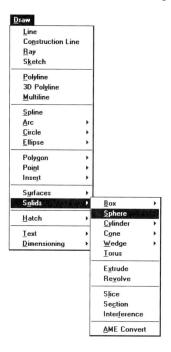

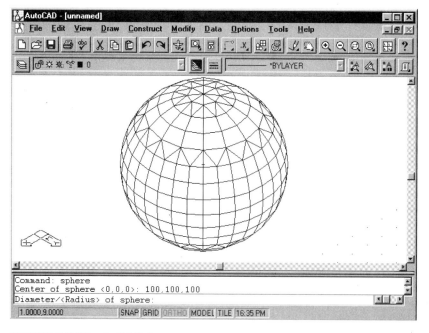

FIGURE 35-13 Solid Sphere

Constructing a solid sphere is very simple. You first designate the center point of the sphere, then either the radius or diameter. Let's construct a sphere. Use Figure 35-14 for the points to enter.

Command: **Sphere**
Center of sphere: *(Enter "Point 1".)*
Diameter/<Radius> of sphere: *(Enter "Point 2".)*

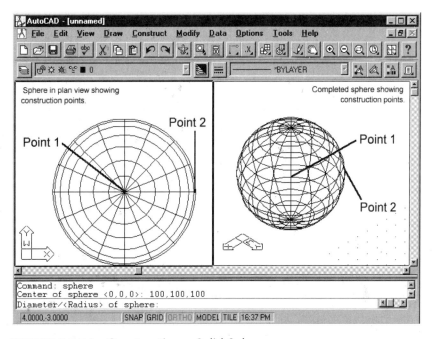

FIGURE 35-14 Constructing a Solid Sphere

The sphere is constructed with the center point you entered on the X,Y axis of the current UCS. The circle described by the two entered points is then rotated about the center point to construct the sphere. The vertical axis of the sphere is perpendicular to the X,Y plane of the UCS.

Constructing a Solid Torus

The Torus command is used to create a solid torus. A torus is a circle rotated about a point to create a tube, like a doughnut. The Torus allows some variations of the traditional torus object. Figure 35-15 shows a solid torus.

35-14 USING AUTOCAD

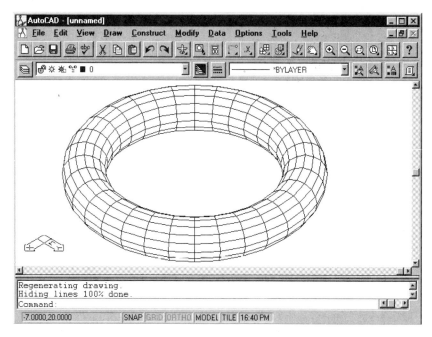

FIGURE 35-15 Solid Torus

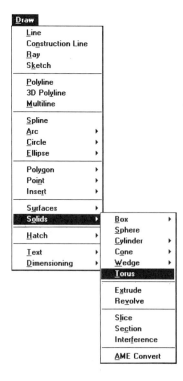

Before we construct a solid torus, let's look at the components of a torus. Figure 35-16 shows plan views of a torus with the components listed.

CONSTRUCTING SOLID PRIMITIVES **35-15**

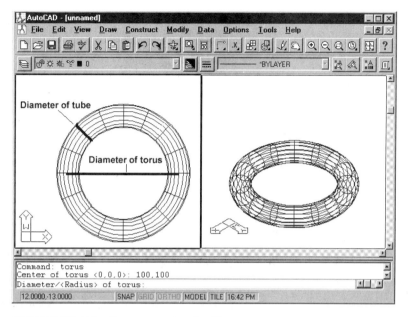

FIGURE 35-16 Components of a Torus

Let's construct a standard torus. Use the following command sequence and Figure 35-17.

Command: **Torus**
Center of torus: *(Enter "Point 1".)*
Diameter/<Radius> of torus: *(Enter "Point 2".)*
Diameter/<Radius> of tube: *(Enter "Point 3".)*

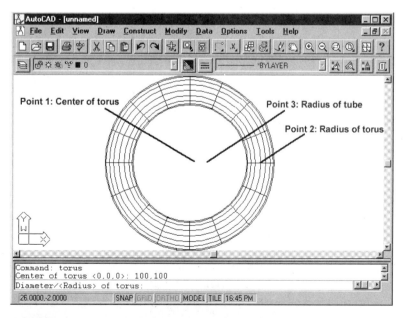

FIGURE 35-17 Constructing a Torus

It is possible to construct a torus with a tube radius that exceeds the radius of the torus. Take a moment to create a torus that has a radius of 2 and a tube radius of 5. View the torus in 3D with the Vpoint command. Notice how the tubes intersect around the center point of the torus.

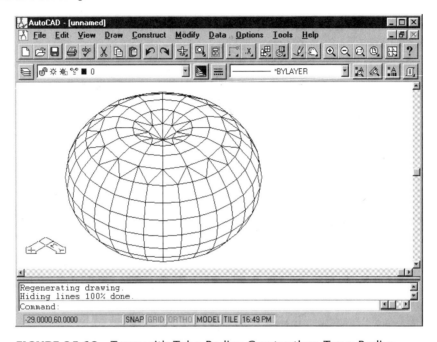

FIGURE 35-18 Torus with Tube Radius Greater than Torus Radius

You can also use a negative value for the torus radius. The tube radius, however, must be a positive number of greater value. For example, if the torus radius is –3, the tube radius must be greater than 3. Construct a torus with a torus radius of –3 and a tube radius of 4.

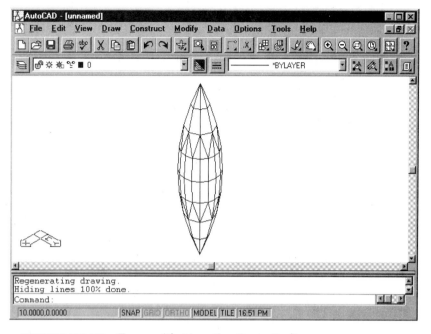

FIGURE 35-19 Torus with Negative Torus Radius

If you have used the Vpoint command to view your torus, select the Plan command to return to the plan view and Zoom All before proceeding.

Constructing a Solid Wedge

The Wedge command is used to create a solid wedge.

35-18 USING AUTOCAD

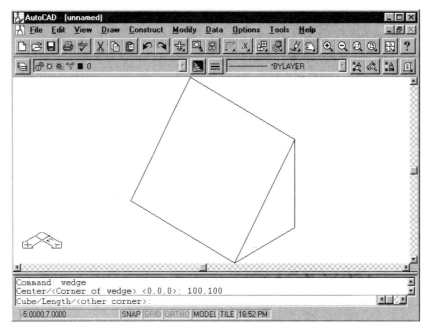

FIGURE 35-20 Solid Wedge

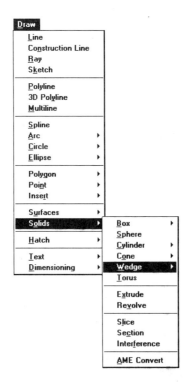

The wedge is constructed by specifying the dimensions of the base, then the height of the end of the wedge. You may alternately enter the length, width, and height in numeric values. Let's construct a wedge. Refer to Figure 35-21.

Command: **Wedge**
Center/<Corner of wedge>: *(Enter "Point 1".)*
Cube/Length/<other corner>: *(Enter "Point 2".)*

Notice how the base of the wedge is "rubber banded" from the first point. Let's continue in the command sequence.

Height: **2**

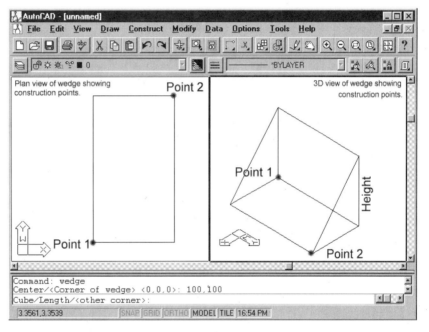

FIGURE 35-21 Constructing a Wedge

If you view the wedge in 3D, you will notice the height is applied to the end described by the first point entered, and the "point" of the wedge is placed at the end of the base described by the second point.

Length Option of Constructing a Solid Wedge. If you wish to construct a wedge by entering the actual dimensions as numeric values, simply select the "Length" option. The following command sequence shows the prompts displayed when you select Length.

Command: **Wedge**
Center/<Corner of wedge>: *(Enter a point.)*
Length/<Other corner>: **LENGTH**
Length: *(Enter a numeric value.)*
Width: *(Enter a numeric value.)*
Height: *(Enter a numeric value.)*

Center and Cube Option of Constructing a Solid Wedge. You can create a wedge centered on a point. As well, the wedge can have equal sides (the Cube option):

Command: **Wedge**
Center/<Corner of wedge>: **CENTER**
Center of wedge <0,0,0>: *(Pick.)*
Cube/Length/<Corner of wedge>: **CUBE**
Length:

CHAPTER REVIEW

1. What are some solid primitive objects?

2. What are the ways to access the AutoCAD solid modeling program?

3. What are the two types of solid cone bases that can be constructed?

4. What is a *torus*?

CHAPTER 36

CREATING CUSTOM SOLIDS

In the previous chapter, you learned how to construct solid primitives. In this chapter, you will practice the techniques necessary to create solid objects of your own design. After completing this chapter, you will be able to:

- Manipulate the Extrude and Revolve commands to create custom solid objects.
- Comprehend the rules and limitations of the commands.

CREATING SOLIDS

You can use AutoCAD commands to create virtually any custom solid object you wish. The following is a listing of ACIS commands and their functions that we will explore in this chapter.

Extrude command: Creates solid extrusions from circles and polylines. You may also create tapered extrusions.

Revolve command: Creates solid revolutions by revolving a polyline around an axis.

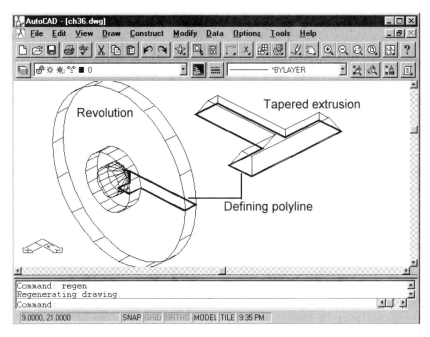

FIGURE 36-1 Solid Revolution and Solid Tapered Extrusion

Let's use these commands to create some solids. First, start a new drawing of any name. Let's get going and create some interesting solids!

Creating Solid Extrusions

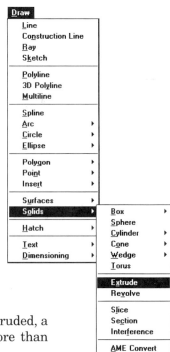

The Extrude command is used to create solid objects by extruding existing polylines and circles. The following objects can be extruded:

Polylines	Closed Splines
Polygons	Doughnuts
Circles	Regions
Ellipses	

The effect of the Extrude command is similar to extruding an object in 3D by specifying a thickness. The difference, of course, is that Extrusion creates a solid extrusion as opposed to a basic 3D extrusion. The object created is extruded perpendicular to its X,Y plane.

Solid Extrusion Rules and Limitations. In order to be extruded, a polyline must have at least three vertices, and no more than 500 vertices.

CREATING CUSTOM SOLIDS 36-3

The polyline must either be closed or be capable of being closed without crossing over another polyline. If the polyline is not closed, AutoCAD will attempt to close it by adding a segment between the endpoints. If that (or any existing) segment crosses over another segment, the polyline can not be extruded.

You can extrude several polylines in one operation. If any selected polylines are not capable of being extruded, they are ignored. If a polyline with a non-zero width is selected, it is extruded with a zero width at the center line of the polyline.

Extrusions are created perpendicular to the X,Y plane of the object.

Creating an Extrusion. Use the Pline command to draw an object similar to the one shown in Figure 36-2. Be sure to close with either the Close option or with object snap ENDpoint.

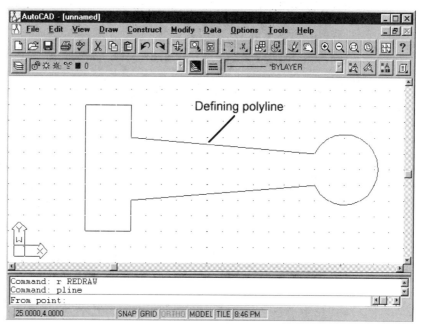

FIGURE 36-2 Drawing the Object

Now use the Extrude command to create an extrusion that is 3 units "thick". Use the following command sequence.

Command: **Extrude**
Select objects: *(Select the object to extrude.)*
1 selected 1 found
Select objects: ([ENTER])
Path/<Height of extrusion>: **3**
Extrusion taper angle: ([ENTER] *to accept 0.)*

Now use the Vpoint command to view the object in 3D. When you are finished, use the Plan command to return to the plan view and Zoom All.

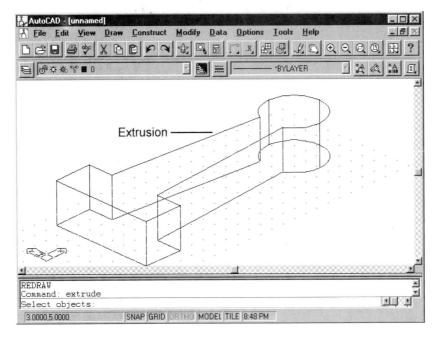

FIGURE 36-3 Extruded Object in 3D

Creating a Tapered Extrusion. Let's use Extrude to create a tapered extrusion. Use the Pline command to draw an object similar to the one shown in Figure 36-4.

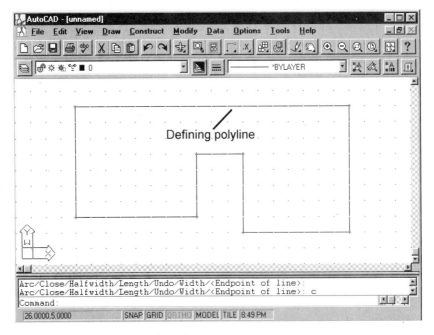

FIGURE 36-4 Drawing the Object

Now use the Extrude command to create an extrusion 5 units "thick", with a taper of 30 degrees.

> Command: **Extrude**
> Select objects:
> 1 selected, 1 found
> Select objects: (ENTER)
> Path/<Height of extrusion>: **0.5**
> Extrusion taper angle <0>: **30**

Now use the Vpoint command to view the object as before. Notice how the sides of the polyline segments taper 30 degrees.

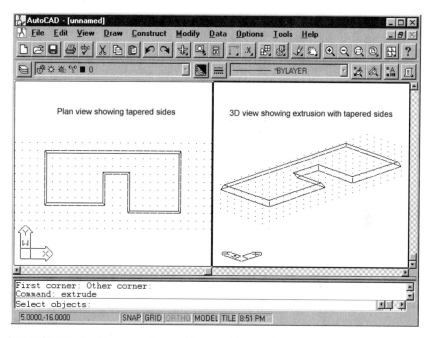

FIGURE 36-5 Constructing a Tapered Extrusion

The degree of taper specified must be greater than zero, but less than 90 degrees. You cannot use negative values to create an "outward" taper. Extrusions can only taper inward from the original object.

Creating a Solid Revolution

The Revolve command is used to create solid revolutions. This is similar to the 3D Revsurf command.

36-6 USING AUTOCAD

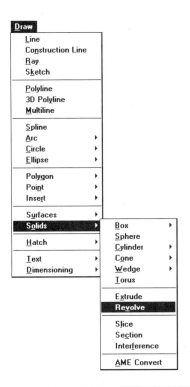

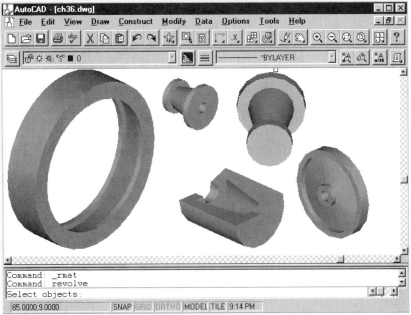

FIGURE 36-6 Solid Revolutions

Revolved solids are created by revolving an object about a specified axis. The following objects may be revolved:

Polylines	Closed Splines
Polygons	Doughnuts
Circles	Regions
Ellipses	

Revolved Solids Rules and Limitations. In order to be revolved, a polyline must have at least three vertices, but no more than 500 vertices. Polylines with non-zero width are changed to zero width. You can only revolve one object at a time.

The closed polyline rules apply as with the Extrusion command. You must be able to close the polyline without crossing another polyline segment.

Creating a Revolved Solid. Let's use the Revolve command to create a revolved solid. Use the Pline command to draw an object similar to the one shown in Figure 36-7. Omit the notations shown.

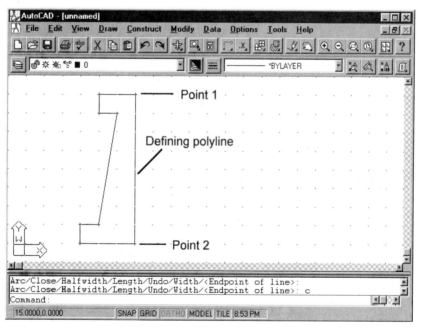

FIGURE 36-7 Creating a Revolved Solid

Use Figure 36-7 and the following command sequence to create the solid.

Command: **Revolve**
Select objects: *(Select the object.)*
1 selected, 1 found
Select objects: (ENTER)
Axis of revolution - Object/X/Y/<Start point of axis>: **ENDpoint**
of *(Select "Point 1".)*
Endpoint of axis: *(Select "point 2".)*
Angle of revolution <full circle>: (ENTER *to accept a full circle revolution.)*

Now use the Vpoint command to view the object in 3D.

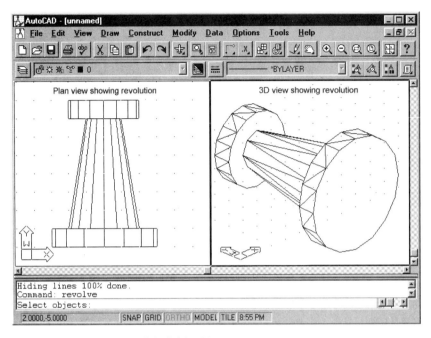

FIGURE 36-8 Revolved Solid in 3D

After you have viewed the solid object, use Undo to return to the original polyline. Try a new revolved solid, using 180 degrees as the included angle.

Revolving Around an Entity. You can use a separate entity as the axis to revolve a polyline around. Use Undo to return the drawing to the original polyline. Draw a line in the position shown in Figure 36-9.

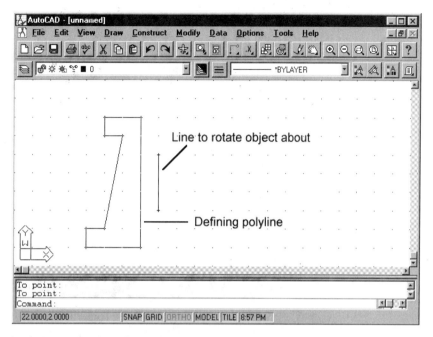

FIGURE 36-9 Revolving Around an Entity

Now let's create a solid object with an open shaft along its central axis.

Command: **Revolve**
Select objects: *(Select the object.)*
1 selected, 1 found
Select objects: (ENTER)
Axis of revolution - Object/X/Y/<Start point of axis>: **OBJECT**
Pick object to revolve about: *(Select the line.)*
Angle of revolution <full circle>: (ENTER *to accept a full circle revolution.)*

View the object in 3D and notice the central shaft created by the location of the line from the polyline.

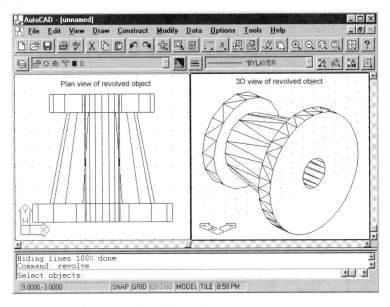

FIGURE 36-10 Revolved Object in 3D

X and Y Rotation Options. The X and Y options of the Revolve command use either the X-axis or Y-axis of the current UCS as the axis of revolution. Figure 36-11 shows the effects of using each with a full circle revolution.

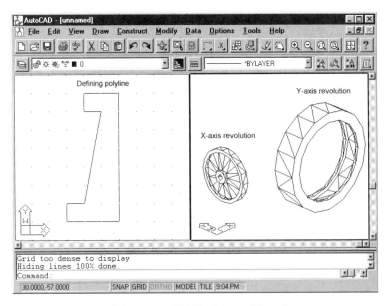

FIGURE 36-11 Using X and Y Options of Revolve

TUTORIAL

The Tube: An Extrusion Along a Path

We have seen how AutoCAD creates ACIS solid objects by extruding up from a closed object or by revolving a closed object around an axis. Release 13 adds the ability to extrude a closed object — such as a circle — along a complex path, such as a spline. This is the way to model tubes, handrails, and piping with bends. Let's see how to draw a tube:

1. Specify the path for the tube by drawing a spline curve:

    Command: **Spline**
    Object/<Enter first point>: *(Pick a point.)*
    Enter point: *(Pick a point.)*
    Close/Fit Tolerance/<Enter point>: *(Pick a point.)*
    Close/Fit Tolerance/<Enter point>: *(Pick a point.)*
    Enter start tangent: *(Pick a point.)*
    Enter end tangent: *(Pick a point.)*

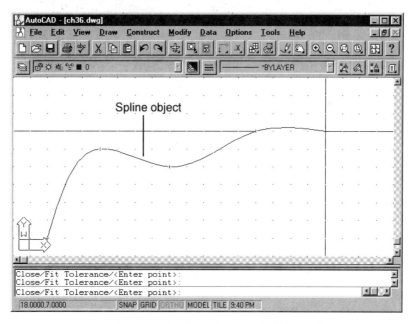

FIGURE 36-12 Drawing the Spline Path

2. Change the UCS so that you are looking at the end of the spline object. That lets you draw the object that defines the cross-sectional shape of the tube:

    Command: **UCS**
    Origin/ZAxis/3point/OBject/View/X/Y/Z/Prev/Restore/Save/Del/?/<World>: **y**
    Rotation angle about Y axis <0>: **90**

3. Set the view to match the UCS with the Plan command:

   Command: **Plan**
   <Current UCS>/Ucs/World: *(Press ENTER.)*
   Regenerating drawing.

   You may want to use the Zoom Window command to make the objects the right size.

4. Draw a circle to define the cross-sectional shape of the tube. (It doesn't need to be a circle; it can be any closed shape, such as an ellipse, a polygon, a rectangle, or closed polyline shape.)

   Command: **Circle**
   3P/2P/TTR/<Center point>: *(Pick point.)*
   Diameter/<Radius>: *(Pick point.)*

   Don't draw the circle too big, otherwise the extrusion will fail.

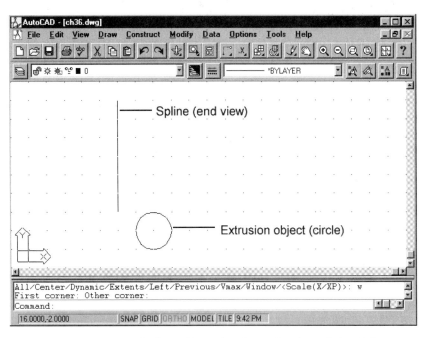

FIGURE 36-13 Circle Defines the Cross Section of the Tube

5. Finally, you are ready to apply the Extrude command:

   Command: **Extrude**
   Select objects: *(Pick the circle.)*
   1 found Select objects: *(Press ENTER.)*
   Path/<Height of Extrusion>: **Path**
   Select path: *(Pick spline curve.)*
   Path was moved to the center of the profile.
   Profile was oriented to be perpendicular to the path.

   It takes a few minutes for AutoCAD to extrude the circle along the path. Particularly complex paths and objects take a long time to extrude.

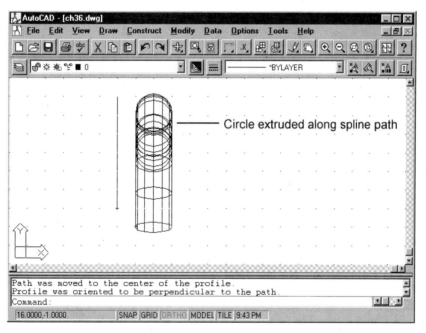

FIGURE 36-14 Circle Extruded Along the Spline Path

36-14 USING AUTOCAD

6. Once AutoCAD is finished, change the viewpoint so that you can see the tube:

   Command: **Vpoint**
   *** Switching to the WCS ***
   Rotate/<View point> <1.0000,0.0000,0.0000>: **-1,-1,.7**
   *** Returning to the UCS ***
   Regenerating drawing.

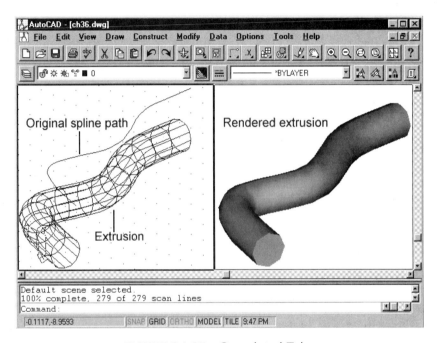

FIGURE 36-15   Completed Tube

7. Use the Hide or Shade command to see the tube with hidden lines removed:

   Command: **Hide**
   Regenerating drawing.
   Hiding lines 100% done.

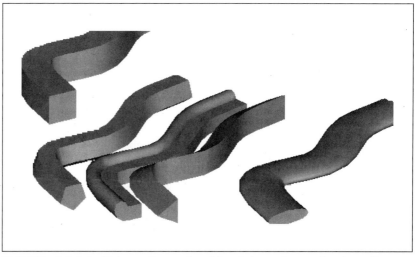

FIGURE 36-16 Tubes Created with Other Cross Section Shapes: Square, Pentagon, Closed Polyline, Triangle, Ellipse

CHAPTER REVIEW

1. What is an *extrusion*?

2. What objects can be extruded?

3. What is created with the Revolve command?

CHAPTER 37

MODIFYING SOLID OBJECTS

To create solid objects of any design, you must be able to modify basic solid shapes. After completing this chapter, you will be able to:

- Perform the methods of editing solid shapes.
- Manipulate the commands to modify existing solid models.

MODIFYING SOLID SHAPES

Solid models can be constructed by the combination of custom and primitive shapes. You can intersect, combine, and subtract shapes to achieve the model you desire. Then you can chamfer and fillet edges, move, rotate, and change the properties of the model and its parts. Solid modifier commands are used to modify solid models in these ways.

Let's look at the commands used to modify solid shapes.

Creating Solid Intersections

The Intersect command is used to create a solid from the intersection of two solids. The resulting solid is created from the common volume occupied by both solids. Figure 37-1 shows the result of using the Intersect command with a box and a sphere.

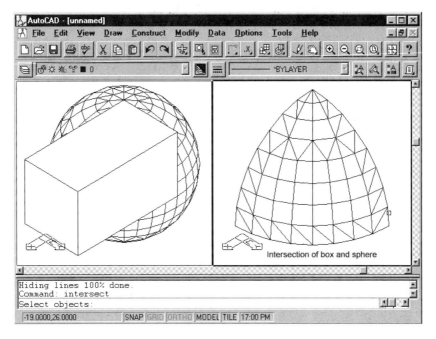

FIGURE 37-1 Creating Solid Intersections

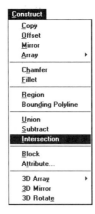

When you select the Intersect command, you are prompted to choose the objects that are to be used for the operation.

Command: **Intersect**
Select objects:

Subtracting Solids

The Subtract command is used to subtract one solid shape from another, creating a new solid. Figure 37-2 shows a box with four cylinders. The Subtract routine has been used to subtract the cylinders from the box.

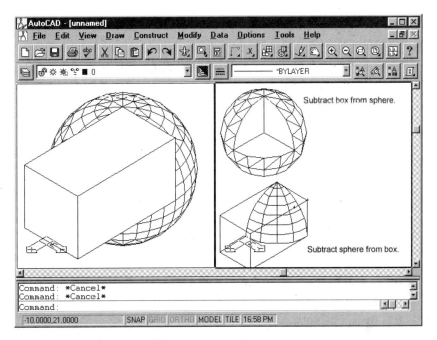

FIGURE 37-2 Subtracting Solids

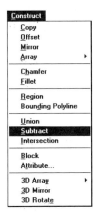

Let's look at how the Subtract command prompts for the solids you will modify. Subtract first asks for the objects that will be subtracted from them. The subtracted object(s) can be one or more objects. When you select the subtracted object(s), AutoCAD automatically performs a union of the selected objects.

Command: **Subtract**
Select solids and regions to subtract from...
Select objects:

AutoCAD then prompts for the solid objects to be subtracted from the source object:

Select solids and regions to subtract...
Select objects:

AutoCAD then subtracts these objects from the source object, creating a new solid.

Joining Solid Objects

The Union command joins two solid objects together as a single object. The new solid encloses both objects and their common space (if they overlap).

Command: **Union**
Select objects:

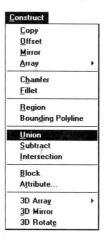

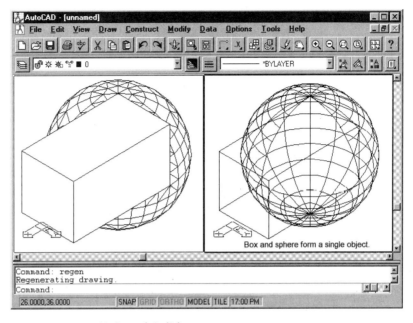

FIGURE 37-3 Unioned Solids

Chamfering a Solid

The Chamfer command is used to create bevel edges on existing solid objects.

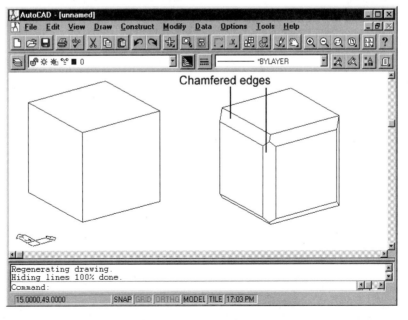

FIGURE 37-4 Solid Object with Chamfers

The chamfer is achieved by first selecting a "base surface," then the edges of the base surface to be chamfered. Finally, the chamfer dimension is specified. Let's look at how the Chamfer command sequence is used.

Command: **Chamfer**
Polyline/Distance/Angle/Trim/Method/<Select first line>: *(Select solid.)*
Pick base surface:
Next/<OK>:

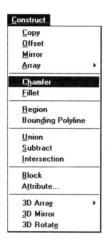

When you select a base surface, one of the surfaces around the edge you selected is highlighted. If this is the correct surface, press [ENTER] to accept the default <OK>. If you want to designate another surface, respond with Next. Adjacent surfaces will blink to show the selected surface. When the correct surface is selected, press [ENTER]. Next, you will choose the edges common with the base surface that are to be chamfered. Let's continue with the command sequence.

Enter base surface distance <0.0000>:
Enter other surface distance <0.0000>:
(You may enter a numeric value or show AutoCAD the distance by picking two points on the drawing.)
Loop/<select edge>:

Select the edges to be chamfered as the response to this prompt. If the edge you select is not adjacent to the base surface, it will not be chamfered. You may select as many edges as you wish.

MODIFYING SOLID OBJECTS **37-7**

Filleting a Solid

The Fillet command is used to create a filleted edge on a solid object.

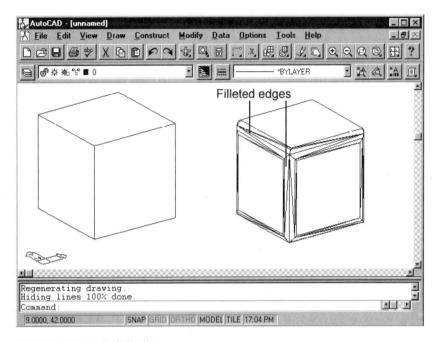

FIGURE 37-5 Solid Fillet

Fillet can create concave or convex fillets. The fillet is created by selecting the edges to be filleted.

Command: **Fillet**
Polyline/Radius/Trim/<select first object>: *(Pick a solid.)*
Chain/Radius/<select edge>:
Enter radius:

The radius or diameter can be specified by either entering a numeric value or showing AutoCAD the distance by entering two points on the drawing.

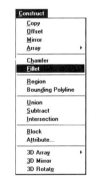

CHAPTER REVIEW

1. Describe the function of the Subtract command.

2. Which command is used to create a new solid from the common area described by two overlapping solids?

3. How would you join two existing solids?

4. When chamfering a solid object, what is meant by the *base surface*?

CHAPTER 38

CREATING COMPOSITE SOLID MODELS

To effectively use AutoCAD's ACIS modeler, you must practice using the commands in actual modeling conditions. After completing this chapter, you will be able to:

- Demonstrate the modeling commands through use with an actual solid model.
- Draw and edit a solid model.

DRAWING A COMPOSITE SOLID MODEL

Let's start a solid model drawing. We will construct the model, edit it, and display it as both a hidden line and solid object. Figure 38-1 shows the finished model.

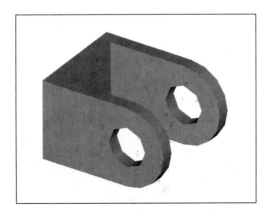

FIGURE 38-1 Completed Solid Model Drawing

Beginning the Model

Let's start by beginning a new drawing. Use the New command to start a new drawing. Enter "MODEL1" as the drawing name. This tutorial uses the default prototype drawing settings. If this file has been changed, the actual results or display may vary from those shown.

Figure 38-2 is a dimensioned drawing of the object we will draw. You may want to refer to it as we are constructing the base model from solid primitives.

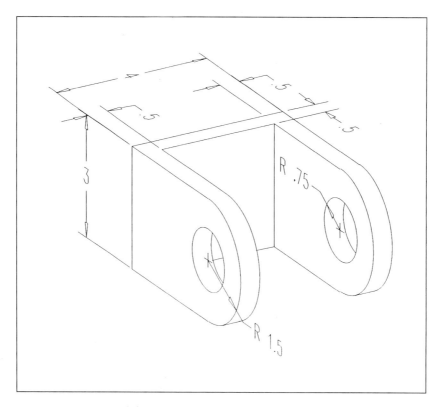

FIGURE 38-2 Model 1 Dimensions

Let's begin our drawing by using these commands to construct our basic model shape. Figure 38-3 shows the building block components we will use to "assemble" our model.

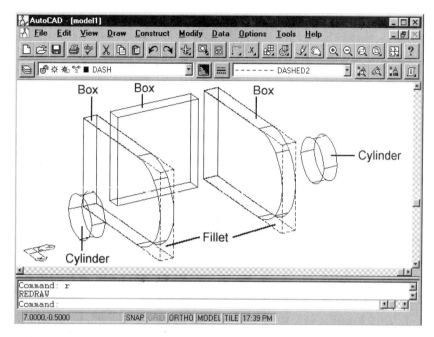

FIGURE 38-3 Building Block Component

Before starting any drawing, you should analyze the object to determine the best primitives to use. We are going to use both solid commands and some 3D principles to construct our model.

Let's get started. Set Snap to 0.5.

Creating the First Box. Select the Box command from the Draw|Solids menu and enter the points indicated in the following command sequence. We will discuss each step as we perform it.

   Command: **Box**

Select the first corner of the box.

   Center/<Corner of box>: **3,2**

Next, define the opposite corner of the box.

   Cube/Length/<Other corner>: **@5,0.5**

Now define the height of the box.

   Height: **3**

Type the Box command again. Refer to Figure 38-4 for the points to enter.

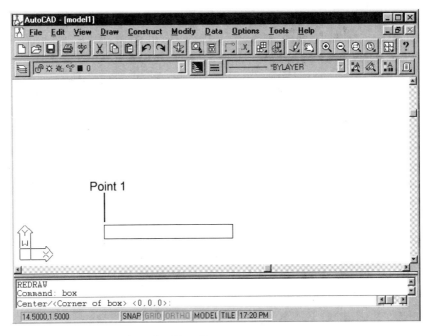

FIGURE 38-4 Drawing the First Box

Command: **Box**
Corner of box: *(Select point "1".)*
Cube/Length/<Other corner>: **@0.5,3**
Height: **3**

Creating the Second Box. Let's use the Box command one more time. Refer to Figure 38-5.

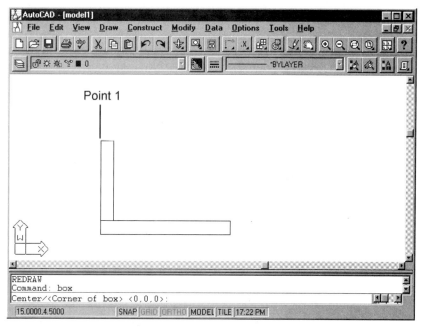

FIGURE 38-5 Drawing the Second Box

Command: **Box**
Corner of box: *(Select point "1".)*
Cube/Length/<Other corner>: **@5,0.5**
Height: **3**

Your drawing should look like the one in Figure 38-6.

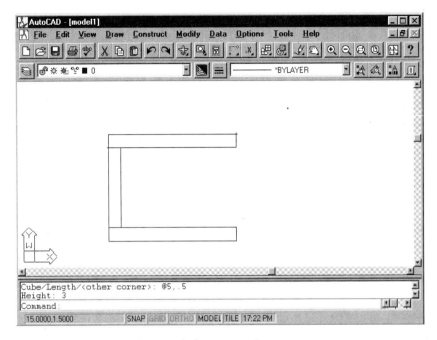

FIGURE 38-6 Completion of Three Box Components

Rotating the View. Let's use the Vpoint command to rotate the view so we can complete our work.

Command: **Vpoint**
Rotate/<View point> <0.0000,0.0000,1.0000>: **R**
Enter angle in X-Y plane from X axis <270>: **290**
Enter angle from X-Y plane <90>: **23**

Let's use a zoom factor to "zoom out" the display of our drawing.

Command: **Zoom**
All/Center/Dynamic/Extents/Left/Previous/Vmax/Window/<Scale (X/XP)>: **.6X**

Your drawing should look similar to Figure 38-7.

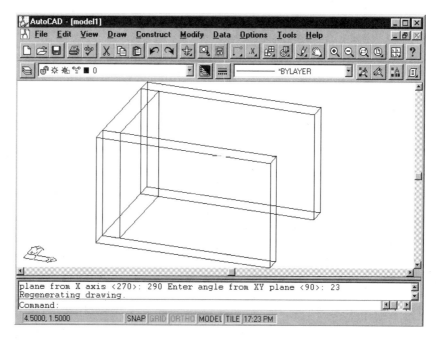

FIGURE 38-7 Box Components in 3D

Adding the Drill Holes. We want to add a drill hole to the side of the object. Let's begin by setting our UCS icon so we can see the origin of the UCS we will be working in.

Command: **Ucsicon**
ON/OFF/All/Noorigin/ORigin <ON>: **OR**

Now let's change the UCS. Refer to the following command sequence and Figure 38-8.

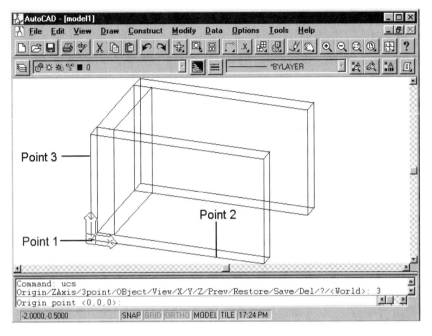

FIGURE 38-8   Setting the UCS

Command: **UCS**
Origin/ZAxis/3point/OBject/View/X/Y/Z/Prev/Restore/Save/Del/?/<World>: **3**
Origin point <0,0,0>: *(Select point "1".)*
Point on positive portion of the X-axis <1.0000,0.0000,0.0000>: **NEA**
of *(Select point "2".)*
Point on positive-Y portion of the UCS X-Y plane <0.0000,1.0000, 0.0000>: **NEA**
of *(Select point "3".)*

It is now time to add the drill hole. We will construct this with the Cylinder command. Let's walk through each step of this command. Refer to Figure 38-9.

Command: **Cylinder**

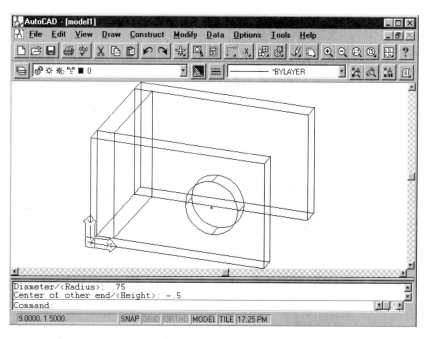

FIGURE 38-9 Drawing the Drill Hole

Next, specify the center point of the cylinder. Note that the absolute coordinates we use are relative to the origin of the new UCS.

Elliptical/<Center point>: **3.5,1.5**

Now specify the radius of the cylinder.

Diameter/<Radius>: **.75**

Finally, we will designate the extrusion height of the cylinder. Since we want the extrusion to extend in a negative direction, the value will be negative.

Center of other end/<Height>: **-0.5**

38-10 USING AUTOCAD

Adding the Second Drill Hole. Let's use the Vpoint command to rotate the view so we can add the second drill hole.

Command: **Vpoint**
Rotate/<View point> <default>: **R**
Enter angle in X-Y plane from X axis <290>: **330**
Enter angle from X-Y plane <23>: (ENTER)

Next, move the UCS to the opposite side of the object. Refer to the following command sequence and Figure 38-10.

Command: **UCS**
Origin/ZAxis/3point/Object/View/X/Y/Z/Prev/Restore/Save/Del/?/<World>: **O**
Origin point <0,0,0>: **END**
of *(Select point "1".)*

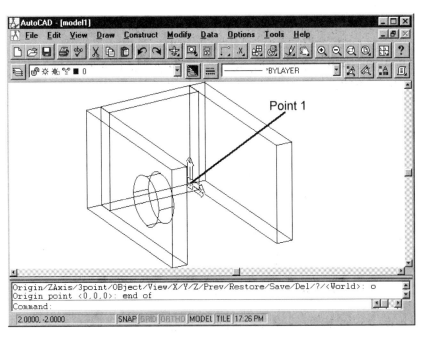

FIGURE 38-10 Relocating the UCS

Let's draw this cylinder the same way as above. (Figure 38-11).

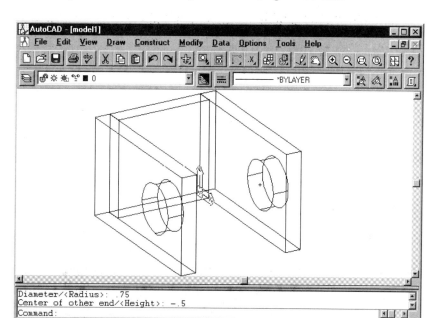

FIGURE 38-11 Drawing the Second Drill Hole

Converting to a Composite Solid. Let's now create a composite solid. To do this, we must change the three boxes into a single solid object, then subtract the cylinders that make the drill holes. We will first use the Union command to combine the boxes. Use the following command sequence.

Command: **Union**
Select objects: *(Select each of the three boxes and press* ENTER *.)*

AutoCAD will now go to work and perform all the necessary calculations to combine the boxes into a single solid object. You will notice when the drawing is redisplayed that the edge lines between the boxes are no longer a part of the object.

Subtracting the Cylinders. Let's continue and subtract the cylinders from the object. We will use the Subtract command.

Command: **Subtract**
Select solids and regions to subtract from...
Select objects: *(Select the boxes.)*
Select objects: (**ENTER**)
1 solid selected

Select solids and regions to subtract...
Select objects: *(Select each of the cylinders, then press **ENTER**.)*
Select objects: (**ENTER**)

AutoCAD will now evaluate the object and remove the cylinders from the solid object created by the union of the three boxes. Note that you will not notice a visual difference after this step. Your drawing should now look similar to Figure 38-12.

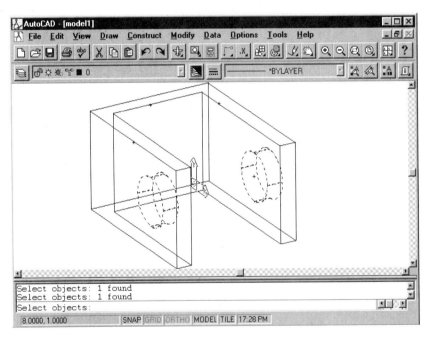

FIGURE 38-12 Subtracting Cylinders to Make Holes

Filleting the Corners. Next we will use the Fillet command to fillet the corners. Refer to the following command sequence and Figure 38-13. We will discuss each step as we proceed.

Command: **Fillet**
Chain/Radius/<select edge>: *(Select point "1", then press [ENTER].)*
Enter radius: **1.5**
Chain/Radius/<select edge>: *(Select point "2")*
Chain/Radius/<select edge>: *(Press [ENTER].)*

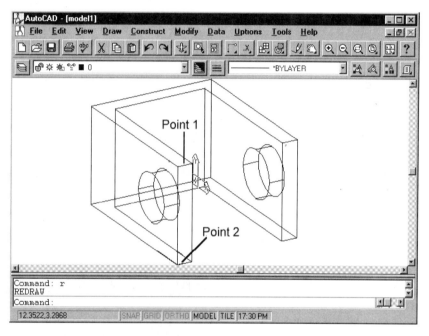

FIGURE 38-13   Filleting the Box

Your drawing should look like Figure 38-14. You may need to use the Regen command to clean up the view.

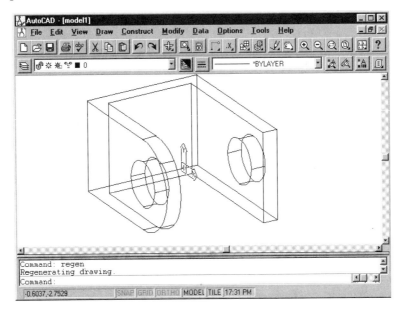

FIGURE 38-14 Filleted Box

Now use the Fillet command again to fillet the back leg of the object in the same manner. When completed, your drawing should look like Figure 38-15.

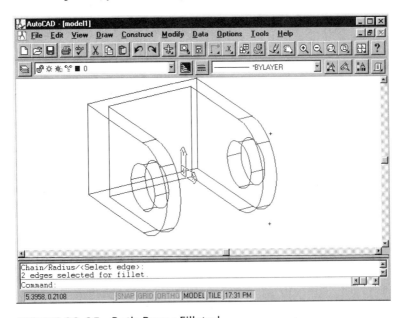

FIGURE 38-15 Both Boxes Filleted

Displaying Your Model

We can produce hidden line and shaded models of our model.

Command: **Hide**

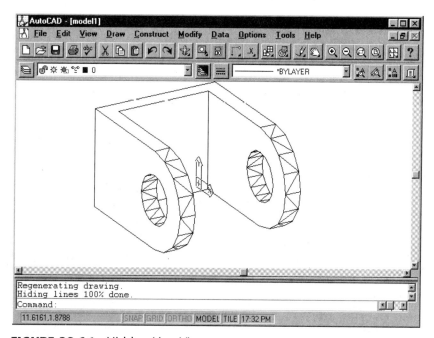

FIGURE 38-16 Hidden Line View

You may want to use the Shade command to produce a shaded model.

Command: **Shade**

SUMMARY

There you have it. Your "Model1" drawing should look like Figure 38-17.

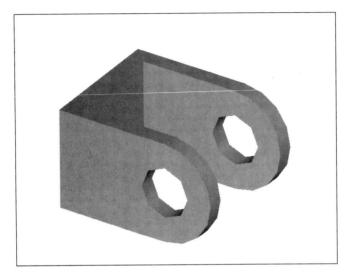

FIGURE 38-17 Completed "Model1" Drawing

CHAPTER

39

REALISTIC RENDERING

To create a realistic rendering of a 3D model, you must know how to place materials and lights. After completing this chapter, you will be able to:

- Use the Render command for quick and advanced renderings of 3D objects.
- Place lights and create scenes.
- Apply material definitions to objects.
- Save the rendering to a file on disk.

THE RENDER COMMAND

The Render command is a more flexible and advanced version of the Shade command (Chapter 31). The Render command creates much more realistic images than does the Shade command. For example, Render smooths out all of the facets you see on a 3D sphere, creating a round ball.

As with the Shade command, you can create a rendering by simply typing the Render command. Or you can set many options to create a complex rendering: you can set up many lights, apply materials to objects, place a background image, and save the rendering in one of several common file formats.

However, Render requires that your graphics board be capable of displaying at least 256 colors. For the most realistic renderings, your computer system should be set up with a graphics board that displays 16.7 million colors (also known as "24 bits"). To display that many colors, your graphics board needs 4MB of RAM, a feature that is becoming more common and less expensive today.

Unlike the DOS version, Release 13 for Windows does not need a special device driver to display high-quality renderings. As long as your graphics board works with Windows, you can use the Render command.

Before continuing this chapter, you should use the Windows Control Panel to change the Display to 16.7 million colors, if possible.

Your First Rendering

Let's try using the Render command on the Model1 solid model created in the previous chapter.

1. Start AutoCAD and open the Model1.Dwg file.

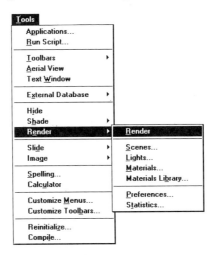

2. From the Tools menu, select Render|Render. AutoCAD loads the rendering module. If this is the first time that the Render command has been used with this computer, AutoCAD displays some rendering configuration information in the Text window. After this first time, you never see this information again.

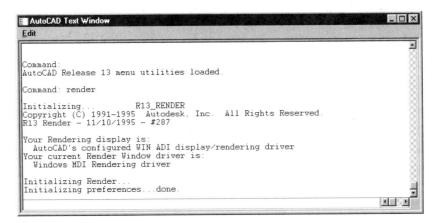

FIGURE 39-1 Render Configuration Information

3. When AutoCAD displays the Render dialog box, ignore all the options for now and click on the OK button. In 10 seconds or so — depending on the speed of your computer — you see a beautifully rendered image of the Model1 drawing appear.

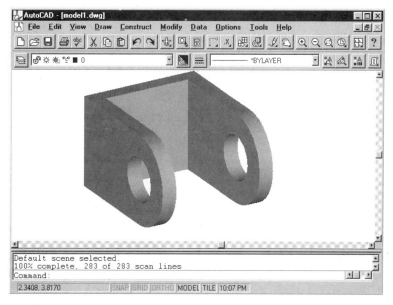

FIGURE 39-2 Model1 Drawing Rendered

Advanced Renderings

For most renderings, you probably won't need to know anything more about the Render command. For total control, though, it is useful to know all about Render's options. Let's start the Render command again and examine its dialog box.

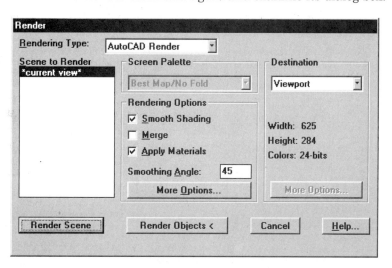

FIGURE 39-3 Render Dialog Box

Rendering Type. Unless you have installed another rendering program that runs inside AutoCAD (such as Autodesk AutoVision), the only option is AutoCAD Render.

Scene to Render. With your first rendering, there is only one scene defined: the current view. Later, we will use the Scene command to create named scenes whose names appear in this part of the Render dialog box.

Screen Palette. If your computer has a graphics board displaying 64,000 colors (16-bit) or better, this dialog box is greyed out. If the graphics board displays 32,000 or 256 colors (15- or 8-bit), you choose how you want the palette (the range of colors) dealt with. Unless you have a reason for changing this option, leave it at the default: Best Map/No Fold.

Rendering Options. This is the core of the Render command. For the most part, you leave the options as they are set by default. If you want to change any, here is a summary of the effect they have on a rendering:

> **Smooth Shading:** When on, AutoCAD creates a smooth transition through faceted areas of the 3D model. This makes the model look more realistic; the cutoff angle for smoothing facets is specified by the Smooth Angle option (default = 45 degrees). You would only turn off the Smooth Shading option for faster renderings.
>
> **Merge:** When off, AutoCAD clears the screen before displaying the rendering. You would turn Merge on for two reasons: (1) to re-render a small portion of the model (using the Render Objects option) without rendering the entire model; and (2) to display a background image (via the Replay) command.
>
> **Apply Materials:** When on, AutoCAD renders the objects using the materials defined by the RMat command. This helps the object look more realistic. You would turn off this option for faster renderings.

Figure 39-4 shows the effect of toggling the three primary options: smooth shading off, merge on, and apply materials off. Compare the lower quality of the rendering with Figure 39-2, but note the addition of the clouds as a background image.

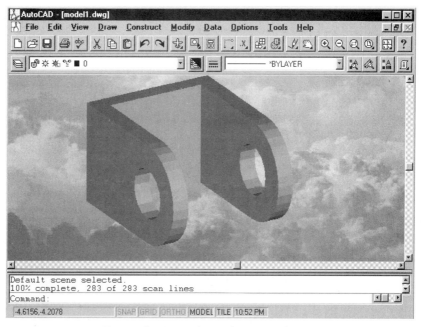

FIGURE 39-4 A Faceted, Merged Rendering with No Material Applied

More Options: Click on this button to display the AutoCAD Render Options dialog box.

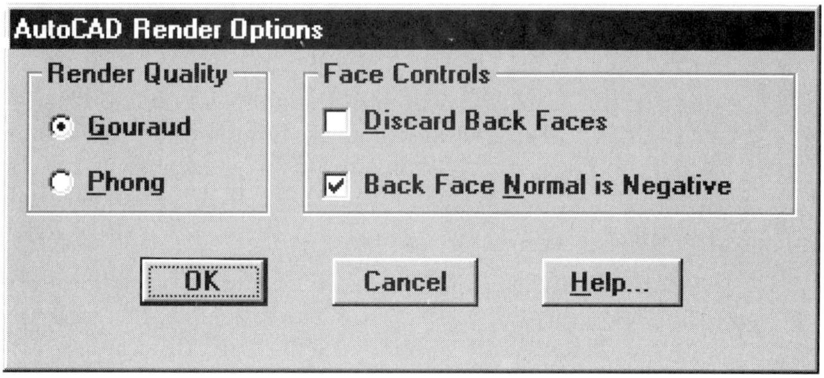

FIGURE 39-5 AutoCAD Render Options Dialog Box

Render Quality: Gouraud and Phong are the names of the algorithms used by AutoCAD Render to create its images. Each is named after the computer scientist who came up with the rendering algorithms. Phong is more realistic when your model has lights, but it takes longer to render than does Gouraud.

Face Controls: AutoCAD uses the concept of "face normals" to determine which parts of the model are the front and which are the back. A "face" is the 3D polygon that makes up the surface of the model; a "normal" is a vector that points out at right angles from the face. AutoCAD determines the direction of the normal by applying the right-hand rule to the order in which the face vertices were drawn. A "positive" normal points toward you; a "negative" normal points away from you. To save time in rendering, negative face normals are ignored, since they won't be seen in the rendering.

You would turn on Discard Face Normals for faster rendering; turn off the option when mistakes appear in the rendering.

CREATING LIGHTS — LIGHT

Whereas the Shade command is limited to a single light source, the Render command has four different kinds of light sources: distant, point, spot, and ambient. Except for ambient light, you can place as many of each of these lights as you want. All lights can emit any color at any level of brightness. The names of the lights have special meaning:

Point lights emit light in all directions with varying intensity. The best example of a point light is the light bulb on the ceiling of your room.

Spotlights emit light beams in the shape of a cone. The best example of a spotlight is a high-intensity desk lamp or a vehicle-mounted spotlight. When you place spotlights in a drawing, you specify the "hotspot" of the light (where the light is brightest) and the "falloff", where the light diminishes in intensity.

Distant lights emit parallel light beams of constant intensity. The best examples of distant lights are the sun and moon. Typically, you want to place a single distant light to simulate the sun. To simulate a setting sun, you would change the color of the light to an orange-red color.

Ambient light is an omnipresent light source that ensures that every object in the scene has illumination. There is a single ambient light in every rendering; you would turn off the ambient light to simulate a nighttime scene.

REALISTIC RENDERING **39-7**

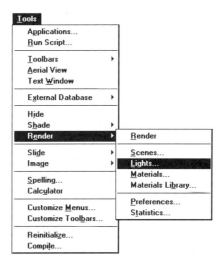

The Light command places lights into the drawing. (Note: The Render command does not use any lights defined in a drawing until they are included in a scene definition via the Scene command.) Let's see how to place a sun into the Model1 drawing:

1. Select Render | Lights from the Tools menu. AutoCAD displays the Lights dialog box.

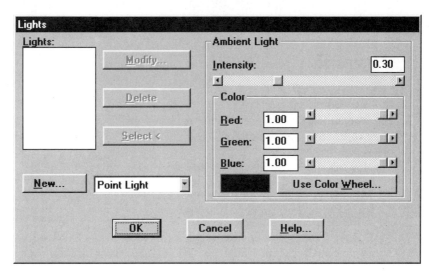

FIGURE 39-6 Lights Dialog Box

Initially there are no lights defined, other than a single light source located at your eye.

2. To place a new light, you need to decide on the type of light: spotlight, point light, or distant light. For now, select Distant light from the popdown list box because a sun-like light is the easiest to work with (spotlights are the hardest lights to work with).

3. Click on the New button to give the light a name and to specify its parameters.

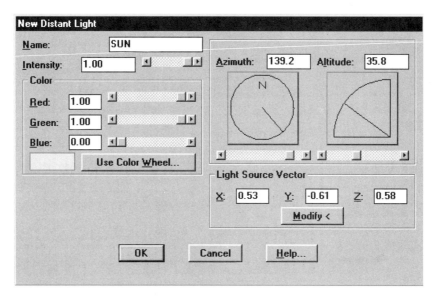

FIGURE 39-7 New Distant Light Dialog Box

4. The dialog box for each of the three light types is roughly similar, depending on their characteristics. Common to all three are:

   Name: Give the light a convenient name, such as "Sun."

   Intensity: The brighter the light, the higher the intensity value. An intensity of zero turns off the light.

   Color: You can select any color for each light. Clicking on the Use Color Wheel button gives you three choices: RGB (select varying amounts of red, green, and blue); HLS (select varying amounts of hue, lightness, and saturation), or ACI (the AutoCAD color index). ACI is the easiest option to use. Select the color yellow for the color of the light and click "OK".

5. The New Distant Light dialog box also lets you define the Sun's azimuth (how far around in the day) and altitude (height in the sky). Since it doesn't matter how far the Sun is from the object, AutoCAD places the distant light in the drawing for you.

6. Click OK. Notice that an image of a light appears in the drawing.

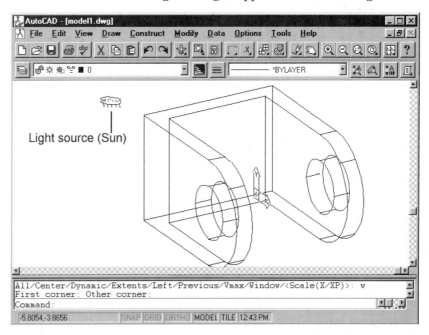

FIGURE 39-8 New Distant Light in Drawing

When you place lights in the drawing, AutoCAD actually places a block on a special layer named "AutoShade" (named after an earlier, now-obsolete rendering program no longer sold by Autodesk). Except for ambient light, each type of light has a unique block shape, as shown in Figure 39-9.

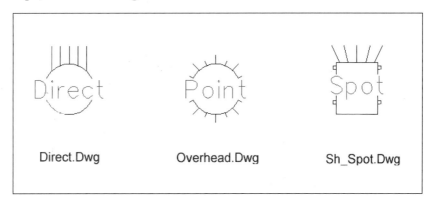

FIGURE 39-9 The Three Light Blocks

You have defined a single distant light called the Sun, but you cannot use it . . . yet.

39-10 USING AUTOCAD

COLLECTING LIGHTS INTO SCENES — SCENE

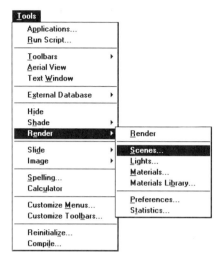

As mentioned earlier, the Render command will not use a light until it has been included in a scene definition. The Scene command does only two things: (1) lets you decide which lights should be used in a rendering, and (2) lets you specify the name of the view for that rendering. The lights and the view are collected into a named scene. Let's see how the Scene command works.

1. Select Render | Scenes from the Tools menu. AutoCAD displays the Scenes dialog box.

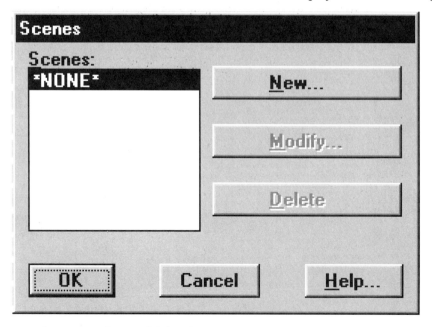

FIGURE 39-10 Scenes Dialog Box

2. As the word *NONE* indicates, this drawing has no scenes defined yet. Click on the New button. AutoCAD displays the New Scene dialog box.

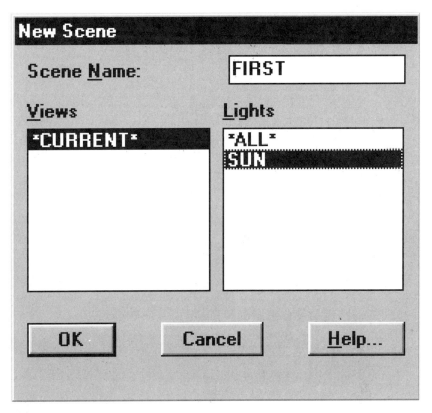

FIGURE 39-11 New Scene Dialog Box

3. Type a name for this scene, such as "First".
4. Then select the Sun light. (If you had used the View Save command earlier to define named views, these would be listed under Views.)
5. Click the OK button twice to dismiss both dialog boxes.
6. Let's try rendering with the distant light. This time, the Render command's dialog box displays the name of the scene you just defined: FIRST. Select the FIRST scene, then click on the Render Scene button. After a moment, AutoCAD displays the rendering tinted with the yellow color you selected for the sun light.

APPLYING MATERIALS AND BACKGROUNDS — RMAT AND REPLAY

To enhance the realism of the rendering, AutoCAD lets you apply material definitions to objects (with the RMat command) and place a background image with the Replay command.

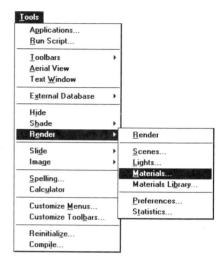

The material definition uses four parameters to define the surface characteristics of objects: color, reflection, roughness, and ambient reflection. AutoCAD comes with two dozen predefined material definitions stored in the Render.Mli file. Another 144 definitions are available on the AutoCAD CD-ROM in subdirectory \Autovis\Avis_sup in file AutoVis.Mli.

Let's use the RMat command to load and apply a material definition to the Model1 drawing.

1. From the Tools menu, select the Render | Materials command. AutoCAD displays the Materials dialog box, which lists no materials.

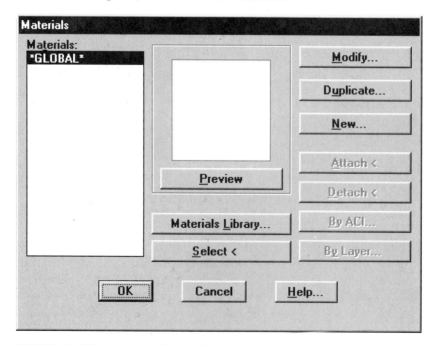

FIGURE 39-12 Materials Dialog Box

2. Before you can attach a material, you have to load its definition into AutoCAD, much like loading a linetype from Acad.Lin. Click on the Materials Library button. AutoCAD displays the Materials Library dialog box (you can get directly to this dialog box with the MatLib command).

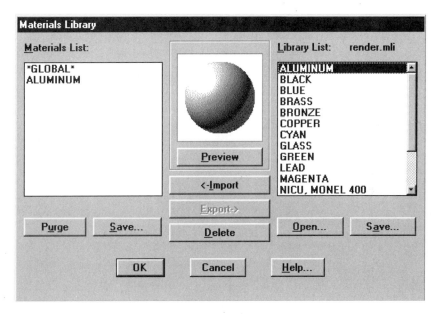

FIGURE 39-13 Materials Library Dialog Box

3. The dialog box lists 14 materials defined in the Render.Mli file. Click on Aluminum, then click on the Import button. This adds Aluminum to the materials list.
4. For a preview of how a material looks on a sphere, click on the Preview button. AutoCAD quickly renders a sphere using the selected material. (This does not work when more than one material is selected.)
5. Click on the OK button to return to the Materials dialog box. Aluminum is listed in the Materials list box.
6. To attach the aluminum material to the Model1 object, click on the Attach button.
7. AutoCAD clears the dialog box and prompts you:

Select objects to attach "ALUMINUM" to: *(Pick the 3D model.)*
1 found Select objects: *(Press ENTER.)*
Updating drawing...done

The dialog box returns. In addition to attaching the material to individual objects, you can attach a material definition to all objects of a specific color (click on the By ACI button) or to all objects on a layer (click on the By Layer button).

8. Click on the OK button and use the Render command to re-render the model. Ensure that the Apply Materials option is turned on. The newly rendered model will look lighter and shinier than before due to the aluminum material definition.

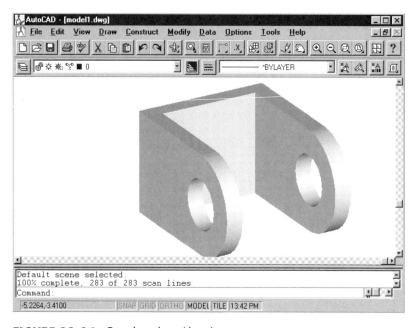

FIGURE 39-14 Rendered as Aluminum

9. Try applying other material definitions to see the difference they make in color and shininess. Note that the Glass material does not make the object transparent since AutoCAD Render is unable to simulate transparency.

10. Now, let's place an image behind the object with the Replay command. This is useful for enhancing the rendering. For example, if you have designed a 3D house in AutoCAD, you could place a landscape image behind the rendering.

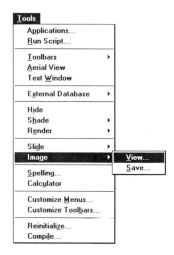

To place the background image, the image must be in TIFF, Targa, or GIF format. Note that the Replay command cannot read TIFF files saved with the popular LZW compression format; AutoCAD says "Compression algorithm does not support random access" and does not load the image.

AutoCAD does not come with any sample background images. You can use scenic images provided on clip-art CD-ROMs or make your own with a low-cost video grabber (such as Snappy), an electronic camera, or by scanning a picture.

To load the background image, select Image | View from the Tools menu.

REALISTIC RENDERING **39-15**

11. AutoCAD displays the Replay dialog box. Select a GIF, TIFF, or TGA file.
12. AutoCAD displays the Image Specifications dialog box. This lets you specify a smaller size and the placement of the image on the screen. In most cases, there is no need to change anything. Click on the OK button.

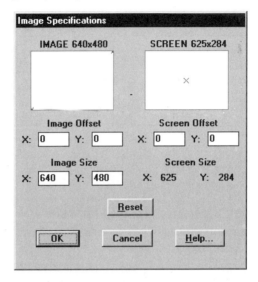

FIGURE 39-15 Image Specifications Dialog Box

13. AutoCAD loads and displays the image.

FIGURE 39-16 Background Image

39-16 USING AUTOCAD

14. Now re-render the Model1 drawing with the Render command. This time, ensure that the Merge option is turned on. AutoCAD renders the object on top of the background image.

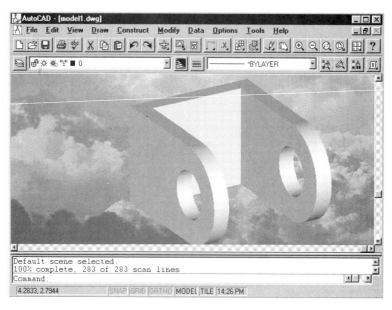

FIGURE 39-17 Model Rendered with Background Image

SAVING RENDERINGS — SAVEIMG

There are three ways to save the result of the rendering:

1. When the Render dialog box appears, select File from the Destination option. Then click on the More Options button. AutoCAD displays the File Output Configuration dialog box.

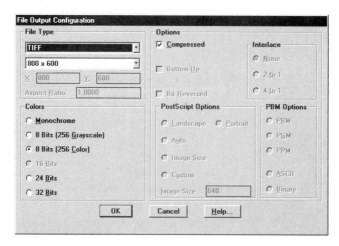

FIGURE 39-18 File Output Configuration Dialog Box

REALISTIC RENDERING   **39-17**

Here you can choose to save the rendering in one of 12 file formats and a variety of parameters, such as number of colors and compression schemes. There are two drawbacks to using the File option of the Render command: (1) you do not see the rendering before it is saved, and (2) the background (Replay) image is not merged with the rendering.

2. Use the SaveImg command to save the rendering to a file. This command is used after the Render command has created the rendering in a viewport. Select Image | Save from the Tools menu. As you see from Figure 39-18, you have fewer choices for the output file format than with the File option of the Render command — just three: Targa, TIFF, and GIF. The Options button lets you select the type of image compression.

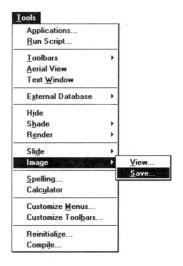

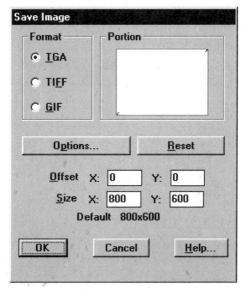

FIGURE 39-19   Save Image Dialog Box

3. Press the [Alt]+[PrtScr] command. This captures the entire AutoCAD window and rendering to the Windows Clipboard.

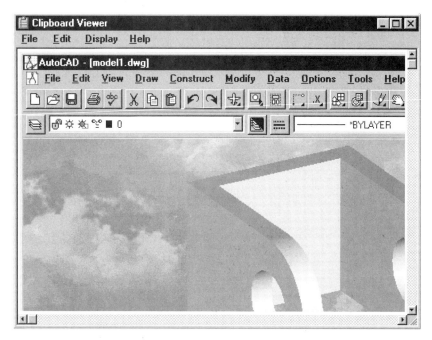

FIGURE 39-20 Clipboard Viewer

IMPROVING THE SPEED OF RENDERING

The time it takes for AutoCAD to create a rendering depends on many factors. These include:

- **Options selected for the Render command.** Setting the following options results in a rendering that is at least two and one half times faster:

  Smooth shading: **Off**
  Merge: **Off**
  Apply materials: **Off**
  Render quality: **Gouraud**
  Discard back faces: **On**
  Destination: small viewport or 320x200 file
  Lights: **None**

REALISTIC RENDERING **39-19**

- **Speed and memory of the computer.** The faster the computer's CPU and the larger the computer's RAM, the faster the rendering.
- **Complexity of the model.** A complex model with many parts takes longer to render than a simple model with only a few parts.
- **Amount of the model being rendered.** Rendering all of the model takes longer than rendering a small portion.
- **Size of the viewport being rendered.** Rendering to a very small viewport greatly decreases the rendering time.

While the Render command is an improvement over the Shade command, there are many rendering functions Render is unable to accomplish. For example, Render cannot simulate transparency (glass), fog effects, or cast shadows. If you wish to add these effects to your rendered image, you will need to purchase an add-on rendering package, such as Autodesk AutoVision.

THE RENDERUNLOAD COMMAND

When you are finished with your rendering tasks, use the Renderunload command to unload the Render module. This frees up memory for AutoCAD and Windows. Render will automatically reload the next time you use one of its commands.

CHAPTER REVIEW

1. What is a scene?

   _____

   _____

   _____

2. What are the two ways of defining a viewpoint for a rendering?

   _____

   _____

3. Name three differences between the Shade and Render commands.

   _____

   _____

   _____

4. What are the four types of lights in AutoCAD Render?

5. List five ways to speed up a rendering.

CHAPTER 40

INTRODUCTION TO AUTOLISP

AutoLISP is a powerful programming tool that is a part of AutoCAD. After completing this chapter, you will be able to:

- Comprehend the nature and purpose of the AutoLISP programming language.
- Load and use existing AutoLISP routines.

USING AUTOLISP

AutoCAD is a powerful drawing and design program. While many people use this program for various purposes, most have implemented CAD for a specific discipline. While the "generic" form of AutoCAD contains all the command routines needed to construct any type of drawing, most users can benefit from the custom programming capabilities within AutoCAD.

In Chapter 29 we learned how to customize menus and write macros to combine several steps into one. A more powerful method of customization is the use of the LISP programming language embedded within the AutoCAD package. This feature, referred to as AutoLISP, allows you to set and recall points, mix mathematical routines within a list of instructions, and perform many other routines.

Why Use AutoLISP?

AutoLISP allows the AutoCAD user to customize new commands that perform one or many functions. For example, third-party programmers have used AutoLISP to create packages that can automatically create a 3D contour map from site data, create "unfolded" patterns from three-dimensional objects, and construct a drawing from a list of dimensions that describes an object (parametric drawing construction).

Using AutoLISP to customize routines for your work creates a more efficient drawing system. Whether you continue to increase your knowledge of LISP programming and write extensive routines, or just gain a general understanding and write simple time-saving routines, you will find that AutoLISP will enhance your AutoCAD work!

Using an AutoLISP Program

You don't have to be a LISP programmer to use AutoLISP routines. AutoCAD contains many LISP routines for your use. Let's look at one of these programs, then learn how to load and use it.

The electronic work disk contains a Lisp file named "AXROT.LSP". This is an AutoLISP program used to rotate an object around any of its axes. Before starting, make sure this program is in the directory in which you have placed your AutoCAD files. Table 40-1 is a listing of the program.

```
;;; ------------------------------------------------------------
;;; AXROT.LSP
;;;     Copyright (C) 1990 by Autodesk, Inc.
;;;
;;;     Permission to use, copy, modify, and distribute this software and
;;;     documentation for any purpose and without fee is hereby granted.
;;;
;;;     THIS SOFTWARE IS PROVIDED "AS IS" WITHOUT EXPRESS OR IMPLIED WAPP
;;;     ALL IMPLIED WARRANTIES OF FITNESS FOR ANY PARTICULAR PURPOSE AND
;;;     MERCHANTABILITY ARE HEREBY DISCLAIMED.
;;;
;;;     By Jan S. Yoder                                        May 11,
;;;     Modified for AutoCAD Rel 11                            July
;;; ------------------------------------------------------------
;;; DESCRIPTION
;;;
;;;     A routine to do 3 axis rotation of a selection set.
;;;
;;; ------------------------- Main program -----------------------
(defun c:axrot (/ axerr s olderr obpt oce ogm ohl oucsf ssel kwd dr bpt
  (if (and (= (getvar "cvport") 1) (= (getvar "tilemode") 0))
    (progn
      (prompt "\n *** Command not allowed in Paper space ***\n")
      (princ)
    )
    (progn
      ;; Internal error handler

      (defun axerr (s)                  ; If an error (such as CTRL-C) oc
        ;; while this command is active...
        (if (/= s "Function cancelled")
          (princ (strcat "\nError: " s))
        )
        (setq *error* olderr)           ; restore old *error* handler
        (setvar "gridmode" ogm)         ; restore saved modes
        (setvar "highlight" ohl)
        (setvar "ucsfollow" oucsf)
        (command "ucs" "restore" "_AXROT_")
        (command "ucs" "del" "_AXROT_")
        (command "undo" "e")            ; complete undo group
        (setvar "cmdecho" oce)
        (princ)
      )

      (setq olderr *error*
            *error* axerr)
      (setq oce (getvar "cmdecho")
            ogm (getvar "gridmode")
            ohl (getvar "highlight")
            oucsf (getvar "ucsfollow")
      )
      (setvar "cmdecho" 0)
      (command "undo" "group")
      (command "ucs" "save" "_AXROT_")
      (setvar "gridmode" 0)
      (setvar "ucsfollow" 0)
      (setq ssel (ssget))

      (if ssel
        (progn
          (setvar "highlight" 0)
          (initget 1 "X Y Z")
          (setq kwd (getkword "\nAxis of rotation X/Y/Z: "))
          (setq dr (getreal "\nDegrees of rotation <0>: "))
          (if (null dr)
            (setq dr 0)
          )
          (setq bpt (getpoint "\nBase point <0,0,0>: "))
          (if (null bpt)
            (setq bpt (list 0 0 0))
          )
          (setq bpt (trans bpt 1 0))
          (cond ((= kwd "X") (command "ucs" "Y" "90"))
                ((= kwd "Y") (command "ucs" "X" "-90"))
                ((= kwd "Z") (command "ucs" "Z" "0"))
          )
          (setq bpt (trans bpt 0 1))
          (command "rotate" ssel "" bpt dr)
          (command "ucs" "p")            ;restore previous ucs
          (command "'redrawall")
        )
        (princ "\nNothing selected. ")
      )
      (setvar "gridmode" ogm)           ;restore saved modes
      (setvar "highlight" ohl)
      (setvar "ucsfollow" oucsf)
      (command "ucs" "del" "_AXROT_")
      (command "undo" "e")              ;complete undo group
      (setvar "cmdecho" oce)
      (princ)
    )
  )
)
(princ "\n\tC:AXROT loaded.  Start command with AXROT.")
(princ)
;;; ------------------------------------------------------------
```

TABLE 40-1

40-4 USING AUTOCAD

Drawing the Box. Let's start a new drawing and try out this program. Start a new drawing named "LISPTST". Use AutoCAD's 3D capabilities to draw a cube similar to the one shown in Figure 40-1. You can do this by using the BOX command under the screen menu path, or the Draw/Surfaces.../3D Objects pull-down menus (the last is an icon box). As you draw the cube, you will notice that it appears only as a square in plan.

FIGURE 40-1 Solid Cube in Plan

Command: **AI_Box**
Corner of box: *(Enter a point on the screen.)*
Length: *(Move the crosshairs to show a length and press [ENTER].)*
Cube/<Width>: **C**
Rotation angle about Z axis: **0**

Using the Axrot.LSP Program. Now let's use the Axrot Lisp program. The first thing we have to do is load the program. You can load a Lisp program in different ways. We will load this one from the command line. Enter the following exactly as listed, including the quotes and parentheses.

Command: **(LOAD "AXROT")**

If you entered the line correctly, you will see a short message that the program was loaded. Now that it is loaded, we can use the program as a command. Let's do this now.

Command: **Axrot**
Select objects: *(Select the cube.)*
Select objects: 1 selected, 1 found
Select objects: *(Press [ENTER].)*
Axis of rotation X/Y/Z: **X**
Degrees of rotation <0>: **30**
Base point <0,0,0>: *(Select a point about the center of the cube.)*

The box has rotated 30 degrees about the X-axis. Let's perform one more rotation. Press [ENTER] to repeat the command.

Command: **Axrot**
Select objects: *(Select the cube.)*
Select objects: 1 selected, 1 found
Select objects: *(Press [ENTER].)*
Axis of rotation X/Y/Z: **Y**
Degrees of rotation <0>: **30**
Base point <0,0,0>: *(Select a point about the center of the cube.)*

Your cube should now look similar to the one in Figure 40-2. Notice how you can see the rotations performed around the X- and Y-axes.

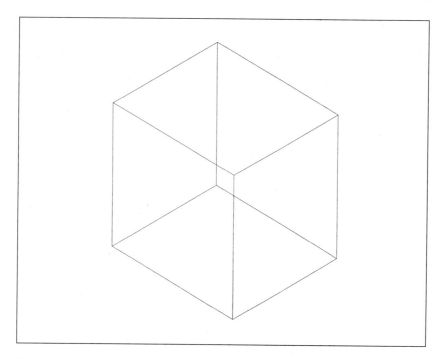

FIGURE 40-2   3D Cube After Using AXROT.LSP

Now perform a Hide to see the cube more clearly.

CHAPTER REVIEW

1. Why would you program in AutoCAD with AutoLISP?

2. What are some uses for AutoLISP?

3. How would you load a LISP file named 3DARRAY.LSP while within an AutoCAD drawing?

4. Describe an AutoLISP program that would be useful for you.

CHAPTER

41

PROGRAMMING IN AUTOLISP

To use AutoLISP, you must learn the available AutoLISP functions and the basics of programming in AutoLISP. After completing this chapter, you will be able to:

- Perform the basics of programming in AutoLISP.
- Analyze a LISP program file.
- Use a LISP file inside of AutoCAD.

AUTOLISP BASICS

AutoLISP programming can range from very simple to very complex functions. A LISP program is nothing more than a list of instructions. The LISP program embedded in AutoCAD contains many functions that can be used to perform the tasks you desire. Let's take a look at some of these.

Arithmetic Functions

You can use LISP to perform math functions. Start an AutoCAD program of any name. We are going to enter some LISP routines directly at the command line.

At the command line, enter the following exactly. Be sure to include the parentheses.

 Command: (+ 5 4)
 9

Notice how AutoCAD "returns" the value of 9. This simple routine adds the values of 5 and 4, then prints the sum of these to the screen. Continue and enter the following and notice the results of addition, subtraction, multiplication, and division.

Command: (- 100 40)

Command: (* 5 6)

Command: (/ 100 20)

Notice how the LISP routine is entered at the command line using parentheses. The opening and closing parentheses are used to tell AutoLISP what is an AutoCAD or AutoLISP expression. If AutoCAD detects an opening parenthesis, the following expression is passed to AutoLISP. The expression ends with a closing parenthesis. The expressions you entered at the command line were started with an opening parenthesis, so AutoCAD knew it was an AutoLISP expression. The closing parenthesis denoted the end of the expression. Note how you are able to use a space between parts of the expression. If you were in AutoCAD, you could not do that.

If you get a "1>" return from AutoCAD, enter a closing parenthesis, then [ENTER]. This code means that there is not a closing parenthesis to match an opening parenthesis. There must always be an equal number of opening and closing parentheses. If you want to test this situation, re-enter one of the previous expressions, leaving off the closing parenthesis.

AutoLISP and AutoCAD Commands

An AutoCAD command can be used within an AutoLISP routine. To place a command with the routine, use the COMMAND function, followed by the AutoCAD command name in parentheses, then the proper "arguments." The arguments are the responses to the command prompts.

If the command requires user input, a pause can be placed within the routine. For example, if you wanted to draw a circle with the center at the absolute coordinate of 5,5 and then drag the circle to the desired radius, the following routine could be used.

(command "circle" "5,5" pause)

If you try this routine, you will notice that the Circle command's prompts are "echoed" (displayed) to the command line. Notice also that the coordinate of 5,5 is placed in quotes. This is because it is the actual value and not a variable. There may be times that you do not wish to display these prompts. For example, if the routine contains all the inputs for the prompts, it is useless to show them.

The prompts can be suppressed by setting the CMDECHO variable within the expression. The following line can be included within a LISP routine to do this.

(setvar "cmdecho" 0)

This line sets the AutoCAD variable Cmdecho to zero, which suppresses the command prompts. Appendix C contains a listing of AutoCAD variables.

SETQ Function

The SETQ function is used to assign a value. For example, we could assign values for A and B as follows.

   Command: (setq A 10)

Continue and assign a value for B.

   Command: (setq B 25)

Now let's use an arithmetic expression to use the values set for A and B.

   Command: (+ A B)
   35

Notice how AutoCAD returned the value 35 for the sum of A and B.

Getting and Storing Points in AutoLISP

You can use AutoLISP to store a point that you enter as a coordinate or as a point selected on the drawing screen. For example, you may want to store a point with the Setq function for later use in the routine. You can do this with the GETPOINT function. We will use this function in an example later.

You can also get and store a distance. The GETDIST function is used to obtain a distance. The distance can either be entered as a numeric value or by showing AutoCAD two points on the drawing screen.

Placing Command Prompts within AutoLISP

If you write your own LISP command routines, you may wish to create your own prompts. Prompts are placed within quotes (") inside the LISP routine. For example, the following line would prompt REFERENCE POINT: at the command line.

   (setvar "lastpoint" (getpoint "REFERENCE POINT:"))

Using Notes within a Routine

As you write LISP routines, you may want to include notes for your own purposes. A line that starts with a semicolon will not be executed as a part of the routine.

It is sometimes helpful to include a heading that describes the routine you are writing. We will use a description when we write a LISP routine later in this chapter.

There are many more functions available in AutoLISP, but we will stop here and use the ones we have learned about. If you want to learn all the available functions, the AutoLISP *Programmer's Reference* contains a complete listing.

WRITING AND USING LISP

It's time to write, store, and use a simple LISP routine. Let's assume that you are a machine parts designer. You frequently draw parts that have drill holes of different diameters at a certain position on a plate. The position is always relative to a corner of the plate. You want to write a custom AutoCAD command to streamline your drawing process. The following LISP routine could be used. Let's look at the routine and analyze the functions.

```
;--------------------------------------------------------
;DPCIR.LSP
;--------------------------------------------------------
;DESCRIPTION
;
;Constructs a circle of a specified radius at a designated
;point from a reference position.
;--------------------------------------------------------
;
   (defun C:DPCIR()
     (setvar "CMDECHO" 0)
     (setvar "lastpoint" (getpoint "Select Corner of box:"))
     (setq p1 (getpoint "\nEnter X,Y distance (using @):"))
     (setq p2 (getdist "\nEnter radius:"))
     (command "CIRCLE" p1 p2))
```

DPCIR.LSP

This routine is a simple Ascii text file. The file was arbitrarily named "DPCIR.LSP" for Designated Point Circle. Note that all Lisp files have an .LSP file extension. The file is simply a collection of lines of Lisp instructions. Let's discuss each section of the routine.

The first nine lines are preceded by a semicolon. This means they will not be processed as part of the routine. We used these lines to print a description of the file. This might be useful if we looked through a library of LISP routines later.

We then used Define Function (Defun) to define a new command named DPCIR. This will allow us to use the routine as we would a regular AutoCAD command.

The next line sets the variable Cmdecho to zero. We did this because we do not want the prompts for the Circle command, which occurs later, to be displayed.

Next we used the system variable LASTPOINT (see Appendix C for explanations of this and all system variables) to set a reference point to locate our circle. If you remember, this reference point is to be the corner of the plate. We used the Getpoint function to obtain this point and included the prompt: "Select Corner Of Plate:".

We next used the Setq function to store a value of a point we arbitrarily named "p1". The point for p1 was obtained with the Getpoint function and we included a prompt for a relative coordinate.

Next we used Setq again to store a value for the point named "p2". This time we used the Getdist function to obtain this distance. We also included a prompt for the radius distance.

Finally, we used the Circle command. The values p1 and p2 are the arguments for the prompts that the Circle command issues (center point and radius).

If we wanted to enhance the routine, we could have used the Setvar routine to set the object snap to INTersection when we captured the plate corner, then set the object snap back to NONE. The great part of LISP programming is that you can write the command to suit yourself!

Using the LISP Routine

Let's try the LISP program we just looked at. The first step is to create a file named "DPCIR.LSP". To do this, use a text processor in "non-document" mode. This means that the file must be all ASCII text. You can also use Edlin. Name the file "DPCIR.LSP". Type each line exactly as it is printed, then save the file and copy it to the directory where you have stored the AutoCAD program.

Next, enter AutoCAD and start a drawing of any name. Let's now load the LISP program.

   Command: (LOAD "DPCIR")

If you get an error message, go back and check the contents of your file to verify that you copied it exactly as written. If the loading was successful, AutoCAD should return "C:DPCIR".

Now let's use the program. Draw a square that is 5 units on each side. Now let's place our drill hole. Use your Dpcir routine as a regular command.

   Command: **Dpcir**
   Select corner of box: *(Select the lower left corner.)*
   Enter X,Y distance (using @): **@2,2**
   Enter radius: **.5**

SUMMARY

Writing routines in AutoLISP takes time and practice to learn. The effective use of the LISP programming techniques can be further explored in many excellent books dedicated to this purpose.

CHAPTER REVIEW

1. Write a LISP routine that will multiply two times four.

2. What does "n>" mean when **n** is a number?

3. How can an AutoCAD command be used within a LISP routine?

4. How would you suppress the command prompts from an AutoCAD command contained within a LISP file?

5. What does the Setq function do?

CHAPTER

42

PROGRAMMING TOOLBAR MACROS

Toolbar macros are simple, easy-to-create "mini-programs" that help reduce the number of keystrokes and menu picks. After completing this chapter, you will be able to:

- Understand the nature of macros.
- Write your own macros.

USING MACROS

Behind every icon button on every toolbox is a "macro." A macro is a collection of one or more commands and options, such as "Zoom Window" or "QSave Zoom Extents Plot." By clicking on an icon button, you execute the macro. AutoCAD runs the commands in the macro faster than you can type them or select them from the menus.

Toolbar macros are unique to the Windows version of AutoCAD; you cannot create toolbar macros in the DOS version. You should not confuse toolbar macros with menu macros; toolbar and menu macros share the same name, but they are programmed somewhat differently.

Why Use Macros?

Toolbar macros have a number of advantages and disadvantages over using AutoLISP in AutoCAD.

The advantages of macros are:

- Faster to write than AutoLISP programs; the programming environment is inside AutoCAD, not in an external text editor as with AutoLISP.
- Always available when AutoCAD is running (AutoLISP programs must be specifically loaded or you must make special arrangements to have AutoLISP programs load automatically).
- Easier to start the macro than an AutoLISP routine: just click on the correct button.

The disadvantages of macros are:

- Macros are limited to 255 characters in length; AutoLISP programs can be thousands and thousands of characters in length.
- Macros are limited to what you can type at the keyboard; however, AutoLISP can be used to extend the functionality of macros.
- Both toolbar macros and AutoLISP are limited to running within AutoCAD. If you want to automate steps inside and outside AutoCAD, you must use a Windows-based macro recorder, such as the Recorder provided with Windows v3.1.

You can think of toolbar macros as more user-friendly versions of script files. In short, you would use a toolbar macro when you want to execute one or more commands automatically. For anything more complex — such as performing calculations or parametric drafting — you need to use AutoLISP.

Toolbar Macro Basics

As mentioned, a macro is simply one or more AutoCAD commands in a row. For example, when you receive a drawing from a co-worker or client, you want to set up the four engineering views: top, front, side, and isometric.

To set up the four views by entering the commands, you type:

Command: **vports 4**
Command: *(Set side view.)* **vpoint 0,-1,0**
Command: **zoom e zoom .9x**
Command: *(Pick lower left viewport for front view.)* **vpoint -1,0,0**
Command: **zoom .9x**
Command: *(Pick upper left viewport for top view.)* **vpoint 0,0,0**
Command: **zoom .9x**
Command: *(Pick upper right viewport for iso view.)* **vpoint 1,-1,1**
Command: **zoom .9x**

That's a total of 108 keystrokes.

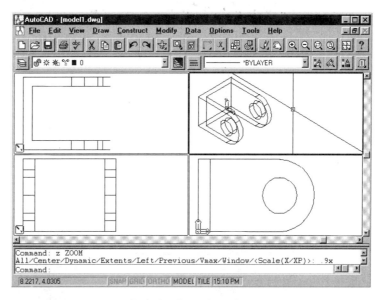

FIGURE 42-1 Four Standard Engineering Viewpoints

Or, you do this by selecting items from the menus:

View|Tiled Viewports|4 Viewports
View|3D Viewpoint Presets|Left

> **NOTE:** Whether you select Left or Front viewpoint depends on how the UCS is set up for the current drawing; similarly, you may need to select an isometric view other than SE.

View|Zoom|Scale
(Type .9x and press ENTER.)
(Click on lower right viewport.)
View|3D Viewpoint Presets|Front
View|Zoom|Scale
(Type .9x and press ENTER.)
(Click on upper right viewport.)
View|3D Viewpoint Presets|Top
View|Zoom|Scale
(Type .9x and press ENTER.)
(Click on upper left viewport.)
View|3D Viewpoint Presets|SE Isometric
View|Zoom|Scale
(Type .9x and press ENTER.)

That's 33 menu and screen picks, along with 16 typed characters.

42-4 USING AUTOCAD

After doing this for a few drawings, you find it tedious to set up each drawing this way. You decide to combine the many keystrokes and/or menu picks into a single toolbar macro. Here's how to do this:

1. Start AutoCAD.
2. Right-click on any icon. AutoCAD displays the Toolbars dialog box.

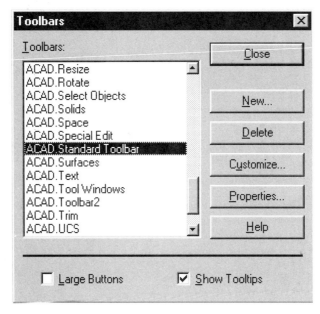

FIGURE 42-2 Toolbars Dialog Box

3. Click the Customize button. AutoCAD displays the Customize Toolbars dialog box.

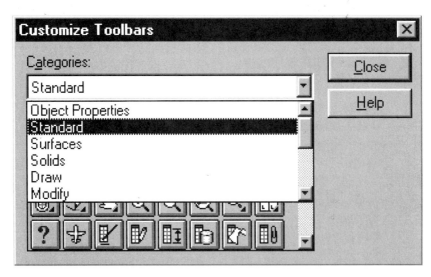

FIGURE 42-3 Customize Toolbars Dialog Box

4. Select Standard from the Categories list box.

5. Scroll through the many icons until you see the Tile Model Space icon without the small black triangle (you'll find it in the very last row).

6. Drag the icon out of the dialog box and onto the AutoCAD drawing area.

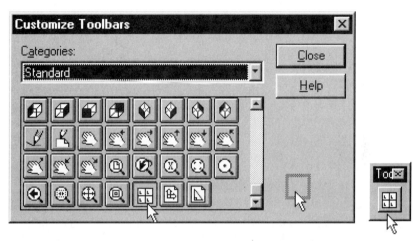

FIGURE 42-4 Dragging the Icon

7. Right-click on the icon. The Button Properties dialog box appears. Notice that the dialog box has four areas of interest.

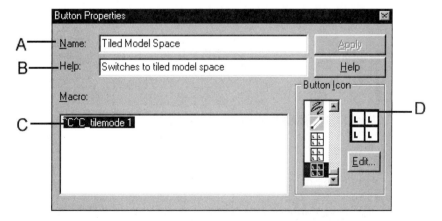

FIGURE 42-5 Button Properties Dialog Box

8. The Name area (A). This is the name that appears in the Tooltip when the cursor lingers over the button. Currently, the name reads "Tiled Model Space". Change the name text to read:

Name: Engineering Views

9. The Help area (B). This is the text that appears on AutoCAD's status bar when you select the button. Currently, the help line reads "Switches to tiled model space." Change the help text to read:

Help: Creates four viewports with the standard engineering views.

10. The macro area (C). These are the commands that are executed when you click on the button. Currently, the macro reads, "^C^C_tilemode 1". You probably recognize the Tilemode command, but what about the other characters? Here's what they mean:

   ^C^C The ^ (caret) symbol is equivalent to holding down the [CTRL] key; the ^C is the same as pressing [CTRL] [C], which is the old AutoCAD method of canceling a command (C is short for Cancel). Two ^C^C in a row cancel out of nested commands, like PEdit.

   _tilemode The _ (underscore) symbol "internationalizes" the command. AutoCAD is available in many languages other than English: German, Spanish, French, Swedish, and so on. In these international versions of AutoCAD, the commands have been "localized." However, to ensure that any language version of AutoCAD understands the same macro, the underscore allows English language command names to work.

   1 Finally, the number 1 is the value of Tilemode to create tiled viewports; 0 is the value for overlapping viewports.

Change the macro to read:

^C^C_vports 4 _cvport 2 _vpoint 0,-1,0 _z e _z .9x _cvport 3 _vpoint -1,0,0 _z .9x _cvport 4 _vpoint 0,0,0 _z .9x _cvport 5 _vpoint 1,-1,1 _z .9x

Much of this macro is similar to the commands we typed at the beginning of the chapter to create the four viewports and their viewpoints. Some characters you may not be familiar with are:

   _z This is the alias for the Zoom command. By using the alias, we reduce the number of characters in the macro. Recall that the macro can have a maximum of 255 characters. Recall, too, that the underscore prefix internationalizes the Zoom command.

   _cvport This is the system variable for switching viewports.

The dialog box should now look like the one in Figure 42-6.

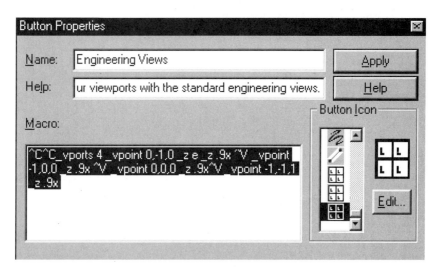

FIGURE 42-6 Changes Made to Dialog Box

11. The icon customization area (D) (Figure 42-5). Here you can select a different icon or create a new one. Currently, the icon shows four window panes. Let's modify it to better represent the four engineering views. Click on the Edit button. AutoCAD displays the Button Editor dialog box.

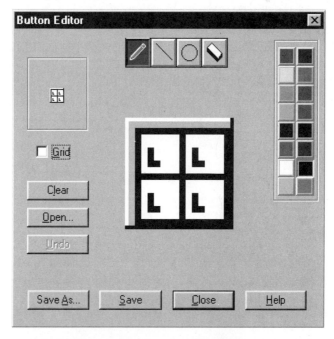

FIGURE 42-7 Button Editor Dialog Box

12. Use the drawing tools to modify the icon. It is generally easier to use the paint tool for everything: to erase, paint over in another color.

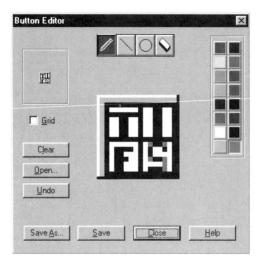

FIGURE 42-8 Edited Button

13. When done, click on the Save button.
14. Click on the Close button to dismiss the Button Editor.
15. Click the Apply button. The icon is modified. AutoCAD gives the toolbar the generic title, "Toolbar 1".

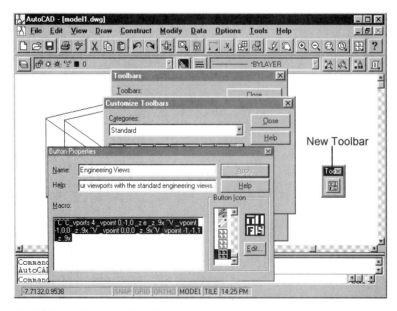

FIGURE 42-9 Modified Toolbar

PROGRAMMING TOOLBAR MACROS **42-9**

16. Dismiss any remaining dialog boxes and AutoCAD updates its menu files to store your changes.

17. To test the macro, open the Model1 drawing from Chapter 38. It will probably appear in the 3D view we left it in.

18. Select the icon. The four engineering views should appear. If AutoCAD reports an error, check the wording of the macro carefully. A command might be misspelled; a comma or other character may be missing.

SPECIAL CHARACTERS

In addition to ^C and ^V, toolbar macros are allowed to use other special characters. The complete list includes:

Character	Meaning
^B	Toggle snap mode between on and off, like CTRL B
^C	Cancel current command, like CTRL C
^D	Change coordinate display mode, like CTRL D
^E	Switch to next isometric plane, like CTRL E
^G	Toggle grid display, like CTRL G
^H	Backspace, like CTRL H
^O	Toggle ortho mode, like CTRL O
^V	Switch to next viewport, like CTRL V
\n	Starts a new line
\t	Tab; has same effect as space
\nnn	Use ASCII character *nnn*
\\	Allows use of the \ character

TABLE 42-1 Special Characters

SUMMARY

Toolbar macros are a quick and easy way to automate a number of command steps in AutoCAD.

CHAPTER REVIEW

1. What is a macro?

2. When is it better to use a toolbar macro? An AutoLISP routine?

3. What does the ^ (caret) symbol mean?

4. Write a toolbar macro for preparing a drawing before plotting: Zoom to extents, save the drawing, then start Plot.

APPENDIX A

PROFESSIONAL CAD TECHNIQUES

Studying and learning the commands in this book are only a start toward becoming a professional CAD operator. Using the proper techniques is just as important. The following is a list of items which should be followed to obtain the most from your expertise.

Learn every command. Most CAD operators tend to learn just enough to get by. If this is your intention, stick with your pencil. You will not realize the full benefits of time savings and more professional results. Force yourself to use each command. This is best accomplished by introducing yourself to each new command and using it until you are comfortable with it. Some of the parts of Auto-CAD that appear the hardest to learn become easy with just a little practice!

Practice, Practice, Practice! Efficiency on a CAD system comes with practice. (Doesn't it also take practice to perform good board drafting?) Never pass up the opportunity to spend time with a system. After you are comfortable with the commands you are using, try new ones. Before long, you will be constructing drawings with ease!

Plan your approach. Never just start drawing. There are hundreds of approaches to every drawing when you are using CAD. The time savings you will realize will come from your technique, not your speed. Drawing faster doesn't mean working harder or faster, it means working smarter. This program has some amazing capabilities, let it work for you!

Use layers. If you are drawing a set of house plans, don't draw the floor plan more than once. Use the layering system to use the plan for each drawing that requires it. The electrical, plumbing, HVAC, lighting, framing, dimensions, furniture, and other plans can exist on each layer.

Use blocks. Never draw anything twice! If you have drawn it once, writeblock it. Before long, you will have a nice library to draw from. The drawings may be transferred onto a library disk for later use. If you build a large enough library of the drawings of your trade, you will soon be assembling more than you will be drawing.

Customize toolbars. Set up a toolbar that contains all the commands that you normally use in your work. Be sure to include any blocks that are normally used. This allows you to choose the commands and drawing parts quickly.

Buy proper equipment. Don't try to save money on brain surgery and CAD equipment. You will purchase a CAD station to save time (i.e., money). If the equipment doesn't suit the job, you won't obtain the savings and may end up not using it at all. It is cheaper to buy the proper equipment than it is to replace it later!

Set up a proper workstation. Like any other type of job, the proper work setup is essential. If you have purchased the right equipment, plan a proper setting for it. A little planning will make your work days (or those of your employees) more efficient.

Be patient. Your first few months with CAD won't save you all the time that you had anticipated. It will come! After a while, you will notice that your speed has increased dramatically. This time period will vary with each individual and each task. It is normal to experience a learning curve, but don't get discouraged; it shouldn't last too long.

Designate a CAD manager. If you are using a multistation CAD setup, assign one person the task of coordinating all operations. This person will emerge soon after you install the system. Some people just seem to gravitate to the position. This will soon become a very valuable employee.

Keep pushing! Don't ever stop experimenting with ways to streamline your graphics procedures. Every task is different, with many ways to approach it. One of the advantages of a CAD system is the ability to customize it to your specific needs. With the ability of AutoCAD, your real challenge is to discover the ways it can serve you best!

Use Aerial View. Zoom in to a drawing to perform detail work. Don't attempt to draw complicated parts when the display shows the work in small scale.

Run check plots. Run periodic check plots to check your work. You will also obtain a feel for the scale of the finished work. Use the laser printer for inexpensive plots. Use a pen plotter with a felt-tip pen for a "dress rehearsal" before plotting the final work.

Use both CAD and traditional techniques. You will soon learn which work CAD is best for and which work traditional methods are best for. Use a combination of both by hand drafting over plots. Sometimes it is best to lay out your drawing in rough form by hand, then digitize the base drawing in and finish with CAD.

Use Copy and Array. Use the Copy and Array commands for repeated work. Never, never draw the same thing twice!

Standardize. Standardize your own procedures and those of your office for CAD use. Create standards for base drawing layers, text sizes, fonts, linetypes, and other operations. Not only will your work be consistent, but your drawing operations will be easier to perform.

APPENDIX B

COMMAND SUMMARY

The following list of commands are documented by Autodesk as being available in AutoCAD Release 13 for DOS. (Windows contains additional commands not found in the DOS version.) Commands prefixed with ' (apostrophe) are transparent commands, which can be executed during another command.

A complete reference of AutoCAD Release 13 commands and how to access and use them can be found in *The Illustrated AutoCAD Quick Reference* by Ralph Grabowski, available through Delmar Publishers.

'About
Displays AutoCAD information dialog box that includes version and serial numbers, acad.msg text, and other information.

AcisIn
Imports an ASCII-format ACIS file into the drawing and creates a 3D solid, 2D region, or body object.

AcisOut
Exports an AutoCAD 3D solid, 2D region, or body as an ASCI-format ACIS file (file extension .SAT).

Ai-Box
Collection of commands to draw 3D primitives from surface meshes: Ai_Box, Ai_Cone, Ai_Dish, Ai_Dome, Ai_mesh, Ai_Pyramid, Ai_Torus, and Ai_Wedge.

Align
Uses three pairs of 3D points to move and rotate (align) 3D objects.

AmeConvert
Converts drawings made with AME v2.0 and v2.1 into ACIS solid models.

'Aperture
Adjusts the size of the target box used with object snap.

'AppLoad
Displays a dialog box that lets you list AutoLISP, ADS, and ARX program names for easy loading into AutoCAD.

Arc
Permits drawing of arcs using different parameters.

Area
Computes area and perimeter of a polygon.

Array
Makes multiple copies of an object or group of objects.

AseAdmin
AutoCAD SQL (structured query language) Extension command: displays a dialog box to execute administrative functions.

AseExport
Exports data and tables in SDF, CDF, or native database formats.

AseLinks
Displays a dialog box that displays database link information, letting you edit and delete the info.

AseRow
Displays a dialog box that displays database table information, letting you edit table data, create links, and selection sets.

AseSelect
Displays a dialog box that lets you create a selection set (union, intersection, or difference) from two selection sets (text or graphics) from linked database rows.

AseSqlEd
Displays a dialog box that lets you enter and execute SQL statements.

AseUnload
Unloads ASE from memory to free up memory for AutoCAD.

Attdef
Creates an attribute definition.

'Attdisp
Controls whether the attributes are displayed.

Attedit
Used for editing attributes.

Attext
Allows attribute entities to be removed from a drawing and written to another disk for use with another program.

AttReDef
Lets you assign existing attributes to a new block, and new attributes to an existing block.

Audit
A diagnostic command used to examine and/or correct errors in a drawing file.

'Base
Specifies point of origin for insertion into another drawing.

Bhatch
Used to fill an automatically defined boundary with a hatch pattern; the use of a dialog box allows preview and adjustment without starting over.

'Blipmode
Controls whether a marker blip is displayed on the screen when picking a point.

Block
Forms a complex object from a group of separate entities in a drawing.

BmpOut
Exports selected objects from the current viewport to a raster BMP file.

Boundary
Draws a closed boundary polyline.

Box
An ACIS command that creates a 3D solid box or cube.

Break
Erases part of an entity or breaks it into two entities.

'Cal
The geometry calculator that evaluates integer, real, and vector expressions.

Chamfer
Trims two intersecting lines and connects the two trimmed lines with a chamfer line.

Change
Permits modification of an entity's characteristics.

Chprop
Similar to the Change command, except only the properties (linetype, color, etc.) of the entity are affected.

Circle
Draws any size circles.

'Color (or 'Colour)
Used to set a new color for all subsequently drawn entities.

Compile
Used to compile shape and font files.

Cone
An ACIS command that creates a 3D solid cone.

Config
Allows configuration of peripheral devices and operating parameters.

Copy
Copies selected objects.

CopyClip
Copies selected objects to the Windows Clipboard in several formats.

CopyHist
Copies Text window text to the Windows Clipboard.

CopyLink
Copies all objects in the current viewport to the Windows Clipboard in several formats.

CutClip
Cuts selected objects from the drawing to the Windows Clipboard in several formats.

Cylinder
An ACIS command that creates a 3D solid cylinder.

Dblist
Provides database information of a drawing.

Ddattdef
Displays Attribute Definition dialog box.

Ddatte
Displays Attribute Editing dialog box.

Ddattext
Displays Attribute Extraction dialog box.

Ddchprop
Displays Change Properties dialog box.

'DdColor
Displays a dialog box that lets you set the new working color.

Ddedit
Displays Attribute Edit dialog box.

'Ddemodes
Displays Current Properties dialog box.

'Ddgrips
Displays Grips dialog box.

Ddim
Displays a series of dimension control dialog boxes.

Ddinsert
Displays Insert dialog box.

'Ddlmodes
Displays Layer Control dialog box.

'DdLtype
Displays a dialog box that lets you load, set, and create linetypes.

DdModify
Displays a dialog box that lets you modify most properties of the selected object.

'Ddosnap
Displays Running Object Snap dialog box.

'DdPType
Displays a dialog box that lets you choose the style of point.

Ddrename
Displays Rename dialog box.

'Ddrmodes
Displays Drawing Aids and Modes dialog box.

'Ddselect
Displays Entity Selection dialog box.

DdStyle
Displays a dialog box for creating text styles from font files (not present in versions of Release 13 prior to c4).

Dducs
Displays UCS dialog box.

DdUcsP
Displays a dialog box that lets you choose from preset UCS orientations.

'Ddunits
Displays Units Control dialog box.

DdView
Displays a dialog box that lets you create, restore, and delete named views.

DdVPoint
Displays a dialog box that lets you set a new 3D viewpoint.

'Delay
Allows for delay between operations in a script file.

Dim
Semi-automatic dimensioning capabilities.

Dim1
Execute a single dimension command.

DimAligned
Draws a linear dimension aligned to an object.

DimAngular
Draws an angular dimension.

DimBaseLine
Draws a linear, angular, or ordinate dimension that continues from a baseline.

DimCenter
Draws the center mark on circles and arcs.

DimContinue
Draws a linear, angular, or ordinate dimension that continues from the last dimension.

DimDiameter
Draws diameter dimension on circles and arcs.

DimEdit
Edits the text and extension lines of associative dimensions.

DimLinear
Draws linear dimensions.

DimOrdinate
Draw ordinate dimensions in the X and Y directions.

DimOverride
Overrides current dimension variables to change the look of selected dimensions.

DimRadius
Draws radial dimensions for circles and arcs.

DimStyle
Creates, names, modifies, applies named dimension styles.

DimTEdit
Moves and rotates text in dimensions.

'Dist
Computes the distance between two points.

Divide
Divides an entity into an equal number of parts and places either a specified block or a point entity at the division points on the entity.

Donut (or Doughnut)
Constructs solid filled circles and doughnuts.

'Dragmode
Permits dynamic dragging of an entity to the desired position on the display.

'DsViewer
Opens the Aerial View window.

Dtext
Allows you to place text on the screen and view it in place as it is entered.

Dview
Used to display a 3D view dynamically.

Dxbin
Creates binary drawing interchange files.

Dxfin
Converts an interchange file into an AutoCAD file.

Dxfout
Allows an AutoCAD file to be reformatted for use with another program.

Edge
Changes the visibility of 3D face edges.

Edgesurf
Draws an edge-defined surface.

'Elev
Sets current elevation and thickness.

Ellipse
Used to construct ellipses.

End
Saves the drawing file and exits the drawing editor.

Erase
Removes entities from a drawing.

Explode
Breaks down a block into the individual entities from which it was constructed; breaks down a Polyline into lines and arcs.

Export
Displays a dialog box to export the drawing in a variety of file formats.

Extend
Used to extend objects in a drawing to meet a boundary object.

Extrude
Extrudes a 2D closed object into a 3D solid object.

FileOpen
Opens a drawing file without displaying a dialog box.

'Files
Allows user to perform file utility tasks while in the drawing editor.

'Fill
Creates solid fill in a closed polygon.

Fillet
Connects two lines with an arc.

'Filter
Creates a selection set of objects based on their properties.

'GifIn
Imports GIF format raster files as a block made of solids.

'Graphscr
Displays control function for single-screen systems that allows user to toggle to graphics screen.

'Grid
Displays grid of specified spacing.

Group
Creates a named selection set of objects.

Hatch
Performs non-associative crosshatching of an area with a specified pattern.

HatchEdit
Edits existing associative hatch patterns (created by BHatch).

'Help (or ?)
Displays a list of AutoCAD commands with detailed information available.

Hide
Removes hidden lines from the currently displayed view.

HpMPlot
A special plot command for HPGL/2-compatible plotters.

'Id
Displays the position of a point in *X,Y* coordinates.

Import
Displays a dialog box to import a variety of file formats into the drawing.

Insert
Inserts a block or another drawing into the current drawing.

InsertObj
The command-line version of the Edit | Insert Object menu selection.

Interfere
Determines the interference of two or more 3D solids.

Intersect
Creates a 3D solid (or 2D region) from the intersection of two or more 3D solids (or 2D regions).

'Isoplane
Allows user to select another isoplane.

'Layer
Creates or switches drawing layers and assigns linetypes and colors to them.

Leader
Draws a leader dimension.

Lengthen
Lengthens or shortens open objects.

Light
Creates, names, places, and deletes "lights" used by the Render command.

'Limits
Sets the drawing boundaries.

Line
Draws a straight line.

'Linetype
Lists, creates, or modifies linetype definitions or loads them for use in a drawing.

List
Displays database information for a single entity in a drawing.

Load
Loads a shape file into a drawing.

LogFileOff
Closes the log file Acad.Log.

LogFileOn
Writes the text of the Command: prompt area to the log file Acad.Log.

'Ltscale
Specifies a scale for all linetypes in a drawing.

MakePreview
Creates a preview image the current view, saved to disk in BMP Bitmap format.

MassProp
Calculates and displays the mass properties of 3D solids and 2D regions.

MatLib
Imports material-look definitions; used by the Render command.

Measure
Places point entities at a specified distance on an object.

Menu
Loads a menu of AutoCAD commands into the menu area.

MenuLoad
Loads a partial menu file.

MenuUnLoad
Unloads part of the menu file.

Minsert
Used to make multiple inserts of a block.

Mirror
Creates a mirror image of an entity.

Mirror3D
Creates a mirror image of objects that can be rotated about a plane.

MlEdit
Displays a dialog box that lets you perform limited editing of multilines.

MLine
Draws multiple parallel lines (up to 16 parallel lines).

MlStyle
Displays a dialog box that lets you define named mline styles, including color, linetype, and endcapping.

Move
Moves an entity from one location to another.

Mslide
Creates a slide of the current display.

Mspace
Used to switch to model space.

MText
Draws paragraph text that fits inside a rectangular boundary; requires an external text editor.

MtProp
Sets and changes the properties of paragraph text, including text style, height, direction, attachment, width, and rotation.

Multiple
When used before a command, causes the command to repeat after each use.

Mview
Used in paper space to create and manipulate viewports.

MvSetup
Places a drawing border and sets up standard engineering views.

New
Used to create a new drawing.

Offset
Constructs a parallel copy to an entity or constructs a larger or smaller image of the entity through a point.

OleLinks
The command-line version of the Edit | Links menu selection.

Oops
Restores entities that were accidentally erased with the previous command.

Open
Used to open an existing drawing.

'Ortho
Causes all lines to be drawn orthogonally with the set snap rotation angle.

'Osnap
Allows geometric points of existing objects to be easily located.

'Pan
Moves the display window for viewing a different part of the drawing without changing the magnification.

PasteClip
Pastes an object from the Windows Clipboard into the upper left corner of the drawing.

PasteSpec
Provides control over the format of the object pasted into the drawing.

'PcxIn
Imports a PCX-format raster file and places it as a block (made of solids) in the drawing.

Pedit
Permits editing of polylines.

Pface
Constructs a polygon mesh that is defined by the location of each vertex in the mesh.

Plan
Returns to the current UCS plan view.

Pline
(Polylines) Lines of specified width that can be manipulated.

Plot
Plots a drawing with a pen printer plotter.

Point
Draws a specified point.

Polygon
Used to draw a regular polygon with a specified number of sides.

Preferences
Allows you to set some AutoCAD settings, in addition to the Config command and system variables.

Psdrag
Controls the scale and position of an imported PostScript image that is being dragged into place.

Psfill
Allows 2D polylines to be filled with PostScript fill patterns.

Psin
Used to import EPS (Encapsulated PostScript) files.

Psout
Used to export the current view of a drawing to an EPS (Encapsulated PostScript) file.

Pspace
Used to switch to paper space.

Purge
Selectively deletes unused blocks, layers, or linetypes.

Qsave
Command that saves the drawing without requesting a file name.

Qtext
Permits text entities to be drawn as rectangles.

Quit
Exits the drawing editor without saving the updated drawing.

Ray
Draws a semi-infinite line.

RConfig
Configures the setup for the Render command.

Recover
Used to attempt recovery of corrupted or damaged files.

Rectang
Draws a rectangle.

Redefine
Restores AutoCAD's definition of a command.

Redo
Restores operations deleted by the Undo command.

'Redraw
Cleans up the display.

'Redrawall
Performs a redraw in all viewports.

Regen
Causes the entire drawing to be regenerated and redraws the screen.

Regenall
Performs a regeneration in all viewports.

'Regenauto
Allows the user to control whether the drawing is automatically regenerated.

Region
Creates a 2D region object from existing closed objects.

Reinit
Reinitializes the I/O ports, digitizer, display, plotter, and PGP file.

Rename
Allows name changes of blocks, linetypes, layers, or text style.

Render
Creates a rendering of 3D objects in the drawing.

'RenderUnLoad
Unloads Render to free up memory.

Replay
Displays a dialog box that lets AutoCAD display a GIF, TGA, or TIFF raster file.

'Resume
Continues playing back a script file that had been interrupted by the [ESC] or [BACKSPACE] keys.

Revolve
Creates a 3D solid by revolving a 2D closed object around an axis.

Revsurf
Draws a revolved surface.

RMat
Displays a dialog box that lets you define, load, create, attach, detach, and modify material-look definitions; used by the Render command.

Rotate
Rotates an entity or a group of entities around a specified center point.

Rotate3D
Rotates objects about a 3D axis.

RPref
Displays a dialog box that lets you set your preferences for renderings.

'Rscript
Forces a script to be restarted from the beginning.

Rulesurf
Draws a ruled surface.

Save
Saves an updated drawing without exiting out of the drawing editor.

Saveas
Saves the current drawing under a different specified name.

SaveAsR12
Saves the drawing in Release 12 format, eliminating or converting R13-specific objects into R12 objects.

SaveImg
Saves the current rendering in GIF, TGA, or TIFF format.

Scale
Changes the scale of an entity or entities.

Scene
Creates, modifies, and deletes named scenes; used by the Render command.

'Script
Allows user to invoke a script file while in the drawing editor.

Section
Creates a 2D region from a 3D solid by intersecting a plane through the solid.

Select
Used to preselect objects to be edited.

'Setvar
Used to change AutoCAD's system variables.

Shade
Creates a fast shade of a 3D object, similar to the fast shade option in AutoShade.

Shape
Places shapes from a shape file into a drawing.

ShowMat
Reports the material definition assigned to the selected object.

Sketch
Allows freehand sketching as a part of the drawing.

Slice
Slices a 3D solid with a plane.

'Snap
Allows user to turn snap on or off, change the snap resolution, set different spacing for the X and Y axis, or rotate the grid, and set isometric mode.

Solid
Draws filled in polygons.

Spell
Checks the spelling of text in the drawing.

Sphere
Draws a 3D sphere.

Spline
Draws a NURBS (spline) curve.

SplinEdit
Edits a spline edit.

Stats
Displays a dialog box that lists information about current state of rendering.

'Status
Displays a status screen containing information about the current drawing.

StlOut
Exports 3D solids to an SLA file, in ASCII or binary format.

Stretch
Move selected objects while allowing their connections to other objects in the drawing to remain unchanged.

'Style
Creates and modifies text styles.

Subtract
Creates a new 3D solid (or 2D region) by subtracting one object from a second object.

SysWindows
Controls the size of the drawing window.

Tablet
Permits alignment of digitizer with existing drawing coordinates.

Tabsurf
Draws a tabulated surface.

Text
Allows text to be entered into a drawing.

'Textscr
Display control function for single-screen systems that allows user to toggle to text screen.

'TiffIn
Imports a TIFF raster file as a block (made of solids) into the drawing.

'Time
Keeps track of time functions for each drawing.

Tolerance
Displays a dialog box that lets you select tolerance symbols.

Toolbar
Controls the display of toolboxes.

Torus
Draws a doughnut-shaped 3D solid.

Trace
Draws lines of a specified width.

'Treestat
Displays drawing information on the current spatial index.

Trim
Trims objects in a drawing by defining other objects as cutting edges then specifying the part of the object to be cut from between them.

U
Used to undo the most recent command.

UCS
Used to manipulate the User Coordinate System.

UCSicon
Controls the on-screen display of the UCS icon indicator.

Undefine
Disables a command.

Undo
Used to undo several command moves in a single operation.

Union
Creates a new 3D solid (or 2D region) from two solids (or regions).

'Units
Allows the user to select display format and precision of that format.

'View
Saves the display as a view or displays a named view.

Viewres
Controls the fast zoom mode and resolution for circle and arc regenerations.

VlConv
Converts Visual Link rendering data into AutoVision format.

Vplayer
Controls the visibility of the individual viewport layers.

Vpoint
Sets the view point from which the user will see the drawing.

Vports (or Viewports)
Sets the number and configuration of viewports displayed on the screen.

Vslide
Allows user to view a slide file.

Wblock
Writes entities to a new drawing file.

Wedge
Draws a 3D solid wedge.

WmfIn
Imports a WMF file into the drawing as a block.

WmfOpts
Controls how a WMF file is imported.

WmfOut
Exports the drawing as a WMF file.

Xbind
Binds externally referenced drawings, converting them to blocks in the master drawing.

XLine
Draws an infinite line.

'Xplode
Explodes more than one object at a time.

Xref
Used to place an externally referenced drawing into a master drawing.

XrefClip
Inserts and clips an externally referenced drawing.

'Zoom
Allows user to increase or decrease the size of the display for viewing purposes.

3D
Draws 3D surface objects out of polygon meshes: box, cone, dish, dome, mesh, pyramid, sphere, torus, and wedge.

3dArray
Creates a 3D array.

3Dface
Creates a solid face in a defined plane.

3Dmesh
Draws a 3D mesh.

3Dpoly
Draws a 3D polyline.

3dsIn
Imports 3D Studio geometry and rendering data; cannot import procedural materials and smoothing groups.

3dsOut
Exports AutoCAD geometry and rendering data.

APPENDIX C

SYSTEM VARIABLES

Courtesy of Autodesk Inc.

Variable	Characteristics	Description
ACADPREFIX	(Read-only) Type: String Not saved	Stores the directory path, if any, specified by the ACAD environment variable, with path separators appended if necessary.
ACADVER	(Read-only) Type: String Not saved	Stores the AutoCAD version number, which can have values like 13 or 13a. Note that this variable differs from the DXF file $ACADVER header variable, which contains the drawing database level number.
AFLAGS	Type: Integer Not saved Initial value: 0	Sets attribute flags for the ATTDEF command bit-code. It is the sum of the following: 0 No attribute mode selected 1 Invisible 2 Constant 4 Verify 8 Preset
ANGBASE	Type: Real Saved in: Drawing Initial value: 0.0000	Sets the base angle 0 with respect to the current UCS.
ANGDIR	Type: Integer Saved in: Drawing Initial value: 0	Sets the angle from angle 0 with respect to the current UCS. 0 Counterclockwise 1 Clockwise
APERTURE	Type: Integer Saved in: Config Initial value: 10	Sets object snap target height, in pixels.
AREA	(Read-only) Type: Real Not saved	Stores the last area computed by AREA, LIST, or DBLIST. Use the SETVAR command to access this system variable.
ATTDIA	Type: Integer Saved in: Drawing Initial value: 0	Controls whether INSERT uses a dialog box for attribute value entry. 0 Issues prompts on the command line. 1 Uses a dialog box.
ATTMODE	Type: Integer Saved in: Drawing Initial value: 1	Controls Attribute Display mode. 0 Off 1 Normal 2 On

Variable	Characteristics	Description
ATTREQ	Type: Integer Saved in: Drawing Initial value: 1	Determines whether INSERT uses default attribute settings during insertion of blocks. 0 Assumes the defaults for the values of all attributes. 1 Enables prompts or dialog box for attribute values, as selected by ATTDIA.
AUDITCTL	Type: Integer Saved in: Config Initial value: 0	Controls whether AutoCAD creates an *.adt* file (audit report). 0 Disables or prevents the writing of *.adt* files. 1 Enables the writing of *.adt* files by AUDIT.
AUNITS	Type: Integer Saved in: Drawing Initial value: 0	Sets Angular Units mode. 0 Decimal degrees 1 Degrees/minutes/seconds 2 Gradians 3 Radians 4 Surveyor's units
AUPREC	Type: Integer Saved in: Drawing Initial value: 0	Sets angular units decimal places.
BACKZ	(Read-only) Type: Real Saved in: Drawing	Stores the back clipping plane offset from the target plane for the current viewport, in drawing units. Meaningful only if the back clipping bit in VIEWMODE is on. The distance of the back clipping plane from the camera point can be found by subtracting BACKZ from the camera-to-target distance.
BLIPMODE	Type: Integer Saved in: Drawing Initial value: 1	Controls whether marker blips are visible. 0 Turns off marker blips. 1 Turns on marker blips.
CDATE	(Read-only) Type: Real Not saved	Sets calendar date and time.
CECOLOR	Type: String Saved in: Drawing Initial value: "BYLAYER"	Sets the color of new objects.
CELTSCALE	Type: Real Saved in: Drawing Initial value: 1.0000	Sets the current global linetype scale for objects.

Variable	Characteristics	Description
CELTYPE	Type: String Saved in: Drawing Initial value: "BYLAYER"	Sets the linetype of new objects.
CHAMFERA	Type: Real Saved in: Drawing Initial value: 0.0000	Sets the first chamfer distance.
CHAMFERB	Type: Real Saved in: Drawing Initial value: 0.0000	Sets the second chamfer distance.
CHAMFERC	Type: Real Saved in: Drawing Initial value: 0.0000	Sets the chamfer length.
CHAMFERD	Type: Real Saved in: Drawing Initial value: 0.0000	Sets the chamfer angle.
CHAMMODE	Type: Integer Not saved Initial value: 0	Sets the input method by which AutoCAD creates chamfers. 0 Requires two chamfer distances. 1 Requires one chamfer length and an angle.
CIRCLERAD	Type: Real Not saved Initial value: 0.0000	Sets the default circle radius. A zero sets no default.
CLAYER	Type: String Saved in: Drawing Initial value: "0"	Sets the current layer.
CMDACTIVE	(Read-only) Type: Integer Not saved	Stores the bit-code that indicates whether an ordinary command, transparent command, script, or dialog box is active. It is the sum of the following: 1 Ordinary command is active. 2 Ordinary command and a transparent command are active. 4 Script is active. 8 Dialog box is active.

Variable	Characteristics	Description
CMDDIA	Type: Integer Saved in: Config Initial value: 1	Controls whether dialog boxes are enabled for more than just PLOT and external database commands. 0 Disables dialog boxes. 1 Enables dialog boxes.
CMDECHO	Type: Integer Not saved Initial value: 1	Controls whether AutoCAD echoes prompts and input during the AutoLISP (**command**) function. 0 Disables echoing. 1 Enables echoing.
CMDNAMES	(Read-only) Type: String Not saved	Displays the name (in English) of the currently active command and transparent command. For example, LINE'ZOOM indicates that the ZOOM command is being used transparently during the LINE command.
CMLJUST	Type: Integer Saved in: Config Initial value: 0	Specifies multiline justification. 0 Top 1 Middle 2 Bottom
CMLSCALE	Type: Real Saved in: Config Initial value: 1.0000	Controls the overall width of a multiline. A scale factor of 2.0 produces a multiline that is twice as wide as the style definition. A zero scale factor collapses the multiline into a single line. A negative scale factor flips the order of the offset lines (that is, the smallest or most negative is placed on top when the multiline is drawn from left to right).
CMLSTYLE	Type: String Saved in: Config Initial value: ""	Sets the name of the multiline style that AutoCAD uses to draw the multiline.
COORDS	Type: Integer Saved in: Drawing Initial value: 1	Controls when coordinates are updated. 0 Coordinate display is updated on pick points only. 1 Display of absolute coordinates is continuously updated. 2 Distance and angle from last point are displayed when a distance or angle is requested.

Variable	Characteristics	Description
CVPORT	Type: Integer Saved in: Drawing Initial value: 2	Sets the identification number of the current viewport. You can change this value, thereby changing the current viewport, if the following conditions are met: • The identification number you specify is that of an active viewport. • A command in progress has not locked cursor movement to that viewport. • Tablet mode is off.
DATE	(Read-only) Type: Real Not saved	Stores the current date and time represented as a Julian date and fraction in a real number: <Julian date>.<Fraction> For example, on January 29, 1993, at 2:29:35 in the afternoon, the DATE variable would contain 2446460.603877364. Your computer clock provides the date and time. The time is represented as a fraction of a day. To compute differences in time, subtract the times returned by DATE. To extract the seconds since midnight from the value returned by DATE, use AutoLISP expressions: (setq s (getvar "DATE")) (setq seconds (* 86400.0 (- s (fix s)))) The DATE system variable returns a true Julian date only if the system clock is set to UTC/Zulu (Greenwich Mean Time). TDCREATE and TDUPDATE have the same format as DATE, but their values represent the creation time and last update time of the current drawing.
DBMOD	(Read-only) Type: Integer Not saved	Indicates the drawing modification status using bitcode. It is the sum of the following: 1 Object database modified 2 Symbol table modified 4 Database variable modified 8 Window modified 6 View modified

Variable	Characteristics	Description
DCTCUST	Type: String Saved in: Config Initial value: ""	Displays the current custom spelling dictionary path and file name.
DCTMAIN	Type: String Saved in: Config Initial value: ""	Displays the current main spelling dictionary file name. The full path is not shown because this file is expected to reside in the \support directory. You can specify a default main spelling dictionary using the SETVAR command. When prompted for a new value for DCTMAIN, you can enter one of the following keywords: Keyword Language name enu American English ena Australian English ens British English (ise) enz British English (ize) ca Catalan cs Czech da Danish nl Dutch (primary) nls Dutch (secondary) fi Finnish fr French (unaccented capitals) fra French (accented capitals) de German (Scharfes s) ded German (Dopple s) it Italian no Norwegian (Bokmal) non Norwegian (Nynorsk) pt Portuguese (Iberian) ptb Portuguese (Brazilian) ru Russian (infrequent io) ru i Russian (frequent io) es Spanish (unaccented capitals) esa Spanish (accented capitals) sv Swedish
DELOBJ	Type: Integer Saved in: Drawing Initial value: 1	Controls whether objects used to create other objects are retained or deleted from the drawing database. 0 Objects are deleted. 1 Objects are retained.

Variable	Characteristics	Description
DIASTAT	(Read-only) Type: Integer Not saved	Stores the exit method of the most recently used dialog box. 0 Cancel 1 OK
DIMALT	Type: Switch Saved in: Drawing Initial value: Off	When turned on, enables alternate units dimensioning. See also DIMALTD, DIMALTF, DIMALTZ (DIMALTTZ, DIMALTTD), and DIMAPOST.
DIMALTD	Type: Integer Saved in: Drawing Initial value: 2	Controls alternate units decimal places. If DIMALT is enabled, DIMALTD governs the number of decimal places displayed in the alternate measurement.
DIMALTF	Type: Real Saved in: Drawing Initial value: 25.4000	Controls alternate units scale factor. If DIMALT is enabled, DIMALTF multiplies linear dimensions by a factor to produce a value in an alternate system of measurement.
DIMALTTD	Type: Integer Saved in: Drawing Initial value: 2	Sets the number of decimal places for the tolerance values of an alternate units dimension. DIMALTTD sets this value when entered on the command line or set in the Alternate Units section of the Annotation dialog box.
DIMALTTZ	Type: Integer Saved in: Drawing Initial value: 0	Toggles suppression of zeros for tolerance values. DIMALTTZ sets this value when entered on the command line or set in the Alternate Units section of the Annotation dialog box. DIMALTTZ also affects real-to-string conversions performed by the AutoLISP **rtos** and **angtos** functions.
DIMALTU	Type: Integer Saved in: Drawing Initial value: 2	Sets the units format for alternate units of all dimension style family members except angular. 1 Scientific 2 Decimal 3 Engineering 4 Architectural 5 Fractional DIMALTU sets this value when entered on the command line or set in the Alternate Units section of the Annotation dialog box.

Variable	Characteristics	Description
DIMALTZ	Type: Integer Saved in: Drawing Initial value: 0	Toggles suppression of zeros for alternate unit dimension values. 0 Turns off suppression of zeros. 1 Turns on suppression of zeros. DIMALTZ sets this value when entered on the command line or set in the Alternate Units section of the Annotation dialog box. DIMALTZ also affects real-to-string conversions performed by the AutoLISP **rtos** and **angtos** functions.
DIMAPOST	Type: String Saved in: Drawing Initial value: ""	Specifies a text prefix or suffix (or both) to the alternate dimension measurement for all types of dimensions except angular. For instance, if the current Units mode is Architectural, DIMALT is enabled, DIMALTF is 25.4, DIMALTD is 2, and DIMAPOST is set to "mm," a distance of 10 units would be edited as 10"[254.00mm]. To disable an established prefix or suffix (or both), set it to a single period (.).
DIMASO	Type: Switch Saved in: Drawing Initial value: On	Controls the creation of associative dimension objects. Off No association between the dimension and points on the object. The lines, arcs, arrowheads, and text of a dimension are drawn as separate objects. On Creates an association between the dimension and definition points located on a feature of the object, such as an intersection of two lines. If the feature is moved, so must the definition point be. The elements are formed into a single object associated with the geometry used to define it. The DIMASO value is not stored in a dimension style.
DIMASZ	Type: Real Saved in: Drawing Initial value: 0.1800	Controls the size of dimension line and leader line arrowheads. Also controls the size of hook lines. Multiples of the arrowhead size determine whether dimension lines and text are to fit between the extension lines. Also used to scale arrowhead blocks if set by DIMBLK. DIMASZ has no effect when DIMTSZ is other than zero.

Variable	Characteristics	Description
DIMAUNIT	Type: Integer Saved in: Drawing Initial value: 0	Sets the angle format for angular dimensions. 0 Decimal degrees 1 Degrees/minutes/seconds 2 Gradians 3 Radians 4 Surveyor's units DIMAUNIT sets this value when entered on the command line or set from the Primary Units section of the Annotation dialog box.
DIMBLK	Type: String Saved in: Drawing Initial value: ""	Sets the name of a block to be drawn instead of the normal arrowhead at the ends of the dimension line or leader line. To disable an established block name, set it to a single period (.).
DIMBLK1	Type: String Saved in: Drawing Initial value: ""	If DIMSAH is on, DIMBLK1 specifies user-defined arrowhead blocks for the first end of the dimension line. This variable contains the name of a previously defined block. To disable an established block name, set it to a single period (.).
DIMBLK2	Type: String Saved in: Drawing Initial value: ""	If DIMSAH is on, DIMBLK2 specifies user-defined arrowhead blocks for the second end of the dimension line. This variable contains the name of a previously defined block. To disable an established block name, set it to a single period (.).
DIMCEN	Type: Real Saved in: Drawing Initial value: 0.0900	Controls drawing of circle or arc center marks and center lines by the DIMCENTER, DIMDIAMETER, and DIMRADIUS dimensioning commands. 0 No center marks or lines are drawn. <0 Center lines are drawn. >0 Center marks are drawn. The absolute value specifies the size of the mark portion of the center line. DIMRADIUS and DIMDIAMETER draw the center mark or line only if the dimension line is placed outside the circle or arc.

Variable	Characteristics	Description
DIMCLRD	Type: Integer Saved in: Drawing Initial value: 0	Assigns colors to dimension lines, arrowheads, and dimension leader lines. Also controls the color of leader lines created with the LEADER command. The color can be any valid color number or special color label BYBLOCK or BYLAYER. Using the SETVAR command, supply the color number. Integer equivalents for BYBLOCK and BYLAYER are 0 and 256, respectively. From the Command prompt, set the color values by entering **dimclrd** and then a standard color name or **BYBLOCK** or **BYLAYER**.
DIMCLRE	Type: Integer Saved in: Drawing Initial value: 0	Assigns colors to dimension extension lines. The color can be any valid color number or the special color label BYBLOCK or BYLAYER. See DIMCLRD.
DIMCLRT	Type: Integer Saved in: Drawing Initial value: 0	Assigns colors to dimension text. The color can be any valid color number or the special color label BYBLOCK or BYLAYER. See DIMCLRD.
DIMDEC	Type: Integer Saved in: Drawing Initial value: 4	Sets the number of decimal places for the tolerance values of a primary units dimension. DIMDEC stores this value when entered on the command line or set in the Primary Units section of the Annotation dialog box.
DIMDLE	Type: Real Saved in: Drawing Initial value: 0.0000	Extends the dimension line beyond the extension line when oblique strokes are drawn instead of arrowheads.
DIMDLI	Type: Real Saved in: Drawing Initial value: 0.3800	Controls the dimension line spacing for baseline dimensions. Each baseline dimension is offset by this amount, if necessary, to avoid drawing over the previous dimension.
DIMEXE	Type: Real Saved in: Drawing Initial value: 0.1800	Determines how far to extend the extension line beyond the dimension line.
DIMEXO	Type: Real Saved in: Drawing Initial value: 0.0625	Determines how far extension lines are offset from origin points. If you point directly at the corners of an object to be dimensioned, the extension lines stop just short of the object.

Variable	Characteristics	Description	
DIMFIT	Type: Integer Saved in: Drawing Initial value: 3	Controls the placement of text and arrowheads inside or outside extension lines based on the available space between the extension lines.	
		0	Places text and arrowheads between the extension lines if space is available. Otherwise places both text and arrowheads outside extension lines.
		1	If space is available, places text and arrowheads between the extension lines. When enough space is available for text, places text between the extension lines and arrowheads outside them. When not enough space is available for text, places both text and arrowheads outside extension lines.
		2	If space is available, places text and arrowheads between the extension lines. When space is available for the text only, AutoCAD places the text between the extension lines and the arrowheads outside. When space is available for the arrowheads only, AutoCAD places them between the extension lines and the text outside. When no space is available for either text or arrowheads, AutoCAD places them both outside the extension lines.
		3	Places whatever best fits between the extension lines.
		4	Creates leader lines when there is not enough space for text between extension lines. Horizontal justification controls whether the text is drawn to the right or the left of the leader. For more information, see DIMJUST.

Variable	Characteristics	Description
DIMGAP	Type: Real Saved in: Drawing Initial value: 0.0900	Sets the distance around the dimension text when you break the dimension line to accommodate dimension text. Also sets the gap between annotation and a hook line created with the LEADER command. A negative DIMGAP value creates basic dimensioning—dimension text with a box around its full extents. AutoCAD also uses DIMGAP as the minimum length for pieces of the dimension line. When calculating the default position for the dimension text, it positions the text inside the extension lines only if doing so breaks the dimension lines into two segments at least as long as DIMGAP. Text placed above or below the dimension line is moved inside if there is room for the arrowheads, dimension text, and a margin between them at least as large as DIMGAP: 2 * (DIMASZ + DIMGAP).
DIMJUST	Type: Integer Saved in: Drawing Initial value: 0	Controls horizontal dimension text position. 0 Center-justifies the text between the extension lines. 1 Positions the text next to the first extension line. 2 Positions the text next to the second extension line. 3 Positions the text above and aligned with the first extension line. 4 Positions the text above and aligned with the second extension line.

Variable	Characteristics	Description
DIMLFAC	Type: Real Saved in: Drawing Initial value: 1.0000	Sets a global scale factor for linear dimensioning measurements. All linear distances measured by dimensioning (including radii, diameters, and coordinates) are multiplied by the DIMLFAC setting before being converted to dimension text.
		DIMLFAC has no effect on angular dimensions, and it is not applied to the values held in DIMTM, DIMTP, or DIMRND.
		If you are creating a dimension in paper space and DIMLFAC is nonzero, AutoCAD multiplies the distance measured by the absolute value of DIMLFAC. In model space, negative values are ignored, and the value 1.0 is used instead. AutoCAD computes a value for DIMLFAC if you try to change DIMLFAC from the Dim prompt while in paper space and you select the Viewport option.
		Dim: **dimlfac** Current value <1.0000> New value (Viewport): **v** Select viewport to set scale:
		AutoCAD calculates the scaling of model space to paper space and assigns the negative of this value to DIMLFAC.
DIMLIM	Type: Switch Saved in: Drawing Initial value: Off	When turned on, generates dimension limits as the default text. Setting DIMLIM on forces DIMTOL to be off.
DIMPOST	Type: String Saved in: Drawing Initial value: ""	Specifies a text prefix or suffix (or both) to the dimension measurement. For example, to establish a suffix for millimeters, set DIMPOST to mm; a distance of 19.2 units would be displayed as 19.2mm.
		If tolerances are enabled, the suffix is applied to the tolerances as well as to the main dimension.
		To separate DIMPOST values into prefix and suffix parts of the dimension text, use the < > mechanism; this allows AutoCAD to use the DIMPOST values as text. Use this mechanism for angular dimensions.

Variable	Characteristics	Description
DIMRND	Type: Real Saved in: Drawing Initial value: 0.0000	Rounds all dimensioning distances to the specified value. For instance, if DIMRND is set to 0.25, all distances round to the nearest 0.25 unit. If you set DIMRND to 1.0, all distances round to the nearest integer. Note that the number of digits edited after the decimal point depends on the precision set by DIMDEC. DIMRND does not apply to angular dimensions.
DIMSAH	Type: Switch Saved in: Drawing Initial value: Off	Controls use of user-defined arrowhead blocks at the ends of the dimension line. Off Normal arrowheads or user-defined arrowhead blocks set by DIMBLK are used. On User-defined arrowhead blocks are used. DIMBLK1 and DIMBLK2 specify different user-defined arrowhead blocks for each end of the dimension line.
DIMSCALE	Type: Real Saved in: Drawing Initial value: 1.0000	Sets the overall scale factor applied to dimensioning variables that specify sizes, distances, or offsets. It is not applied to tolerances or to measured lengths, coordinates, or angles. Also affects the scale of leader objects created with the LEADER command. 0.0 AutoCAD computes a reasonable default value based on the scaling between the current model space viewport and paper space. If you are in paper space, or in model space and not using the paper space feature, the scale factor is 1.0. >0 AutoCAD computes a scale factor that leads text sizes, arrowhead sizes, and other scaled distances to plot at their face values.
DIMSD1	Type: Switch Saved in: Drawing Initial value: Off	When turned on, suppresses drawing of the first dimension line.
DIMSD2	Type: Switch Saved in: Drawing Initial value: Off	When turned on, suppresses drawing of the second dimension line.
DIMSE1	Type: Switch Saved in: Drawing Initial value: Off	When turned on, suppresses drawing of the first extension line.

Variable	Characteristics	Description
DIMSE2	Type: Switch Saved in: Drawing Initial value: Off	When turned on, suppresses drawing of the second extension line.
DIMSHO	Type: Switch Saved in: Drawing Initial value: On	When turned on, controls redefinition of dimension objects while dragging. Associative dimensions recompute dynamically as they are dragged. Radius or diameter leader length input uses dynamic dragging and ignores DIMSHO. On some computers, dynamic dragging can be very slow, so you can set DIMSHO to off to drag the original image instead. The DIMSHO value is not stored in a dimension style.
DIMSOXD	Type: Switch Saved in: Drawing Initial value: Off	When turned on, suppresses drawing of dimension lines outside the extension lines. If the dimension lines would be outside the extension lines and DIMTIX is on, setting DIMSOXD to on suppresses the dimension line. If DIMTIX is off, DIMSOXD has no effect.
DIMSTYLE	(Read-only) Type: String Saved in: Drawing	Sets the current dimension style by name. To change the dimension style, use DDIM or the DIMSTYLE dimensioning command.
DIMTAD	Type: Integer Saved in: Drawing Initial value: 0	Controls vertical position of text in relation to the dimension line. 0 Centers the dimension text between the extension lines. 1 Places the dimension text above the dimension line except when the dimension line is not horizontal and text inside the extension lines is forced horizontal (DIMTIH = 1). The distance from the dimension line to the baseline of the lowest line of text is the current DIMGAP value. 2 Places the dimension text on the side of the dimension line farthest away from the defining points. 3 Places the dimension text to conform to a JIS representation.

Variable	Characteristics	Description
DIMTDEC	Type: Integer Saved in: Drawing Initial value: 4	Sets the number of decimal places for the tolerance values for a primary units dimension.
DIMTFAC	Type: Real Saved in: Drawing Initial value: 1.0000	Specifies a scale factor for text height of tolerance values relative to the dimension text height as set by DIMTXT. The relative sizes of numbers in stacked fractions also are based on DIMTFAC. $$\text{DIMTFAC} = \frac{\text{Tolerance Height}}{\text{Text Height}}$$ For example, if DIMTFAC is set to 1.0, the text height of tolerances is the same as the dimension text. If DIMTFAC is set to 0.75, the text height of tolerances is three-quarters the size of dimension text. Use DIMTFAC for plus and minus tolerance strings when DIMTOL is on and DIMTM is not equal to DIMTP, or when DIMLIM is on.
DIMTIH	Type: Switch Saved in: Drawing Initial value: On	Controls the position of dimension text inside the extension lines for all dimension types except ordinate dimensions. Off Aligns text with the dimension line. On Draws text horizontally.
DIMTIX	Type: Switch Saved in: Drawing Initial value: Off	Draws text between extension lines. Off The result varies with the type of dimension. For linear and angular dimensions, AutoCAD places text inside the extension lines if there is sufficient room. For radius and diameter dimensions, setting DIMTIX off forces the text outside the circle or arc. On Draws dimension text between the extension lines even if AutoCAD would ordinarily place it outside those lines.

Variable	Characteristics	Description
DIMTM	Type: Real Saved in: Drawing Initial value: 0.0000	When DIMTOL or DIMLIM is on, sets the minimum (or lower) tolerance limit for dimension text. AutoCAD accepts signed values for DIMTM. If DIMTOL is on and DIMTP and DIMTM are set to the same value, AutoCAD draws a ± symbol followed by the tolerance value. If DIMTM and DIMTP values differ, the upper tolerance is drawn above the lower, and a plus sign is added to the DIMTP value if it is positive. For DIMTM, AutoCAD uses the negative of the value you enter (adding a minus sign if you specify a positive number and a plus sign if you specify a negative number). No sign is added to a value of zero.
DIMTOFL	Type: Switch Saved in: Drawing Initial value: Off	When turned on, draws a dimension line between the extension lines even when the text is placed outside the extension lines. For radius and diameter dimensions (while DIMTIX is off), draws a dimension line and arrowheads inside the circle or arc and places the text and leader outside.
DIMTOH	Type: Switch Saved in: Drawing Initial value: On	When turned on, controls the position of dimension text outside the extension lines. 0 Aligns text with the dimension line. 1 Draws text horizontally.
DIMTOL	Type: Switch Saved in: Drawing Initial value: Off	When turned on, appends dimension tolerances to dimension text. Setting DIMTOL on forces DIMLIM off.
DIMTOLJ	Type: Integer Saved in: Drawing Initial value: 1	Sets the vertical justification for tolerance values relative to the nominal dimension text. 0 Bottom 1 Middle 2 Top
DIMTP	Type: Real Saved in: Drawing Initial value: 0.0000	When DIMTOL or DIMLIM is on, sets the maximum (or upper) tolerance limit for dimension text. AutoCAD accepts signed values for DIMTP. If DIMTOL is on and DIMTP and DIMTM are set to the same value, AutoCAD draws a ± symbol followed by the tolerance value. If DIMTM and DIMTP values differ, the upper tolerance is drawn above the lower and a plus sign is added to the DIMTP value if it is positive.

Variable	Characteristics	Description
DIMTSZ	Type: Real Saved in: Drawing Initial value: 0.0000	Specifies the size of oblique strokes drawn instead of arrowheads for linear, radius, and diameter dimensioning. 0 Draws arrows. >0 Draws oblique strokes instead of arrows. Size of oblique strokes is determined by this value multiplied by the DIMSCALE value. Also determines if dimension lines and text fit between extension lines.
DIMTVP	Type: Real Saved in: Drawing Initial value: 0.0000	Adjusts the vertical position of dimension text above or below the dimension line. AutoCAD uses the DIMTVP value when DIMTAD is off. The magnitude of the vertical offset of text is the product of the text height and DIMTVP. Setting DIMTVP to 1.0 is equivalent to setting DIMTAD to on. AutoCAD splits the dimension line to accommodate the text only if the absolute value of DIMTVP is less than 0.7.
DIMTXSTY	Type: String Saved in: Drawing Initial value: "STANDARD"	Specifies the text style of the dimension.
DIMTXT	Type: Real Saved in: Drawing Initial value: 0.1800	Specifies the height of dimension text, unless the current text style has a fixed height.
DIMTZIN	Type: Integer Saved in: Drawing Initial value: 0	Toggles suppression of zeros for tolerance values. DIMZIN stores this value when entered on the command line or set in the Primary Units section of the Annotation dialog box.
DIMUNIT	Type: Integer Saved in: Drawing Initial value: 2	Sets the units format for all dimension style family members except angular. 1 Scientific 2 Decimal 3 Engineering 4 Architectural 5 Fractional

Variable	Characteristics	Description
DIMUPT	Type: Switch Saved in: Drawing Initial value: Off	Controls cursor functionality for User Positioned Text. 0 Cursor controls only the dimension line location. 1 Cursor controls the text position as well as the dimension line location.
DIMZIN	Type: Integer Saved in: Drawing Initial value: 0	Controls the suppression of the inches portion of a feet-and-inches dimension when the distance is an integral number of feet, or the feet portion when the distance is less than one foot. 0 Suppresses zero feet and precisely zero inches. 1 Includes zero feet and precisely zero inches. 2 Includes zero feet and suppresses zero inches. 3 Includes zero inches and suppresses zero feet. DIMZIN stores this value when entered on the command line or set in the Primary Units section of the Annotation dialog box. DIMZIN also affects real-to-string conversions performed by the AutoLISP **rtos** and **angtos** functions.
DISPSILH	Type: Integer Saved in: Drawing Initial value: 0	Controls the display of silhouette curves of body objects in wire-frame mode. 0 Off 1 On
DISTANCE	(Read-only) Type: Real Not saved	Stores the distance computed by the DIST command.
DONUTID	Type: Real Not saved Initial value: 0.5000	Sets the default for the inside diameter of a donut.
DONUTOD	Type: Real Not saved Initial value: 1.0000	Sets the default for the outside diameter of a donut. Must be nonzero. If DONUTID is larger than DONUTOD, the two values are swapped by the next command.
DRAGMODE	Type: Integer Saved in: Drawing Initial value: 2	Sets Object Drag mode during editing operations. 0 No dragging 1 On (if requested) 2 Auto

Variable	Characteristics	Description
DRAGP1	Type: Integer Saved in: Config Initial value: 10	Sets regen-drag input sampling rate.
DRAGP2	Type: Integer Saved in: Config Initial value: 25	Sets fast-drag input sampling rate.
DWGCODEPAGE	(Read-only) Type: String Saved in: Drawing	Stores the drawing code page. This variable is set to the system code page when you create a new drawing; otherwise, AutoCAD does *not* maintain it. It should reflect the code page of the drawing. You can set it to any of the values used by the SYSCODEPAGE system variable or set it as undefined. It is saved in the header.
DWGNAME	(Read-only) Type: String Not saved	Stores the drawing name as entered by the user. If the drawing hasn't been named yet, DWGNAME reports that it is unnamed. If the user specified a drive/directory prefix, it is included as well.
DWGPREFIX	(Read-only) Type: String Not saved	Stores the drive/directory prefix for the drawing.
DWGTITLED	(Read-only) Type: Integer Not saved	Indicates whether the current drawing has been named. 0 The drawing has *not* been named. 1 The drawing has been named.
DWGWRITE	Type: Integer Not saved Initial value: 1	Controls the initial state of the read-only toggle in the Open Drawing dialog box of the OPEN command. 0 Opens the drawing for reading only. 1 Opens the drawing for reading and writing.
EDGEMODE	Type: Integer Not saved Initial value: 0	Controls how the TRIM and EXTEND commands determine cutting and boundary edges. 0 Uses the selected edge without an extension. 1 Extends or trims the object to an imaginary extension of the cutting or boundary object.
ELEVATION	Type: Real Saved in: Drawing Initial value: 0.0000	Stores the current 3D elevation relative to the current UCS for the current space.

Variable	Characteristics	Description
EXPERT	Type: Integer Not saved Initial value: 0	Controls the issuance of certain prompts. 0 Issues all prompts normally. 1 Suppresses "About to regen, proceed?" and "Really want to turn the current layer off?" 2 Suppresses the preceding prompts and "Block already defined. Redefine it?" (BLOCK) and "A drawing with this name already exists. Overwrite it?" (SAVE or WBLOCK). 3 Suppresses the preceding prompts and those issued by LINETYPE if you try to load a linetype that's already loaded or create a new linetype in a file that already defines it. 4 Suppresses the preceding prompts and those issued by UCS Save and VPORTS Save if the name you supply already exists. 5 Suppresses the preceding prompts and those issued by the DIMSTYLE Save option and DIMOVERRIDE if the dimension style name you supply already exists (the entries are redefined). When a prompt is suppressed by EXPERT, the operation in question is performed as though you entered *y* at the prompt. The setting of EXPERT can affect scripts, menu macros, AutoLISP, and the command functions.
EXPLMODE	Type: Integer Saved in: Drawing Initial value: 1	Controls whether the EXPLODE command supports non-uniformly scaled (NUS) blocks. 0 Does not explode NUS blocks. 1 Explodes NUS blocks.
EXTMAX	(Read-only) Type: 3D Point Saved in: Drawing	Stores the upper-right point of drawing extents. Expands outward as new objects are drawn, shrinks only with ZOOM All or ZOOM Extents. Reported in World coordinates for the current space.
EXTMIN	(Read-only) Type: 3D Point Saved in: Drawing	Stores the lower-left point of drawing extents. Expands outward as new objects are drawn, shrinks only with ZOOM All or ZOOM Extents. Reported in World coordinates for the current space.
FACETRES	Type: Real Saved in: Drawing Initial value: 0.5	Further adjusts the smoothness of shaded and hidden line-removed objects. Valid values are from 0.01 to 10.0.

APPENDIX C: SYSTEM VARIABLES   **C-23**

Variable	Characteristics	Description
FFLIMIT	Type: Integer Saved in: Config Initial value: 0	Limits the number of PostScript and TrueType fonts in memory. Valid values are from 0 to 100. If set to 0, there is no limit.
FILEDIA	Type: Integer Saved in: Config Initial value: 1	Suppresses the display of the file dialog boxes. 0 Disables file dialog boxes. You can still request a file dialog box to appear by entering a tilde (~) in response to the command's prompt. The same is true for AutoLISP and ADS functions. 1 Enables file dialog boxes. However, if a script or AutoLISP/ADS program is active, an ordinary prompt appears.
FILLETRAD	Type: Real Saved in: Drawing Initial value: 0.0000	Stores the current fillet radius.
FILLMODE	Type: Integer Saved in: Drawing Initial value: 1	Specifies whether objects created with SOLID are filled in. 0 Objects are not filled. 1 Objects are filled.
FONTALT	Type: String Saved in: Config Initial value: ""	Specifies the alternate font to be used when the specified font file cannot be located. If an alternate font is not specified, AutoCAD displays a warning.
FONTMAP	Type: String Saved in: Config Initial value: ""	Specifies the font mapping file to be used when the specified font cannot be located. A font mapping file contains one font mapping per line where the original font and the substitute font are separated by a semicolon (;). For example, to substitute romans with the Times TrueType font, you would have a line in your mapping file that reads: **romans;c:\windows\system\times.ttf**
FRONTZ	(Read-only) Type: Real Saved in: Drawing	Stores the front clipping plane offset from the target plane for the current viewport, in drawing units. Meaningful only if the front clipping bit in VIEWMODE is on and the front clip not at eye bit is also on. The distance of the front clipping plane from the camera point is found by subtracting FRONTZ from the camera-to-target distance.

Variable	Characteristics	Description
GRIDMODE	Type: Integer Saved in: Drawing Initial value: 0	Specifies whether the grid is turned on or off. 0 Turns the grid off. 1 Turns the grid on.
GRIDUNIT	Type: Real Saved in: Drawing Initial value: 0.0000,0.0000	Specifies the grid spacing (X and Y) for the current viewport. Changes to the grid spacing are not reflected in the displayed grid until you use the REDRAW or REGEN command. AutoCAD does *not* perform automatic redraws or regens when variables are changed.
GRIPBLOCK	Type: Integer Saved in: Config Initial value: 0	Controls the assignment of grips in blocks. 0 Assigns grip only to the insertion point of the block. 1 Assigns grips to objects within the block.
GRIPCOLOR	Type: Integer Saved in: Config Initial value: 5	Controls the color of nonselected grips (drawn as a box outline). The valid range is 1–255.
GRIPHOT	Type: Integer Saved in: Config Initial value: 1	Controls the color of selected grips (drawn as a filled box). The valid range is 1–255.
GRIPS	Type: Integer Saved in: Config Initial value: 1	Allows the use of selection set grips for the Stretch, Move, Rotate, Scale, and Mirror grip modes. 0 Disables grips. 1 Enables grips. To adjust the size of the grips and the effective selection area used by the cursor when you snap to a grip, use the GRIPSIZE system variable.
GRIPSIZE	Type: Integer Saved in: Config Initial value: 3	Sets the size of the box drawn to display the grip in pixels. The valid range is 1–255.
HANDLES	(Read-only) Type: Integer Saved in: Drawing	Reports that object handles are enabled and can be accessed by applications.

Variable	Characteristics	Description
HIGHLIGHT	Type: Integer Not saved Initial value: 1	Controls object highlighting; does not affect objects selected with grips. 0 Disables object selection highlighting. 1 Enables object selection highlighting.
HPANG	Type: Real Not saved Initial value: 0.0000	Specifies the hatch pattern angle.
HPBOUND	Type: Real Saved in: Drawing Initial value: 1	Controls the object type created by the BHATCH and BOUNDARY commands. 0 Creates a polyline. 1 Creates a region.
HPDOUBLE	Type: Integer Not saved Initial value: 0	Specifies hatch pattern doubling for "U" user-defined patterns. 0 Disables hatch pattern doubling. 1 Enables hatch pattern doubling.
HPNAME	Type: String Not saved Initial value: " "	Sets default hatch pattern name of up to 34 characters, no spaces allowed. Returns " " if there is no default. Enter a period (.) to set no default.
HPSCALE	Type: Real Not saved Initial value: 1.0000	Specifies the hatch pattern scale factor; must be nonzero.
HPSPACE	Type: Real Not saved Initial value: 1.0000	Specifies the hatch pattern line spacing for "U" user-defined simple patterns; must be nonzero.
INSBASE	Type: 3D point Saved in: Drawing Initial value: 0.0000, 0.0000, 0.0000	Stores insertion base point set by BASE command, expressed in UCS coordinates for the current space.
INSNAME	Type: String Not saved Initial value: " "	Sets default block name for DDINSERT or INSERT. The name must conform to symbol naming conventions. Returns " " if no default. Enter a period (.) to set no default.

Variable	Characteristics	Description
ISAVEBAK	Type: Integer Saved in: Config Initial value: 1	Improves the speed of incremental saves, especially for larger drawings in Windows. ISAVEBAK controls the creation of a backup file (.bak). In Windows, copying the file data to create a backup file for large drawings takes a major portion of the incremental save time. 0 No backup file is created (even for a full save). 1 A backup file is created. **Warning** In the case of a contingency (such as a power failure in the middle of a save), it's possible that drawing data can be lost.
ISAVEPERCENT	Type: Integer Saved in: Config Initial value: 50	Determines the amount of wasted space tolerated in a drawing file. The value of ISAVEPERCENT is an integer from 0 to 100. The default value of 50 means that the estimate of wasted space within the file doesn't exceed 50% of the total file size. Wasted space is eliminated by periodic full saves. When the estimate exceeds 50%, the next save will be a full save. This resets the wasted space estimate to 0. If ISAVEPERCENT is set to 0, a save always results in a full save.
ISOLINES	Type: Integer Saved in: Drawing Initial value: 4	Specifies the number of isolines per surface on objects. Valid integer values are from 0 to 2047.
LASTANGLE	(Read-only) Type: Real Not saved	Stores the end angle of the last arc entered, relative to the *XY* plane of the current UCS for the current space.
LASTPOINT	Type: 3D point Saved in: Drawing Initial value: 0.0000,0.0000 0.0000	Stores the last point entered, expressed in UCS coordinates for the current space; referenced by @ during keyboard entry.
LENSLENGTH	(Read-only) Type: Real Saved in: Drawing	Stores the length of the lens (in millimeters) used in perspective viewing for the current viewport.

Variable	Characteristics	Description
LIMCHECK	Type: Integer Saved in: Drawing Initial value: 0	Controls object creation outside the drawing limits. 0 Enables object creation. 1 Disables object creation.
LIMMAX	Type: 2D point Saved in: Drawing Initial value: 12.0000,9.0000	Stores upper-right drawing limits for the current space expressed in World coordinates.
LIMMIN	Type: 2D point Saved in: Drawing Initial value: 0.0000,0.0000	Stores lower-left drawing limits for the current space expressed in World coordinates.
LOCALE	(Read-only) Type: String Not saved Initial value: "en"	Displays the ISO language code of the current AutoCAD version you're running.
LOGINNAME	(Read-only) Type: String Not saved	Displays the user's name as configured or input when AutoCAD is loaded.
LONGFNAME	(Read-only) Type: Integer Saved in: Config Initial value: 1	Indicates whether long file name support is enabled or disabled. 0 Disables long file name support 1 Enables long file name support LONGFNAME is for use on platforms that allow long file name support, such as Windows NT or Windows 95.
LTSCALE	Type: Real Saved in: Drawing Initial value: 1	Sets global linetype scale factor.
LUNITS	Type: Integer Saved in: Drawing Initial value: 2	Sets Linear Units mode. 1 Scientific 2 Decimal 3 Engineering 4 Architectural 5 Fractional

Variable	Characteristics	Description
LUPREC	Type: Integer Saved in: Drawing Initial value: 4	Sets linear units decimal places or denominator.
MAXACTVP	Type: Integer Not saved Initial value: 16	Sets maximum number of viewports to regenerate at one time.
MAXOBJMEM	Type: Integer Not saved Default value: 0	Controls the object pager and specifies how much virtual memory AutoCAD allows the drawing to use before it starts paging the drawing out to disk into the object pagers swap files. The default value 0 disables the object pager. When this variable is set to a negative value or the value 2,147,483,647, the object pager is also disabled. When set to any other value, the object pager is enabled and the specified value is used as the upper limit for the object pagers virtual memory. The environment variable CADMAXOBJMEM also controls the object pager. **Warning** If you restart your system without exiting AutoCAD, these swap files are not deleted. You should delete them. Do not delete them from within AutoCAD.
MAXSORT	Type: Integer Saved in: Config Initial value: 200	Sets maximum number of symbol names or file names to be sorted by listing commands. If the total number of items exceeds this number, no items are sorted.
MENUCTL	Type: Integer Saved in: Config Initial value: 1	Controls the page switching of the screen menu. 0 Screen menu does *not* switch pages in response to keyboard command entry. 1 Screen menu switches pages in response to keyboard command entry.

Variable	Characteristics	Description
MENUECHO	Type: Integer Not saved Initial value: 0	Sets menu echo and prompt control bits. It is the sum of the following: 1 Suppresses echo of menu items (^P in a menu item toggles echoing). 2 Suppresses display of system prompts during menu. 4 Disables ^P toggle of menu echoing. 8 Displays input/output strings; debugging aid for DIESEL macros.
MENUNAME	(Read-only) Type: String Not saved	Stores the name and path of the currently loaded base menu file.
MIRRTEXT	Type: Integer Saved in: Drawing Initial value: 1	Controls how MIRROR reflects text. 0 Retains text direction. 1 Mirrors the text.
MODEMACRO	Type: String Not saved Initial value: " "	Displays a text string on the status line, such as the name of the current drawing, time/date stamp, or special modes. Use MODEMACRO to display a string of text, or use special text strings written in the DIESEL macro language to have AutoCAD evaluate the macro from time to time and base the status line on user-selected conditions.
MTEXTED	Type: String Saved in: Config Initial value: " "	Sets the name of the program to use for editing mtext objects.

Variable	Characteristics	Description
OFFSETDIST	Type: Real Not saved Initial value: −1.0000	Sets the default offset distance. <0 Changes to Through mode. >0 Sets the default offset distance.
ORTHOMODE	Type: Integer Saved in: Drawing Initial value: 0	Controls orthogonal display of lines or polylines. 0 Turns off Ortho mode. 1 Turns on Ortho mode.
OSMODE	Type: Integer Saved in: Drawing Initial value: 0	Sets running Object Snap modes using the following bit-codes. To specify more than one object snap, enter the sum of their values. For example, entering 3 specifies the Endpoint (1) and Midpoint (2) object snaps. 0 NONe 1 ENDpoint 2 MIDpoint 4 CENter 8 NODe 16 QUAdrant 32 INTersection 64 INSertion 128 PERpendicular 256 TANgent 512 NEArest 1024 QUIck 2048 APPint
PDMODE	Type: Integer Saved in: Drawing Initial value: 0	Sets Point Object Display mode. For information on values to enter, see the POINT command.
PDSIZE	Type: Real Saved in: Drawing Initial value: 0.0000	Sets point object display size. 0 Creates a point at 5% of the graphics area height. >0 Specifies an absolute size. <0 Specifies a percentage of the viewport size.
PELLIPSE	Type: Integer Saved in: Drawing Initial value: 0	Controls the ellipse type created with ELLIPSE. 0 Creates a true ellipse object. 1 Creates a polyline representation of an ellipse.
PERIMETER	(Read-only) Type: Real Not saved	Stores the last perimeter value computed by AREA, LIST, or DBLIST.

Variable	Characteristics	Description
PFACEVMAX	(Read-only) Type: Integer Not saved	Sets the maximum number of vertices per face.
PICKADD	Type: Integer Saved in: Config Initial value: 1	Controls additive selection of objects. 0 Disables PICKADD. The objects most recently selected, either by an individual pick or windowing, become the selection set. Previously selected objects are removed from the selection set. Add more objects to the selection set by holding down [Shift] while selecting. 1 Enables PICKADD. Each object selected, either individually or by windowing, is added to the current selection set. To remove objects from the set, hold down [Shift] while selecting.
PICKAUTO	Type: Integer Saved in: Config Initial value: 1	Controls automatic windowing when the Select objects prompt appears. 0 Disables PICKAUTO. 1 Draws a selection window (both window and crossing window) automatically at the Select objects prompt.
PICKBOX	Type: Integer Saved in: Config Initial value: 3	Sets object selection target height, in pixels.
PICKDRAG	Type: Integer Saved in: Config Initial value: 0	Controls the method of drawing a selection window: 0 Draws the selection window by clicking the mouse or digitizer at one corner and then at the other corner. 1 Draws the selection window by clicking at one corner, holding down the mouse or digitizer button, dragging, and releasing the mouse or digitizer button at the other corner.
PICKFIRST	Type: Integer Saved in: Config Initial value: 1	Controls the method of object selection so that you select objects first and then use an edit or inquiry command. 0 Disables PICKFIRST. 1 Enables PICKFIRST.

Variable	Characteristics	Description
PICKSTYLE	Type: Integer Saved in: Drawing Initial value: 3	Controls group selection and associative hatch selection. 0 No group selection or associative hatch selection 1 Group selection 2 Associative hatch selection 3 Group selection and associative hatch selection
PLATFORM	(Read-only) Type: String Not saved	Indicates which platform of AutoCAD is in use. One of the following strings may appear: Microsoft Windows Sun4/SPARCstation 386 DOS Extender DECstation Apple Macintosh Silicon Graphics Iris Indigo
PLINEGEN	Type: Integer Saved in: Drawing Initial value: 0	Sets the linetype pattern generation around the vertices of a 2D polyline. Does *not* apply to polylines with tapered segments. 0 Polylines are generated to start and end with a dash at each vertex. 1 Generates the linetype in a continuous pattern around the vertices of the polyline.
PLINEWID	Type: Real Saved in: Drawing Initial value: 0.0000	Stores the default polyline width.
PLOTID	Type: String Saved in: Config Initial value: ""	Changes the default plotter, based on its assigned description, and retains the text string of the current plotter description. Change to another configured plotter by entering its full or partial description. For example, if you have a plotter named "UNIX plotter at 10 X 10," enter **UNIX**. Applications can use PLOTTER to step through the available plotter descriptions retained by PLOTID and thus control the default plotter.

Variable	Characteristics	Description
PLOTROTMODE	Type: Integer Saved in: Drawing Initial value: 1	Controls the orientation of plots. 0 Rotates the effective plotting area so that the corner with the Rotation icon aligns with the paper at the lower-left for 0, top-left for 90, top-right for 180, and lower-right for 270. 1 Aligns the lower-left corner of the effective plotting area with the lower-left corner of the paper.
PLOTTER	Type: Integer Saved in: Config Initial value: 0	Changes the default plotter, based on its assigned integer, and retains an integer number that AutoCAD assigns for each configured plotter. This number can be in the range of 0 up to the number of configured plotters. You may configure up to 29 plotters. Change to another configured plotter by entering its valid, assigned number. For example, if you configured four plotters, the valid numbers are 0 through 3. Note AutoCAD does *not* permanently assign a number to a given plotter. If you delete a plotter configuration, AutoCAD assigns new numbers to each of the configured plotters. AutoCAD also updates the value of PLOTTER. Applications can use PLOTTER to step through the available plotter descriptions retained by PLOTID and thus control the default plotter.
POLYSIDES	Type: Integer Not saved Initial value: 4	Sets the default number of sides for POLYGON. The range is 3–1024.
POPUPS	(Read-only) Type: Integer Not saved	Displays the status of the currently configured display driver. 0 Does *not* support dialog boxes, the menu bar, pull-down menus, and icon menus. 1 Supports these features.
PROJMODE	Type: Integer Saved in: Config Initial value: 1	Sets the current Projection mode for Trim or Extend operations. 0 True 3D mode (no projection) 1 Project to the *XY* plane of the current UCS 2 Project to the current view plane

Variable	Characteristics	Description
PSLTSCALE	Type: Integer Saved in: Drawing Initial value: 1	Controls paper space linetype scaling. 0 No special linetype scaling. Linetype dash lengths are based on the drawing units of the space (model or paper) in which the objects were created, scaled by the global LTSCALE factor. 1 Viewport scaling governs linetype scaling. If TILEMODE is set to 0, dash lengths are based on paper space drawing units, even for objects in model space. In this mode, viewports can have varying magnifications, yet display linetypes identically. For a specific linetype, the dash lengths of a line in a viewport are the same as the dash lengths of a line in paper space. You can still control the dash lengths with LTSCALE. When you change PSLTSCALE or use a command such as ZOOM with PSLTSCALE set to 1, objects in viewports are not automatically regenerated. Use REGEN or REGENALL to update the linetypes in the viewports.
PSPROLOG	Type: String Saved in: Config Initial value: ""	Assigns a name for a prologue section to be read from the *acad.psf* file when using PSOUT.
PSQUALITY	Type: Integer Saved in: Drawing Initial value: 75	Controls the rendering quality of PostScript images and whether they are drawn as filled objects or as outlines. 0 Disables PostScript image generation. <0 Sets the number of pixels per AutoCAD drawing unit for the PostScript resolution. >0 Sets the number of pixels per drawing unit, but uses the absolute value; causes AutoCAD to show the PostScript paths as outlines and does not fill them.

Variable	Characteristics	Description
QTEXTMODE	Type: Integer Saved in: Drawing Initial value: 0	Controls Quick Text mode. 0 Turns off Quick Text mode; displays characters. 1 Turns on Quick Text mode; displays a box in place of text.
RASTERPREVIEW	Type: Integer Saved in: Drawing Initial value: 0	Controls whether drawing preview images are saved with the drawing and sets the format type. 0 BMP only 1 BMP and WMF 2 WMF only 3 No preview image created
REGENMODE	Type: Integer Saved in: Drawing Initial value: 1	Controls automatic regeneration of the drawing. 0 Turns off REGENAUTO. 1 Turns on REGENAUTO.
RE-INIT	Type: Integer Not saved Initial value: 0	Reinitializes the I/O ports, digitizer, display, plotter, and *acad.pgp* file using the following bit-codes: 0 No reinitialization 1 Digitizer port reinitialization 2 Plotter port reinitialization 4 Digitizer reinitialization 8 Display reinitialization 16 PGP file reinitialization (reload) To specify more than one reinitialization, enter the sum of their values, for example, **3** to specify both digitizer port (1) and plotter port (2) reinitialization. For more information, see the REINIT command.
RIASPECT	Type: Real Not saved Initial value: 0.0000	Changes the image aspect ratio for imported raster images. GIF and TIFF images specify the pixel aspect ratio to prevent circles from displaying as ellipses when images are transported to differently shaped displays. The ratio stored in RIASPECT overrides any specification in the GIF and TIFF file you import. PCX files contain no aspect ratio. A useful RIASPECT setting is 0.8333, which is the ratio for importing VGA or MCGA images in 320 x 200 mode.

Variable	Characteristics	Description
RIBACKG	Type: Integer Not saved Initial value: 0	Specifies the background color number for imported raster images. Areas of the image equal to the background are not converted to solid objects in the block. If you have a different screen background, set RIBACKG to the AutoCAD color number corresponding to your screen background. For example, if you use a white screen background, specify **ribackg 7**. Use RIBACKG to reduce the size of the imported image by setting the background color as the color that makes up most of your raster image.
RIEDGE	Type: Integer Not saved Initial value: 0	Controls the edge detection feature: 0 Disables edge detection. 1–255 Sets the threshold for RIEDGE detection. GIFIN, PCXIN, and TIFFIN use the value for detecting features in the drawing. To trace over an imported raster image, use RIEDGE to locate the edges of the features of the image and include only these edges in the drawing. To import an image for viewing and not for tracing edges, set RIBACKG to 0. If you're importing an image for a monochrome display, use RIEDGE or the brightness threshold setting RITHRESH. AutoCAD performs edge detection using the Pythagorean sum of two crossed Sobel gradient operators as described in *Digital Image Processing*, by Gonzalez and Wintz.

Variable	Characteristics	Description
RIGAMUT	Type: Integer Not saved Initial value: 256	Controls the number of colors GIFIN, PCXIN, and TIFFIN use when they import a color image. If you're using a display with less than 256 colors, setting RIGAMUT restricts the colors used by these commands. Common settings for RIGAMUT are 8 and 16. RIGAMUT specifies only the number of colors. The colors themselves are the standard AutoCAD shades. For example, setting RIGAMUT to 2 does not produce black-and-white images; It creates black-and-red images because the only colors allowed are 0 (black) and 1 (red). To import an image for a monochrome display, use the edge detection variable RIEDGE or the brightness threshold variable RITHRESH, not RIGAMUT. Limiting the color gamut often reduces the size of an imported image in the drawing database, because importing raster images groups pixels that would otherwise be distinct to the same AutoCAD shade. Doing this compresses the image into fewer AutoCAD objects.
RIGREY	Type: Integer Not saved Initial value: 0	Imports an image as a gray-scale image: 0 Disables gray-scale image importing. >0 Converts each pixel in the image to a gray-scale value. Because AutoCAD has relatively few gray shades, importing an image using RIGREY reduces the size of the imported image in the drawing database while preserving essential details. Using the edge detection variable RIEDGE achieves similar effects. The gray-scale value is based on the human eye response function defined for NTSC television (the YIQ color system).

Variable	Characteristics	Description
RITHRESH	Type: Integer Not saved Initial value: 0	Controls importing an image based on luminance (brightness): 0 Turns off RITHRESH. >0 Rasterin uses a brightness threshold filter so only pixels with a luminance value greater than the RITHRESH value are included in the drawing. The default value is 0, which turns off the brightness threshold feature. If you have an image with a dark, "noisy" background and a light foreground, you can use RITHRESH to drop out the background and import just the brighter foreground material. In general, if you're importing an image for a monochrome display, it's best to use RIEDGE (edge detection) or RITHRESH, not RIGAMUT (number of colors).
SAVEFILE	(Read-only) Type: String Saved in: Config	Stores current auto-save file name.
SAVEIMAGES	Type: Integer Saved in: Drawing Initial value: 0	Controls writing graphics metafiles for application-defined objects, solids, bodies, and regions. SAVEIMAGES is used to specify whether to save the images of application-defined objects, solids, bodies, and regions with the drawing. 0 The application's definition of the objects controls whether to save the graphical descriptions of objects. Solids, bodies and regions are not saved. 1 Always saves images. 2 Never saves images. If 1, or for certain classes if 0, then application-defined objects are visible if the drawing is reloaded without the supporting application present. The image displayed is the image that was last described by the supporting application. If 2, or for certain classes and solids, bodies, and regions if 0, then no image data is saved for these application-defined objects if the drawing is reloaded without the supporting applications. The ability to display application-defined objects without their application present is very desirable. However, when graphical metafiles are saved, your drawing size is larger and the save time is longer.

Variable	Characteristics	Description
SAVENAME	(Read-only) Type: String Not saved	Stores the file name you save the drawing to.
SAVETIME	Type: Integer Saved in: Config Initial value: 120	Sets automatic save interval, in minutes. 0 Disables automatic save. >0 Automatically saves the drawing at intervals specified by the nonzero integer. The SAVETIME timer starts as soon as you make a change to a drawing. It is reset and restarted by a manual SAVE, SAVEAS, or QSAVE. The current drawing is saved to *auto.sv$*.
SCREENBOXES	(Read-only) Type: Integer Saved in: Config	Stores the number of boxes in the screen menu area of the graphics area. If the screen menu is disabled (configured off), SCREENBOXES is zero. On platforms that permit the AutoCAD graphics window to be resized or the screen menu to be reconfigured during an editing session, the value of this variable might change during the editing session.
SCREENMODE	(Read-only) Type: Integer Saved in: Config	Stores a bit-code indicating the graphics/text state of the AutoCAD display. It is the sum of the following bit values: 0 Text screen is displayed. 1 Graphics mode is displayed. 2 Dual-screen display is configured.
SCREENSIZE	(Read-only) Type: 2D point Not saved	Stores current viewport size in pixels (*X* and *Y*).
SHADEDGE	Type: Integer Saved in: Drawing Initial value: 3	Controls shading of edges in rendering. 0 Faces shaded, edges not highlighted 1 Faces shaded, edges drawn in background color 2 Faces not filled, edges in object color 3 Faces in object color, edges in background color
SHADEDIF	Type: Integer Saved in: Drawing Initial value: 70	Sets ratio of diffuse reflective light to ambient light (in percent of diffuse reflective light).

Variable	Characteristics	Description
SHPNAME	Type: String Not saved Initial value: ""	Sets default shape name. Must conform to symbol naming conventions. If no default is set, it returns "". Enter a period (.) to set no default.
SKETCHINC	Type: Real Saved in: Drawing Initial value: 0.1000	Sets SKETCH record increment.
SKPOLY	Type: Integer Saved in: Drawing Initial value: 0	Determines whether SKETCH generates lines or polylines. 0 Generates lines. 1 Generates polylines.
SNAPANG	Type: Real Saved in: Drawing Initial value: 0	Sets snap/grid rotation angle (UCS-relative) for the current viewport. Note Changes to this variable are not reflected in the displayed grid until a redraw is performed. AutoCAD does *not* perform automatic redraws when variables are changed.
SNAPBASE	Type: 2D point Saved in: Drawing Initial value: 0.0000,0.0000	Sets snap/grid origin point for the current viewport (in UCS X,Y coordinates). Note Changes to this variable are not reflected in the displayed grid until a redraw is performed. AutoCAD does *not* perform automatic redraws when variables are changed.
SNAPISOPAIR	Type: Integer Saved in: Drawing Initial value: 0	Controls current isometric plane for the current viewport. 0 Left 1 Top 2 Right
SNAPMODE	Type: Integer Saved in: Drawing Initial value: 0	Controls the Snap mode. 0 Snap off 1 Snap on for current viewport
SNAPSTYL	Type: Integer Saved in: Drawing Initial value: 0	Sets snap style for current viewport. 0 Standard 1 Isometric
SNAPUNIT	Type: 2D point Saved in: Drawing Initial value: 1.0000,1.0000	Sets snap spacing (X and Y) for current viewport. Note Changes to this variable are not reflected in the displayed grid until a redraw operation is performed. AutoCAD does *not* perform automatic redraws when variables are changed.

Variable	Characteristics	Description
SORTENTS	Type: Integer Saved in: Config Initial value: 96	Controls the display of object sort order operations using the following codes: 0 Disables SORTENTS. 1 Sorts for object selection. 2 Sorts for object snap. 4 Sorts for redraws. 8 Sorts for MSLIDE slide creation. 16 Sorts for REGENs. 32 Sorts for plotting. 64 Sorts for PostScript output. To select more than one, enter the sum of their codes. For example, enter 3 to specify code 1 and code 2.
SPLFRAME	Type: Integer Saved in: Drawing Initial value: 0	Controls display of spline-fit polylines. 0 Does not display the control polygon for spline fit polylines. Displays the fit surface of a polygon mesh, not the defining mesh. Does not display the invisible edges of 3D faces or polyface meshes. 1 Displays the control polygon for spline-fit polylines. Only the defining mesh of a surface-fit polygon mesh is displayed (not the fit surface). Invisible edges of 3D faces or polyface meshes are displayed.
SPLINESEGS	Type: Integer Saved in: Drawing Initial value: 8	Sets the number of line segments to be generated for each spline.
SPLINETYPE	Type: Integer Saved in: Drawing Initial value: 6	Sets the type of spline curve to be generated by PEDIT Spline. 5 Quadratic B-spline 6 Cubic B-spline
SURFTAB1	Type: Integer Saved in: Drawing Initial value: 6	Sets the number of tabulations to be generated for RULESURF and TABSURF. Also sets the mesh density in the M direction for REVSURF and EDGESURF.
SURFTAB2	Type: Integer Saved in: Drawing Initial value: 6	Sets the mesh density in the N direction for REVSURF and EDGESURF.

Variable	Characteristics	Description
SURFTYPE	Type: Integer Saved in: Drawing Initial value: 6	Controls the type of surface-fitting to be performed by PEDIT Smooth. 5 Quadratic B-spline surface 6 Cubic B-spline surface 8 Bezier surface
SURFU	Type: Integer Saved in: Drawing Initial value: 6	Sets the surface density in the *M* direction.
SURFV	Type: Integer Saved in: Drawing Initial value: 6	Sets the surface density in the *N* direction.
SYSCODEPAGE	(Read-only) Type: String Saved in: Drawing	Indicates the system code page specified in *acad.xmf*. Codes are as follows: ascii dos860 dos932 iso8859-8 big5 dos861 iso8859-1 iso8859-9 dos437 dos863 iso8859-2 johab dos850 dos864 iso8859-3 ksc5601 dos852 dos865 iso8859-4 mac-roman dos855 dos866 iso8859-6 dos857 dos869 iso8859-7
TABMODE	Type: Integer Not saved Initial value: 0	Controls the use of Tablet mode. 0 Disables Tablet mode. 1 Enables Tablet mode.
TARGET	(Read-only) Type: 3D point Saved in: Drawing	Stores location (in UCS coordinates) of the target point for the current viewport.
TDCREATE	(Read-only) Type: Real Saved in: Drawing	Stores time and date of drawing creation.
TDINDWG	(Read-only) Type: Real Saved in: Drawing	Stores total editing time.
TDUPDATE	(Read-only) Type: Real Saved in: Drawing	Stores time and date of last update/save.
TDUSRTIMER	(Read-only) Type: Real Saved in: Drawing	Stores user elapsed timer.

Variable	Characteristics	Description
TEMPPREFIX	(Read-only) Type: String Not saved	Contains the directory name (if any) configured for placement of temporary files, with a path separator appended.
TEXTEVAL	Type: Integer Not saved Initial value: 0	Controls method of evaluation of text strings. 0 All responses to prompts for text strings and attribute values are taken literally. 1 Text starting with "(" or "!" is evaluated as an AutoLISP expression, as for nontextual input. Note The DTEXT command takes all input literally, regardless of the setting of TEXTEVAL.
TEXTFILL	Type: Integer Saved in: Drawing Initial value: 1	Controls the filling of Bitstream, TrueType, and Adobe Type 1 fonts. 0 Displays text as outlines. 1 Displays text as filled images.
TEXTQLTY	Type: Real Saved in: Drawing Initial value: 50	Sets the resolution of Bitstream, TrueType, and Adobe Type 1 fonts. Lower values decrease resolution and increase display and plotting speed. Higher values increase resolution and decrease display and plotting speed. Valid values are 0 to 100.0.
TEXTSIZE	Type: Real Saved in: Drawing Initial value: 0.2000	Sets the default height for new text objects drawn with the current text style (meaningless if the style has a fixed height).
TEXTSTYLE	Type: String Saved in: Drawing Initial value: STANDARD	Contains the name of the current text style.
THICKNESS	Type: Real Saved in: Drawing Initial value: 0.0000	Sets the current 3D thickness.
TILEMODE	Type: Integer Saved in: Drawing Initial value: 1	Controls access to paper space, as well as the behavior of AutoCAD viewports. 0 Enables paper space and viewport objects (uses MVIEW). AutoCAD clears the graphics area and prompts you to create one or more viewports. 1 Enables Release 10 Compatibility mode (uses VPORTS). AutoCAD returns to Tiled Viewport mode, restoring the most recently active tiled-viewport configuration. Paper space objects—including viewport objects—are not displayed, and the MVIEW, MSPACE, PSPACE, and VPLAYER commands are disabled.

Variable	Characteristics	Description
TOOLTIPS	(Windows only) Type: Integer Saved in: Config Initial value: 1	Controls the display of ToolTips. 0 Turns off display of ToolTips. 1 Turns on display of ToolTips.
TRACEWID	Type: Real Saved in: Drawing Initial value: 0.0500	Sets default trace width.
TREEDEPTH	Type: Integer Saved in: Drawing Initial value: 3020	Specifies maximum depth, that is, the number of times the tree-structured spatial index may divide into branches. 0 Suppresses the spatial index entirely, eliminating the performance improvements it provides in working with large drawings. This setting assures that objects are always processed in database order, making it unnecessary ever to set the SORTENTS system variable. >0 Enables TREEDEPTH. An integer of up to four digits is valid. The first two digits refer to model space, and the second two digits refer to paper space. <0 Model-space objects are treated as two-dimensional (Z coordinates are ignored), as is always the case with paper space objects. Such a setting is appropriate for 2D drawings and results in more efficient use of memory without loss of performance. **Note** You cannot use TREEDEPTH transparently.

Variable	Characteristics	Description
TREEMAX	Type: Integer Saved in: Config Initial value: 10000000	Limits memory consumption during drawing regeneration by limiting the maximum number of nodes in the spatial index (oct-tree).
		The value for TREEMAX is stored in the AutoCAD configuration file. By imposing a fixed limit with TREEMAX, you can load drawings created on systems with more memory than your system and with a larger TREEDEPTH than your system can handle. These drawings, if left unchecked, have an oct-tree large enough to eventually consume more memory than is available to your computer. TREEMAX also provides a safeguard against experimentation with inappropriately high TREEDEPTH values.
		The initial default for TREEMAX is 10000000 (ten million), a value high enough to effectively disable TREEMAX as a control of TREEDEPTH. The value to which you should set TREEMAX depends on your system's available RAM. You get about 15,000 oct-tree nodes per megabyte of RAM.
		If you want an oct-tree to use up to, but no more than, 2 megabytes of RAM, set TREEMAX to 30000 (2 x 15,000). If AutoCAD runs out of memory allocating oct-tree nodes, restart AutoCAD, set TREEMAX to a smaller number, and try loading the drawing again.
		AutoCAD might occasionally run into the limit you set with TREEMAX. Follow the resulting prompt instructions. The ability to increase TREEMAX depends on your computer's available memory.
TRIMMODE	Type: Integer Not saved Initial value: 1	Controls whether AutoCAD trims selected edges for chamfers and fillets. 0 Leaves selected edges intact. 1 Trims selected edges to the endpoints of chamfer lines and fillet arcs.

Variable	Characteristics	Description
UCSFOLLOW	Type: Integer Saved in: Drawing Initial value: 0	Generates a plan view whenever you change from one UCS to another. You can set UCSFOLLOW separately for each viewport. If UCSFOLLOW is on for a particular viewport, AutoCAD generates a plan view in that viewport whenever you change coordinate systems. Once the new UCS has been established, you can use VPOINT, DVIEW, PLAN, or VIEW to change the view of the drawing. It will change to a plan view again the next time you change coordinate systems. 0 UCS does *not* affect the view. 1 Any UCS change causes a change to plan view of the new UCS in the current viewport. The setting of UCSFOLLOW is maintained separately for both spaces and can be accessed in either space, but the setting is ignored while in paper space (it is always treated as if set to 0). Although you can define a non-World UCS in paper space, the view remains in plan view to the World Coordinate System.
UCSICON	Type: Integer Saved in: Drawing Initial value: 1	Displays the Coordinate System icon using bit-code for the current viewport. It is the sum of the following: 1 On; icon display is enabled. 2 Origin; if icon display is enabled, the icon floats to the UCS origin if possible.
UCSNAME	(Read-only) Type: String Saved in: Drawing	Stores the name of the current coordinate system for the current space. Returns a null string if the current UCS is unnamed.
UCSORG	(Read-only) Type: 3D point Saved in: Drawing	Stores the origin point of the current coordinate system for the current space. This value is always returned in World coordinates.
UCSXDIR	(Read-only) Type: 3D point Saved in: Drawing	Stores the X direction of the current UCS for the current space.
UCSYDIR	(Read-only) Type: 3D point Saved in: Drawing	Stores the Y direction of the current UCS for the current space.

Variable	Characteristics	Description
UNDOCTL	(Read-only) Type: Integer Not saved	Stores a bit-code indicating the state of the UNDO feature. It is the sum of the following values: 0 UNDO is disabled 1 UNDO is enabled. 2 Only one command can be undone. 4 Auto-group mode is enabled. 8 A group is currently active.
UNDOMARKS	(Read-only) Type: Integer Not saved	Stores the number of marks that have been placed in the UNDO control stream by the Mark option. The Mark and Back options are unavailable if a group is currently active.
UNDOONDISK	Type: Integer Saved in: Drawing Initial value: 1	Controls whether the Undo file is kept on the hard disk or in RAM. A setting of 1 is typically appropriate for systems with less than 48 MB of RAM. Writing the Undo information to disk is slower than recording it in RAM; however, the additional available RAM should increase the performance of other operations, such as editing and object creation. 0 The Undo file is kept in RAM. This requires less available disk space but greater memory and swap space. 1 The Undo file is kept on disk and requires greater available disk space but less memory and swap space. **Warning** AutoCAD might fail if it runs out of disk space while writing to the Undo file.
UNITMODE	Type: Integer Saved in: Drawing Initial value: 0	Controls the units display format. 0 Displays fractional, feet-and-inches, and surveyor's angles as previously set. 1 Displays fractional, feet-and-inches, and surveyor's angles in input format.
USERI15	Type: Integer Saved in: Drawing	USERI1, USERI2, USERI3, USERI4, and USERI5 are used for storage and retrieval of integer values.
USERR15	Type: Real Saved in: Drawing	USERR1, USERR2, USERR3, USERR4, and USERR5 are used for storage and retrieval of real numbers.
USERS15	Type: String Saved in: Not saved	USERS1, USERS2, USERS3, USERS4, and USERS5 are used for storage and retrieval of text string data.

Variable	Characteristics	Description
VIEWCTR	(Read-only) Type: 3D point Saved in: Drawing	Stores the center of view in the current viewport, expressed in UCS coordinates.
VIEWDIR	(Read-only) Type: 3D vector Saved in: Drawing	Stores the viewing direction in the current viewport expressed in UCS coordinates. This describes the camera point as a 3D offset from the target point.
VIEWMODE	(Read-only) Type: Integer Saved in: Drawing	Controls Viewing mode for the current viewport using bit-code. The value is the sum of the following bit values: 0 Disabled 1 Perspective view active. 2 Front clipping on. 4 Back clipping on. 8 UCS Follow mode on. 16 Front clip not at eye. If on, the front clip distance (FRONTZ) determines the front clipping plane. If off, FRONTZ is ignored, and the front clipping plane is set to pass through the camera point (vectors behind the camera are not displayed). This flag is ignored if the front clipping bit (2) is off.
VIEWSIZE	(Read-only) Type: Real Saved in: Drawing	Stores height of view in current viewport, expressed in drawing units.
VIEWTWIST	(Read-only) Type: Real Saved in: Drawing	Stores view twist angle for the current viewport.

Variable	Characteristics	Description
VISRETAIN	Type: Integer Saved in: Drawing Initial value: 0	Controls visibility of layers in xref files. 0 The xref layer definition in the current drawing takes precedence over these settings: On/Off, Freeze/Thaw, color, and linetype settings for xref-dependent layers. 1 On/Off, Freeze/Thaw, color, and linetype settings for xref-dependent layers take precedence over the xref layer definition in the current drawing.
VSMAX	(Read-only) Type: 3D point Saved in: Drawing	Stores the upper-right corner of the current viewport's virtual screen, expressed in UCS coordinates.
VSMIN	(Read-only) Type: 3D point Saved in: Drawing	Stores the lower-left corner of the current viewport virtual screen, expressed in UCS coordinates.
WORLDUCS	(Read-only) Type: Integer Not saved	Indicates whether the UCS is the same as the World Coordinate System. 0 Current UCS is different from the World Coordinate System. 1 Current UCS is the same as the World Coordinate System.
WORLDVIEW	Type: Integer Saved in: Drawing Initial value: 1	Controls whether UCS changes to WCS during DVIEW or VPOINT. 0 Current UCS remains unchanged. 1 Current UCS is changed to the WCS for the duration of DVIEW or VPOINT. DVIEW and VPOINT command input is relative to the current UCS.
XREFCTL	Type: Integer Saved in: Config Initial value: 0	Controls whether AutoCAD writes .xlg files (external reference log files). 0 Xref log (.xlg) files are *not* written. 1 Xref log (.xlg) files are written.

APPENDIX D

ACAD PROTOTYPE DRAWING SETTINGS

The setup for each new drawing is determined by the settings in a "prototype" drawing. (See Chapter 7.) The default drawing is named "ACAD". Following is a listing of the settings for the ACAD drawing file. See Chapter 7 for an explanation of how to create custom prototype drawings.

APERTURE	10 pixels			
Attributes	Visibility controlled individually, entry of values during DDINSERT or INSERT permitted (using prompts rather than dialog box)			
BASE	Insertion base point (0.0, 0.0, 0.0)			
Blipmode	On			
CHAMFER	Distance 0.0			
COLOR	Entity color "BYLAYER"			
Coordinate display	Updated continuously			
DIM variables				
	DIMALT	Off	DIMCLRD	0
	DIMALTD	2	DIMCLRE	0
	DIMALTF	25.4000	DIMCLRT	0
	DIMAPOST	(None)	DIMDLE	0.0000
	DIMASO	On	DIMDLI	0.3800
	DIMASZ	0.1800	DIMEXE	0.1800
	DIMBLK	(None)	DIMEXO	0.0625
	DIMBLK1	(None)	DIMGAP	0.0900
	DIMBLK2	(None)	DIMLFAC	1.0000
	DIMCEN	0.0900	DIMLIM	Off

DIM variables *(continued)*

DIMPOST	(None)	DIMTIH	On
DIMRND	0.0000	DIMTIX	Off
DIMSAH	Off	DIMTM	0.0000
DIMSCALE	1.0000	DIMTOFL	Off
DIMSE1	Off	DIMTOH	On
DIMSE2	Off	DIMTOL	Off
DIMSHO	On	DIMTP	0.0000
DIMSOXD	Off	DIMTSZ	0.0000
DIMSTYLE	"Standard"	DIMTVP	0.0000
DIMTAD	0	DIMTXT	0.18
DIMTFAC	1.0000	DIMZIN	0

DRAGMODE	Auto
ELEV	Elevation 0.0, thickness 0.0
FILL	On
FILLET	Radius 0.0
GRID	Off, spacing (0.0, 0.0)
HANDLES	Enabled
Highlighting	Enabled
ISOPLANE	Left
LAYER	Current/only layer is "0", On, with color 7 (white) and linetype "CONTINUOUS"
LIMITS	Off, drawing limits (0.0, 0.0) to (12.0, 9.0)
LINETYPE	Entity linetype "BYLAYER", no loaded linetypes other than "CONTINUOUS"
LTSCALE	1.0
MENU	"acad"
MIRROR	Text mirrored with other entities
Object selection	Pick box size 3 pixels
ORTHO	Off
OSNAP	None
PLINE	Line-width 0.0
POINT	Display mode 0, size 0
QTEXT	Off
REGENAUTO	On
SKETCH	Record increment 0.10, producing lines
SHADE	Rendering type 3, percent diffuse reflection 70
SNAP	Off, spacing (1.0, 1.0)
SNAP/GRID	Standard style, base point (0.00, 0.00), rotation 0.0°
SPACE	Model
Spline curves	Frame off, segments 8, spline type = cubic
STYLE	One defined text style ("STANDARD"), using font file "txt", with variable height, width factor 1.0, horizontal orientation.
Surfaces	6 tabulations in M and N directions, 6 segments for smoothing in U and V directions, smooth surface type = cubic B-spline
TABLET	Off

TEXT	Style "STANDARD", height 0.20, rotation 0.0°
TILEMODE	On
TIME	User elapsed timer on
TRACE	Width 0.05
UCS	Current UCS equivalent to World, auto plan view off, coordinate system icon on (at origin)
UNITS (angular)	Decimal degrees, 0 decimal places, angle 0 direction is to the right angles increase counterclockwise
UNITS (linear)	Decimal, 4 decimal places
Viewing modes	One active viewport, plan view, perspective off, target point (0, 0, 0), front and back clipping off, lens length 50mm, twist angle 0.0, fast zoom on, circle zoom percent 100, worldview 0
ZOOM	To drawing limits

APPENDIX E

HATCH PATTERNS

Courtesy of Autodesk Inc.

APPENDIX E: HATCH PATTERNS **E-3**

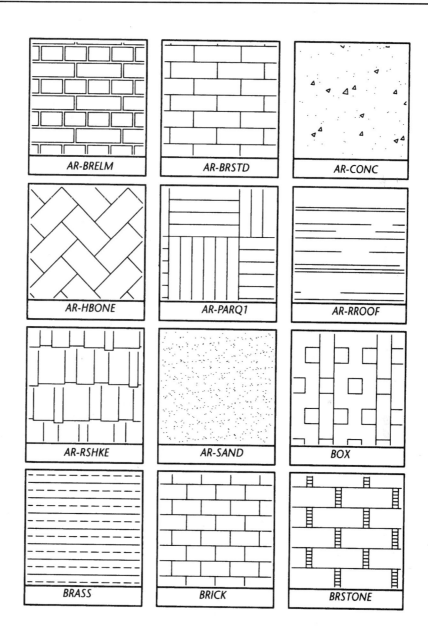

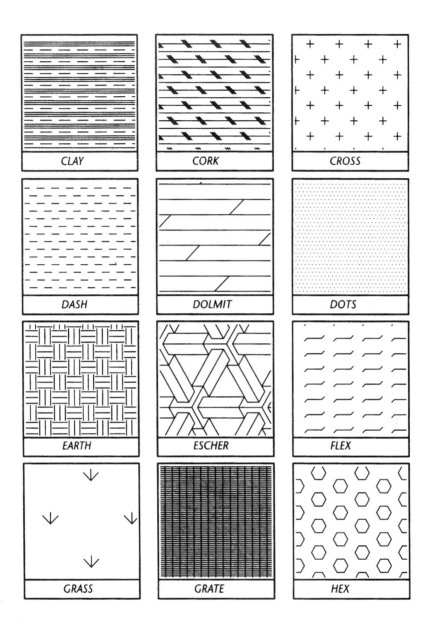

APPENDIX E: HATCH PATTERNS **E-5**

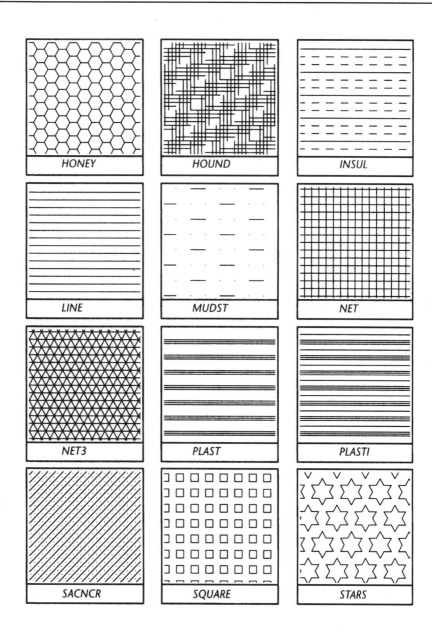

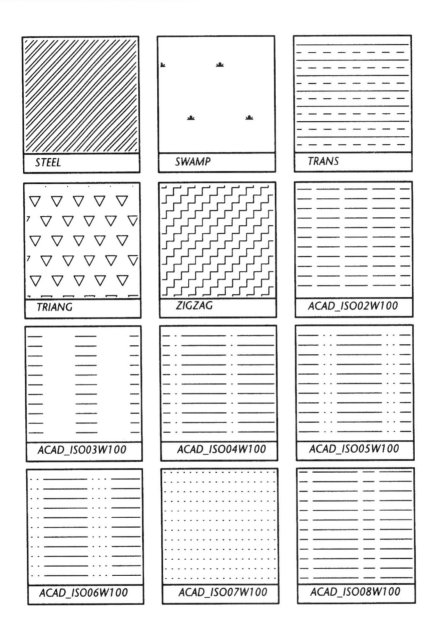

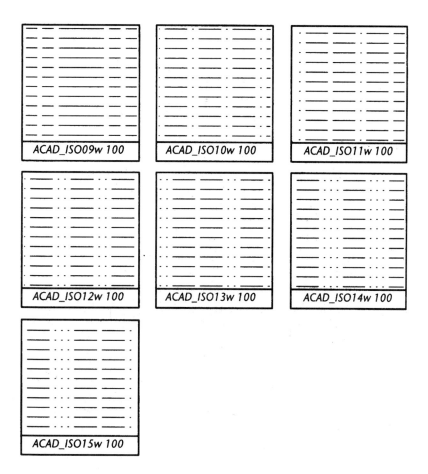

APPENDIX F

AUTOCAD LINETYPES

Courtesy of Autodesk Inc.

Standard Linetypes

AutoCAD provides a library of standard linetypes in the *acad.lin* file.

Name	Pattern
Border	— — . — — . — — . —
Border2	— — . — — . — — . — — . —
BorderX2	—— —— . ——
Center	— — — — — —
Center2	— . — . — . — . —
CenterX2	—— — ——
Dashdot	— . — . — . — . —
Dashdot2	-.-.-.-.-.-.-.-.-
DashdotX2	—— . —— . ——
Dashed	— — — — — — —
Dashed2	- - - - - - - - - - -
DashedX2	—— —— —— —— ——
Divide	— . . — . . — . . —
Divide2	-..-..-..-..-..-
DivideX2	—— . . —— . . ——
Dot	. . . .
Dot2	....
DotX2	. . . .
Hidden	— — — — — — — —
Hidden2	- - - - - - - - - - - - -
HiddenX2	— — — — — —
Phantom	—— — — —— — — ——
Phantom2	— - - — - - — - - —
PhantomX2	—— — — ——
ACAD_ISO02W100	— — — — —
ACAD_ISO03W100	— —— —— —— —
ACAD_ISO04W100	—— . —— . —— . ——
ACAD_ISO05W100	—— . . —— . . ——
ACAD_ISO06W100	—— . . . —— . . . ——
ACAD_ISO07W100	. . . .
ACAD_ISO08W100	—— —— — —— ——
ACAD_ISO09W100	—— — — —— — — ——
ACAD_ISO10W100	—— . —— . —— . ——
ACAD_ISO11W100	—— — —— — —— — ——
ACAD_ISO12W100	—— . . —— . . ——
ACAD_ISO13W100	—— — . —— — . ——
ACAD_ISO14W100	—— . . —— . . ——
ACAD_ISO15W100	—— . . . —— . . . ——

Complex Linetypes

AutoCAD provides some sample complex linetypes in the *ltypeshp.lin* file.

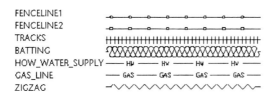

FENCELINE1
FENCELINE2
TRACKS
BATTING
HOW_WATER_SUPPLY
GAS_LINE
ZIGZAG

APPENDIX G

TABLES

INCHES TO MILLIMETRES

in.	mm	in.	mm	in.	mm	in.	mm
1	25.4	26	660.4	51	1295.4	76	1930.4
2	50.8	27	685.8	52	1320.8	77	1955.8
3	76.2	28	711.2	53	1346.2	78	1981.2
4	101.6	29	736.6	54	1371.6	79	2006.6
5	127.0	30	762.0	55	1397.0	80	2032.0
6	152.4	31	787.4	56	1422.4	81	2057.4
7	177.8	32	812.8	57	1447.8	82	2082.8
8	203.2	33	838.2	58	1473.2	83	2108.2
9	228.6	34	863.6	59	1498.6	84	2133.6
10	254.0	35	889.0	60	1524.0	85	2159.0
11	279.4	36	914.4	61	1549.4	86	2184.4
12	304.8	37	939.8	62	1574.8	87	2209.8
13	330.2	38	965.2	63	1600.2	88	2235.2
14	355.6	39	990.6	64	1625.6	89	2260.6
15	381.0	40	1016.0	65	1651.0	90	2286.0
16	406.4	41	1041.4	66	1676.4	91	2311.4
17	431.8	42	1066.8	67	1701.8	92	2336.8
18	457.2	43	1092.2	68	1727.2	93	2362.2
19	482.6	44	1117.6	69	1752.6	94	2387.6
20	508.0	45	1143.0	70	1778.0	95	2413.0
21	533.4	46	1168.4	71	1803.4	96	2438.4
22	558.8	47	1193.8	72	1828.8	97	2463.8
23	584.2	48	1219.2	73	1854.2	98	2489.2
24	609.6	49	1244.6	74	1879.6	99	2514.6
25	635.0	50	1270.0	75	1905.0	100	2540.0

The above table is exact on the basis: 1 in. = 25.4 mm

MILLIMETRES TO INCHES

mm	in.	mm	in.	mm	in.	mm	in.
1	0.039370	26	1.023622	51	2.007874	76	2.992126
2	0.078740	27	1.062992	52	2.047244	77	3.031496
3	0.118110	28	1.102362	53	2.086614	78	3.070866
4	0.157480	29	1.141732	54	2.125984	79	3.110236
5	0.196850	30	1.181102	55	2.165354	80	3.149606
6	0.236220	31	1.220472	56	2.204724	81	3.188976
7	0.275591	32	1.259843	57	2.244094	82	3.228346
8	0.314961	33	1.299213	58	2.283465	83	3.267717
9	0.354331	34	1.338583	59	2.322835	84	3.307087
10	0.393701	35	1.377953	60	2.362205	85	3.346457
11	0.433071	36	1.417323	61	2.401575	86	3.385827
12	0.472441	37	1.456693	62	2.440945	87	3.425197
13	0.511811	38	1.496063	63	2.480315	88	3.464567
14	0.551181	39	1.535433	64	2.519685	89	3.503937
15	0.590551	40	1.574803	65	2.559055	90	3.543307
16	0.629921	41	1.614173	66	2.598425	91	3.582677
17	0.669291	42	1.653543	67	2.637795	92	3.622047
18	0.708661	43	1.692913	68	2.677165	93	3.661417
19	0.748031	44	1.732283	69	2.716535	94	3.700787
20	0.787402	45	1.771654	70	2.755906	95	3.740157
21	0.826772	46	1.811024	71	2.795276	96	3.779528
22	0.866142	47	1.850394	72	2.834646	97	3.818898
23	0.905512	48	1.889764	73	2.874016	98	3.858268
24	0.944882	49	1.929134	74	2.913386	99	3.897638
25	0.984252	50	1.968504	75	2.952756	100	3.937008

The above table is approximate on the basis: 1 in. = 25.4 mm, 1/25.4 = 0.039370078740+

From Goetsch, Nelson, and Chalk, *Technical Drawing*, 3rd edition, copyright 1994 by Delmar Publishers Inc.

INCH/METRIC – EQUIVALENTS

Fraction	Decimal Equivalent Customary (in.)	Metric (mm)	Fraction	Decimal Equivalent Customary (in.)	Metric (mm)
1/64	.015625	0.3969	33/64	.515625	13.0969
1/32	.03125	0.7938	17/32	.53125	13.4938
3/64	.046875	1.1906	35/64	.546875	13.8906
1/16	.0625	1.5875	9/16	.5625	14.2875
5/64	.078125	1.9844	37/64	.578125	14.6844
3/32	.09375	2.3813	19/32	.59375	15.0813
7/64	.109375	2.7781	39/64	.609375	15.4781
1/8	.1250	3.1750	5/8	.6250	15.8750
9/64	.140625	3.5719	41/64	.640625	16.2719
5/32	.15625	3.9688	21/32	.65625	16.6688
11/64	.171875	4.3656	43/64	.671875	17.0656
3/16	.1875	4.7625	11/16	.6875	17.4625
13/64	.203125	5.1594	45/64	.703125	17.8594
7/32	.21875	5.5563	23/32	.71875	18.2563
15/64	.234375	5.9531	47/64	.734375	18.6531
1/4	.250	6.3500	3/4	.750	19.0500
17/64	.265625	6.7469	49/64	.765625	19.4469
9/32	.28125	7.1438	25/32	.78125	19.8438
19/64	.296875	7.5406	51/64	.796875	20.2406
5/16	.3125	7.9375	13/16	.8125	20.6375
21/64	.328125	8.3384	53/64	.828125	21.0344
11/32	.34375	8.7313	27/32	.84375	21.4313
23/64	.359375	9.1281	55/64	.859375	21.8281
3/8	.3750	9.5250	7/8	.8750	22.2250
25/64	.390625	9.9219	57/64	.890625	22.6219
13/32	.40625	10.3188	29/32	.90625	23.0188
27/64	.421875	10.7156	59/64	.921875	23.4156
7/16	.4375	11.1125	15/16	.9375	23.8125
29/64	.453125	11.5094	61/64	.953125	24.2094
15/32	.46875	11.9063	31/32	.96875	24.6063
31/64	.484375	12.3031	63/64	.984375	25.0031
1/2	.500	12.7000	1	1.000	25.4000

From Nelson, *Drafting for Trades and Industry—Basic Skills.* Delmar Publishers Inc.

METRIC EQUIVALENTS

LENGTH

U.S. to Metric

1 inch = 2.540 centimetres
1 foot = .305 metre
1 yard = .914 metre
1 mile = 1.609 kilometres

Metric to U.S.

1 millimetre = .039 inch
1 centimetre = .394 inch
1 metre = 3.281 feet or 1.094 yards
1 kilometre = .621 mile

AREA

1 inch^2 = 6.451 centimetre^2
1 foot^2 = .093 metre^2
1 yard^2 = .836 metre^2
1 acre^2 = 4,046.873 metre^2

1 millimetre^2 = .00155 inch^2
1 centimetre^2 = .155 inch^2
1 metre^2 = 10.764 foot^2 or 1.196 yard^2
1 kilometre^2 = .386 mile^2 or 247.04 acre^2

VOLUME

1 inch^3 = 16.387 centimetre^3
1 foot^3 = .028 metre^3
1 yard^3 = .764 metre^3
1 quart = .946 litre
1 gallon = .003785 metre^3

1 centimetre^3 = 0.61 inch^3
1 metre^3 = 35.314 foot^3 or 1.308 yard^3
1 litre = .2642 gallons
1 litre = 1.057 quarts
1 metre^3 = 264.02 gallons

WEIGHT

1 ounce = 28.349 grams
1 pound = .454 kilogram
1 ton = .907 metric ton

1 gram = .035 ounce
1 kilogram = 2.205 pounds
1 metric ton = 1.102 tons

VELOCITY

1 foot/second = .305 metre/second
1 mile/hour = .447 metre/second

1 metre/second = 3.281 feet/second
1 kilometre/hour = .621 mile/second

ACCELERATION

1 inch/second^2 = .0254 metre/second^2
1 foot/second^2 = .305 metre/second^2

1 metre/second^2 = 3.278 feet/second^2

FORCE

N (newton) = basic unit of force, kg-m/s^2. A mass of one kilogram (1 kg) exerts a gravitational force of 9.8 N (theoretically 9.80665 N) at mean sea level.

From Goetsch, Nelson, and Chalk, *Technical Drawing*, 3rd edition, copyright 1994 by Delmar Publishers Inc.

MULTIPLIERS FOR DRAFTERS

Multiply	By	To Obtain	Multiply	By	To Obtain
Acres	43,560	Square feet	Degrees/sec.	0.002778	Revolutions/sec.
Acres	4047	Square metres	Fathoms	6	Feet
Acres	1.562×10^{-3}	Square miles	Feet	30.48	Centimetres
Acres	4840	Square yards	Feet	12	Inches
Acre–feet	43,560	Cubic feet	Feet	0.3048	Metres
Atmospheres	76.0	Cms. of mercury	Foot–pounds	1.286×10^{-3}	British Thermal Units
Atmospheres	29.92	Inches of mercury	Foot–pounds	5.050×10^{-7}	Horsepower–hrs.
Atmospheres	33.90	Feet of water	Foot–pounds	3.241×10^{-4}	Kilogram–calories
Atmospheres	10,333	Kgs./sq. metre	Foot–pounds	0.1383	Kilogram–metres
Atmospheres	14.70	Lbs./sq. inch	Foot–pounds	3.766×10^{-7}	Kilowatt–hrs.
Atmospheres	1.058	Tons/sq. ft.	Foot–pounds/min.	1.286×10^{-3}	B.T.U./min.
Board feet	144 sq. in. × 1 in.	Cubic inches	Foot–pounds/min.	0.01667	Foot–pounds/sec.
British Thermal Units	0.2520	Kilogram–calories	Foot–pounds/min.	3.030×10^{-5}	Horsepower
British Thermal Units	777.5	Foot–lbs.	Foot–pounds/min.	3.241×10^{-4}	Kg.–calories/min.
British Thermal Units	3.927×10^{-4}	Horsepower–hrs.	Foot–pounds/min.	2.260×10^{-5}	Kilowatts
British Thermal Units	107.5	Kilogram–metres	Foot–pounds/sec.	7.717×10^{-2}	B.T.U./min.
British Thermal Units	2.928×10^{-4}	Kilowatt–hrs.	Foot–pounds/sec.	1.818×10^{-3}	Horsepower
B.T.U./min.	12.96	Foot–lbs./sec.	Foot–pounds/sec.	1.945×10^{-2}	Kg.–calories/min.
B.T.U./min.	0.02356	Horsepower	Foot–pounds/sec.	1.356×10^{-3}	Kilowatts
B.T.U./min.	0.01757	Kilowatts	Gallons	3785	Cubic centimetres
B.T.U./min.	17.57	Watts	Gallons	0.1337	Cubic feet
Cubic centimetres	3.531×10^{-5}	Cubic feet	Gallons	231	Cubic inches
Cubic centimetres	6.102×10^{-2}	Cubic inches	Gallons	3.785×10^{-3}	Cubic metres
Cubic centimetres	10^{-6}	Cubic metres	Gallons	4.951×10^{-3}	Cubic yards
Cubic centimetres	1.308×10^{-6}	Cubic yards	Gallons	3.785	Litres
Cubic centimetres	2.642×10^{-4}	Gallons	Gallons	8	Pints (liq.)
Cubic centimetres	10^{-3}	Litres	Gallons	4	Quarts (liq.)
Cubic centimetres	2.113×10^{-3}	Pints (liq.)	Gallons–Imperial	1.20095	U.S. gallons
Cubic centimetres	1.057×10^{-3}	Quarts (liq.)	Gallons–U.S.	0.83267	Imperial gallons
Cubic feet	2.832×10^{4}	Cubic cms.	Gallons water	8.3453	Pounds of water
Cubic feet	1728	Cubic inches	Horsepower	42.44	B.T.U./min.
Cubic feet	0.02832	Cubic metres	Horsepower	33,000	Foot–lbs./min.
Cubic feet	0.03704	Cubic yards	Horsepower	550	Foot–lbs./sec.
Cubic feet	7.48052	Gallons	Horsepower	1.014	Horsepower (metric)
Cubic feet	28.32	Litres	Horsepower	10.70	Kg.–calories/min.
Cubic feet	59.84	Pints (liq.)	Horsepower	0.7457	Kilowatts
Cubic feet	29.92	Quarts (liq.)	Horsepower	745.7	Watts
Cubic feet/min.	472.0	Cubic cms./sec.	Horsepower–hours	2547	B.T.U.
Cubic feet/min.	0.1247	Gallons/sec.	Horsepower–hours	1.98×10^{6}	Foot–lbs.
Cubic feet/min.	0.4720	Litres/sec.	Horsepower–hours	641.7	Kilogram–calories
Cubic feet/min.	62.43	Pounds of water/min.	Horsepower–hours	2.737×10^{5}	Kilogram–metres
Cubic feet/sec.	0.646317	Millions gals./day	Horsepower–hours	0.7457	Kilowatt–hours
Cubic feet/sec.	448.831	Gallons/min.	Kilometres	10^{5}	Centimetres
Cubic inches	16.39	Cubic centimetres	Kilometres	3281	Feet
Cubic inches	5.787×10^{-4}	Cubic feet	Kilometres	10^{3}	Metres
Cubic inches	1.639×10^{-5}	Cubic metres	Kilometres	0.6214	Miles
Cubic inches	2.143×10^{-5}	Cubic yards	Kilometres	1094	Yards
Cubic inches	4.329×10^{-3}	Gallons	Kilowatts	56.92	B.T.U./min.
Cubic inches	1.639×10^{-2}	Litres	Kilowatts	4.425×10^{4}	Foot–lbs./min.
Cubic inches	0.03463	Pints (liq.)	Kilowatts	737.6	Foot–lbs./sec.
Cubic inches	0.01732	Quarts (liq.)	Kilowatts	1.341	Horsepower
Cubic metres	10^{6}	Cubic centimetres	Kilowatts	14.34	Kg.–calories/min.
Cubic metres	35.31	Cubic feet	Kilowatts	10^{3}	Watts
Cubic metres	61.023	Cubic inches	Kilowatt–hours	3415	B.T.U.
Cubic metres	1.308	Cubic yards	Kilowatt–hours	2.655×10^{6}	Foot–lbs.
Cubic metres	264.2	Gallons	Kilowatt–hours	1.341	Horsepower–hrs.
Cubic metres	10^{3}	Litres	Kilowatt–hours	860.5	Kilogram–calories
Cubic metres	2113	Pints (liq.)	Kilowatt–hours	3.671×10^{5}	Kilogram–metres
Cubic metres	1057	Quarts (liq.)	Lumber Width (in.) × Thickness (in.) / 12	Length (ft.)	Board feet
Degrees (angle)	60	Minutes			
Degrees (angle)	0.01745	Radians	Metres	100	Centimetres
Degrees (angle)	3600	Seconds	Metres	3.281	Feet
Degrees/sec.	0.01745	Radians/sec.	Metres	39.37	Inches
Degrees/sec.	0.1667	Revolutions/min.			

From Goetsch, Nelson, and Chalk, *Technical Drawing*, 3rd edition, copyright 1994 by Delmar Publishers Inc.

MULTIPLIERS FOR DRAFTERS (cont'd)

Multiply	By	To Obtain	Multiply	By	To Obtain
Metres	10^{-3}	Kilometres	Pounds (troy)	373.24177	Grams
Metres	10^{3}	Millimetres	Pounds (troy)	0.822857	Pounds (avoir.)
Metres	1.094	Yards	Pounds (troy)	13.1657	Ounces (avoir.)
Metres/min.	1.667	Centimetres/sec.	Pounds (troy)	3.6735×10^{-4}	Tons (long)
Metres/min.	3.281	Feet/min.	Pounds (troy)	4.1143×10^{-4}	Tons (short)
Metres/min.	0.05468	Feet/sec.	Pounds (troy)	3.7324×10^{-4}	Tons (metric)
Metres/min.	0.06	Kilometres/hr.	Quadrants (angle)	90	Degrees
Metres/min.	0.03728	Miles/hr.	Quadrants (angle)	5400	Minutes
Metres/sec.	196.8	Feet/min.	Quadrants (angle)	1.571	Radians
Metres/sec.	3.281	Feet/sec.	Radians	57.30	Degrees
Metres/sec.	3.6	Kilometres/hr.	Radians	3438	Minutes
Metres/sec.	0.06	Kilometres/min.	Radians	0.637	Quadrants
Metres/sec.	2.237	Miles/hr.	Radians/sec.	57.30	Degrees/sec.
Metres/sec.	0.03728	Miles/min.	Radians/sec.	0.1592	Revolutions/sec.
Microns	10^{-6}	Metres	Radians/sec.	9.549	Revolutions/min.
Miles	5280	Feet	Radians/sec./sec.	573.0	Revs./min./min.
Miles	1.609	Kilometres	Radians/sec./sec.	0.1592	Revs./sec./sec.
Miles	1760	Yards	Reams	500	Sheets
Miles/hr.	1.609	Kilometres/hr.	Revolutions	360	Degrees
Miles/hr.	0.8684	Knots	Revolutions	4	Quadrants
Minutes (angle)	2.909×10^{-4}	Radians	Revolutions	6.283	Radians
Ounces	16	Drams	Revolutions/min.	6	Degrees/sec.
Ounces	437.5	Grains	Square yards	2.066×10^{-4}	Acres
Ounces	0.0625	Pounds	Square yards	9	Square feet
Ounces	28.349527	Grams	Square yards	0.8361	Square metres
Ounces	0.9115	Ounces (troy)	Square yards	3.228×10^{-7}	Square miles
Ounces	2.790×10^{-5}	Tons (long)	Temp. (°C.) + 273	1	Abs. temp. (°C.)
Ounces	2.835×10^{-5}	Tons (metric)	Temp. (°C.) + 17.78	1.8	Temp. (°F.)
Ounces (troy)	480	Grains	Temp. (°F.) + 460	1	Abs. temp. (°F.)
Ounces (troy)	20	Pennyweights (troy)	Temp. (°F.) − 32	5/9	Temp. (°C.)
Ounces (troy)	0.08333	Pounds (troy)	Watts	0.05692	B.T.U./min.
Ounces (troy)	31.103481	Grams	Watts	44.26	Foot–pounds/min.
Ounces (troy)	1.09714	Ounces (avoir.)	Watts	0.7376	Foot–pounds/sec.
Ounces (fluid)	1.805	Cubic inches	Watts	1.341×10^{-3}	Horsepower
Ounces (fluid)	0.02957	Litres	Watts	0.01434	Kg.–calories/min.
Ounces/sq. inch	0.0625	Lbs./sq. inch	Watts	10^{-3}	Kilowatts
Pounds	16	Ounces	Watt–hours	3.415	B.T.U.
Pounds	256	Drams	Watt–hours	2655	Foot–pounds
Pounds	7000	Grains	Watt–hours	1.341×10^{-3}	Horsepower–hrs.
Pounds	0.0005	Tons (short)	Watt–hours	0.8605	Kilogram–calories
Pounds	453.5924	Grams	Watt–hours	367.1	Kilogram–metres
Pounds	1.21528	Pounds (troy)	Watt–hours	10^{-3}	Kilowatt–hours
Pounds	14.5833	Ounces (troy)	Yards	91.44	Centimetres
Pounds (troy)	5760	Grains	Yards	3	Feet
Pounds (troy)	240	Pennyweights (troy)	Yards	36	Inches
Pounds (troy)	12	Ounces (troy)	Yards	0.9144	Metres

From Goetsch, Nelson, and Chalk, *Technical Drawing*, 3rd edition, copyright 1994 by Delmar Publishers Inc.

CIRCUMFERENCES AND AREAS OF CIRCLES
From 1/64 to 50, Diameter

Dia.	Circum.	Area	Dia.	Circum.	Area	Dia.	Circum.	Area	Dia.	Circum.	Area
1/64	.04909	.00019	8	25.1327	50.2655	17	53.4071	226.980	26	81.6814	530.929
1/32	.09818	.00077	8 1/8	25.5254	51.8485	17 1/8	53.7998	230.330	26 1/8	82.0741	536.047
1/16	.19635	.00307	8 1/4	25.9181	53.4562	17 1/4	54.1925	233.705	26 1/4	82.4668	541.188
1/8	.39270	.01227	8 3/8	26.3108	55.0883	17 3/8	54.5852	237.104	26 3/8	82.8595	546.355
3/16	.58905	.02761	8 1/2	26.7035	56.7450	17 1/2	54.9779	240.528	26 1/2	83.2522	551.546
1/4	.78540	.04909	8 5/8	27.0962	58.4262	17 5/8	55.3706	243.977	26 5/8	83.6449	556.761
5/16	.98175	.07670	8 3/4	27.4889	60.1321	17 3/4	55.7633	247.450	26 3/4	84.0376	562.002
3/8	1.1781	.11045	8 7/8	27.8816	61.8624	17 7/8	56.1560	250.947	26 7/8	84.4303	567.266
7/16	1.3744	.15033	9	28.2743	63.6173	18	56.5487	254.469	27	84.8230	572.555
1/2	1.5708	.19635	9 1/8	28.6670	65.3967	18 1/8	56.9414	258.016	27 1/8	85.2157	577.869
9/16	1.7671	.24850	9 1/4	29.0597	67.2007	18 1/4	57.3341	261.587	27 1/4	85.6084	583.207
5/8	1.9635	.30680	9 3/8	29.4524	69.0292	18 3/8	57.7268	265.182	27 3/8	86.0011	588.570
11/16	2.1598	.37122	9 1/2	29.8451	70.8822	18 1/2	58.1195	268.803	27 1/2	86.3938	593.957
3/4	2.3562	.44179	9 5/8	30.2378	72.7597	18 5/8	58.5122	272.447	27 5/8	86.7865	599.369
13/16	2.5525	.51849	9 3/4	30.6305	74.6619	18 3/4	58.9049	276.117	27 3/4	87.1792	604.806
7/8	2.7489	.60132	9 7/8	31.0232	76.5886	18 7/8	59.2976	279.810	27 7/8	87.5719	610.267
15/16	2.9452	.69029				19	59.6903	283.529	28	87.9646	615.752
1	3.1416	.78540	10	31.4159	78.5398	19 1/8	60.0830	287.272	28 1/8	88.3573	621.262
1 1/8	3.5343	.99402	10 1/8	31.8086	80.5156	19 1/4	60.4757	291.039	28 1/4	88.7500	626.797
1 1/4	3.9270	1.2272	10 1/4	32.2013	82.5159	19 3/8	60.8684	294.831	28 3/8	89.1427	632.356
1 3/8	4.3197	1.4849	10 3/8	32.5940	84.5408	19 1/2	61.2611	298.648	28 1/2	89.5354	637.940
1 1/2	4.7124	1.7671	10 1/2	32.9867	86.5902	19 5/8	61.6538	302.489	28 5/8	89.9281	643.548
1 5/8	5.1051	2.0739	10 5/8	33.3794	88.6641	19 3/4	62.0465	306.354	28 3/4	90.3208	649.181
1 3/4	5.4978	2.4053	10 3/4	33.7721	90.7626	19 7/8	62.4392	310.245	28 7/8	90.7135	654.838
1 7/8	5.8905	2.7612	10 7/8	34.1648	92.8856	20	62.8319	314.159	29	91.1062	660.520
2	6.2832	3.1416	11	34.5575	95.0332	20 1/8	63.2246	318.099	29 1/8	91.4989	666.226
2 1/8	6.6759	3.5466	11 1/8	34.9502	97.2053	20 1/4	63.6173	322.062	29 1/4	91.8916	671.957
2 1/4	7.0686	3.9761	11 1/4	35.3429	99.4020	20 3/8	64.0100	326.051	29 3/8	92.2843	677.713
2 3/8	7.4613	4.4301	11 3/8	35.7356	101.623	20 1/2	64.4027	330.064	29 1/2	92.6770	683.493
2 1/2	7.8540	4.9087	11 1/2	36.1283	103.869	20 5/8	64.7954	334.101	29 5/8	93.0697	689.297
2 5/8	8.2467	5.4119	11 5/8	36.5210	106.139	20 3/4	65.1881	338.163	29 3/4	93.4624	695.127
2 3/4	8.6394	5.9396	11 3/4	36.9137	108.434	20 7/8	65.5808	342.250	29 7/8	93.8551	700.980
2 7/8	9.0321	6.4918	11 7/8	37.3064	110.753				30	94.2478	706.858
3	9.4248	7.0686	12	37.6991	113.097	21	65.9735	346.361	30 1/8	94.6405	712.761
3 1/8	9.8175	7.6699	12 1/8	38.0918	115.466	21 1/8	66.3662	350.496	30 1/4	95.0332	718.689
3 1/4	10.2102	8.2958	12 1/4	38.4845	117.859	21 1/4	66.7589	354.656	30 3/8	95.4259	724.640
3 3/8	10.6029	8.9462	12 3/8	38.8772	120.276	21 3/8	67.1516	358.841	30 1/2	95.8186	730.617
3 1/2	10.9956	9.6211	12 1/2	39.2699	122.718	21 1/2	67.5442	363.050	30 5/8	96.2113	736.618
3 5/8	11.3883	10.3206	12 5/8	39.6626	125.185	21 5/8	67.9369	367.284	30 3/4	96.6040	742.643
3 3/4	11.7810	11.0447	12 3/4	40.0553	127.676	21 3/4	68.3296	371.542	30 7/8	96.9967	748.693
3 7/8	12.1737	11.7932	12 7/8	40.4480	130.191	21 7/8	68.7223	375.825	31	97.3894	754.768
4	12.5664	12.5664	13	40.8407	132.732	22	69.1150	380.133	31 1/8	97.7821	760.867
4 1/8	12.9591	13.3640	13 1/8	41.2334	135.297	22 1/8	69.5077	384.465	31 1/4	98.1748	766.990
4 1/4	13.3518	14.1863	13 1/4	41.6261	137.886	22 1/4	69.9004	388.821	31 3/8	98.5675	773.139
4 3/8	13.7445	15.0330	13 3/8	42.0188	140.500	22 3/8	70.2931	393.203	31 1/2	98.9602	779.311
4 1/2	14.1372	15.9043	13 1/2	42.4115	143.139	22 1/2	70.6858	397.608	31 5/8	99.3529	785.509
4 5/8	14.5299	16.8002	13 5/8	42.8042	145.802	22 5/8	71.0785	402.038	31 3/4	99.7456	791.731
4 3/4	14.9226	17.7206	13 3/4	43.1969	148.489	22 3/4	71.4712	406.493	31 7/8	100.1383	797.977
4 7/8	15.3153	18.6655	13 7/8	43.5896	151.201	22 7/8	71.8639	410.972			
5	15.7080	19.6350	14	43.9823	153.938	23	72.2566	415.476	32	100.5310	804.248
5 1/8	16.1007	20.6290	14 1/8	44.3750	156.699	23 1/8	72.6493	420.004	32 1/8	100.9237	810.543
5 1/4	16.4934	21.6476	14 1/4	44.7677	159.485	23 1/4	73.0420	424.557	32 1/4	101.3164	816.863
5 3/8	16.8861	22.6906	14 3/8	45.1604	162.295	23 3/8	73.4347	429.134	32 3/8	101.7091	823.208
5 1/2	17.2788	23.7580	14 1/2	45.5531	165.130	23 1/2	73.8274	433.736	32 1/2	102.1018	829.577
5 5/8	17.6715	24.8505	14 5/8	45.9458	167.989	23 5/8	74.2201	438.363	32 5/8	102.4945	835.971
5 3/4	18.0642	25.9672	14 3/4	46.3385	170.873	23 3/4	74.6128	443.014	32 3/4	102.8872	842.389
5 7/8	18.4569	27.1085	14 7/8	46.7312	173.782	23 7/8	75.0055	447.689	32 7/8	103.2799	848.831
6	18.8496	28.2743	15	47.1239	176.715	24	75.3982	452.389	33	103.6726	855.299
6 1/8	19.2423	29.4647	15 1/8	47.5166	179.672	24 1/8	75.7909	457.114	33 1/8	104.0653	861.791
6 1/4	19.6350	30.6796	15 1/4	47.9094	182.654	24 1/4	76.1836	461.863	33 1/4	104.4580	868.307
6 3/8	20.0277	31.9191	15 3/8	48.3020	185.661	24 3/8	76.5763	466.637	33 3/8	104.8507	874.848
6 1/2	20.4204	33.1831	15 1/2	48.6947	188.692	24 1/2	76.9690	471.435	33 1/2	105.2434	881.413
6 5/8	20.8131	34.4716	15 5/8	49.0874	191.748	24 5/8	77.3617	476.258	33 5/8	105.6361	888.003
6 3/4	21.2058	35.7847	15 3/4	49.4801	194.828	24 3/4	77.7544	481.106	33 3/4	106.0288	894.618
6 7/8	21.5985	37.1223	15 7/8	49.8728	197.933	24 7/8	78.1471	485.977	33 7/8	106.4215	901.257
7	21.9912	38.4845	16	50.2655	201.062	25	78.5398	490.874	34	106.8142	907.920
7 1/8	22.3839	39.8712	16 1/8	50.6582	204.216	25 1/8	78.9325	495.795	34 1/8	107.2069	914.609
7 1/4	22.7765	41.2825	16 1/4	51.0509	207.394	25 1/4	79.3252	500.740	34 1/4	107.5996	921.321
7 3/8	23.1692	42.7183	16 3/8	51.4436	210.597	25 3/8	79.7179	505.711	34 3/8	107.9923	928.058
7 1/2	23.5619	44.1787	16 1/2	51.8363	213.825	25 1/2	80.1106	510.705	34 1/2	108.3850	934.820
7 5/8	23.9546	45.6636	16 5/8	52.2290	217.077	25 5/8	80.5033	515.724	34 5/8	108.7777	941.607
7 3/4	24.3473	47.1730	16 3/4	52.6217	220.353	25 3/4	80.8960	520.768	34 3/4	109.1704	948.417
7 7/8	24.7400	48.7069	16 7/8	53.0144	223.654	25 7/8	81.2887	525.836	34 7/8	109.5631	955.253

From Goetsch, Nelson, and Chalk, *Technical Drawing*, 3rd edition, copyright 1994 by Delmar Publishers Inc.

ARCHITECTURAL

FINAL PLOT SCALE	SHEET SIZE				
	A 11 x 8½	B 17 x 11	C 24 x 18	D 36 x 24	E 48 x 36
1/16	176', 136'	272', 176'	384', 288'	576', 384'	768', 576'
3/32	132', 102'	204', 132'	288', 216'	432', 288'	576', 432'
1/8	88', 68'	136', 88'	192', 144'	288', 192'	384', 288'
3/16	66', 51'	102', 66'	144', 108'	216', 144'	288', 216'
1/4	44', 34'	68', 44'	96', 72'	144', 96'	192', 144'
3/8	29'-4", 22'-8"	45'-4", 29'-4"	64', 48'	96', 64'	128', 96'
1/2	22', 17'	34', 22'	48', 36'	72', 48'	96', 72'
3/4	14'-8", 11'-4"	22'-8", 14'-8"	32', 24'	48', 32'	64', 48'
1	11', 8'-6"	17', 11'	24', 18'	36', 24'	48', 36'
1½	7'-4", 5'-8"	11'-4", 7'-4"	16', 12'	24', 16'	32', 24'
3	3'-8", 2'-10"	5'-8", 3'-8"	8', 6"	12', 8'	16,' 12'

ENGINEERING

FINAL PLOT SCALE	SHEET SIZE				
	A 11 x 8½	B 17 x 11	C 24 x 18	D 36 x 24	E 48 x 36
10	110, 85	170, 110	240, 180	360, 240	480, 360
20	220, 170	340, 220	480, 360	720, 480	960, 720
30	330, 255	510, 330	720, 540	1080, 720	1440, 1080
40	440, 340	680, 440	960, 720	1440, 960	1920, 1440
50	550, 425	850, 550	1200, 900	1800, 1200	2400, 1800
60	660, 510	1020, 660	1440, 1080	2160, 1440	2880, 2160
100	1100, 850	1700, 1100	2400, 1800	3600, 2400	4800, 3600
Full Size	11, 8.5	17, 11	24, 18	36, 24	48, 36

APPENDIX H

GLOSSARY

Absolute coordinates	Points designated by a specific X, Y, and Z distance from a fixed origin.
Alphanumeric	Numbers, letters, and special characters.
ANSI	American National Standards Institute. ANSI is a professional organization which sets standards.
ASCII	American Standard Code for Information Interchange. ASCII is a standard computer data communications code.
Aspect ratio	The height-to-width ratio of an image on an output device.
Attribute	Information associated with a graphic object.
Baud	Data transmission rate of a computer. The baud rate is the number of bits sent per second.
Bill of materials	A listing of the parts required to assemble an object. Bills of materials may be extracted from a database.
Bit	The smallest unit of information of a binary system.
Buffer	Memory reserved for temporary storage of data.
Bug	A malfunction or design error in computer hardware or software.
Byte	Collection of eight bits that represent one letter.
CAD	Computer-aided design or drafting.

CAM	Computer-aided manufacturing.
Command	A specific word or other entry to provide an instruction.
Configuration	Providing setup for peripheral hardware by installing proper software drivers in a program.
Coordinate	A point located by X, Y, and Z directions and represented by real numbers.
CPU	Central processing unit. The part of a computer where the arithmetic and logic functions are performed.
Crosshair	Horizontal and vertical crossed lines on the display screen that designate the current working point of the drawing.
Crosshatching	Filling in of an area with a design.
CRT	Cathode ray tube. The display screen on which the computer display is shown. A CRT is similar to a television screen.
Cursor	Display device, such as a flashing bar or box, that represents the current working point on the display.
Database	The collection of information that is used to form a drawing or used to perform program functions.
Data extraction	Retrieval of data from the database.
Default	A predetermined value for a computer function.
Digitize	Electronically tracing a drawing.
Digitizer	An electronically sensitive pad used to translate movement of a hand-held device to cursor movement on a screen. May also be used to trace drawings.
Directory	A file listing of specific files or other data.
Disk	A circular piece of plastic with a magnetic coating used to store computer data.
Display	Area on a CRT where graphic or text images are displayed.
Dot matrix	A dot grid that is used to display images by displaying specific patterns.
Dragging	Moving of an entity or entities relative to the displacement of an input device.
Drum plotter	A pen plotter that moves both the paper and pen to achieve a plot.
Dual display	Use of two display devices simultaneously.
Entity	A single drawing element such as a line, circle, or arc.
Extension	The last three letters that follow a period in a file name, e.g., .DWG or .EXE.
Fillet	The corner fit or radius between two nonparallel lines.

Flatbed plotter	A pen plotter in which the paper remains stationary and the pen moves in both directions to create the plot.
Floppy disk	A thin plastic disk with a magnetic coating used to store computer data.
Fonts	The design of an alphabet.
Function key	Programmable keyboard key used for a specific purpose by a software program.
Gigabyte (GB)	One gigabyte (1 KB × 1 MB).
Grid	A series of dots arranged in a designated X and Y spacing used for reference purposes when drawing.
Hard copy	The printed or plotted copy of a drawing or data.
Hardware	The physical computer equipment such as the computer, printer, mouse, etc.
Hatching	Filling an area with a pattern.
Input device	Any device used to enter data into a computer. Examples of input devices are keyboard, mouse, and digitizing pad.
Kilobyte (KB)	A unit of 1024 bytes.
Layer	An overlay used to store specific data on. A drawing may contain several overlays or layers with data on each.
Light pen	An input device that enters points directly onto the display screen by emitting a light beam that is detected by the CRT.
Macro	Several keystroke operations executed by a single entry.
Mainframe computer	The largest and most powerful computer. Mainframes are used where large amounts of data are stored, such as by government and insurance companies.
Megabyte (MB)	Unit of storage equal to 1,048,576 bytes (1KB × 1KB).
Menu	A screen listing of options in a computer program.
Microcomputer	A small desktop computer. Usually referred to as a personal computer.
Minicomputer	A mid-sized computer of a size and capability between those of a microcomputer and a mainframe.
Mirroring	Reversal of a graphic image around a specified axis.
Modem	Modulator-demodulator. A modem is used to send computer data over telephone lines.
Mouse	A small, hand-held input device that is moved around on a surface. This movement is sensed by the computer and a relative displacement of the cursor or crosshair is made.
Network	The linking together of several computer systems.

OLE	Object linking and embedding.
Operating system	Computer program that acts as a translator between the computer and applications programs.
Pan	Movement around a drawing while in a specified magnification.
Pen plotter	Output device used to record a CAD drawing by moving a pen over a drawing surface to obtain a hard-copy drawing.
Peripheral	Any device used in conjunction with a computer, e.g., printer, light pen, etc.
Pixel	Dots that make up the display. The arrangement of displayed pixels determines the image displayed.
Plotter	Device used to obtain a hard copy of a graphic display. Plotters may be pen, electrostatic, or dot matrix.
Polar	Coordinate system used to specify distance and angle, defined in reference to a specified starting origin.
Program	The set of instructions that the computer uses to perform tasks. The program is also referred to as software.
Prompt	A message or symbol displayed by the computer that tells the operator that the computer is ready for input.
Puck	The hand-held input device used with a digitizing tablet to move the crosshairs on the display.
RAM	Random access memory. Temporary memory storage area in a computer.
Relative coordinates	Points that are located relative to a specified point.
Resolution	The measure of precision and clarity of a display.
ROM	Read-only memory. Permanent memory storage area in a computer.
Rubberbanding	"Stretching" a line from a fixed point to the current location of the crosshairs.
Scrolling	Vertical movement of text lines.
Snap	Drawing aid that allows entered points to "snap" to the closest point of an imaginary grid.
Software	Set of instructions used by the computer to perform a task. Software is also referred to as the program.
Stylus	A pen-like input device used with a digitizing pad in the same manner as a puck.
Windowing	Enlargement or reduction of graphic screen images.

INDEX

Absolute coordinates,
  drawing lines using,
  9-5—9-6
Add
  areas, area calculations
    and, 19-8
  using shift to, 10-11
Advanced rendering,
  39-3—39-6
Aerial view, 9-43—9-45
Aligned
  dimension, 15-12—15-13
  dimension drawing and,
    15-7
  text, placing, 13-6
Alternate units, dimension
  annotation and,
  16-16—16-17
Ambient light, 39-6
Angle
  dimensioning,
    15-16—15-17
  measurement, setting,
    8-4—8-7
  option, Xline and, 9-52
  rotating,
    by reference, 22-19
    by specifying, 22-19
  surfaces, dimensioning
    and, 15-12—15-14
Angular, dimension drawing
  and, 15-7
Annotation (text), dimen-
  sion, 16-14—16-20
Aperture, 28-3
  size, setting, 28-3
APParent intersection,
  object snap and, 9-25
Architectural units, 8-3

Architecture, CAD and, 2-3,
  2-4
Arcs, 9-1
  break and, 10-25
  constructing, elliptical,
    20-6
  dimensioning, 15-17—
    15-18, 17-7—17-8
  radius of, 15-19
  drawing, 9-16—9-22
    center, start, 9-21
    end point, 9-21
    included angle, 9-21
    length of chord, 9-21
    line/arc continuation,
      9-22
    manual entry method,
      9-22
  polylines and, 24-3—24-5
  start, center
    end, 9-17, **9-18**
    end of chord, 9-19
  start, end
    included angle, 9-20
    radius, 9-20
    starting direction, 9-21
  three point, 9-17
  filleting, 10-27
Area, 19-1, 19-6—19-7
Arrays, 22-7—22-9
  objects, drawings and,
    22-7—22-9
Arrow blocks, dimensioning,
  15-27—15-28
Arrowheads, dimensioning,
  16-6—16-8
Aspect
  grid and, 9-46
  snap and, 9-48

Attach
  option, MText and, 13-13
  Xref and, 27-3—27-4
Attedit, 26-7, 26-10
Attributes
  defining, 26-2
  described, 26-1
  display control of, 26-6—
    26-7
  editing, 26-7—26-10
  extractions, 26-10—26-15
  global editing and, 26-10
  stored as block, 20-23
  suppression of prompts,
    26-1
  tag, 26-3—26-5
  tutorial, 26-2—26-5
AU, object selection with,
  10-6
AutoCAD
  3D,
    described, 30-1
    drawing in, 33-1—33-2
    extruded entities, 33-2—
      33-6
  aperture size, setting,
    28-3
  blip marks, displaying,
    28-3
  camera/targeting position,
    31-7—31-13
  commands, 41-2
    redefining, 28-5
  creating sectional views
    with, 12-5—12-20
  dimensioning in, 15-1
  drawing, 1-2
    regenerations, control-
      ling, 28-4

X-1

dynamic viewing and, 31-4—31-7
getting started, 1-14
macros, 42-1—42-2
object linking and embedding and, 27-9—27-16
overview of, 6-1—6-2
starting, 1-1—1-2
system variables, setting, 28-1—28-3
terminology used by, 6-2—6-3
text components and, 13-2—13-3
viewports used with, 25-1—25-5
view resolution, setting, 28-4—28-5
AutoLISP
  arithmetic functions of, 41-1—41-2
  command prompt placement in, 41-3
  commands, 41-2
  getting/storing points in, 41-3
  notes within routines, 41-3
  SETQ function, 41-3
  using, 40-1—40-5
  writing and, 41-4—41-5
Automatic option, object selection with, 10-6
Auxiliary views
  construction of, 11-11
  tutorial, 11-12—11-13
Axis, view creation by, 31-2—31-3

Base, 20-22
Baseline
  dimensioning, 15-8, 15-38
  dimensions, 15-14—15-15
Baseplate, drawing, 1-16
Bhatch, using, 12-6—12-8
Binary files, drawing interchange, 24-23

Bind, external reference drawings, 27-5
Bisector option, Xline and, 9-52
Blank (spacebar), plotting and, 18-18
Blip marks, 1-9
  displaying, 28-3
Blocks, inserts, whole drawings as, 20-22
Block, 20-16
Blocks, 20-15
  advantages of, 20-23
  arrow, dimensioning, 15-27—15-28
  combining entries into, 20-16—20-17
  dividing with, 22-11
  drawing file creation from, 20-17—20-18
  exploding, 22-12—22-13
  inserting, 20-18—20-23
    with dialog box, 20-23—20-26
  inserts, redefining, 20-22
  layers and, 20-17
  multiple insertions and, 20-26—20-28
  separate, 15-28
Box
  3D, 33-29—33-30
  selecting objects with a, 10-6
  solid, drawing, 35-3—35-5
Break, 10-11
  affect on entities, 10-25
  erasing with, 10-24—10-25
Business, CAD and, 2-4
Buttons, in dialog boxes, 7-13—7-14
BYBLOCK, 23-2
BYLAYER, 23-2

CAD
  applications of, 2-3—2-7
  benefits of, 2-2

components of, 3-1—3-6
Calculating areas, 19-6—19-8
CAmera, 31-6
  positioning, 31-7—31-13
^C^C, 42-6
CD, 5-7
Center
  lines, dimension, 15-6
  marks,
    dimension, 15-6
    dimension drawing and, 15-8
    dimensioning, 15-19—15-20
Centered text, placing, 13-6
Central processing unit, 3-2
Chamfering
  lines, 20-11—20-12
  polylines, 20-11—20-12
  solid, 37-5—37-6
Chamfers, dimensioning, 17-19
Change, 22-1—22-7
Check button, 7-13
Choose, defined, 6-4
Chprop, 22-7
Circles, 9-1
  break and, 10-25
  calculating areas of, 19-7
  changing entry point of, 22-5, **22-6**
  copying, 10-31
  dimensioning, 15-17—15-18
  radius of, 15-19
  drawing, 9-11—9-15
    with center and diameter, 9-12
    with center and radius, 9-12
    designating three points, 9-13, **9-14**
    designating two points, 9-13
    isometric, 20-5—20-6

INDEX **X-3**

solid-filled, 20-7—20-8
tangent three point, 9-15
tangent to objects, 9-14
filleting, 10-27, 10-28
isometric, 21-7
lower, removing, 10-29
moving, 10-30
trimming, 22-15
Circumscribed polygons, 20-14
Clipping planes
  3D, 30-4—30-5
  setting, 31-14—31-16
Close, line, 9-3
CN (color number), plotting and, 18-18
Color
  changing, 22-2, 22-3
  setting current, 23-1—23-2
  setting grip, 22-27—22-28
  toolbar and, 7-3
Commands
  defined, 1-2
  editing, 7-3—7-7
  line entry keys, 7-8
  nomenclature, 6-4—6-5
  prompt area, 7-3
  repeating, 7-9
Composite, solids, 34-3
Computer-aided drafting
  applications of, 2-3—2-7
  benefits of, 2-2
  computers and, 3-1—3-3
Computers
  CAD and, 3-1—3-3
  categories of, 3-2
  components of, 3-2—3-3
  directory listings, 5-5—5-6
  disk drives, 4-1
  display devices, 3-5
  file,
    directories, 4-5
    names, 5-5—5-6
  floppy disks, 4-2
  hard drives, 4-2
  input devices, 3-5—3-6

paths, 4-5
peripheral hardware, 3-3—3-4
printers, 3-4—3-5
wild-cards, 5-6
write-protecting data, 4-4
Cones
  dimensioning, 17-6
  solid, 35-6—35-9
Construction
  line, **9-2**
  ray, **9-2**
Contin, 9-22
Continue
  dimension drawing and, 15-8
  line, 9-3
Control key, described, 6-4
Coordinates
  3D viewpoint and, 31-2
  defined, 6-2—6-3
  drawing lines with, 9-4—9-9
  absolute, 9-5—9-6
  polar, 9-8, **9-9**
  relative, 9-7
  screen, 7-2
Coordinate system
  3D, 30-2—30-4
  drawing in, 33-6
  user, 32-1—32-3
    changing, 32-4—32-9, 32-13
    head-on indicator, 32-3—32-4
    icon, 32-3
    listing information, 32-13
    preset orientations, 32-10
    UCSFOLLOW variable, 32-10
Copy, 10-11
Copying
  disks, 5-10—5-11
  file, 5-9—5-10
Corners
  filleting, 1-22

radiusing the, 10-32
Crosshatching, sectional views, 12-4—12-5
Crossing polygon window, selecting objects with, 10-7, 10-22
Cube, solid, 35-5, **35-6**
Cursor, 3-5
Curves
  dimensioning, 17-7—17-8
  spline, constructing, 24-14—24-15
Custom menus, 29-1
Cvport, 42-6
Cylinders
  dimensioning, 17-6
  solid, 35-10—35-11

Data, write-protecting, 4-4
Database information, listing, 19-3
Dblist, 19-1
DDATTDEF, 7-11, 26-3—26-5
DDATTE, 7-11, 26-9
DDATTEST, 7-11
DDCHPROP, 7-11
DDCOLOR, 7-11
DDEDIT, 7-11, 13-18
DDEMODES, 7-11
DDGRIPS, 7-11
DDIM, 7-11, 16-2
DDINSERT, 7-11
DDLMODES, 7-11
DDLTYPE, 7-11, 11-15
DDMODIFY, 7-11, 13-17
DDOSNAP, 7-11
DDPTYPE, 7-11
DDRENAME, 7-11
DDRMODES, 7-11
DDSELECT, 7-1, 10-9—10-12
DDSTYLE, 7-11
DDUCS, 7-11, 32-12
DDUCSP, 7-11
DDUNITS, 7-11

DDVIEW, 7-11
DDVPOINT, 7-11
Decimal
  degree, angle settings and, 8-5
  units, 8-3
Default<Default>, 6-4
Default option, Xline and, 9-51
Defined blocks, 20-17
Definition points, dimensioning, 15-27
Degrees, angle settings and, 8-5
Del, UCS and, 32-8
Delta option, lengthening lines and, 22-17
Detach, external reference drawings, 27-5
Dialog boxes, 7-10—7-14
  attribute editing with, 26-9
  buttons in, 7-13—7-14
  by command, 7-11
  changing properties with, 22-2—22-5
  extract attributes and, 26-11
  inserting blocks with, 20-23—20-26
  layers and, 14-7—14-12
  MText, 13-13—13-16
  plotting and, 18-2
  Rename and, 23-29
  setting view by, 31-3, **31-4**
  text styles by, 13-24—13-26
  UCS, 32-12
  using, 7-12—7-14
Diameter
  dimension drawing and, 15-7
  dimensioning, 15-40—15-41
Digitizers, 3-5—3-6
  screen menus and, 7-5
DimAngular, 15-16—15-17

DimCen, 16-18
Dimcenter, 15-20
Dimcontinue, 15-10—15-12
Dimedit, 15-22—15-24, 15-26
Dimension
  aligned, 15-12—15-13
  annotation (text), 16-14—16-20
  baseline, 15-14—15-15
    creating, 15-14—15-15
  center marks/lines, 15-6
  components,
    arrows, 17-3
    constructing, 17-2—17-5
    extension lines, 17-3
  format, 16-10—16-14
  information, placing on a drawing, 17-4
  limits, 15-4, **15-5**
  line, 15-1—15-2, 16-3—16-4
    angled surfaces, 15-12—15-14
    extension, 15-13, 16-5—16-6
    geometry, 16-3—16-10
    leader, 15-5
    placing a linear, 15-8—15-10
    vertical, 15-12
    rotated, constructing, 15-13—15-14
    status, displaying/changing, 15-27
    string, continuing, 15-10—15-12
    styles, 16-1
    text, 15-3—15-4
      constructing, 17-4—17-5
    tolerances, 15-4
    units, alternate, 15-5
    variables, 15-6, 16-1
      setting, 16-2—16-3
Dimensioning
  3D objects, 17-5—17-8

  angles, 15-16—15-17
  arcs, 15-17—15-18
    radius of, 15-19
  arrow blocks, 15-27—15-28
  arrowheads, 16-6—16-8
  baseline, 15-38
  center marks, placing, 15-19—15-20
  circles, 15-17—15-18
    radius of, 15-19
  commands, 15-7
  components, 15-1—15-3
  definition points, 15-27
  diameter, 15-40—15-41
  edit commands, 15-22—15-26
  extension lines, obliquing, 15-23—15-24
  isometric, 21-8
  leader construction, 15-42
  linear, aligned, 15-40
  mechanical components, 17-9—17-11
  mode, entering, 15-6
  ordinate, 15-21—15-22
  practices, described, 17-1
  radius, 15-41
  text,
    changing, 15-23
    relocating, 15-25—15-26
    restoring to default position, 15-22—15-23
    rotating, 15-26
  tutorial, 15-37—15-42
  utility commands, 15-27
  vertical, 15-39
Dimlinear, 15-8—15-10, 15-14
Dimordinate, 15-22
Dimsho, 16-2
DimTedit, 15-22, 15-26
DIR, 5-4—5-5
  / P, 5-5
  / W, 5-5
  /W/P, 5-5

INDEX **X-5**

Directories
  creating, 5-8
  deleting, 5-8
  displaying files in, 5-7
  listings, 5-5—5-6
  making current, 5-7
Discarding
  drawings, 7-22
    remaining in AutoCAD,
      1-12
Dish, 3D, 33-32
Disk drives, 4-1
Disk operating system.
    See DOS
Disks
  copying, 5-10—5-11
  floppy/hard, 4-2—4-3
  formatting, 5-11
Display
  defined, 6-3
  systems, 3-5
  time command and, 19-11
Dist, 19-1, 19-4—19-6
Distance, 31-6
  computing, 19-4—19-6
Distant lights, 39-6
Divide, 22-11
Dome, 3D, 33-31—33-32
Donut, 20-7—20-8
DOS
  changing active drives,
    5-2—5-3
  displaying files, 5-3—5-4
  described, 5-1—5-2
  directories,
    creating, 5-8
    deleting, 5-8
    displaying files in, 5-7
    making current, 5-7
  displaying files, 5-4—5-5
    in directories, 5-7
    wild-cards and, 5-6
  files, deleting, 5-8—5-9
Dot-matrix printers, 3-4
Doughnuts, drawing, solid-
    filled, 20-7—20-8

DOWN ARROW key, 7-9
Drafting techniques,
    traditional, 2-1
Dragmode, 9-2
  setting, 9-49—9-50
Draw commands
  ellipses, 20-1—20-4
  isometric, circles/ellipses,
    20-5—20-6
Drawing
  arcs, 9-16—9-22
    center, start, 9-21
      end point, 9-21
      included angle, 9-21
      length of chord, 9-21
    line/arc continuation,
      9-22
    manual entry method,
      9-22
    start, center
      end, 9-17, **9-18**
      end of chord, 9-19
    start, end
      included angle, 9-20
      radius, 9-20
      starting direction, 9-22
  calculating areas in, 19-6—
    19-8
  circles, 9-11—9-15
    with center
      and diameter, 9-12
      and radius, 9-12
    designating
      three points, 9-13, **9-14**
      two points, 9-13
    tangent three point, 9-15
    tangent to objects, 9-14
  file,
    diagnosis, 24-24
    recovery, 24-24—24-25
  information,
    database, 19-3
    listing, 19-2—19-3
  interchange,
    binary files, 24-23
    file format, 24-22—24-23

lines, 9-2—9-4
operations,
  redoing, 9-53
  undoing, 9-53
points, 9-9—9-11
polylines, 24-1—24-2
regenerations, controlling,
    28-4
screen, placing a grid on,
    9-46—9-47
starting, 9-1—9-2
Drawings
  3D, viewing, 31-1
  aerial view, 9-43—9-45
  aids, 9-45
  discarding, 7-22
    remaining in AutoCAD,
      1-12
  editing, existing, 7-19—
    7-20
  editor, 7-2—7-3
  exiting, 1-12—1-13
  external reference. *See*
    External reference
    drawings
  files, defined, 6-3
  graphic, text in, 13-1—13-3
  isometric. *See* Isometric
    drawings
  limits, setting, 8-7—8-8
  modes, 9-45
  moving objects in, 10-18
  multiview. *See* Multiview
    drawings
  new, starting, 1-3, 7-16—
    7-18
  objects, purging, 23-7—
    23-8
  panning, 9-41—9-42
    transparent, 9-42
  placing a grid on, 9-46—
    9-47
  regenerating, 9-32
  renaming parts of, 23-8—
    23-9
  saving, 1-12, 7-20—7-22

quit AutoCAD, 1-12
remain in AutoCAD, 1-13
scaling, 8-8—8-9
session, beginning, 7-1
setting up, 8-1—8-7
starting, no prototype, 7-18
status of, checking, 8-10
units, setting the, 8-2—8-3
viewing, 3D, 1-21
views of, storing/
 displaying, 23-3
zooming, 9-32—9-45
Drop-down list box, 7-14
Dtext codes, 13-27
Dview, 31-4—31-7
Dxbin, 24-23
DXF file, using a, 24-23
Dxfin, 24-23
Dxfout file, producing, 24-22
Dynamic
 option, lengthening lines
  and, 22-18
 text, drawing, 13-10—
  13-11
 viewing, 31-4—31-7
 zooms,
  regenerating, 9-38
  using without pointing
   device, 9-38—9-39

Edge-defined surface,
 33-28—33-29
Edge method, polygon con-
 struction and, 20-14
Edgesurf, 33-28—33-29
Edit, box, 7-14
Editing
 attributes, 26-7—26-10
 commands, 7-3—7-7
  Array, 22-7—22-9
  Change, 22-1—22-7
  Chprop, 22-7
  dimensioning, 15-22—
   15-26
  Divide, 22-11
  Explode, 22-12—22-13

Extend, 22-15—22-17
general notes on, 22-25
grips editing and,
 22-26—22-35
Lengthen, 22-17—
 22-18
Measure, 22-12
Mirror, 22-9—22-10
Rotate, 22-18—22-20
Scale, 22-20—22-21
Stretch, 22-21—22-22
Trim, 22-13—22-15
Undo, 22-23—22-25
drawings, existing, 7-19
entities, making copies of,
 10-23—10-24
erased, restoring, 10-17
global, 26-10
items selected, changing,
 10-9
objects,
 erasing, 10-17
 moving, 10-18
 preselecting, 10-13—
  10-16
polylines, 24-5—24-7
using object selection,
 10-2—10-9
Electrostatic plotters, 3-4
Elev, 22-2
Elevation
 extruded entities and,
  33-2—33-3
 changing, 33-6
 tutorial, 33-4—33-6
Ellipses, 20-5—20-6
 drawing, 20-1—20-4
 by axis and eccentricity,
  20-2—20-3
 isometric, 20-5—20-6
 types of, 20-6
Elliptical arc, constructing,
 20-6
END, 7-21
ENDpoint, object snap,
 9-24

Engineering
 CAD and, 2-4
 units, 8-3
Enlarging parts, for dimensioning, 17-10
Enter
 area calculations and, 19-8
 defined, 6-4
 point, 6-4
Entertainment, CAD and,
 2-4
Entities
 changing properties of,
  22-1
 combining, into a block,
  20-16—20-17
 dividing, 22-11
 effect of break on, 10-25
 extruded,
  elevation and, 33-2—33-6
  thickness and, 33-3
 making copies of, 10-23—
  10-24
 polylines, 22-14—22-15
 selecting,
  with a fence, 10-7—10-8
  with polygon window,
   10-6—10-7
 single, object selection
  and, 10-8
 thickness, changing,
  22-5—22-6
 trimming, 22-13—22-15
Entry keys, 7-4
Erase, 10-11
Erasing
 break command and,
  10-24—10-25
 objects, 10-17, 10-21
Exchange file formats,
 24-22—24-23
Explode, 22-12—22-13
 blocks, 22-12—22-13
 groups, 10-16
Exporting, PostScript
 images, 23-10—23-12

INDEX **X-7**

Extend, 22-15—22-17
  polylines and, 22-16—22-17
Extending, objects, 22-15—22-17
Extension lines
  dimensional, 17-3
  dimensioning, obliquing, 15-23—15-24
External reference drawings
  Bind, 27-5
  detach, 27-5
  List, 27-4—27-5
  Overlay, 27-7
  overview of, 27-1—27-3
  Path, 27-6
  Reload, 27-6
  Xbind, 27-7
  Xref and, 27-3—27-8
Extrude, custom solids and, 36-1
Extruded entities, 33-2—33-6
  elevation, setting, 33-6
Extrusions, custom solids and, 36-2—36-5

Fast cursor, 6-4
Fence, selecting entities using, 10-7—10-8
Field utilities, 19-8—19-9
Files, 19-1
  copying, 5-9—5-10
  deleting, 5-8—5-9
  directories, 4-5
  formats, exchange, 24-22—24-23
  names of, 5-5—5-6
  recovery, 24-24—24-25
  renaming, 5-11
Fillet, 10-11
  connecting objects with, 10-26—10-28
Filleting
  arcs, 10-27
  circles, 10-27, 10-28

  corners, 1-22
  lines, 10-27
  polylines, 10-26
  solids, 35-7
  two lines, 10-26
Flatbed plotters, 3-4
Flip screen, 6-4
Floppy disks, 4-2
  care of, 4-3
  write-protecting, 4-4
Flyouts, accessing, 1-10—1-11
Formatting, disks, 5-11
Fractional units, 8-3
Fractions, dimensioning and, 17-5
Full sections, 12-2—12-3
Function keys, 7-8

Global editing, 26-10
Grads, angle settings and, 8-5
Graphic drawings, text in, 13-1—13-3
Grid, 7-2, 9-1
  placing on drawing screen, 9-46—9-47
  snapping to a, 9-47—9-49
  spacing, 9-46
Grips editing, 22-26—22-35
  commands, 22-30—22-35
    Mirror, 22-35
    Move, 22-31—22-32
    Rotate, 22-32—22-33
    Scale, 22-33—22-35
    Stretch, 22-30—22-31
  enabling, 22-26—22-29
  using, 22-29—22-30
Groups, 10-11
  ? list, 10-15
  add to, 10-16
  changing, 10-14—10-15
  create, 10-14, 10-15
  explode, 10-16
  identification, 10-14
  order, 10-15—10-16

  remove from, 10-16
  rename, 10-16
  selectability of, 10-16

Hard drives, 4-2
  care of, 4-3
Hatch
  boundary,
    definitions, 12-10—12-14
    selecting, 12-8—12-10, **12-10**
  defining your own, 12-17
  pattern adjustment, 12-17
  tutorial, 12-18—12-19
  using, 12-14
Hatching
  angle/scale selection, 12-7
  pattern/style selection, 12-6—12-7
  boundary, 12-8—12-9, **12-10**
Head-on indicator, UCS, 32-3—32-4
Hidden lines
  multiview drawings, 11-11, 11-14
  producing, 31-16
  removing, 1-23
Holes
  creating, 1-20
  dimensioning, 17-10, **17-11**
  drawing, 1-18—1-19
HOME key, 7-9
Horizontal option, Xline and, 9-52

Icons
  boxes, designing, 29-12
  menus, 7-15, 29-10—29-12
  paper space, 25-6
  user coordinate system, 32-3
  using, 1-9
ID, 19-1
  screen coordinates, 19-2

Images, shaded, 31-17—31-19
Image tile, 7-14
Importing, PostScript images, 23-12—23-14
Ink jet printers, 3-5
Input devices, 3-5—3-6
Inquiry commands, 19-1
Inscribed polygons, 20-13
Insert
  blocks and, 20-18
  key, 7-9
  object snap, 9-25
  redefining, 20-22
  whole drawings as, 20-22
Inserts, 20-15
INS key, 7-9
Interior design, CAD and, 2-4
INTersection, object snap, 9-25
Intersections, solid, 37-2
Isometric
  circles, 21-7
  dimensioning, 21-8
  drawing in, 21-6
  drawings,
    described, 21-1—21-3
    isoplane, switching the, 21-5—21-6
    modes for, entering, 21-4—21-5
    principles of, 21-3—21-4
  text, 21-7, **21-8**

Justified text, placing, 13-4

Keyboard
  AutoCAD and, 7-8—7-9
    command line entry keys, 7-8
  functions keys, 7-8
  insert key, 7-9
  toggle keys, 7-8
  use of, 7-4
Keys, special, 6-4

Laser printers, 3-4
Last drawn object, selecting the, 10-4
Layer, 22-2
  blocks and, 20-17
  changing the, 22-4
  colors, 14-5
  command, using, 14-3—14-6
  current drawing, setting, 14-4
  described, 14-1—14-2
  dialog box and, 14-7—14-12
  freezing/thawing with, 14-8
  locking/unlocking with, 14-9
  turning on/off with, 14-8
  filtering listings with, 14-10—14-12
  freezing, 14-6
  linetypes and, 14-5
  listing information, 14-3
  locking/unlocking, 14-6
  making a, 14-4
  naming a, 14-4
  setting color with, 14-9—14-10
  linetype setting with, 14-10
  status, toolbar and, 7-3
  thawing, 14-6
  toolbar and, 14-12—14-13
  turning on/off, 14-4—14-5
Leader
  construction dimensioning, 15-42
  dimension drawing and, 15-8
LEFT ARROW key, 7-9
Lengthen, 22-17—22-18
Lights
  collecting into scenes, 39-10—39-12
  rendering, 39-6—39-9

Limits, 8-1
  defined, 6-3
  dimension, 15-4, 15-5
  drawings, setting, 8-7—8-8
Line, 9-1, 9-2
  break and, 10-25
  changing entry point of, 22-5
  dimension, 15-1—15-2
    angled surfaces, 15-12—15-14
    extension, 15-3
    placing a linear, 15-8—15-10
    vertical, 15-12
    extension, dimension, 15-13, 16-5—16-6
  options, 9-3—9-4
  see also Lines
Linear
  dimensioning, 15-7, 15-37—15-38
  aligned, 15-40
Lines
  chamfering, 20-11—20-12
  dimensional, 17-3
    center, 15-6
    leader, 15-5
  drawing, 9-2—9-4
    using coordinates, 9-4—9-9
  filleting, 10-27
  hidden. See Hidden lines
  lengthening, 22-17—22-18
  multiple parallel, constructing, 24-8—24-13
  see also Line
Linetype
  changing the, 22-4—22-5
  command, 11-14—11-15
  layers and, 14-5
  setting with dialog box, 14-10
  loading, 11-15—11-16
  scales, 11-16
  toolbar and, 7-3

INDEX **X-9**

writing, 11-17—11-18
LISP routine, using, 41-5
List, 19-1, 19-2—19-3
  box, 7-14
  external reference drawings, 27-4—27-5
Listing blocks, 20-17
Ltscale, 11-16, 22-2
LType, 22-2

Macros
  toolbar, 42-2—42-9
  using, 42-1—42-2
Mainframe computers, described, 3-2
Manual, using the, 6-4
Manufacturing, CAD and, 2-4
Math coprocessor, 3-3
Measure, 22-9—22-10
Measure command, 22-12
Measurement, angle, setting, 8-4—8-7
Memory, 3-3
Menus
  custom, 29-1
  icon, 7-15, 29-10—29-12
  linking, 29-4—29-5
  loading, 29-27
  multiple, 29-3
    commands in, 29-5—29-6
  pull-down, 7-4, 7-6—7-7, 29-8—29-10
    changing the, 1-3—1-4
    using, 1-5—1-9
  root, 7-6
  screen, 7-4, 7-5
  simple, 29-2
  tablet, 7-4, 29-7—29-8
Mesh
  3D, 33-20—33-21
  polyface, 33-21
  polygon, 33-19—33-20
Mice, 3-6
Middle aligned text, placing, 13-7

MIDpoint, object snap, 9-26
Minsert
  operations, 20-27
  using, 20-28
Minutes, angle settings and, 8-5
Mirror, 22-9
  grips editing and, 22-35
Mirroring
  objects, 22-9—22-10
  text, 22-10
MlEdit, 24-10—24-11
Mline, 24-8—24-13
MlStyle, 24-11—24-13
Mode indicators, 7-2—7-3
Model, 7-2
  space,
    switching to, 25-5
    viewports and, 25-6—25-7
Modeling, solid. *See* Solid modeling
Mouse, screen menus and, 7-5
Move, 10-11
  grips editing and, 22-31—22-32
Mslide, 23-4—23-5
MText, 13-11—13-16
  changing, 13-16—13-19
  characteristics, 13-27—13-29
  dialog box, 13-13—13-16
Mtprop, 13-16
Multilines, 9-2
  parallel, constructing, 24-8—24-13
Multiple option, editing and, 10-3
Multiview drawings
  hidden lines on, 11-11, 11-14
  linetype command and, 11-14—11-15
  orthographic projection, 11-1—11-5

  tutorial, 11-6—11-10
Mview, 25-7—25-9

Name, toolbar and, 7-3
NEArest, object snap, 9-26
Nested blocks, 20-17
NEW, 1-4
NODe and, object snap and, 9-27
Nomenclature, command, 6-4—6-5
Notch, adding a, 10-31—10-32
Note boxes, using, 6-5
Noun, selecting, 10-10—10-12
Numerical factor, changing scale by, 22-20—22-21

Objects
  connecting, with fillet, 10-26—10-28
  creating groups of, 10-13—10-16
  erasing, 10-17, 10-21
  extending, 22-15—22-17
  grouping, 10-13
  linking and embedding, 27-9—27-16
  placing in AutoCAD, 27-12, 27-16—27-18
  method, calculating areas, 19-7—19-8
  mirroring, 22-9—22-10
  moving, 10-18
  picking, boundary selection by, 12-8—12-9
  pointing, editing and, 10-2—10-3
  preselecting, 10-13—10-16
  removing from selection set, 10-22
  rotating, 22-18—22-20
    by dragging, 22-20
  selecting all, 10-8
  stretching, 22-21—22-22

UCS and, 32-6—32-7
Object selection, 10-1
  with automatic option, 10-6
  with a box, 10-6
  editing using, 10-2—10-9
    object pointing and, 10-2—10-3
    with a window, 10-4
    editing with, using object pointing, 10-3
Object snap, 9-1
  APParent intersection and, 9-25
  CENter and, 9-24
  ENDpoint and, 9-24
  INSert and, 9-25
  modes, 9-24
  NEArest, 9-26
  NODe and, 9-27
  PERpendicular and, 9-28
  QUADrant and, 9-27
  QUIck and, 9-28
  TANgent and, 9-28
  using, 9-22—9-31
    methods of, 9-29—9-31
Off
  grid and, 9-46
  snap and, 9-48
  time command and, 19-11
Offset, 20-9, 20-10
  option, Xline and, 9-52
  sections, 12-4
Offsets
  parallel, constructing, 20-9
  "through", constructing, 20-10
OLE. *See* Object linking and embedding
On
  grid and, 9-46
  snap and, 9-48
  time command and, 19-11
Oops, 10-11, 10-17

Ordinate
  dimension drawing and, 15-8
  dimensioning and, 15-21—15-22
Origin, UCS and, 32-4
Ortho, 7-2, 9-2
Orthogonal control, drawing with, 9-49
Orthographic projection
  multiview drawings, 11-1—11-5
  tutorial, 11-6—11-10
Overlay, external reference drawings, 27-7

Pan, 9-1, 31-6
Panning, defined, 6-3
Paper space
  icon, 25-6
  relative scales, and, 25-9
  viewports and, 25-6—25-7
Paragraph text, 13-11—13-16
Parallel
  lines, constructing multiple, 24-8—24-13
  offsets, constructing, 20-9
Path, external reference drawings, 27-6
Paths, computer, 4-5
Pedit, 24-5—24-7
Pen plotters, 3-3—3-4
Percent option, lengthening lines and, 22-18
PERpendicular, object snap and, 9-28
Personal computers, described, 3-2
Pface, 33-21
PgDn key, 7-9
PgUp key, 7-9
Pick box
  size, 10-12
  using, 10-20

Pick points, tutorial, 12-15
Pline, 24-2—24-3
Plotter plots, 18-2
Plotters, 3-3—3-4
Plotting
  adjust area fill and, 18-8
  from command line, 18-17—18-20
  determining equipment for, 18-3—18-4
  dialog boxes and, 18-2
  display and, 18-6
  the drawing, 18-16
  extents and, 18-6
  hidden lines and, 18-8
  limits and, 18-6
  linetype and, 18-5
  overview of, 18-1—18-2
  paper size selection/orientation, 18-9—18-10
  pen assignments, 18-4—18-5
  plot to file, 18-8—18-9
  preview,
    full, 18-16
    partial, 18-14—18-15
  plot, 18-13
  scale, rotation, plot origin, 18-10—18-11
  setting plot scale, 18-11—18-13
  speed and, 18-6
  view and, 18-6—18-7
  width and, 18-6
  window and, 18-7—18-8
Point, 9-1
  lights, 39-6
  method, calculating areas, 19-6—19-7
POints, 31-6
  drawing, 9-9—9-11
  removing, 10-30
  restoring, 10-30
Polar coordinates, drawing lines using, 9-8, **9-9**
Polyface mesh, 33-21

INDEX X-11

Polygon meshes, 3D, 33-19—33-20
Polygons
  constructing, 20-15
  drawing, 20-12, **20-13**
    circumscribed, 20-14
    edge method, 20-14
    inscribed, 20-13
Polygon window
  crossing, selecting objects with, 10-7, 10-22
  selecting entities using a, 10-6—10-7
Polylines
  area calculation and, 19-7
  boundary, 12-20
  break and, 10-25
  chamfering, 20-11—20-12
  drawing, 24-1—24-2
  drawing arcs with, 24-3—24-5
  editing, 24-5—24-7
  entities, 22-14—22-15
  filleting, 10-26
  Pline and, 24-2—24-3
  trimming, 22-14—22-15
  using extend with, 22-16—22-17
  vertex editing and, 24-18—24-22
PostScript
  entities, fill displaying, 23-14—23-16
  images
    exporting, 23-10—23-12
    importing, 23-12—23-14
    producing/using, 23-9—23-16
Precision
  setting, for units, 8-3—8-4
  setting angle measurement and, 8-6
Previous, UCS and, 32-8
Previously designated selection set, reselecting the, 10-4

Primary units, dimension annotation and, 16-15—16-16
Primitives
  solids, 34-3, 35-1—35-2
  drawing, 35-2—35-3
Printer
  plots, 18-1
  plotters, 3-4
Prompt, described, 1-6
Properties, chprop and, 22-7
PRotate, blocks and, 20-21
Prototype drawings, 7-16
  configuring a default, 7-18
  selecting a, 7-16—7-17
PScale, blocks and, 20-21
Psdrag, 23-13
Psfill, 23-14—23-16
Psquality, 23-14
Puck, 3-5
Pull-down menus, 7-4, 7-6—7-7, 29-8—29-10
  changing the, 1-3—1-4
  using, 1-5—1-9
Purge, 23-7—23-8
PXscale, blocks and, 20-21
Pyramids, dimensioning, 17-7
PYscale, blocks and, 20-21
PZscale, blocks and, 20-21

QSAVE, 7-21
Qtext
  dialog box setting with, 13-31
  plotting and, 13-31
QUAdrant, object snap and, 9-27
QUIck, object snap and, 9-28

Radians, angle settings and, 8-5
Radio button, 7-13
Radius, dimensioning, 17-7, 15-41
Ray, 9-50—9-51

Rectangular arrays
  constructing, 22-7—22-8
  rotated, 22-9
Redo, 9-2
  drawing operation, 9-53
Redraw, 9-1
  clearing screen with, 9-31
  transparently, 9-31
  viewports and, 25-5
  zoom and, 9-38
Reference, changing scale by, 22-21
Regen, 9-32
Regenauto, 28-4
Regenerating
  dynamic zooms, 9-38
  using without pointing device, 9-38—9-39
  viewports and, 25-5
Relative coordinates, drawing lines using, 9-7
Reload, external reference drawings, 27-6
Removed sections, 12-4
Rename, 23-8—23-9
Renaming
  drawing parts, 23-8—23-9
  files, 5-11
  copying and, 5-9—5-10
Render, 39-1—39-2
Rendering
  advanced, 39-3—39-6
  first, 39-2—39-3
  lights, 39-6—39-9
  into scenes, 39-10—39-12
  materials/backgrounds, 39-12—39-16
  saving, 39-16—39-18
  speed improvement of, 39-18—39-19
Replay, 39-12—39-16
Restore, UCS and, 32-8
Return, defined, 6-4
Revolution, solid, 36-5—36-10

Revolved
  custom solids, 36-1
  sections, 12-3
  surface, 33-25—33-27
Revsurf, 33-25—33-27
Right aligned text, placing, 13-8
RIGHT ARROW key, 7-9
Right hand rule, 32-8
Rmat, 39-12—39-16
Rollerbed plotters, 3-4
Root menu, 7-6
Rotate, 22-18—22-20
  blocks and, 20-21
  dimension,
    drawing and, 15-7
    text, 15-26
  dimensions, constructing, 15-13—15-14
  grips editing and, 22-32—22-33
  snap and, 9-48
Round off, dimension, 16-20
Rubber banding, described, 1-7
Ruled surface command, 33-22—33-23
Rulesurf, 33-22—33-23
Running mode, 9-29
  object snap and, 9-30

S (show current values), plotting and, 18-19
SAVE, 7-20
SAVEAS, 7-21
Saveimg, 39-16—39-18
Saving, drawings, 1-12, 7-20—7-22
Scale, 22-20—22-21
  blocks and, 20-21
  grips editing and, 22-33—22-35
  relative, creating, 25-9
Scale factors
  block insertion and, 20-19
  corner specification of, 20-20
  negative, 20-20
  preset values, 20-21
  rotation angle and, 20-20
Scaling, 8-1
  drawings, 8-8—8-9
Scientific units, 8-3
Screen
  coordinates, 7-2
  ID, 19-2
  display, 29-2
  menus, 7-4, 7-5
  pointing, 7-9
Scroll bars, 7-12—7-13
  zoom and, 9-42—9-43
Seconds, angle settings and, 8-5
Sectional views, 12-1—12-20
  creating, with AutoCAD, 12-5—12-20
  crosshatching, 12-4—12-5
Sections, types of, 12-2—12-4
Select, 10-11
  defined, 6-4
Selection
  process, canceling, 10-8
  set, previously designated, reselecting the, 10-4
  settings, designating, 10-9—10-12
Setvar, 28-1—28-3
Shaded images, 31-17—31-19
Shaft, drawing, 1-17
Sizes, points, 9-9—9-11
Sketch, 9-2
Slidelib, 23-6
Slides
  libraries of, 23-6
  selecting from, 29-12
  making, 23-4—23-5
  shows with, 23-6—23-7
  storing/displaying, 23-4—23-8
  viewing, 23-5—23-6
Snap, 9-2
  grid and, 9-46
  options, 9-48
  spacing, 9-48
  time command and, 19-11
Solid
  chamfering, 37-5—37-6
  filleting, 35-7
  modeling,
    commands, 34-34
    described, 34-31
    uses of, 34-2
  objects,
    joining, 37-4, **37-5**
    modifying, 37-1—37-7
Solids
  box, 35-3—35-5
  composite, 34-3
    displaying, 38-15
    drawing, 38-1—38-14
  cone, 35-6—35-9
  creating, 12-21—12-22
  cube, 35-5, **35-6**
  custom, 36-1—36-2
    creating, 36-2—36-5
    extrusions, 36-2—36-5
    revolution, 36-5—36-10
    tutorial, 36-11—36-15
  cylinder, 35-10—35-11
  drawing with, 34-2—34-3
  intersections, 37-2
  primitives, 34-3, 35-1—35-2
  drawing, 35-2—35-3
  sphere, 35-12—35-13
  subtracting, 37-3—37-4
  torus, 35-13—35-17
  wedge, 35-17—35-20
Sphere
  3D, 33-33
  solid, 35-12—35-13
Spline, 24-16—24-17

curves, constructing, 24-14—24-15
  edit commands and, 24-15
Splinedit, 24-17—24-18
Spotlights, 39-6
Status, 8-1
Stretch, 22-21—22-22
  grips editing and, 22-30—22-31
  rules, 22-22
String, dimension, 15-10—15-12
Stylus, 3-5
Submenus, 29-4
Subtract, area calculations and, 19-8
Subtracting solids, 37-3—37-4
Surveyor's units, angle settings and, 8-5
System board, 3-2

Tablet menus, 7-4, 29-7—29-8
Tabsurf, 33-23—33-24
Tabulated surface, 33-23—33-24
  constructing, 33-24
TANgent, object snap and, 9-28
Target, 31-6
  positioning, 31-7—31-13
TbConfig, 29-14—29-17
Template files, creating, 26-12
Temporary mode, 9-29
  object snap and, 9-30—9-31
Text
  backwards text, 13-22
  changing entry point of, 22-6
  codes, 13-27
  command, 13-3
  components, AutoCAD and, 13-2—13-3
  dimension, 15-3—15-4
    annotation and, 16-19—16-20
  dimensional,
    changing, 15-23
    constructing, 17-4—17-5
    relocating, 15-25—15-26
    restoring to default position, 15-22—15-23
    rotating, 15-26
  drawing dynamically, 13-10—13-11
  Dtext codes, 13-27
  font file, 13-20—13-21
  graphic drawings and, 13-1—13-10
  height, 13-21
  isometric, 21-7, **21-8**
  mirroring, 22-10
  MText characteristics, 13-27—13-30
  obliquing angle, 13-22
  paragraph, 13-11—13-16
  placing,
    aligned, 13-6
    centered, 13-6
    justified, 13-4
    middle aligned, 13-7
    multiple lines of, 13-9—13-10
    right aligned, 13-8
    in a specific distance, 13-7
  redrawing, 13-30—13-32
  regenerating, 13-30—13-32
  special considerations concerning, 13-26—13-27
  standards, 13-1
  styles, 13-19—13-23
    by dialog box, 13-24—13-26
    creating, 13-20—13-23
    using, 13-26
  tutorial on, 13-5
  upside-down text, 13-22, **13-23**
  vertical, 13-23
  width factor, 13-21, **13-22**
Thickness, 22-2
  extruded entities and, 33-3
Three 3D
  approach to, 30-2
  clipping planes, 30-4—30-5
  coordinate system, 30-2—30-4
  drawing in, 33-6
  described, 30-1
  drawing in, 33-1—33-2
    extruded entities, 33-2—33-6
  drawings, viewing, 31-1
  edge-defined surface, 33-28—33-29
  faces, 33-17—33-19
    creating solid, 33-16
    placing, 33-17—33-18
  mesh, 33-20—33-21
  objects,
    box, 33-29—33-30
    dish, 33-32
    dome, 33-31—33-32
    sphere, 33-33
    torus, 33-34
    wedge, 33-35
  revolved surface, 33-25—33-27
  ruled surfaces, 33-22—33-23
    constructing, 32-23
  tabulated surface, 33-23—33-24, **33-25**
    constructing, 33-24
  tutorial, 33-7—33-16
  versus perspective, 30-4
  viewpoint, setting, 31-2—31-3
Three point
  arcs, 9-17
  UCS and, 32-5
Through offsets, constructing, 20-10
Tile, 7-2

Tilemode, 25-6, 42-6
Time, 7-2, 9-1, 19-10—19-11
Toggle keys, 7-8
Tolerance
  dimension, 15-4
    annotation and, 16-18—16-19
    dimension drawing and, 15-8
Toolbar
  button, using, 1-10
  customizing, 29-13
  indicators, 7-3
  layers and, 14-12—14-13
  macros, 42-2—42-9
  manipulating, 1-11
Torus
  3D, 33-34
  solid, 35-13—35-17
Total option, lengthening lines and, 22-18
Trace, break and, 10-25
Track balls, 3-6
Transparent
  panning, 9-42
  zooms, 9-40—9-41
Trim, 22-13—22-15
Trimming
  circles, 22-15
  entities, 22-13—22-15
  polylines, 22-14—22-15

U (single undo), 9-2
UCS
  changing, 32-4—32-9
  changing current, 32-13
  dialog box, 32-12
  head-on indicator, 32-3—32-4
  icon, 32-3
  listing information, 32-13
  preset orientations, 32-10
UCSFOLLOW system variable, 32-10
UCSICON, 32-11, 38-7—38-9

Underline, 6-4
Undo, 22-23—22-25
  line, 9-3
Undoing, drawing operations, 9-53
Units, 8-1
  defined, 6-3
  dimension, 15-5
  setting the drawing, 8-2—8-3
  setting precision for, 8-3—8-4
UP ARROW key, 7-9
Utility commands, 19-1
  dimensioning, 15-27
  field, 19-8—19-9

Verb, selecting, 10-10—10-12
Vertex editing, polylines, 24-18—24-22
Vertical
  dimensioning, 15-39
  option, Xline and, 9-52
View
  box, zoom and, 9-36, 9-37
  setting by dialog box, 31-3, **31-4**
  twisting, 31-14
  UCS and, 32-7
  zooming, 31-13—31-16
Viewing, dynamic, 31-4—31-7
Viewpoint, 3D, setting, 31-2—31-3
Viewports, 25-23
  AutoCAD and, 25-1—25-5
  current, 25-2
  drawing between, 25-2—25-5
  entities, break and, 10-25
  model space, 25-6—25-7
  switching to, 25-7
  paper space, 25-6—25-7
  relative scales and, 25-9

redraws/regeneration in, 25-5
  tutorial, 25-10—25-12
  working spaces and, 25-5
Viewres, 28-4—28-5
View resolution, setting, 28-4—28-5
Views. *See* Specific type of view
Vmax, zoom, 9-39
Vslide, 23-5—23-6

Wblock, 20-19
Wedge
  3D, 33-35
  dimensioning, 17-5
  solid, 35-17—35-20
Wild-cards, 5-6
Window
  corners, showing points by, 7-9
  erasing with a, 10-21
  implied, 10-11—10-12
  object selection using a, 10-4
  plotting and, 18-7—18-8
  polygon, selecting objects using a, 10-6—10-7
  selecting objects with, defining a crossing, 10-5
  using, 10-20
  zoom, 9-4
Working spaces, viewports and, 25-5
Workstations, described, 3-2
World, UCS and, 32-8
Wpolygon, 10-6—10-7
Write-protecting data, 4-4

X (exit), plotting and, 18-19
Xbind, external reference drawings, 27-7
Xline, 9-51—9-52

Xref, 27-3—27-8
Xscale, blocks and, 20-21
X/Y/Z, UCS and, 32-7

Yacht design, CAD and, 2-4
Yscale, blocks and, 20-21

ZAxis, UCS and, 32-5
Zero-angle direction, angle measurement setting and, 8-6—8-7
Zoom, 31-6, 42-6
   current view window, 9-36
   drawing extents and, 9-36
   extents, 9-39
   generated area, 9-36
   lower left corner, 9-39
   previous, 9-39
   transparent, 9-40—9-41
   using, 9-33—9-40
     all, 9-34
     center, 9-35
     dynamic, 9-35—9-36
     redrawing and, 9-38
     regenerating dynamic and, 9-38
     scale, 9-33—9-34
     using without pointing device, 9-38—9-39
   view box, 9-36
   Vmax, 9-39
   window, 9-40
Zooming
   defined, 6-3
   drawings and, 9-32—9-45
   view, 31-13—31-16
Zscale, blocks and, 20-21